ADVANCES IN GENETICS

VOLUME 27

Genetic Regulatory Hierarchies in Development

Contributors to This Volume

Hubert Amrein
Clare Bergson
Ruth Bryan
José A. Campos-Ortega
Robin Chadwick
Ulrike Gaul
David Glaser
Thomas Jack
Herbert Jäckle
Thomas C. Kaufman
Michael A. Kuziora
Michael Levine
Nadine McGinnis
William McGinnis
Barbara J. Meyer
Rolf Nöthiger
Gary Olsen
Michael Regulski
Christine Rushlow
Mark A. Seeger
Lucille Shapiro
George F. Sprague, Jr.
Monica Steinmann-Zwicky
Paul W. Sternberg
Anne M. Villeneuve

ADVANCES IN GENETICS

Edited by

JOHN G. SCANDALIOS

Department of Genetics
North Carolina State University
Raleigh, North Carolina

THEODORE R. F. WRIGHT

Department of Biology
University of Virginia
Charlottesville, Virginia

VOLUME 27

Genetic Regulatory Hierarchies in Development

Edited by

THEODORE R. F. WRIGHT

Department of Biology
University of Virginia
Charlottesville, Virginia

ACADEMIC PRESS, INC.
Harcourt Brace Jovanovich, Publishers

San Diego   New York   Boston
London   Sydney   Tokyo   Toronto

ACADEMIC PRESS, INC.
San Diego, California 92101

United Kingdom Edition published by
ACADEMIC PRESS LIMITED
24-28 Oval Road, London NW1 7DX

LIBRARY OF CONGRESS CATALOG CARD NUMBER: 47-30313

ISBN   0-12-017627-0   (alk. paper)

PRINTED IN THE UNITED STATES OF AMERICA
90 91 92 93   9 8 7 6 5 4 3 2 1

CONTENTS

Genetic Regulatory Hierarchy in *Caulobacter* Development

RUTH BRYAN, DAVID GLASER, AND LUCILLE SHAPIRO

Combinatorial Associations of Regulatory Proteins and the Control of Cell Type in Yeast

GEORGE F. SPRAGUE, JR.

Genetic Control of Cell Type and Pattern Formation in *Caenorhabditis elegans*

PAUL W. STERNBERG

The Regulatory Hierarchy Controlling Sex Determination and Dosage Compensation in *Caenorhabditis elegans*

ANNE M. VILLENEUVE AND BARBARA J. MEYER

Genetic Control of Sex Determination in *Drosophila*

MONICA STEINMANN-ZWICKY, HUBERT AMREIN, AND ROLF NÖTHIGER

Role of Gap Genes in Early *Drosophila* Development

ULRIKE GAUL AND HERBERT JÄCKLE

Role of the *zerknüllt* Gene in Dorsal–Ventral Pattern Formation in *Drosophila*

CHRISTINE RUSHLOW AND MICHAEL LEVINE

Molecular and Genetic Organization of the Antennapedia Gene Complex of *Drosophila melanogaster*

THOMAS C. KAUFMAN, MARK A. SEEGER, AND GARY OLSEN

Establishment and Maintenance of Position-Specific Expression of the *Drosophila* Homeotic Selector Gene *Deformed*

WILLIAM MCGINNIS, THOMAS JACK, ROBIN CHADWICK,
MICHAEL REGULSKI, CLARE BERGSON, NADINE MCGINNIS,
AND MICHAEL A. KUZIORA

Mechanisms of a Cellular Decision during Embryonic Development of *Drosophila melanogaster*: Epidermogenesis or Neurogenesis

José A. Campos-Ortega

CONTRIBUTORS TO VOLUME 27

Numbers in parentheses indicate the pages on which the authors' contributions begin.

HUBERT AMREIN* (189), *Zoological Institute, University of Zurich, CH-8057 Zurich, Switzerland*

CLARE BERGSON (363), *Departments of Molecular Biophysics and Biochemistry and of Biology, Yale University, New Haven, Connecticut 06511*

RUTH BRYAN (1), *Department of Microbiology, College of Physicians and Surgeons of Columbia University, New York, New York 10032*

JOSÉ A. CAMPOS-ORTEGA (403), *Institut für Entwicklungsphysiologie, Universität zu Köln, D-5000 Köln 41, Federal Republic of Germany*

ROBIN CHADWICK (363), *Departments of Molecular Biophysics and Biochemistry and of Biology, Yale University, New Haven, Connecticut 06511*

ULRIKE GAUL† (239), *Max-Planck-Institut für Entwicklungsbiologie, D-7400 Tübingen, Federal Republic of Germany*

DAVID GLASER‡ (1), *Department of Microbiology, College of Physicians and Surgeons of Columbia University, New York, New York 10032*

THOMAS JACK (363), *Departments of Molecular Biophysics and Biochemistry and of Biology, Yale University, New Haven, Connecticut 06511*

HERBERT JÄCKLE (239), *Institut für Genetik und Mikrobiologie, Universität München, D-8000 München 19, Federal Republic of Germany*

THOMAS C. KAUFMAN (309), *Department of Biology, Program in Genetics, Indiana University, Bloomington, Indiana 47405*

MICHAEL A. KUZIORA (363), *Departments of Molecular Biophysics and Biochemistry and of Biology, Yale University, New Haven, Connecticut 06511*

MICHAEL LEVINE (277), *Department of Biological Sciences, Columbia University, New York, New York 10027*

* Present address: Department of Biochemistry and Molecular Biology, Harvard University, Cambridge, Massachusetts 02138.

† Present address: Howard Hughes Medical Institute and Department of Molecular and Cell Biology, University of California, Berkeley, Berkeley, California 94720.

‡ Present address: HydroQual Inc., Mahwah, New Jersey 07463.

NADINE MCGINNIS (363), *Departments of Molecular Biophysics and Biochemistry and of Biology, Yale University, New Haven, Connecticut 06511*

WILLIAM MCGINNIS (363), *Departments of Molecular Biophysics and Biochemistry and of Biology, Yale University, New Haven, Connecticut 06511*

BARBARA J. MEYER (117), *Department of Biology, Massachusetts Institute of Technology, Cambridge, Massachusetts 02139*

ROLF NÖTHIGER (189), *Zoological Institute, University of Zurich, CH-8057 Zurich, Switzerland*

GARY OLSEN (309), *Department of Biology, Program in Genetics, Indiana University, Bloomington, Indiana 47405*

MICHAEL REGULSKI (363), *Departments of Molecular Biophysics and Biochemistry and of Biology, Yale University, New Haven, Connecticut 06511*

CHRISTINE RUSHLOW (277), *Department of Biological Sciences, Columbia University, New York, New York 10027*

MARK A. SEEGER (309), *Department of Biology, Program in Genetics, Indiana University, Bloomington, Indiana 47405*

LUCILLE SHAPIRO* (1), *Department of Microbiology, College of Physicians and Surgeons of Columbia University, New York, New York 10032*

GEORGE F. SPRAGUE, JR. (33), *Institute of Molecular Biology and Department of Biology, University of Oregon, Eugene, Oregon 97403*

MONICA STEINMANN-ZWICKY (189), *Zoological Institute, University of Zurich, CH-8057 Zurich, Switzerland*

PAUL W. STERNBERG (63), *Howard Hughes Medical Institute, Division of Biology, California Institute of Technology, Pasadena, California 91125*

ANNE M. VILLENEUVE (117), *Department of Biology, Massachusetts Institute of Technology, Cambridge, Massachusetts 02139*

* Present address: Department of Developmental Biology, School of Medicine, Arnold and Mabel Beckman Center for Molecular and Genetic Medicine, Stanford University Medical Center, Stanford, California 94305.

PREFACE

In the last five to ten years, the analysis of the genetic regulation of development has made almost unbelievable progress at the genetic, molecular, and cellular levels. The initial supposition that genes in development would be regulated by activator or repressor proteins binding to specific sequences in cis-acting elements has been extensively substantiated. Garcia-Bellido's [(1977) *Am. Zool.* **17**, 613–629] hypothesis that the regulation of gene activity in development would be hierarchical, with "activator" genes regulating "selector" genes (the homeotic loci in *Drosophila*), which in turn would regulate "realisator" genes, has also been extensively confirmed and extended. So much information on the combinatorial and hierarchical control of genes in development has been accumulated recently in numerous diverse organisms that it would be impossible to cover it all in any detail in a volume such as this one. Instead, this volume presents a small, representative sample of genetic regulatory systems in sufficient detail to permit one to appreciate the genetic, cellular, and molecular approaches that have been used to elucidate gene interactions in development and to permit one to comprehend both the complexity and simplicity of these genetic combinatorial and hierarchical networks.

Many of the exciting advances in our knowledge of the genetic regulation of development have been made in *Drosophila,* and the bulk of this volume, six articles, is devoted to this organism. However, since more than 70 genes are now known to be involved in the establishment of polarity and segmentation in *Drosophila,* no pretense is made at complete coverage of this area of research in this organism.

It is hoped that inclusion of articles on *Caulobacter,* yeast, and *Caenorhabditis elegans* will emphasize that other genetic developmental systems have many advantages and much information to contribute to our ideas on the genetic regulation of development. Since no mammalian systems are included in this volume, the reader is referred to the minireview by Blau [(1988) *Cell* **53**, 673–674] for an entry into hierarchies of regulatory genes in mammalian development.

Finally, one might suggest that the investigation of gene regulation in development has been the easy part, in comparison to the now more difficult task of determining exactly what the protein products of the regulated "realisator" genes actually do at the cellular and molecular level to effect the coherent differentiation and development of different cell types, tissues, and organs.

THEODORE R. F. WRIGHT

GENETIC REGULATORY HIERARCHY IN
Caulobacter DEVELOPMENT

Ruth Bryan, David Glaser,* and Lucille Shapiro†

Department of Microbiology, College of Physicians and Surgeons of Columbia University,
New York, New York 10032

I. Introduction

Regulatory hierarchies orchestrate the expression of large numbers of genes involved in developmental programs. Eukaryotes and pro-karyotes both use cascades of positive and negative trans-acting factors to control the level of expression of many genes in an ordered fashion, for example, mating type expression in yeast, vulva development in the nematode, and formation of the bithorax complex in *Drosophila*. Regu-

---

* Present address: HydroQual Inc., Mahwah, New Jersey 07463.

† Present address: Department of Developmental Biology, School of Medicine, Arnold and Mabel Beckman Center for Molecular and Genetic Medicine, Stanford University Medical Center, Stanford, California 94305.

1

latory hierarchies also control sporulation in *Bacillus subtilis,* fruiting body formation in *Myxococcus xanthus,* and flagellar biogenesis in *Caulobacter crescentus* and *Escherichia coli.* Here we describe the *Caulobacter* regulatory cascade that controls flagellar morphogenesis and then compare the observed mechanisms to those that appear in other prokaryotic developmental systems.

II. Life Cycle of *Caulobacter crescentus*

Caulobacters are dimorphic, gram-negative bacteria which are found in fresh water, salt water, and soils (Poindexter, 1981). This article focuses on the freshwater *Caulobacter crescentus,* which is the best known species of the group (Shapiro, 1985; Newton, 1989). The organism spends part of its cell cycle as a nonmotile stalked cell and part as a motile swarmer cell. The life cycle of *Caulobacter crescentus* is shown in Fig. 1. The predivisional cell is asymmetric: at one pole is a stalk, an appendage consisting in *C. crescentus* of cell wall and membranes, periodically interrupted by disklike crossbands (Poindexter, 1964). At

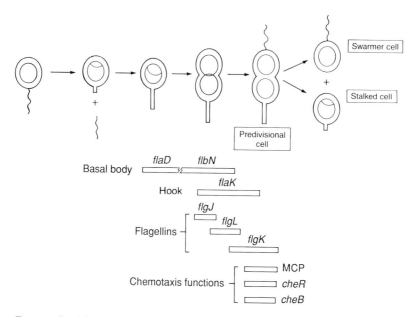

FIG. 1. *Caulobacter* cell cycle. Morphological changes during the cycle are indicated, as well as the timing of expression of flagellar and chemotaxis genes.

the other pole is a single flagellum. Upon division, two different cell types are produced, one flagellated and one stalked, as shown in Fig. 2.

Immediately following division, the daughter stalked cell initiates DNA replication. As it replicates its DNA, it elongates and, in a defined order, produces the proteins needed to build the flagellum at the pole

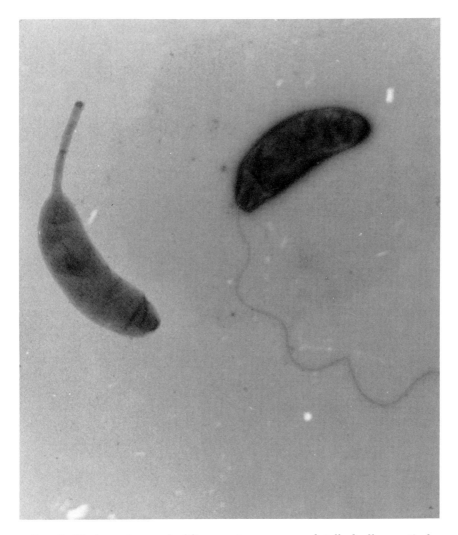

FIG. 2. Electron micrograph of *C. crescentus* swarmer and stalked cells, negatively stained with uranyl acetate. The helical filament on the swarmer cell is approximately 15 nm in diameter. The stalk diameter is about 90 nm.

opposite the stalk. Following completion of DNA replication and the synthesis and assembly of the flagellum at the swarmer pole, the cell divides. The cell begins swimming just prior to the time of cell division.

After cell division, the daughter swarmer cell does not initiate DNA synthesis. It continues to synthesize the 25K flagellin for assembly of the distal portion of the flagellar filament. After one-third of the cell cycle the swarmer cell loses its filament and begins to grow a stalk at the site where the filament had been attached. At this time, the cell initiates DNA replication, and thereafter the cell cycle continues exactly as in the daughter stalked cell. In rich laboratory media, this cell cycle is invariant, with approximately 30% of the life-span of the swarmer daughter cell spent as a swarmer and the remainder as a stalked and predivisional cell.

Caulobacters are well adapted to survival under conditions of limited nutrients. The stalk, with a large surface area to volume ratio, functions in nutrient uptake. The stalked cell attaches to surfaces using the holdfast at the end of the stalk, while the motile cell (swarmer) can undergo chemotaxis, swimming toward a more favorable location (Shaw *et al.,* 1983). Thus, ecologically, *Caulobacter* exploits two survival strategies each generation: one cell remains and one swims away.

III. Structure of the Flagellum

The flagellum of *Caulobacter crescentus* is similar to those of *Salmonella typhimurium* and *E. coli* in size and architecture (Fig. 3). It is composed of three sections, the basal body, the hook, and the filament. The basal body, which forms the most proximal portion of the flagellum (Hahnenberger and Shapiro, 1987; Stallmeyer *et al.,* 1989), is composed of five rings threaded on a central rod and embedded in the cell surface, with the innermost ring in the inner membrane and the outermost ring in the outer membrane. One of the middle rings appears to be associated with the peptidoglycan layer. The basal body of *C. crescentus* is quite similar to those of *E. coli* and *S. typhimurium,* except that it contains five instead of four rings. The extra ring is located at the site of filament separation and may be involved in loss of the flagellum each generation (Stallmeyer *et al.,* 1989).

The hook is a helical assembly of 70-kDa monomers (Wagenknect *et al.,* 1981). It is believed that the hook is assembled outside the cell by the addition of protein monomers which travel out through the hollow center of the basal body rod (Macnab, 1987).

The filament in *C. crescentus* is also similar in structure and size to

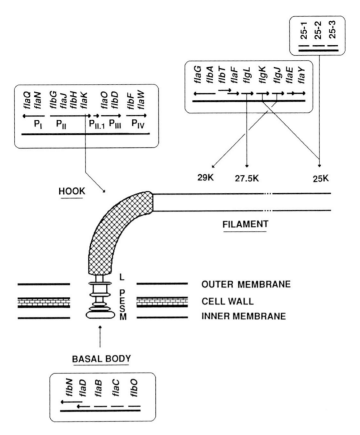

FIG. 3. *Caulobacter crescentus* flagellum, showing genes responsible for the parts of the structure. The genetic structure of the four flagellar gene clusters is also shown. The 29K flagellin occupies a stretch of the filament approximately equal to the length of the hook. The 27.5K flagellin occupies approximately one-quarter of the length of the filament.

those of *E. coli* and *S. typhimurium*. However, the filaments of *E. coli* and *S. typhimurium* are composed of a single filament protein, while the *C. crescentus* filament contains at least three antigenically distinct proteins. The 29K flagellin, encoded by *flgJ*, occupies a small portion of the filament adjacent to the hook (Driks *et al.*, 1989). The 27.5-kDa flagellin, encoded by *flgL*, is located next in the filament and occupies approximately one-quarter of the total filament. The most distally located flagellin, 25 kDa in size, is the most abundant (Weissborn *et al.*, 1982; Koyasu *et al.*, 1981). There are four genes encoding the 25-kDa flagellin. One of these is *flgK*, which is located in the α-flagellin cluster between *flgL* and *flgJ* on the genetic map. *flgK* is expressed (Minnich and Newton, 1987), and at least one of the other three 25K flagellin genes is also expressed (Minnich *et al.*, 1988). Filaments are made even in the presence of deletions or disruptions of *flgJ*, *flgK*, or *flgL*, although mutant strains show decreased swarming on soft agar plates (Driks *et al.*, 1989; Minnich *et al.*, 1988).

IV. Flagellar Genes

Mutations in more than 40 genes lead to reduced motility in *C. crescentus* (Johnson and Ely, 1979; Johnson *et al.*, 1983; Ely *et al.*, 1984; Newton *et al.*, 1989; Schoenlein *et al.*, 1989). Four clusters of genes essential for formation of the flagellum are diagrammed in Fig. 3. Complex patterns of transcription have been found in these clusters, as indicated by arrows over gene designations in Fig. 3. At least nine of the *fla* genes are not located in these clusters (Ely *et al.*, 1984).

The basal body cluster is so named because a mutation in one of the genes, *flaD*, causes a partial basal body to be formed and mutations in the other genes cause complete loss of the basal body (Hahnenberger and Shapiro, 1987). The hook cluster contains the gene for the hook structural protein, *flaK* (Ohta *et al.*, 1984, 1985). Mutations in another gene in this cluster, *flaJ,* lead to an abnormally long hook (Johnson and Ely, 1979). Expression of the other genes in the hook cluster is required for proper expression of the hook operon. Three copies of the gene for the 25K flagellin are located in the β-flagellin cluster (S. Lee and N. Agabian, unpublished). In addition, genes for all three flagellins are located in the α-flagellin cluster (Minnich and Newton, 1987; Gill and Agabian, 1983). Cells containing mutations in any of the six non-flagellin genes in the α-flagellin cluster still have a complete hook but are blocked in filament assembly, suggesting that all these genes are involved in filament assembly (Johnson *et al.*, 1983; Minnich *et al.*, 1988; Kaplan *et al.*, 1989; Schoenlein *et al.*, 1989). Three motility (*mot*) genes have been identified, and seven genes, three in a cluster and four isolated, are essential for chemotaxis (Ely *et al.*, 1986).

V. Timing of Gene Expression

An important question in the construction of subcellular architecture is whether component proteins must be synthesized in the order of their assembly. Analysis of the flagellar genes thus far identified reveals that the time of structural gene expression is correlated with the position of the protein product in the flagellar structure (Figs. 1 and 3). This suggests that assembly of a complex structure may require the ordered synthesis of the components.

Nuclease S1 protection assays and analysis of transcription fusions of flagellar promoters to reporter genes show that basal body, hook, and flagellin mRNAs are synthesized approximately when their protein products are being synthesized (Milhausen and Agabian, 1983; Ohta *et al.*, 1985; Chen *et al.*, 1986; Champer *et al.*, 1987; Loewy *et al.*, 1987; Minnich and Newton, 1987; Hahnenberger and Shapiro, 1988; Dingwall *et al.*, 1990). Thus, the timing of gene expression is controlled at the level of transcription. The genes for the basal body proteins are transcribed shortly after the cell begins to make a stalk (Hahnenberger and Shapiro, 1988; A. Dingwall, unpublished). The hook and the 29K flagellin genes are turned on next, followed by the 27.5K flagellin and then the 25K flagellin (Osley *et al.*, 1977; Lagenaur and Agabian, 1978; Ohta *et al.*, 1985; Loewy *et al.*, 1987). Chemotaxis proteins are synthesized at the time of cell division (Shaw *et al.*, 1983; Gomes and Shapiro, 1984). The expression of the hook gene and of the 27.5K and 29K flagellin genes is turned off at or before cell division, while the 25K flagellin continues to be synthesized in the swarmer cell (Agabian *et al.*, 1979).

Many of the genes involved in flagellar biogenesis, whose expression is also temporally controlled, seem to have purely regulatory roles. We are only beginning to understand the relationship of their timed expression to their function in the control of flagellar biogenesis. In this context, the timing of gene expression and the position of the gene in the regulatory hierarchy are discussed below.

VI. Regulatory Hierarchy Controlling Flagellar Biogenesis

A. METHODS OF STUDYING THE HIERARCHY

Regulatory interactions between flagellar genes are studied by comparing the activity of a given gene in a wild-type genetic background with its activity in a background containing a single mutation in another flagellar gene. If the level of expression is higher in the mutant strain, then we conclude that the mutated gene is required for negative regulation of the assayed gene. If expression is lower in the mutant strain, then the mutated gene is required for positive control. These experiments do not tell us if the effects are direct or indirect.

Gene expression in different backgrounds has been measured by several methods: (1) immunoprecipitation of gene products using antibodies raised against purified flagellar proteins [flagellins, hook or chemotaxis proteins (e.g., Johnson et al., 1983; Bryan et al., 1987; Schoenlein and Ely, 1989)]; (2) measurement of in vivo methylation of methyl-accepting chemotaxis proteins (Bryan et al., 1987; Champer et al., 1987; A. Dingwall, unpublished); (3) nuclease S1 protection assays of flagellar gene transcripts (Mullin et al., 1987); (4) disruption of a flagellar gene with a transposon containing a promoterless reporter gene, followed by transduction of the chimeric gene into various mutant backgrounds and assay of reporter gene activity (Bellofatto et al., 1984; Champer et al., 1985, 1987); and (5) cloning a flagellar promoter, placing it in front of a reporter gene on a plasmid, and then placing the resulting plasmid in various mutant backgrounds (Newton et al., 1989; Xu et al., 1989). The first two methods do not distinguish among transcriptional control, translational control, and modification of the stabilities of either the message or the protein. Methods 3 through 5 yield information about transcriptional activity or message stability. With methods 1 through 4, the gene whose activity is being assayed can be located on the chromosome.

B. POSITIVE AND NEGATIVE INTERACTIONS

The hierarchy shown in Fig. 4 is an operational model summarizing the work of several laboratories. Based on the experimentally defined positive interactions, it appears that there are three major levels of the hierarchy. Synthesis of the 25K and 27.5K flagellin proteins is significantly reduced in cells containing mutations in almost any known *fla* or *flb* gene (Johnson et al., 1983; Ohta et al., 1984; Bryan et al., 1987). Mutations in some genes reduce flagellin synthesis to an undetectable level (Ohta et al., 1984). In addition, transcription from *flgL* (the 27.5K flagellin) and *flgK* (the 25K flagellin in the α-flagellin cluster) is inhibited in *flaO* and *flbD* mutant strains (Minnich and Newton, 1987). *flgL* and *flgK* are therefore considered to be at the bottom of the hierarchy.

Genes in the middle level of the hierarchy include the genes in the basal body cluster (Hahnenberger and Shapiro, 1987) and two operons in the hook cluster, the *flbG* operon (promoter P_{II}) encoding the hook protein and the *flaN* operon (promoter P_I) (Ohta et al., 1984). These genes are required for full expression of the 25K and 27.5K flagellins. The regulation of transcription of these genes was studied using transcription fusions to the reporter genes encoding neomycin phosphotransferase (Champer et al., 1987), β-galactosidase (Newton et al., 1989), or luciferase (Xu et al., 1989). Full expression of the *flbN* promoter in the basal body cluster requires intact copies of *flbO* and *flaS*

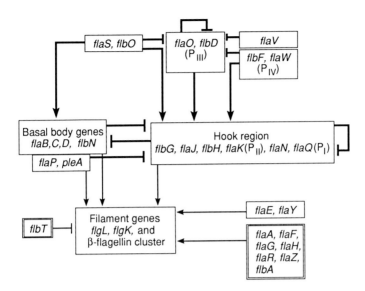

FIG. 4. Genetic regulatory hierarchy in *C. crescentus*. Arrows indicate experiments in which the target gene appears to be under positive influence by the gene at the other end of the arrow. Lines ending in a crossbar indicate negative interactions. Dark lines represent interactions in which the mRNA level was shown to be altered. Light lines indicate that a change took place in the level or rate of synthesis of a protein.

(Xu *et al.*, 1989). Full expression of the *flaG* (P_{II}) promoter and the P_I promoter in the hook cluster requires *flbO* and *flaS* as well as genes in other operons in the hook cluster, *flbF* and *flaW* (P_{IV}) and *flaO* and *flbD* (P_{III}) (Ohta *et al.*, 1984; Champer *et al.*, 1987; Newton *et al.*, 1989; Xu *et al.*, 1989).

Finally, *flaS*, *flbO*, *flaV*, and the genes driven by promoters P_{III} and P_{IV} in the hook cluster are placed in the top level of the hierarchy (Newton *et al.*, 1989; Xu *et al.*, 1989). They are required for expression of some of the genes in the middle level, and no genes have been found which are required for their expression (Newton *et al.*, 1989).

In addition to the positive hierarchy of control, there are negative interactions, defined by the increased expression of a *fla* gene in the presence of another *fla* mutation. Mutations in the basal body cluster and in the hook gene *flaK* lead to increased expression of the hook operon and basal body genes (Newton *et al.*, 1989; Xu *et al.*, 1989). In addition, transcriptional reporter gene insertions in the hook gene *flaK* and in the basal body gene *flbN* prolong the period of transcription of the hook operon and *flbN*, respectively (H. Xu and A. Dingwall, unpublished). This suggests that the observed increases in expression in the presence of mutations may be due in part to a failure to turn off mRNA synthesis (Newton *et al.*, 1989; Xu *et al.*, 1989).

Mutations in several genes at the top of the hierarchy lead to increased levels of synthesis of *flaO* (P_{III}) transcription (Newton *et al.*, 1989; Xu *et al.*, 1989), suggesting negative interactions. Negative regulation has not been observed within the bottom level. For example, mutations in *flgL,* the structural gene for the 27.5K flagellin, do not affect synthesis of the 25K flagellin (Minnich *et al.*, 1988). An interesting example of a negative interaction involves *flbT,* a gene in the α-flagellin cluster. The level of flagellin synthesis is increased in *flbT* mutants, although the cells assemble only a short filament. In a strain containing the wild-type *flbT* gene cloned behind the β-lactamase promoter from pKT230, the levels of flagellins are decreased (Schoenlein and Ely, 1989). In *flbT* mutant strains, about 25% of the cells fail to eject their filaments (A. Driks, unpublished). Turnoff of flagellar synthesis and ejection of the filament may be part of the same developmental pathway.

The positive interactions within the flagellar hierarchy are epistatic to the negative interactions. For example, in a *flaS*$^+$, hook$^-$ background, synthesis from the basal body *flbN* promoter is increased, suggesting that the hook operon exerts negative control over the basal body genes. In a *flaS*$^-$ background, however, synthesis from both the hook and the *flbN* promoter is inhibited (Xu *et al.*, 1989; A. Dingwall, unpublished). This indicates that the enhancement of *flbN* expression by the *flaS* product is a more primary effect than inhibition of *flbN* expression by the hook operon product(s).

A simplified model of the genetic control of flagellar biogenesis in *C. crescentus* places genes in three levels of a hierarchy based on positive interactions (Fig. 4). Within the upper and middle levels there are also negative interactions. This leads to the hypothesis that once the functions of genes at a given level are completed, the genes at that level are turned down and the genes at the next level are turned up, perhaps mediated by a buildup of gene products from the preceding level.

The location of proteins in the flagellar structure is correlated with the time of synthesis of the proteins and to some degree with the apparent position of the genes in the regulatory hierarchy. Hook and basal body genes are above the flagellins in the hierarchy, are synthesized before them during the cell cycle, and occupy a more proximal position in the flagellum. This suggests that proper assembly of the physical structure requires the coordinate regulation of both the timing and the level of gene expression. Note that the epistasis experiments measure the total amount of expression, which depends on both level and duration of expression. The relationship between the modulation of the amount of gene expression and the timing of expression is considered below.

C. HIERARCHY COMPLEXITY

The hierarchy of positive controls described above is divided into discrete levels, that is, no gene is represented as affecting the genes in more than one level below it. This is in part an artificial extension of the data, because not every necessary experiment has been done. There are genes whose location in the hierarchy has been determined by only one method and some that have not been studied at all. In addition, evidence already exists that the entire regulatory network cannot be fully described by so simple a model.

For example, it has been observed that the expression of the regulatory gene *flaO* is not altered in a background containing a mutation in any of several basal body genes (Xu *et al.*, 1989), indicating that *flaO* is indeed above the basal body genes in the hierarchy. However, these experiments have not yet been done with *flaS*. *flaS* was placed near the top of the hierarchy, with the flagellins at the bottom, because *flaS* affects hook and basal body gene expression (Champer *et al.*, 1987; Xu *et al.*, 1989). *flaS* and the hook and basal body genes affect the expression of the flagellin genes (Champer *et al.*, 1987; Newton *et al.*, 1989; Xu *et al.*, 1989), and mutations in the flagellin genes do not block synthesis and assembly of the hook and basal body (Minnich *et al.*, 1988). We do not know if the effect of *flaS* on the flagellins is direct or is mediated through the hook and/or basal body genes. In addition, the effects of mutations in the hook, basal body, or flagellin genes on the expression of *flaS* have not yet been tested.

Evidence indicates that the *flaE* and *flaY* genes are not a simple part of the three-level heirarchy shown in Fig. 4. *flaS*, *flaY*, and *flaE* are all necessary for full expression of the flagellins, but *flaE* and *flaY* do not require *flaS* for their full expression (R. Bryan, unpublished). Cells with mutations in *flaE* and *flaY* lack complete filaments but contain an intact hook, and hook protein synthesis occurs at wild-type levels (Johnson *et al.*, 1983; Schoenlein *et al.*, 1989).

There are additional genes that do not seem to fit neatly into any one level. For example, the genes placed in the double box at the lower right of Fig. 4 are not at the bottom of the hierarchy, because they are clearly required for flagellin synthesis. They differ from the genes of the middle level, because they are not required for synthesis of a complete hook structure (Johnson et al., 1983). They may represent a level that is intermediate between the middle level and the flagellins.

The chemostaxis genes (che) were not placed in the diagram because their position is not clear. Expression of the chemotaxis genes in different mutant backgrounds was assayed by measurement of in vivo methylation of methylated chemotaxis proteins (MCPs). Mutations in many flagellar genes interfere with MCP methylation (Bryan et al., 1987; Champer et al., 1987). However, mutations in flaS and flaV, which are high in the hierarchy, do not prevent methylation of MCPs (Champer et al., 1987). This is surprising because wild-type flaS and flaV alleles are required for expression of the hook gene flaK, and flaK is required for methylation of MCPs. There is evidence that the expression of the chemotaxis proteins, including the methylesterase, methyltransferase, and MCPs, may be coordinately controlled. A Tn5 insertion in the unlinked flaE gene led to decreased levels of all of the above proteins (Bryan et al., 1987).

The 29K flagellin gene flgJ was also not placed in Fig. 4 because its regulation is different from that of any other flagellar gene. Synthesis of the protein is stimulated in almost every known fla mutant (Ohta et al., 1984; Hahnenberger and Shapiro, 1987) while synthesis of the other two flagellins is depressed. Transcription of flgJ is the same in wild-type cells and in flbD mutants (Minnich and Newton, 1987), although protein synthesis is elevated in flbD mutants (Ohta et al., 1984), suggesting that regulation of the level of expression is mediated posttranscriptionally. On the other hand, transcription from flgJ is cell cycle dependent (Loewy et al., 1987), indicating that the production of the 29K protein might be controlled both transcriptionally and posttranscriptionally.

Although a fairly simple three-level hierarchy has emerged, it is not completely nested, and some genes do not fit neatly into any one level. In addition, we do not know if regulatory genes act to control other genes on more than one level. In favor of this idea, the same consensus sequences have been found in front of genes that appear to be on different levels. For example, the II-1 and the 13-bp consensus sequences appear in front of both the hook operon (P_{II}) and the flagellin genes (see below and Table 1). Thus, the three-level model may be a framework on which more complex interactions are overlaid.

Based on the simplified model presented above and on the regulatory hierarchy that controls flagellar biogenesis in E. coli (Komeda, 1982; see below), it is tempting to speculate that a single, nested hierarchy of expression exists, which controls the sequential (temporal) expression of these genes. By this hypothesis, a mutation in an early gene would

TABLE 1
Consensus Sequences in *Caulobacter crescentus*
Flagellar Genes

Gene	II-1	13-mer	Promoter $(\sigma \text{ factor})^a$
Basal body cluster			
flaD		-99	σ^{54} (1)
flbN		-54	σ^{54} (1)
Hook cluster			
P$_\mathrm{I}$ (*flaN*)			σ^{54} (2)
P$_\mathrm{II}$ (*flbG*)	-100	-35	σ^{54} (3)
P$_\mathrm{III}$ (*flaO*)	-49 (2)		
Flagellin cluster			
flgL	$-115, -4$		σ^{54} (4)
flgK	-100		σ^{54} (4)
flgJ	$+38$	-100	σ^{28} (5)
flaE		-140	σ^{28} (6)
flaY		-66	σ^{54} (6)

^{a} References for sequences of the 5' regions are as follows. (*1*) A.
Dingwall unpublished, (*2*) Mullin *et al.* (1987), (*3*) Chen *et al.* (1986),
(*4*) Minnich and Newton (1987), (*5*) Gill and Agabian (1983) and D.
Glaser *et al.* [unpublished (revised sequence of the *flgJ* regulatory
region)], (*6*) Kaplan *et al.* (1989).

lead to loss of synthesis of the flagellins. However, strains with mutations in a variety of genes (*flaE, flaF, flaG, flaK, flaP,* and *flaY*), which
are blocked at various stages of development, all produce the 25K and
27.5K flagellins at the correct time in the cell cycle, but at a greatly
reduced level (Bryan *et al.,* 1987). This suggests that the hierarchical
control of the level of gene expression is not the sole mechanism by
which the cell controls the timing of expression.

D. MECHANISMS OF REGULATION

At least two different types of promoter sequences exist in front of *C.
crescentus* flagellar genes. One class resembles the *ntrA* (σ^{54}) promoter
sites found in front of nitrogen metabolism genes in enterics and in
Rhizobium spp. (Gussin *et al.,* 1986). For this class of promoters, it is
clear that several *C. crescentus fla* genes are transcribed by core RNA
polymerase using purified enteric σ^{54} (Ninfa *et al.,* 1989). In addition,
introduction of these *fla* genes into *E. coli* showed that their expression
is regulated by available nitrogen source (J. Gober, unpublished). The other
class of promoters resembles the σ^{28} -10 promoter sequence found in

front of flagellar genes in *E. coli, S. typhimurium,* and *B. subtilis* (Table 1). In this case it has not yet been demonstrated that a σ^{28}-like factor is required, or that it reads these promoters either *in vitro* or *in vivo*; however, the *E. coli tsr* gene, which has a σ^{28} promoter, is read in a cell cycle-controlled fashion when it is present on the *C. crescentus* chromosome (Frederikse and Shapiro, 1989).

The recognition of flagellar promoters by RNA polymerase containing alternate σ factors does not by itself control timing of gene expression, because *flgK, flgL,* and *flbG* all have σ^{54}-like promoters but are timed differently. Likewise, alternate σ factor binding sequences do not determine the response to trans-acting modulators of the level of *fla* gene expression: expression of the flagellin genes *flgK* and *flgL* is inhibited and expression of the hook operon (promoter P_{II}) is stimulated in cells containing mutations in the hook operon. Thus, flagellar gene transcription in this developmental system is controlled on at least three levels: alternate σ factors, trans-acting factors which modulate the level of expression, and still other factors which contribute to the temporal control of gene expression.

Studies of sequences present in the upstream regions of *Caulobacter fla* genes also suggest that more than one regulatory mechanism can act to control an individual gene. For example, see as follows.

1. Several consensus sequences have been found in *C. crescentus* flagellar genes, including alternate promoters and potential upstream regulatory sites. Table 1 lists known and proposed promoters and upstream consensus sequences found for a subset of *C. crescentus* flagellar genes. There are at least three sites where differential regulation can occur: two consensus sequences (II-1 and 13-mer) and the promoter region. There is evidence from deletion experiments and from site-directed mutagenesis suggesting the importance of the II-1 element (S. Vanway and D. Mullin, unpublished; Mullin and Newton, 1989) and the 13-mer (Xu *et al.*, 1989) in the regulation of transcription.

2. The *flgL* flagellin gene has two copies of the II-1 site, while the *flgK* flagellin gene has only one. The regulatory function of multiple copies of a protein binding site has been studied extensively in the case of the arabinose operon of *E. coli,* involving DNA looping mediated by protein–protein binding (Dunn *et al.*, 1984).

3. Changing the location of a regulatory sequence relative to the start of transcription may provide a mechanism whereby one factor can be used to turn on one set of genes and to turn off another, that is, to act as a genetic switch. Element II-1 is present 100 bp upstream of the transcriptional start site of the P_{II} hook operon and 49 bp upstream of

the P$_{III}$ operon. This difference in location may affect its regulatory role in the two genes (Mullin *et al.*, 1987). A similar phenomenon has been suggested in the case of the λ repressor, in which the distance between the protein binding site and the promoter may control the regulatory effect (Patschne, 1986). The latter distances are smaller than those hypothesized to be important in the case of the *C. crescentus* flagellar genes.

4. At least three sets of *C. crescentus* flagellar genes are organized in an overlapped configuration, such that the predicted promoter and the upstream regulatory region for the second gene in the pair is located within the coding region of the first gene. These gene sets include *flaE* and *flaY* (Kaplan *et al.*, 1989), *flbT* and *flaF* (Schoenlein *et al.*, 1989), and *flaD* and *flbN* (A. Dingwall, unpublished). In the case of the *flaE* and *flaY* genes, the predicted stop codon for *flaE*, promoter for *flaY*, and start codon for *flaY* are all contained in what appears to be a strong invert repeat structure (Kaplan *et al.*, 1989). This is also the case for *flaD* and *flbN*. This structure may be part of a complex regulatory mechanism, the nature of which we do not yet understand.

Biological systems have another degree of freedom, namely, quantitative modulation. The copies of consensus sequences present in different genes are rarely exactly the same, which suggests that several sites may bind the same protein with different strengths. Competition between flagellar promoter regions for binding proteins may be one mechanism whereby differential gene expression is controlled. This is analogous to the functioning of the λ repressor, which binds to its three operator sites with different strengths, a property that is essential for proper functioning of the system (Ptashne, 1986).

E. REGULATION STRATEGIES IN DEVELOPMENTAL SYSTEMS

It is clear that many complex factors modulate the expression of individual genes. Why is gene regulation not simpler, with one factor necessary to activate a given gene at the correct time? We suggest three possible reasons.

1. Having more than one factor acting on individual genes actually increases the flexibility of control, because factors can be mixed and matched. More genes can be uniquely regulated with fewer regulatory elements. For example, a specific binding site can either be present or absent from a given gene. Thus, with one specific binding protein, two genes can be differentially regulated, one with and one without the binding site. With two specific binding proteins, there are four possible

combinations of occupancy of binding sites; therefore, four genes can be differentially regulated, assuming that the binding sites are not over-lapping and that occupancy of one site does not influence occupancy of the other. Alternate σ factors also provide flexibility of control. With two alternative σ factors and two binding proteins, 2^3 possible combinations exist. Thus, just using the three possibilities in Table 1 (two protein binding sites and two alternate promoters), there are eight possible combinations and thus, theoretically, eight ways to differentially regulate genes. In contrast, using a single regulatory factor for each gene, eight factors would be required.

2. Multiple regulatory factors may reflect the many events that must be coordinated to control flagellar gene expression, including synthesis and assembly of earlier structures and expression of regulatory genes. Thus, a given promoter may be expressed only if a full set of signals indicates completion of all the necessary previous steps in the developmental pathway; the promoter, in combination with several trans-acting factors, "senses" the states of several components of the system before "proceeding" with the next step.

3. Precise regulation of expression in a complex genetic system may require multiple imprecise mechanisms acting in concert, as proposed by Ghysen and Farber (1979). An example of two imprecise mechanisms together creating a very precise result is aminoacylation of tRNAs. Isoleucine and valine differ in structure by only a single methyl group. Isoleucyl-tRNA synthetase (isoleucine–tRNA ligase) discriminates between these two amino acids in a two-step process, amino acid–AMP activation and transfer of the activated amino acid to iso-leucine–tRNA. The first process has an error rate of approximately 1/225, that is, isoleucyl-tRNA synthetase forms valyl–AMP once in every 225 reactions. When the complex attempts to bind to tRNA^{Ile}, valyl–AMP complexes are hydrolyzed and removed from the enzyme. This process has an error rate of 1/800, that is, one of every 800 valyl–AMP complexes is not hydrolyzed. Therefore, valine is inserted incorrectly at a frequency of $(1/225)(1/800) = 1/180,000$, which is sufficiently low to ensure accurate protein synthesis (Freifelder, 1983).

In summary, evidence suggests that the *Caulobacter* flagellar hierarchy is not a simple, nested structure, but that each gene is controlled by more than one mechanism. Sequence analysis and, more recently, the isolation of cognate trans-acting proteins (J. Gober, unpublished; D. Mullin, unpublished) support the suggestion that the regulation of *Caulobacter* flagellar gene expression depends on several cis-acting elements. Often there is more than one regulatory sequence controlling each gene, and there may be multiple copies of a single binding sequence. Consensus sequences are found at different positions relative to the different promoters. There is overlap between some genes, and there is probably competition among genes for binding factors. These mechanisms may reflect the multitude of processes necessary to synthesize the components of a complex structure, and to construct it at a specific site on the cell, at a defined time in the cell cycle.

VII. Physical Structure, Timing of Gene Expression, and the Regulatory Hierarchy

The mechanisms of regulation of gene expression that have been proposed in *Caulobacter* are not likely to be unique to *C. crescentus* flagellar genes. Rather, they may be general mechanisms for the control of gene expression during a developmental process. Two obvious characteristics of the developmental events in *C. crescentus* are the construction of a physical structure and the differential timing of gene expression.

Does a clock control the timing of gene expression? A clock may be defined as a mechanism that involves the repeating of a process of precise duration. One obvious clock in *Caulobacter* is DNA replication, which operates at a constant speed (Degnen and Newton, 1972; Dingwall and Shapiro, 1989) and over a significant portion of the cell cycle, including the time of flagellar biogenesis. Sheffery and Newton (1981) have suggested that DNA replication may initiate the regulatory cascade. When DNA replication was inhibited, hook and flagellin gene expression stopped soon thereafter. When DNA replication was temporarily inhibited, hook synthesis was delayed for an equivalent amount of time (Sheffrey and Newton, 1981). According to this hypothesis, flagellar morphogenesis is initiated only in cells that are synthesizing DNA, and perhaps only once a certain master gene has been replicated.

The actual passage of the replication fork through flagellar genes is not what causes initiation of expression of these genes. We know this because map position is not correlated with time of gene expression and because *fla* genes located on plasmids are correctly timed (Ohta *et al.*, 1985; Loewy *et al.*, 1987). In addition, DNA replication does not seem to be necessary for synthesis of all components of the flagellar hierarchy. For example, when phospholipid synthesis was inhibited by precursor deprivation in a glycerol-3-phosphate dehydrogenase mutant, DNA replication was inhibited (Contreras *et al.*, 1979; B. Loewy, unpublished). In the subsequent cell cycle, the synthesis of the 25K and 27.5K flagellins was prevented, but the 29K flagellin was synthesized at the correct time (B. Loewy and G. Marczynski, unpublished). Therefore, DNA synthesis is not an absolute prerequisite for expression of all flagellar genes.

How, then, are the individual promoters separately timed? Perhaps the clock model is inappropriate. Instead, we might view flagellar morphogenesis as a self-regulating process of construction, in which the timing of promoter activity is caused by the delays inherent in building a physical structure. The discovery of negative feedback in the hook and basal body clusters (Newton *et al.*, 1989; Xu *et al.*, 1989) suggests that the cell has a mechanism for sensing the completion of a portion of the structure, perhaps a complete basal body–hook complex. When the structure is completed, one or several components may increase in concentration in the cytoplasm, inhibiting the hook and basal body genes and stimulating the flagellin genes.

Coupling between morphological structure and gene expression has also been proposed for regulation of sporulation in *Bacillus subtilis* (Stragier *et al.*, 1988). The *sigE* (*spoIIBG*) gene encodes pro-σ^E, an inactive precursor to σ^E, a σ factor necessary for expression of some sporulation-specific genes, for example, *spoIID*. Pro-σ^E is processed by the product of *spoIIGA* into its active form. Stragier *et al.* (1988) hypothesized that the *spoIIGA* protein product is fully active only when embedded in one of the two membranes of the sporulation septum. Thus, the septum must be formed before processing activity is maximal. The assembly of the *spoIIGA* product into a processing complex in the septum may explain the 1-hour delay that occurs between the expression of *spoIIGA* and the onset of processing activity.

Many, if not all, genes that have been studied in the *C. crescentus* flagellar developmental pathway seem to have a unique time of expression. Thus, there may be a hierarchy of mechanisms of temporal control: DNA synthesis may be necessary to initiate the process, and the completion of a hook–basal body structure may be a critical point at which early genes are turned off and flagellin genes turned on. Within these sets of genes, other mechanisms may fine tune the timing of gene expression.

VIII. Hierarchies in Other Bacterial Developmental Systems

A. Flagellar Biogenesis in *Escherichia coli*

The single *Caulobacter* flagellum is similar in overall appearance to the peritrichous flagella of *E. coli* and *S. typhimurium*. The *Caulobacter* basal body contains an additional ring (Stallmeyer *et al.*, 1989), and the

Caulobacter filament contains three different flagellins (Driks *et al.*, 1989) whereas the *E. coli* filament contains only one. (For the remainder of this section, we refer to *E. coli,* although the statements made are usually true for *S. typhimurium* as well.) There may be additional structural differences which allow the *C. crescentus* rod to separate into two parts during ejection of the hook–filament structure. It is reasonable to expect that the regulation of flagellar gene expression and assembly pathways in these organisms might be similar. However, the *C. crescentus* flagellar assembly pathway and regulatory hierarchy appear to be somewhat different from those found in *E. coli* and *S. typhimurium*. In addition, the organization of flagellar genes in *Caulobacter crescentus* is somewhat different from that found in *E. coli* and *S. typhimurium.*

Pathways of assembly in *E. coli* (Suzuki and Komeda, 1981) and *S. typhimurium* (Suzuki *et al.*, 1978) were deduced by examination of partial flagellum structures found in the membrane fraction of flagellar mutants in these organisms. Basal body assembly starts with the innermost rings and proximal portion of the rod. The rod is then modified and the outer rings and the distal portion of the rod assembled, prior to hook assembly. At the distal end of the hook, three low-abundance hook-associated proteins (HAPs) are assembled (Homma and Iino, 1985), followed by assembly of the flagellin monomers at the distal tip (Emerson *et al.*, 1970).

As shown in Table 2, at least 34 genes are required in *E. coli* for complete flagellar assembly [the gene designations given are from Iino *et al.* (1988) and replace earlier flagellar gene names]. A mutation in any one of 23 genes prevents the formation of any structure (Macnab, 1987), but 16 of these genes are of unknown function. Four of these genes probably encode proteins which make up the rod (Jones *et al.*, 1989). *fliF* (formerly *flaBI*), encodes the protein which makes up the innermost ring of the basal body (Homma *et al.*, 1987). *flhC* and *flhD* (*flaI* and *flbB*) are regulatory genes required for expression of all other flagellar genes (Komeda, 1982). *fliG, fliM,* and *fliN* (*flaBII, flaAII,* and *motD*) proteins may interact with the motor or chemotaxis apparatus, because some alleles of these genes, or the homologous genes from *S. typhimurium,* produce a Fla⁻ phenotype. Other alleles of the same genes result in the assembly of a normal but paralyzed filament, and other alleles lead to bacteria that have a normal filament but cannot undergo chemotaxis (Yamaguchi *et al.*, 1986).

Mutations in four other genes lead to the assembly of partial basal bodies (Suzuki *et al.*, 1978; Suzuki and Komeda, 1981). Mutations of the hook gene lead to assembly of a basal body but no hook. Mutation of any of the three hook associated proteins (HAPs), the polyhook gene, or the

TABLE 2

Genes Required for Flagellar Assembly in *Escherichia coli*
and *Caulobacter crescentus*

No structure formed	Partial or complete basal body formed	Basal body and hook formed	Basal body, hook, and partial filament formed
Escherichia coli[a]			
Twenty three genes	Five genes	Six genes	
flgB,C,F,G,J (*B,C,F,G* encode rod proteins)	*flgA,D*	*flgK,L* and *fliD* (HAPs)	
flhA,B,C,D (*C* and *D* are regulatory)	*flgE* (hook protein)	*fliC* (flagellin protein)	
fliA,E,F,G, H,I,J,L,M, N,O,P,Q,R (*fliF* encodes M-ring protein)	*flgI* (P-ring protein)	*fliA* (regulatory)	
	flgH (L-ring protein)	*fliK* (polyhook)	
Caulobacter crescentus[b]			
Eleven genes	Three genes	Eight genes	Five genes
flaB,C,I,L, M,P,S,V,W	*flaD* (P-ring protein?)	*flaE,F,G,H, N,Q,R*	*flaA,Y,Z*
flbN,O	*flaK* (hook protein) *flaO*	*flaJ* (polyhook)	*flbA,T*

[a] Suzuki *et al.* (1978), Suzuki and Komeda (1981), Macnab (1987), and Jones *et al.* (1990). Gene designations are from Iino *et al.* (1988).

[b] Johnson and Ely (1979), Johnson *et al.* (1983), Ohta *et al.* (1984), and Hahnenberger and Shapiro (1987).

structural gene for the flagellin leads to assembly of the basal body and hook, but no filament. The HAPs function as adapters between the hook and the filament flagellin monomers. HAPs are probably required because the helical parameters of the hook are different from those of the filament (Macnab, 1987).

Table 2 shows the phenotypes caused by mutations in 27 different *C. crescentus* flagellar genes. These phenotypes were studied in two ways. Membrane fractions of Fla⁻ strains were examined for partial structures (Hahnenberger and Shapiro, 1987). Johnson *et al.* (1983) examined culture fluid for the presence of ejected filaments and hooks. In *C.*

crescentus, mutations in 11 genes have been found to prevent assembly of any structure (Johnson *et al.,* 1983; Hahnenberger and Shapiro, 1987). This is far fewer than the 23 genes required for the initial stages of flagellar assembly in *E. coli.* Mutations in *flaD* lead to the formation of a partial basal body containing the two inner rings and part of the rod. Other partial basal body structures were unstable and fell apart into ring structures during purification (Hahnenberger and Shapiro, 1987); these mutations were included in the list of those which completely blocked assembly. Mutations in the regulatory gene *flaO* and in *flaK,* the structural gene for the hook, have basal bodies but no hook (K. Hahnenberger, unpublished). Strains containing a mutation in any of 13 different genes, not including the flagellin genes themselves, are able to assemble a hook but not a complete filament. *Caulobacter crescentus* apparently requires many more genes for the final stage of filament assembly than does *E. coli.* Four additional genes are not listed in Table 2 because their assembly phenotype has not been determined. *flbF, flbG,* and *flbD* are required for hook synthesis, and *flbH* in the hook operon is required for flagellin synthesis (Ohta *et al.,* 1984, 1985).

As shown in Table 2, the two major differences between flagellar mutations in *E. coli* and in *C. crescentus* are the following. (1) More mutant genes in *E. coli* than in *C. crescentus* block the appearance of even a partial structure. (2) More mutant genes in *C. crescentus* than in *E. coli* block the stage of flagellin assembly into a filament. Three possible reasons for the first difference are the following. (1) There may be additional *C. crescentus* flagellar genes that have not yet been found; there are at least four flagellar genes whose phenotypes have not yet been determined. (2) Six of the twenty three *E. coli* mutants are leaky (Yamaguchi *et al.,* 1986), and there may be leaky mutations in *C. crescentus* which have not been detected (Johnson and Ely, 1979). (3) The initial stage of flagellar assembly in *C. crescentus* may involve laying down a structure which marks the site of assembly, perhaps in keeping with the "organizing center" proposed by Newton (1984). Mutations that disrupt this initial structure in *E. coli* may simply lead to a Fla⁻ phenotype, while similar mutations in *C. crescentus* might disrupt stalk initiation or even cell division.

It is harder to explain why *C. crescentus* has so many genes required at the stage of flagellin assembly. One possibility is that some of the mutants blocked in flagellin assembly in *C. crescentus* may be functionally equivalent to mutants blocked before any assembly in *E. coli.* There may be some specific structure, perhaps part of the motor, required for filament assembly. This structure may be assembled before the basal body in *E. coli* and after hook assembly in *C. crescentus.*

Alternatively, the 13 genes required for flagellin assembly may encode products which are functionally equivalent to the three HAPs in *E. coli*.

The genetic organization of the *fla* genes in *C. crescentus* is somewhat different from that in *E. coli*. In both organisms there are three main regions of flagellar genes. The organization of the genes within clusters differs between these organisms. In *E. coli*, the hook gene and the basal body gene encoding the P-ring protein are in the same operon. In *C. crescentus,* the *flaD* gene, which is thought to encode the P-ring protein, is in a different operon from the hook gene. A caveat, however, is that the functions of all the genes in the hook and basal body clusters have not yet been identified. Several of the operons containing flagellar genes in *E. coli* are larger than the largest flagellar operons found thus far in *C. crescentus*. In addition, several *C. crescentus* flagellar genes lie outside the clusters (Ely *et al.,* 1984).

This dispersed genetic organization may be related to the fact that the genetic regulatory hierarchy controlling *C. crescentus* flagellar development appears more complicated than that described by Komeda for *E. coli*. In *E. coli* two genes, *flhC* and *flhD* (formerly called *flaI* and *flbB*), are essential for transcription of all other flagellar genes and are, therefore, at the top of the hierarchy (Komeda, 1982). These genes are under catabolite control (Silverman and Simon, 1974). They are the only flagellar genes required for expression of the hook gene in *E. coli* (Komeda *et al.,* 1984). In *C. crescentus,* by contrast, hook gene synthesis requires expression of at least six other *fla* genes, at least two of which are also required for basal body gene expression (Fig. 4).

The middle level of the *E. coli* regulatory hierarchy contains the majority of the flagellar genes, including basal body genes, the hook gene, the rod protein genes, and the genes whose products interact with the motor and chemotaxis apparatus. This set of genes includes almost all of the genes required for formation of partial basal body and hook structures. Mutations in the genes at this level lead to loss of transcription of the genes at lower levels but do not affect the amount of transcription of genes on their level. Shutoff of synthesis of genes at lower levels of the hierarchy does not occur in a *flgA⁻* (formerly called *flaU*) background. Thus, *flgA,* whose gene product may be a structural component of the basal body (Suzuki and Komeda, 1981), is required for negative control of gene expression (Komeda, 1986). Another way to look at this interaction is as follows: the other genes in the middle level of the *E. coli* hierarchy are not actually positive regulators, but, in the absence of these genes, the *flgA* gene product exerts a negative effect. Experiments have not been done to see if there are similar cases in *C. crescentus* in which a regulatory effect actually depends on the presence of one wild-type allele and a mutation in a different gene.

Both *E. coli* and *C. crescentus* exhibit negative control of the *fla* genes. There are more examples of negative interactions in *C. crescentus* than in *E. coli*. As described above, mutations in several *C. crescentus* genes lead to increased synthesis of either the hook or basal body genes (Newton *et al.*, 1989; Xu *et al.*, 1989). In *C. crescentus* there is another gene, *flbT*, which is required to turn off expression of flagellin genes. Strains which overproduce *flbT* make decreased levels of flagellins (Schoenlein and Ely, 1989).

A striking similarity between the two organisms is that mutations in almost any *fla* gene cause a decrease in the level of synthesis of the flagellins. In *E. coli* these mutations lead to the loss of detectable flagellin synthesis (Komeda *et al.*, 1984). In *Caulobacter* it appears that some flagellar mutations lead to complete loss of flagellin synthesis while others lead to only partial loss (Johnson *et al.*, 1983; Ohta *et al.*, 1984; Bryan *et al.*, 1987). This difference may reflect a real difference in regulation, or it may reflect differences in methods used to assay flagellin synthesis.

In *C. crescentus,* flagellar gene expression is coordinated with the cell cycle. It is possible that initiating and terminating the expression of several genes in concert with the cell cycle may require additional regulatory elements in *C. crescentus* that might not be needed in *E. coli*. However, the cell cycle might play a role in regulating flagellar gene expression in *E. coli* as well. For example, DNA synthesis is required for regeneration of flagella in *E. coli* (see Iino, 1985). In addition, several *E. coli* temperature-sensitive cell division mutants were unable to synthesize flagellin mRNA or assemble filaments at the restrictive temperature (Nishimura and Hirota, 1989). It is interesting that the *E. coli* chemotaxis receptor gene, *tsr,* is cell cycle controlled in *C. crescentus* (Frederikse and Shapiro, 1989).

Because of the similarities between *E. coli* and *C. crescentus* flagellar organization, it is likely that these organisms diverged from a common, flagellated, precursor. The observed differences in regulation, assembly pathway, and genetic organization must have evolved since that divergence.

B. Sporulation in *Bacillus subtilis*

Bacillus subtilis, a gram-positive rod, is one of several related organisms that can sporulate in response to starvation or other environmental insults, thereby increasing its probability of survival. The genetic regulatory apparatus in *B. subtilis* is able to respond to an environmental signal and begin differentiation, during which the synthesis and assembly of sporulation products is controlled in a coordinated manner.

Thus, one major difference between *B. subtilis* and *C. crescentus* is that the differentiation events in *B. subtilis* are induced by environmental cues. The mother cell follows a program of development that is separate from that of the forespore. The following is a description of some of the genetic regulatory features controlling endospore formation in *B. subtilis* (for recent reviews, see Losick *et al.*, 1986, 1989).

The stages of sporulation are defined by characteristic morphogenetic features. Sporulation mutations are named for the stage at which they block development. Stage 0 of sporulation is defined as vegetative growth. *spoO* mutants are able to grow vegetatively but cannot even begin to sporulate. Many *spoO* genes are expressed during vegetative growth of wild-type cells. Chromosomal condensation occurs during stage 1 of sporulation. No mutants blocked at this stage have been found. Stage 2 occurs about 1 hour after initiation of sporulation. During stage 2, an asymmetric membrane septum forms, dividing the cell into two parts. During stage 3, the forespore is engulfed by the membrane layer of the mother cell, and it is at this stage that the developmental programs of the mother cell and the forespore diverge (Errington *et al.*, 1988). During stage 4, a cortex is layed down between membranes. During stage 5, about 4 hours into the sporulation process, a coat is made outside of the cortex by the mother cell. Stage 6 involves maturation and assembly. Very little protein synthesis takes place, and the spore becomes resistant to the outside world. At stage 7, lysis of the mother cell occurs. Both sporulation and germination of an entire population can be induced synchronously, allowing biochemical characterization of the stages of sporulation.

There are at least 60 *spo* (sporulation), *ger* (germination), or *out* (outgrowth) genetic loci known. There probably are more genes, because many of these loci are actually operons. Most of these genes have a regulatory role. There are many other sporulation genes that encode coat proteins or other structural proteins and do not cause a Spo⁻ phenotype, so they are not included in the 60 loci mentioned above.

Sporulation in *B. subtilis* involves a regulatory cascade of genes that are activated in a defined order and whose protein products interact to form a physical structure. In this respect, sporulation and flagellar biogenesis are similar. The hierarchical nature of the regulatory interactions in *B. subtilis* is evidenced by the tight dependence of expression of *spo* genes. Most *spo* genes are expressed only if all the *spo* genes from earlier stages are expressed. In some cases the mechanisms for this dependency are understood. As in *C. crescentus,* the timing of gene expression is correlated with the role of the protein product in the developing spore and with the location in the regulatory hierarchy.

There are several differences between sporulation in *B. subtilis* and flagellar development in *C. crescentus*. The time scale is longer in *B. subtilis;* sporulation takes about 6 hours at 37°C, while flagellar morphogenesis requires about 1 hour. Differentiation events in *B. subtilis* are induced by the outside environment; the flagellar hierarchy in *C. crescentus* is possibly induced by DNA replication. The genetic hierarchy in *B. subtilis* appears to be fully nested. For example, if genes *A* and *B* require gene *C* for their activation and gene *C* requires gene *D*, then *A* and *B* require *D*. In *C. crescentus* the hierarchy is not nested. For example, the chemotaxis genes require *flaK* for full expression and *flaK* requires *flaV*, but the chemotaxis genes do not require *flaV* (Champer *et al.*, 1987).

A recurrent motif in the control of sporulation in *B. subtilis* is the ordered appearance of new σ factors which confer promoter specificity on RNA polymerase and allow expression of specific sets of sporulation genes. Six σ factors have been identified in sporulation; in comparison, there is evidence for two alternate promoters that would presumably use two different σ factors for flagellar gene expression in *C. crescentus*, and only one in *E. coli, S. typhimurium*, and *B. subtilis* flagellar development. In *B. subtilis* sporulation, the availability of specific σ factors is involved in the control of timing; in turn, elaborate mechanisms control the expression of the σ factors themselves (see Jonas *et al.*, 1988; Stragier *et al.*, 1988, 1989). Evidence accumulated so far suggests that σ factors are not sufficient to control timing in the *Caulobacter* flagellar hierarchy, as discussed in Section VI,D. In *B. subtilis* as well, the timing of transcription of some genes depends on the presence of auxiliary factors (Kroos *et al.*, 1989).

IX. Components of Regulatory Hierarchies

Complex genetic systems can be divided into three parts: (1) a sensing apparatus that receives the signal(s) necessary to initiate the process, (2) a regulatory apparatus that transduces the environmental signal and controls the expression of the structural genes, and (3) genes that encode the enzymes or structural proteins.

The signals that initiate flagellar biogenesis differ in *E. coli, B. subtilis*, and *C. crescentus*. Expression of the flagellar hierarchy in *E. coli* is induced by a decrease in glucose concentration (Macnab, 1987). Similarly, *B. subtilis* flagellar development is induced late in logarithmic growth (Helmann and Chamberlin, 1987). In contrast, no environmental signals have been found to initiate flagellar development in

Caulobacter. It is not clear what signals initiate flagellar development in *Caulobacter,* but DNA replication may be involved (see above).

Escherichia coli and *B. subtilis* share both the signal (nutrient deprivation) that triggers flagellar formation as well as some regulatory mechanisms. All known flagellar genes in *E. coli* and *B. subtilis* are transcribed from promoters that resemble the σ^{28} promoter sequence (Helmann and Chamberlin, 1987; Helmann et al., 1988; Bartlett et al., 1988). In *B. subtilis,* disrupting the gene encoding σ^{28} leads to nonmotile filamentous cells that do not synthesize detectble levels of flagellin but do undergo normal sporulation (Helmann et al., 1988).

In contrast, two types of promoter sequences have been found in front of *C. crescentus* flagellar genes. Some *C. crescentus* flagellar promoters contain sequences that resemble the σ^{28} -10 region, and some contain sequences that resemble promoters from nitrogen metabolism genes in *Rhizobium* and in enteric bacteria, which are read by σ^{54} (Table 1). There is functional evidence that these promoters are indeed what they appear to be, and that there must be σ factors in *C. crescentus* that are homologous to σ^{28} and σ^{54}. For example, the *E. coli* MCP gene *tsr* is transcribed in *C. crescentus* predivisional cells from its σ^{28}-like promoter in a cell cycle-dependent manner (Frederikse and Shapiro, 1989). Three *C. crescentus* flagellar genes containing the σ^{54}-like promoter have been transcribed *in vitro* using purified *E. coli* RNA polymerase containing σ^{54} and NtrC and NtrB proteins (Ninfa et al., 1989). Thus, *C. crescentus* utilizes one set of promoters similar to those for the flagellar genes of *E. coli* and *B. subtilis* and another set of promoters similar to the nitrogen assimilation promoters from *E. coli.*

The flagellar and nitrogen metabolism systems may share other regulatory components. Phosphorylation of regulatory proteins is an essential reaction in both the chemotaxis system and the nitrogen assimilation system of enterics. In *S. typhimurium,* CheA phosphorylates CheY, which interacts with the flagellar motor to control swimming behavior (Wylie et al., 1988). In *E. coli,* NtrB (NR$_{II}$) phosphorylates NtrC (NR$_I$), which then activates nitrogen-regulated promoters (Ninfa and Magasanik, 1986). Ninfa et al. (1988) showed that, *in vitro,* CheA can phosphorylate NtrC, and thereby activate transcription from the nitrogen-regulated *glnA* promoter, and NtrB can phosphorylate CheY. These results suggest that functional similarities and perhaps evolutionary homologies between the nitrogen regulation and flagellar development systems are extensive.

Apparently sensory and regulatory components of various genetic systems can be combined in different ways to regulate the gene products of those systems. For example, see as follows.

1. The same regulatory components can function in systems that respond to different cues. Systems controlled by σ^{28} promoters respond to environmental cues (*E. coli* and *B. subtilis*), while in *C. crescentus* σ^{28}-like promoters exist as part of a regulatory hierarchy that does not seem to respond to environmental cues but may depend on DNA replication. σ^{54}-like promoters are part of systems that respond to fixed nitrogen or to oxygen levels in the environment and control nitrogen assimilation in enterics and in *Rhizobium*. They are a component of flagellar development in *C. crescentus*.

2. Similar structural genes can be controlled by different regulatory apparatuses: flagellar morphogenesis is controlled by σ^{28}-like promoters in *E. coli* and *B. subtilis* and σ^{28}- and σ^{54}-like promoters in *C. crescentus*.

3. One regulatory apparatus can be used to control different structural genes: σ^{54}-like promoters control nitrogen metabolism in enterics and in *Rhizobium;* they control flagellar morphogenesis in *Caulobacter*.

Thus, it appears that, evolutionarily, genetic regulatory systems are quite plastic. These patterns raise an interesting question. Are the relationships we see among the environmental sensors, regulatory elements, and structural genes of different systems the results of chance rearrangements of chromosomes that happened to leave, for example, a σ^{54}-like promoter in front of a flagellar gene in *C. crescentus*? Or was there natural selection for specific combinations of elements?

ACKNOWLEDGMENTS

We thank Paul Greene and Hong Xu for critical reading of the manuscript and Adam Driks for providing the electron micrograph of *C. crescentus*.

REFERENCES

Agabian, N., Evinger, M., and Parker, G. (1979). Generation of asymmetry during development. *J. Cell Biol.* **81**, 123–136.

Bartlett, D. H., Frantz, B. B., and Matsumura, P. (1988). Flagellar transcriptional activators *flbB* and *flaI*: Gene sequences and 5' consensus sequences of operons under *flbB* and *flaI* control. *J. Bacteriol.* **170**, 1575–1581.

Bellofatto, V., Shapiro, L., and Hodgson, D. (1984). Generation of a Tn5 promoter probe and its use in the study of gene expression in *Caulobacter crescentus. Proc. Natl. Acad. Sci. U.S.A.* **81**, 1035–1039.

Bryan, R., Champer, R., Gomes, S., Ely, B., and Shapiro, L. (1987). Separation of temporal control and trans-acting modulation of flagellin and chemotaxis genes in *Caulobacter. Mol. Gen. Genet.* **206**, 300–306.

Champer, R., Bryan, R., Gomes, S. L., Purucker, M., and Shapiro, L. (1985). Temporal and spatial control of flagellar and chemotaxis gene expression during *Caulobacter* cell differentiation. *Cold Spring Harbor Symp. Quant. Biol.* **50,** 831–840.

Champer, R., Dingwall, A., and Shapiro, L. (1987). Cascade regulation of *Caulobacter* flagellar and chemotaxis genes. *J. Mol. Biol.* **194,** 71–80.

Chen, L.-S., Mullin, D., and Newton, A. (1986). Identification, nucleotide sequence, and control of developmentally regulated promoters in the hook operon region of *Caulobacter crescentus. Proc. Natl. Acad. Sci. U.S.A.* **83,** 2860–2864.

Contreras, I., Bender, R. A., Mansour, J., Henry, S., and Shapiro, L. (1979). *Caulobacter crescentus* mutant defective in membrane phospholipid synthesis. *J. Bacteriol.* **140,** 612–619.

Degnen, S. T., and Newton, A. (1972). Chromosome replication during development in *Caulobacter crescentus. J. Mol. Biol.* **64,** 671–680.

Dingwall, A., and Shapiro, L. (1989). Rate, origin and bi-directionality of *Caulobacter* chromosome replication as determined by pulsed-field gel electrophoresis.*Proc.Natl. Acad. Sci. U.S.A.* **86,** 119–123.

Driks, A., Byran, R., Shapiro, L., and DeRosier, D. J. (1989). The organization of the *Caulobacter crescentus* flagellar filament. *J. Mol. Biol.* **206,** 627–636.

Dunn, T. M., Hahn, S., Ogden, S., and Schleif, R. F. (1984). An operator at −280 base pairs that is required for repression of araBAD operon promoter: Addition of DNA helical turns between the operator and promoter cyclically hinders repression. *Proc. Natl. Acad. Sci. U.S.A.* **81,** 5017–5020.

Ely, B., Croft, R. H., and Gerardot, C. J. (1984). Genetic mapping of genes required for motility in *Caulobacter crescentus. Genetics* **108,** 523–532.

Ely, B., Gerardot, C. J., Fleming, D. L., Gomes, S. L., Frederikse, P., and Shapiro, L. (1986). General nonchemotactic mutants of *Caulobacter crescentus. Genetics* **114,** 717–730.

Emerson, S. U., Tokuyasu, K., and Simon, M. I. (1970). Bacterial flagella: Polarity of elongation. *Science* **169,** 190–192.

Errington, J., Cutting, S., and Mandelstam, J. (1988). Branched pattern of regulatory interactions between late sporulation genes in *Bacillus subtilis. J. Bacteriol.* **170,** 796–801.

Frederikse, P. H., and Shapiro, L. (1989). An *E. coli* chemoreceptor gene is temporally controlled in *Caulobacter. Proc. Natl. Acad. Sci. U.S.A.* **86,** 4061–4065.

Freifelder, D. (1983). "Molecular Biology." Jones and Bartlett, Boston, Massachusetts.

Ghysen, A., and Farber, A. (1979). A model for the achievement of accuracy in biology and economy. *In* "Kinetic Logic: A Boolean Approach to the Analysis of Complex Regulatory Systems" (R. Thomas and S. Levin, eds.), Vol. 29, pp. 464–472. Springer-Verlag, Berlin and New York.

Gill, P. R., and Agabian, N. (1983). The nucleotide sequence of the $M_r = 28,500$ flagellin gene of *Caulobacter crescentus. J. Biol. Chem.* **258,** 7395–7401.

Gomes, S. L., and Shapiro, L. (1984). Differential expression and positioning of chemotaxis methylation proteins in *Caulobacter. J. Mol. Biol.* **177,** 551–568.

Gussin, G. N., Ronson, C. W., and Ausubel, F. M. (1986). Regulation of nitrogen fixation genes. *Annu. Rev. Genet.* **20,** 567–591.

Hahnenberger, K., and Shapiro, L. (1987). Identification of a gene cluster involved in flagellar basal body biogenesis in *Caulobacter crescentus. J. Mol. Biol.* **194,** 91–103.

Hahnenberger, K. M., and Shapiro, L. (1988). Organization and temporal expression of a flagellar basal body gene in *Caulobacter crescentus. J. Bacteriol.* **170,** 4119–4124.

Helmann, J. D., and Chamberlin, M. J. (1987). DNA sequence analysis suggests that expression of flagellar and chemotaxis genes in *Escherichia coli* and *Salmonella typhimurium* is controlled by an alterate sigma factor. *Proc. Natl. Acad. Sci. U.S.A.* **84**, 6422–6424.

Helmann, J. D., Marquez, L. M., and Chamberlin, M. J. (1988). Cloning, sequencing, and distribution of the *Bacillus subtilis* σ^{28} gene. *J. Bacteriol.* **170**, 1568–1574.

Homma, M., and Iino, T. (1985). Locations of hook-associated proteins in flagellar structures of *Salmonella typhimurium*. *J. Bacteriol.* **162**, 183–189.

Homma, M., Aizawa, S.-I., Dean, G. E., and Macnab, R. M. (1987). Identification of the M-ring protein of the flagellar motor of *Salmonella typhimurium*. *Proc. Natl. Acad. Sci. U.S.A.* **84**, 7483–7487.

Iino, T. (1985). Structure and assembly of flagella. *In* "Molecular Cytology of *Escherichia coli*" (N. Nanninga, ed.), pp. 9–37. Academic Press, London.

Iino, T., Komeda, Y., Kutsukake, K., Macnab, R. M., Matsumura, P., Parkinson, J., Simon, M. I., and Yamaguchi, S. (1988). New unified nomenclature for the flagellar genes of *Escherichia coli* and *Salmonella typhimurium*. *Microbiol. Rev.* **52**, 533–535.

Johnson, R. C., and Ely, B. (1979). Analysis of non-motile mutants of the dimorphic bacterium *Caulobacter crescentus*. *J. Bacteriol.* **137**, 627–635.

Johnson, R. C., Feber, D. M., and Ely, B. (1983). Synthesis and assembly of flagellar components by *Caulobacter crescentus* motility mutants. *J. Bacteriol.* **154**, 1137–1144.

Jonas, R. M., Weaver, E. A., Kenney, T. J., Moran, C. P., Jr., and Haldenwang, W. G. (1988). The *Bacillus subtilis spoIIG* operon encodes both σ^E and a gene necessary for σ^E activation. *J. Bacteriol.* **170**, 507–511.

Jones, C. J., Macnab, R. M., Okino, H., and Aizawa, S.-I. (1990). Stoichiometric analysis of the flagellar hook–basal body complex of *Salmonella typhimurium*. *J. Bacteriol.* (in press).

Kaplan, J. B., Dingwall, A., Bryan, R., Champer, R., and Shapiro, L. (1989). Temporal regulation and overlap organization of two *Caulobacter* flagellar genes. *J. Mol. Biol.* **205**, 71–83.

Komeda, Y. (1982). Fusions of flagellar operons to lactose genes on a Mu *lac* bacteriophage. *J. Bacteriol.* **150**, 16–26.

Komeda, Y. (1986). Transcriptional control of flagellar genes on *Escherichia coli* K12. *J. Bacteriol.* **168**, 1315–1318.

Komeda, Y., Ono, N., and Kagawa, H. (1984). Synthesis of flagellin and hook subunit protein in flagellar mutants of *Escherichia coli* K12. *Mol. Gen. Genet.* **194**, 49–51.

Koyasu, S., Asada, M., Fukuda, A., and Okada, Y. (1981). Sequential polymerization of flagellin A and flagellin B into *Caulobacter* flagella. *J. Mol. Biol.* **153**, 471–475.

Kroos, L., Kunkel, B., and Losick, R. (1989). Switch protein alters specificity of RNA polymerase containing a compartment-specific sigma factor. *Science* **243**, 526–529.

Lagenaur, C., and Agabian, N. (1978). *Caulobacter* flagella organelle: Synthesis, compartmentation, and assembly. *J. Bacteriol.* **135**, 1062–1069.

Loewy, Z. G., Bryan, R. A., Reuter, S. H., and Shapiro, L. (1987). Control of synthesis and positioning of a *Caulobacter crescentus* flagellar protein. *Genes Dev.* **1**, 625–635.

Losick, R., Youngman, P., and Piggot, P. J. (1986). Genetics of endospore formation in *Bacillus subtilus*. *Annu. Rev. Genet.* **20**, 625–669.

Losick, R., Kroos, L., Errington, J., and Youngman, P. (1989). Pathways of developmentally regulated gene expression in the spore-forming bacterium *Bacillus subtilis*. *In* "Genetics of Bacterial Diversity" (K. A. Chater and D. A. Hopwood, eds.). Academic Press, London.

Macnab, R. (1987). Flagella. *In* "*Escherichia coli* and *Salmonella typhimurium* Cellular and Molecular Biology" (F. C. Neidhardt, ed.), Vol. 1, pp. 70–83. Am. Soc. Microbiol., Washington, D.C.

Milhausen, M., and Agabian, N. (1983). *Caulobacter* flagellin mRNA segregated asymmetrically at cell division. *Nature (London)* **302**, 630–632.

Minnich, S. A., and Newton, A. (1987). Promoter mapping and cell cycle regulation of flagellin gene transcription in *Caulobacter crescentus*. *Proc. Natl. Acad. Sci. U.S.A.* **84**, 1142–1146.

Minnich, S. A., Ohta, N., Taylor, N., and Newton, A. (1988). Role of the 25-, 27- and 29-kDa flagellins of *Caulobacter crescentus* in cell motility; a method for the construction of the Tn5 insertion and deletion mutants by gene replacement. *J. Bacteriol.* **170**, 3953–3960.

Mullin, D., and Newton, A. (1989). Ntr-like promoters and upstream regulatory sequence *ftr* are required for transcription of a developmentally regulated *Caulobacter crescentus* flagellar gene. *J. Bacteriol.* **171**, 3218–3227.

Mullin, D., Minnich, S., Chen, L.-S., and Newton, A. (1987). A set of positively regulated flagellar gene promoters in *Caulobacter crescentus* with sequence homology to the *nif* gene promoters of *Klebsiella pneumoniae*. *J. Mol. Biol.* **195**, 939–943.

Newton, A. (1984). Temporal and spatial control of the *Caulobacter* cell cycle. *In* "The Microbial Cell Cycle" (P. Nurse and E. Streiblova, eds.). CRC Press, Boca Raton, Florida.

Newton, A. (1989). Differentiation in *Caulobacter* flagellum development, motility and chemotaxis. *In* "Genetics of Bacterial Diversity" (K. Chater and D. A. Hopwood, eds.), pp. 199–220. Academic Press, London.

Newton, A., Ohta, N., Ramakrishnan, G., Mullin, D., and Raymond, G. (1989). Genetic switching in the flagellar gene hierarchy requires negative as well as positive regulation of transcription. *Proc. Natl. Acad. Sci. U.S.A.* **86**, 6651–6655.

Ninfa, A. J., and Magasanik, B. (1986). Covalent modification of the *glnG* product, NRI, by the *glnL* product, NRII, regulates the transcription of the *glnALG* operon in *Escherichia coli*. *Proc. Natl. Acad. Sci. U.S.A.* **83**, 5909–5913.

Ninfa, A. J., Ninfa, E. G., Lupas, A. N., Stock, A., Magasanik, B., and Stock, J. (1988). Crosstalk between bacterial chemotaxis signal transduction proteins and regulators of transcription of the Ntr regulon: Evidence that nitrogen assimilation and chemotaxis are controlled by a common phosphotransfer mechanism. *Proc. Natl. Acad. Sci. U.S.A.* **85**, 5492–5496.

Ninfa, A. J., Mullin, D. A., Ramakrishnan, G., and Newton, A. (1989). *Escherichia coli* σ^{54} RNA polymerase recognizes *Caulobacter crescentus flbG* and *flaN* flagellar gene promoters *in vitro*. *J. Bacteriol.* **171**, 383–391.

Nishimura, A., and Hirota, Y. (1989). A cell division mechanism controls the flagellar regulon in *Escherichia coli*. *Mol. Gen. Genet.* **216**, 340–346.

Ohta, N., Swanson, E., Ely, B., and Newton, A. (1984). Physical mapping and complementation analysis of transposon Tn5 mutations in *Caulobacter crescentus*: Organization of transcriptional units in the hook gene cluster. *J. Bacteriol.* **158**, 897–904.

Ohta, N., Chen, L.-S., Swanson, E., and Newton, A. (1985). Transcriptional regulation of a periodically controlled flagellar gene operon in *Caulobacter crescentus*. *J. Mol. Biol.* **186**, 107–115.

Osley, M. A., Sheffery, M., and Newton, A. (1977). Regulation of flagellin synthesis in the cell cycle of *Caulobacter*: Dependence on DNA replication. *Cell* **12**, 393–400.

Poindexter, J. (1964). Biological properties and classification of the *Caulobacter* group. *Bacteriol. Rev.* **28**, 231–295.

Poindexter, J. (1981). The caulobacters: Ubiquitous unusual bacteria. *Microbiol. Rev.* **45**, 123–179.

Ptashne, M. (1986). "A Genetic Switch." Blackwell, Cambridge, Massachusetts.

Schoenlein, P. V., and Ely, B. (1989). Characterization of strains containing mutations in the contiguous *flaF*, *flbT*, or *flbA–flaG* transcription unit and identification of a novel Fla phenotype in *Caulobacter crescentus*. *J. Bacteriol.* **171**, 1554–1561.

Schoenlein, P. V., Gallman, L. S., and Ely, B. (1989). Organization of the *flaFG* gene cluster and identification of two additional genes involved in flagellum biogenesis in *Caulobacter crescentus*. *J. Bacteriol.* **171**, 1544–1553.

Shapiro, L. (1985). Generation of polarity during *Caulobacter* cell differentiation. *Annu. Rev. Cell Biol.* **1**, 173–207.

Shaw, P., Gomes, S. L., Sweeney, K., Ely, B., and Shapiro, L. (1983). Methylation involved in chemotaxis is regulated during *Caulobacter* differentiation. *Proc. Natl. Acad. Sci. U.S.A.* **80**, 5261–5265.

Sheffery, M., and Newton, A. (1981). Regulation of periodic protein synthesis in the cell cycle: Control of initiation and termination of flagellar gene expression. *Cell* **24**, 49–57.

Silverman, M., and Simon, M. (1974). Characterization of *Escherichia coli* flagellar mutants that are insensitive to catabolite repression. *J. Bacteriol.* **120**, 1196–1203.

Stallmeyer, M. J. B., Hahnenberger, K., Sosinsky, G. E., Shapiro, L., and DeRosier, D. (1989). Image reconstruction of the flagellar basal body of *Caulobacter crescentus*. *J. Mol. Biol.* **205**, 511–518.

Stragier, P., Bonamy, C., and Karmazyn-Campelli, C. (1988). Processing of sporulation sigma factor in *Bacillus subtilis*: How morphological structure could control gene expression. *Cell* **52**, 697–704.

Stragier, P., Kunkel, B., Kroos, L., and Losick, R. (1989). Chromosomal rearrangement generating a composite gene for a developmental transcription factor. *Science* **243**, 507–512.

Suzuki, T., and Komeda, Y. (1981). Incomplete flagellar structures in *Escherichia coli* mutants. *J. Bacteriol.* **145**, 1036–1041.

Suzuki, T., Iino, T., Horiguchi, T., and Yamaguchi, S. (1978). Incomplete flagellar structures in nonflagellate mutants of *Salmonella typhimurium*. *J. Bacteriol.* **133**, 904–915.

Wagenknecht, T., DeRosier, D., Shapiro, L., and Weissborn, A. (1981). Three-dimensional reconstruction of the flagellar hook from *Caulobacter crescentus*. *J. Mol. Biol.* **151**, 439–465.

Weissborn, A., Steinmann, H. M., and Shapiro, L. (1982). Characterization of the proteins of the *Caulobacter crescentus* flagellar filament: Peptide analysis and filament organization. *J. Biol. Chem.* **257**, 2066–2074.

Wylie, D., Stock, A., Wong, C. Y., and Stock, J. (1988). Sensory transduction in bacterial chemotaxis involves phosphotransfer between *che* proteins. *Biochem. Biophys. Res. Commun.* **151**, 891–896.

Xu, H., Dingwall, A., and Shapiro, L. (1989). Negative transcriptional regulation in the *Caulobacter* flagellar hierarchy. *Proc. Natl. Acad. Sci. U.S.A.* **86**, 6656–6660.

Yamaguchi, S., Fujita, H., Ishihara, A., Aizawa, S.-I., and Macnab, R. M. (1986). Subdivision of flagellar genes of *Salmonella typhimurium* into regions responsible for assembly, rotation, and switching. *J. Bacteriol.* **166**, 187–193.

COMBINATORIAL ASSOCIATIONS OF REGULATORY PROTEINS AND THE CONTROL OF CELL TYPE IN YEAST

George F. Sprague, Jr.

Institute of Molecular Biology and Department of Biology,
University of Oregon, Eugene, Oregon 97403

I. Introduction

Cell specialization is a feature common to all eukaryotic organisms. Multicellular creatures are comprised of a number of different types of cells, 200 or so in the case of mammalian species (Alberts *et al.*, 1983), each of which carries out a particular function for the organism as a whole. By and large, each cell type owes its unique phenotype to the transcription of a specific set of genes encoding the proteins that endow the cells with their special properties. How this cell-type-specific transcription is achieved in molecular terms is not known, but recent studies with the yeast *Saccharomyces cerevisiae* point to a strategy that is simple and elegant. In this strategy, the combinatorial association of several proteins creates regulatory complexes with distinct DNA binding specificities: two complexes serve as transcription activators and two as transcription repressors. The sum of their activities is to gener-

33

ate three yeast cell types, each of which transcribes a unique set of genes. Whether this strategy is directly applicable to the generation of cell types in other eukaryotes is, of course, not known. However, at least one of these yeast regulatory proteins is homologous to a transcription factor from mammalian species, suggesting that combinatorial association of proteins to give complexes of unique activity may be a general theme in eukaryotic gene regulation. In this article, I review the mechanism by which the three yeast cell types are generated and maintained, focusing on the role of the regulatory protein complexes in achieving cell-type-specific transcription. Other recent reviews emphasize other facets of the yeast life cycle and offer different perspectives (Sprague *et al.*, 1983a; Nasmyth and Shore, 1987; Cross *et al.*, 1988; Herskowitz, 1988).

II. Yeast Cell Types and the Mating Process

Yeast exhibits three specialized cell types in the course of its sexual reproductive cycle. Two of the cell types, called **a** and α, are haploid and are capable of vegetative growth by a mitotic cell division cycle. Hence, large pure populations of **a** or α cells can be maintained. However, when **a** and α cells are cocultured they exit the cell cycle and undergo a mating process that culminates in the formation of a third cell type, an **a**/α diploid. Like the haploid cells, **a**/α cells can reproduce by mitosis but, unlike **a** or α cells, cannot mate. Instead they have a new property, the ability to undergo meiosis and sporulation when nutrients are limiting, thereby regenerating the two haploid cell types. Thus, **a** and α cells are specialized for mating whereas **a**/α cells are specialized for sporulation.

The mating process requires that **a** and α cells first communicate with each other and then interact physically so that cell and nuclear fusion can occur. Communication is achieved via the cell-type-specific production of secreted mating factors (pheromones) and receptors for those pheromones. In particular, only α cells secrete the 13-residue polypeptide, α-factor pheromone (Bucking-Throm *et al.*, 1973), which binds to a receptor present only on the surface of **a** cells (Jennesss *et al.*, 1983; Blumer *et al.*, 1988; Konopka *et al.*, 1988; Marsh and Herskowitz, 1988; Reneke *et al.*, 1988). Likewise, **a** cells secrete a polypeptide, **a**-factor pheromone (Wilkinson and Pringle, 1974), which interacts with a receptor present on the surface of α cells (MacKay and Manney, 1974; Bender and Sprague, 1986; Hagen *et al.*, 1986). The crucial role of the pheromones and receptors in mating is revealed by the finding that mutations in their structural genes lead to a nonmating phenotype. For example, α strains harboring mutations in the a-factor receptor structural gene, *STE3*, have a mating efficiency of 10^{-6} compared with wild-type α cells (MacKay and Manney, 1974; Hagen *et al.*, 1986). As expected, the mating defect is limited to α cells; **a** cells do not require this receptor to mate. [*Note:* Italicized symbols (e.g., *STE3*) refer to genes, with uppercase letters indicating the wild-type allele and lowercase letters indicating recessive mutant alleles. Roman symbols (e.g., STE3) refer to the protein product of the gene.]

Binding of pheromone to receptor activates an intracellular signal transduction pathway that is shared by α and **a** cells; that is, it does not involve proteins present only in α or only in **a** cells (Bender and Sprague, 1986; Nakayama *et al.*, 1987). Propagation of a signal along this pathway elicits physiological changes in the responding cell, including increased transcription of a number of genes whose products participate in mating (Hagen and Sprague, 1984; Hartig *et al.*, 1986; McCaffrey *et al.*, 1987; Trueheart *et al.*, 1987), preparation of both the cell surface and the nucleus for fusion (Betz *et al.*, 1978; Fehrenbacher *et al.*, 1978; Rose *et al.*, 1986), and arrest of the cell division cycle in the G_1 phase (Bucking-Throm *et al.*, 1973; Wilkinson and Pringle, 1974). Thus, the cell cycles of the mating pair are synchronized, and proteins that catalyze cell and nuclear fusion are poised to act; mating to form an **a**/α zygote can proceed efficiently.

Steps in this signal transduction pathway and targets of the signal have been identified largely by the isolation of mutants defective in mating or in pheromone response (MacKay and Manney, 1974; Hartwell, 1980). The mutations present in these strains affect mating of both α and **a** cells because a process common to both cell types is defective. Genetic and physiological studies indicate that an early step in the pathway involves a heterotrimeric GTP-binding protein (G protein); the α subunit is encoded by *GPA1* (*SCG1*) (Dietzel and Kurjan, 1987; Miyajima *et al.*, 1987; Jahng *et al.*, 1988), the β subunit by *STE4* (Whiteway *et al.*, 1989), and the γ subunit by *STE18* (Whiteway *et al.*, 1989). By analogy with mammalian systems (for review, see Stryer and Bourne, 1986), it is thought that in the absence of pheromone, the α, β, and γ subunits of the G protein are complexed and inactive, and the α subunit is bound with GDP. When pheromone binds receptor, GTP is exchanged for GDP, and the $\beta\gamma$ subunits are released as a complex. In contrast to the usual situation in mammalian systems, in yeast it is thought that the $\beta\gamma$ complex, rather than the α subunit bound with GTP, then propagates the signal that pheromone is bound to receptor by interacting with an as yet unidentified effector or second messenger generating system. In the context of the involvement of a G protein in yeast signal transduction, it is interesting to note that the yeast receptors are believed to contain seven hydrophobic, membrane-spanning segments in the amino-terminal portion of the protein, followed by a cytoplasmic hydrophilic, serine- and threonine-rich carboxy-terminal segment (Burkholder and Hartwell, 1985; Nakayama *et al.*, 1985; Hagen *et al.*, 1986). Thus, the yeast receptors are members of a family of receptors that includes the rhodopsins, the β-adrenergic receptors, and the muscarinic acetylcholine receptors, all of which share the serpentine membrane topology and function in signal transduction via G proteins (Nathans and Hogness, 1984; Dixon *et al.*, 1986; Kubo *et al.*, 1986).

Other components of the signal transduction pathway may include the products of the *STE5, STE7, STE11,* and *STE12* genes, each of which is required for pheromone response and mating. The role of STE5 is not known. STE7 and STE11 likely are protein kinases (Teague *et al.*, 1986; D. Chaleff and B. Errede, personal communication), suggesting that phosphorylation events are crucial in yeast signal transduction. STE12 is a DNA-binding protein that is part of protein–DNA complexes that form on a DNA sequence known to be responsible for enhanced transcription of pheromone-responsive genes (Dolan *et al.*, 1989; Errede and Ammerer, 1989). In this sense, STE12 appears to be a terminal step in the signal transduction pathway. Perhaps the activity of STE12 is affected by pheromone-induced activation of the pathway; for instance, the phosphorylation state of STE12 may change.

A high degree of specificity characterizes the mating process: **a** cells will mate only with α cells even though in a typical mating mix they may be in contact with other **a** cells as well as α cells. An initial expectation might be that a number of cell-type-specific proteins are required to confer this specificity. However, other than the pheromones and receptors, the only known **a** cell- or α cell-specific proteins that participate in the mating process are **a**-agglutinin and α-agglutinin, cell surface glycoproteins that enable the two cell types to adhere tightly (Betz *et al.*, 1978; Fehrenbacher *et al.*, 1978). The agglutinins apparently have only a subtle role in mating as structural gene mutations cause only a modest reduction in mating efficiency (Suzuki and Yanagishima, 1985; Lipke *et al.*, 1989). Rather, the specificity of mating is dictated largely by the species of pheromone and receptor that a cell synthesizes. For example, an **a** cell can be made to mate with other **a** cells if its capacity to produce the normal (**a** cell) pheromone and receptor set is destroyed by structural gene mutations and instead it is made to express the inappropriate (α cell) pheromone and receptor set (Bender and Sprague, 1989). Apparently, the act of sending and receiving pheromone signals establishes directional cues that identify mating partners. Thus, a generic haploid yeast cell (i.e., one that does not synthesize any **a**- or α-specific proteins) can be converted to one of two specialized cell types with very different biological potential simply by conferring the generic cell with the capacity to produce two or three unique proteins. Moreover, the features that distinguish **a** from α cells are conferred by proteins present at the cell surface, an appropriate setting given that this is where the initial interaction between the mating pair will occur.

III. Control of Cell Type

How is the differential gene expression that gives **a**, α, and **a**/α cells their unique properties achieved? What regulatory mechanisms dictate that each cell type will synthesize a set of unique proteins? The three cell types differ in genetic content only at one locus, the mating-type locus. Genetic and physiological studies have led to the view that the mating-type locus alleles, *MAT***a** and *MAT*α, encode regulatory proteins that govern transcription of four gene sets: α-specific genes, **a**-specific genes, haploid-specific genes, and diploid-specific genes (the $\alpha1-\alpha2$ hypothesis of Strathern *et al.*, 1981; Fig. 1). This picture has been verified using cloned DNA segments corresponding to the structural genes for various cell-type-specific proteins. These latter experiments also demonstrate that regulation is at the level of transcription. The regulatory circuitry by which the mating-type locus products generate cell-type-specific transcription is described below.

A. α CELLS

MATα encodes two proteins, MATα1 and MATα2, that control expression of known α-specific and **a**-specific genes. MATα1 (hereafter referred to as α1) activates transcription of the α-specific gene set (Table 1; Sprague *et al.*, 1983b; Fields and Herskowitz, 1985; Lipke *et al.*, 1989). This set includes *STE3*, the **a**-factor receptor structural gene; *MF*$\alpha1$ and *MF*$\alpha2$, the α-factor structural genes; and *AG*$\alpha1$, the α-agglutinin structural gene. In contrast, MATα2 (hereafter referred to as α2) represses transcription of the **a**-specific gene set (Table 1; Wilson and Herskowitz, 1984; Hartig *et al.*, 1986; Kronstad *et al.*, 1987; Michaelis and Herskowitz, 1988). This gene set includes *STE2*, the α-factor receptor structural gene; *MF***a***1* and *MF***a***2*, the **a**-factor structural genes;

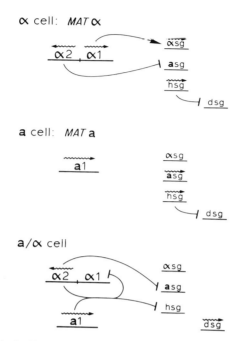

FIG. 1. Control of cell-type-specific genes by the mating-type locus. Expression of α-specific genes (αsg), **a**-specific genes (**a**sg), haploid-specific genes (hsg), and diploid-specific genes (dsg) is shown for each cell type (see Strathern *et al.*, 1981, and other references cited in the text). Wavy lines indicate gene transcription, lines with arrowheads indicate stimulation of transcription, and lines with crossbars indicate inhibition of transcription. See the text for details concerning the roles of the mating-type locus products in regulating transcription.

STE6, a gene required for maturation of **a**-factor; and *BAR1,* the structural gene for an α-factor protease. As a consequence of the action of α1 and α2, the α-specific gene set is transcribed in α cells, but the **a**-specific gene set is not. The cells therefore have the α phenotype.

In addition to the α-specific gene set, α cells also transcribe the haploid-specific gene set. As described below, **a**/α cells have a unique repressor activity that blocks transcription of these genes. Because this repressor is absent from α and **a** cells, both these cell types transcribe the haploid-specific gene set. Many of the genes that are members of this set encode proteins required for mating by both **a** and α cells. For example, *GPA1 (SCG1), STE4,* and *STE18* are haploid-specific genes, which as noted above specify a G protein that functions at an early step in the pheromone signal transduction pathway. *STE5* and *STE12,* two

TABLE 1
Gene–Function Relationships

Gene	Function[a]
α-Specific genes	
STE3	a-Factor receptor
MFα1	α-Factor
MFα2	α-Factor
AGα1	α-Agglutinin
a-Specific genes	
STE2	α-Factor receptor
MFa1	a-Factor
MFa2	a-Factor
STE6	Unknown; a-factor maturation or secretion
BAR1	α-Factor protease
Haploid-specific genes	
GPA1 (SCG1)	α Subunit of G protein
STE4	β Subunit of G protein
STE18	γ Subunit of G protein
STE5	Unknown; response to pheromone
STE12	DNA-binding protein
FUS1	Unknown; cell fusion
RME1	Unknown; inhibitor of meiosis and sporulation
Diploid-specific genes	
IME1	Unknown; inducer of meiosis and sporulation

[a] References for the assignment of these functions are as follows: STE3 (MacKay and Manney, 1974; Bender and Sprague, 1986; Hagen et al., 1986), MFα1 and MFα2 (Kurjan and Herskowitz, 1982; Singh et al., 1983; Kurjan, 1985), AGα1 (Suzuki and Yanagishima, 1985; Lipke et al., 1989), STE2 (MacKay and Manney, 1974; Hartwell, 1980; Jenness et al., 1983; Blumer et al., 1988; Konopka et al., 1988; Marsh and Herskowitz, 1988), MFa1 and MFa2 (Michaelis and Herskowitz, 1988), STE6 (Rine, 1979; Wilson and Herskowitz, 1984), BAR1 (Sprague and Herskowitz, 1981; Manney, 1983; MacKay et al., 1988), GPA1 (Dietzel and Kurjan, 1987; Miyajima et al., 1987; Nakafuku et al., 1987; Jahng et al., 1988), STE4 and STE18 (Whiteway et al., 1989), STE5 (MacKay and Manney, 1974; Hartwell, 1980), STE12 (Dolan et al., 1989; Errede and Ammerer, 1989), FUS1 (McCaffrey et al., 1987; Trueheart et al., 1987), RME1 (Kassir and Simchen, 1976; Rine et al., 1981; Mitchell and Herskowitz, 1986), and IME1 (Kassir et al., 1988).

other players in signal transduction, also show haploid-specific transcription (Fields and Herskowitz, 1987; J. Thorner, personal communication). Similarly, FUS1, a gene required for cell fusion by the mating pair, is transcribed in a and α cells, not in a/α cells (McCaffrey et al., 1987; Trueheart et al., 1987). For simplicity these genes are referred to as haploid-specific although in fact their expression is not sensitive to ploidy: a/a or α/α diploids transcribe this gene set. Strictly speaking, the gene set should be called a/α-inhibited genes, but the haploid-specific terminology is retained because it is parallel to α-specific and a-specific.

Finally, diploid (a/α)-specific genes are not transcribed in α cells because they are repressed by one or more haploid-specific gene products (see below).

B. a CELLS

The *MAT*a allele encodes a single regulator, a1, but this regulator does not have a role in controlling expression of α- or a-specific genes. Rather, the appropriate pattern of expression of these genes is the simple consequence of the absence of the two *MAT*α-encoded regulators: there is no α1 to activate transcription of α-specific genes, so they are not expressed, and there is no α2 to repress transcription of a-specific genes, so they are expressed. As expected given this view, *mat*α1 *mat*α2 double mutants mate as if they were a cells. In fact, this was a central observation that led to the α1–α2 hypothesis (Strathern *et al.*, 1981). Haploid-specific genes are transcribed and diploid-specific genes are not transcribed in a cells for the reasons discussed above for α cells.

C. a/α CELLS

*MAT*a/*MAT*α diploids express only the diploid-specific gene set. This pattern of gene expression is the consequence of two repressor activities. First, the repressor α2 blocks expression of a-specific genes, as it does in α cells. Second, in concert with a1, α2 exhibits a different DNA binding specificity (Strathern *et al.*, 1981; Miller *et al.*, 1985; Goutte and Johnson, 1988). This new activity, a1–α2, leads to repression of both α-specific and haploid-specific genes. In some cases repression is direct, the consequence of binding of a1–α2 to the upstream regions of particular genes (Goutte and Johnson, 1988). For example, two presumptive a1–α2 binding sites exist within the upstream region of *STE5* (Miller *et al.*, 1985), a haploid-specific gene required for response to pheromone and for mating. In other cases, repression by a1–α2 is indirect. For example, at least part of the explanation for the failure to transcribe α-specific genes in a/α cells is that the *MAT*α1 gene, which is required for transcription of this gene set, is itself repressed by the binding of a1–α2 to a portion of its promoter (Klar *et al.*, 1981; Nasmyth *et al.*, 1981; Siliciano and Tatchell, 1984; Miller *et al.*, 1985; Goutte and Johnson, 1988). A second example of indirect repression by a1–α2 is provided by the haploid-specific *FUS1* gene: *FUS1* requires other haploid-specific genes, such as *STE5* and *STE12*, for its transcription but does not contain an **a**1–α2 site within its promoter (McCaffrey *et al.*, 1987). Thus, in **a**/α cells, α2 and **a**1–α2 together block expression of three gene sets, namely, α-specific genes, **a**-specific genes, and haploid-specific genes.

The activation of diploid-specific genes is also an indirect consequence of the action of **a**1–α2. Genes such as *IME1* must be expressed to allow meiosis and sporulation to proceed in **a**/α diploids (Kassir *et al.*, 1988). Transcription of *IME1* requires that a cell contain both the **a**1 and α2 products and also be subjected to the appropriate nutritional conditions. Rather than functioning to activate directly the transcription of *IME1*, the role of **a**1–α2 is to repress transcription of an inhibitor of meiosis, *RME1* (Mitchell and Herskowitz, 1986), apparently by binding to the gene's promoter (Goutte and Johnson, 1988; A. Mitchell, personal communication). Thus, *RME1* is a haploid-specific gene. The presence of high levels of RME1 prevents entry into meiosis by haploid cells. Diminution of *RME1* transcription by **a**1–α2 in diploid cells, coupled with the appropriate nutritional signals, allows transcription of *IME1* and other genes required for sporulation and thereby permits initiation of meiosis.

D. GENE EXPRESSION AND CELL SPECIALIZATION

In summary, yeast cells can be specialized either for mating or for meiosis and sporulation. The regulatory circuitry outlined here highlights the fact that transcriptional control of four sets of genes gives the α, **a,** and **a**/α cell types their unique properties.

1. Mating

Mating-competent cells exhibit two tiers of specialization that differentiate them from **a**/α cells. First, mating-competent cells express either the α-specific or the **a**-specific gene set. Second, both α and **a** cells transcribe the haploid-specific gene set. Expression of this set allows the signal transduction pathway to operate (*GPA1, STE4, STE18, STE5, STE12*) and allows cell fusion to occur (*FUS1*). Thus, haploid **a** or α cells are specialized for mating because only these cells make the appropriate cell surface components to initiate mating and the appropriate intracellular machinery to interpret the information generated by the binding of pheromone to receptor. The specialization of haploid cells for mating is reinforced because these cells syntehsize RME1, an inhibitor of meiosis.

2. *Meiosis and Sporulation*

Diploid **a**/α cells are specialized for meiosis and sporulation by a regulatory logic that is the converse of haploid cells. In particular, **a**/α cells do not express the pheromone or receptor structural genes nor genes that encode key components of the signal transduction pathway, so mating is precluded. Rather, meiosis and sporulation are allowed because transcription of *RME1* is repressed. Release from RME1 inhibition of meiosis permits transcription of *IME1* and other genes required for sporulation.

IV. Molecular Basis of Transcriptional Control of Cell-Type-Specific Genes

The foregoing discussion provided partial insight as to how the mating-type locus products determine cell type. In molecular terms, how do the mating-type locus products achieve the transcriptional regulation that differentiates **a**, α, and **a**/α cells? I focus on studies that address the roles of α1 and α2. These studies revealed an unexpected connection in the transcription of α- and **a**-specific genes and further demonstrated that differential transcription of these two gene sets is achieved via an elegant example of combinatorial control. α1 and α2 form complexes with a transcription factor/regulator, PRTF (pheromone and receptor transcription factor, so named because these are the known genes that utilize this factor; alternatively, P-box recognition transcription factor), present in all three cell types. The α1–PRTF complex acts as a transcription activator of α-specific genes, whereas the α2–PRTF complex acts as a transcription repressor of **a**-specific genes. Moreover, in some situations PRTF acts as an unregulated transcription activator. This regulatory strategy has implications for the control of expression of genes in mammalian cells, particularly since PRTF appears to be a homolog of a mammalian transcription factor, the serum response factor (SRF).

A. MATα1 AND CONTROL OF α-SPECIFIC GENES

The model for the mechanism whereby α1 activates transcription of α-specific genes is derived from two kinds of experiments. First, deletion analysis was used to define the DNA sequences from α-specific genes that are required for their expression *in vivo* and that can impart α-specific expression to reporter genes when substituted for the upstream activation sequences (UAS) of the reporter genes. Second, the proteins that bind to the α-specific UAS elements *in vitro*, as well as the conditions that must be met for binding to be detected, were defined. Both approaches led to the same picture of how regulated transcription activation is achieved. I first present this picture and then present some of the experimental results that support this picture.

The UAS of α-specific genes contain a sequence to which α1 protein binds in conjunction with a transcription factor, PRTF, that is present in all three cell types. Transcription of α-specific genes requires PRTF, and the role of α1 is to assist the binding of PRTF to α-specific UAS elements. The UAS can be divided into two parts (Fig. 2). One part, termed the P box, is a degenerate version of a 16-bp symmetric dyad; the second part is an adjacent 10-bp Q box. PRTF binds efficiently to symmetric versions of the P box but binds only poorly to the degenerate versions of the P box actually present in α-specific UAS elements. However, cooperative interaction with α1 allows PRTF to bind efficiently to the QP sequence of α-specific genes. In one view, α1 makes contact with the Q-box DNA and also with PRTF and thereby compensates for the degenerate nature of the P box (Fig. 3). In another view, α1 does not contact a DNA directly, but rather causes a conformational change in PRTF that enables it to make contact with the Q-box segment as well as the degenerate P sequence. In either case, the role of α1 is to recruit a general transcription factor to bind to UAS elements that it cannot bind to alone. Thus, transcription of α-specific genes is limited to α cells because only these cells contain α1 to facilitate binding of the transcription factor (Fig. 3).

As noted above, this model is based on an analysis both of the UAS elements of α-specific genes and of the proteins that bind to those elements. Deletion analysis of the upstream regions of two α-specific genes, *STE3* (Jarvis *et al.*, 1988) and *MFα1* (Inokuchi *et al.*, 1987; Flessel *et al.*, 1989), delimited short stretches of DNA that could func-

α TTTCCTAATTAGTNCN TCAATGNCAG

P (PAL)   CTTCCTAATTAGGAAG

a   ANCATGTAAA TTTCCTTATTNGGTAA TTACATG

FIG. 2. P-box consensus sequences, showing the synthetic sequence P(PAL) and the consensus P boxes from α-specific and a-specific genes. The Q-box sequence found adjacent to α-specific P boxes is shown in smaller letters. Likewise, the α2 sites flanking a-specific P boxes are shown in smaller letters. [Modified from Bender and Sprague (1987) and Jarvis *et al.* (1988) and reprinted with the permission of Cell Press and the American Society for Microbiology.]

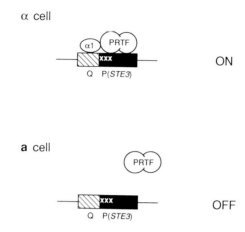

FIG. 3. Model for regulation of α-specific genes. The upstream activation sequences (UAS) of α-specific genes contain two sequence elements, a P box that is a degenerate form of a palindromic P box and an adjacent Q box. The degenerate region of the P box is denoted by ×××. The transcription activator protein PRTF does not bind efficiently to the UAS unless α1 protein is also present. Hence, α-specific genes are OFF in **a** cells, where there is no α1, but ON in α cells, which contain α1. The P box is denoted by the solid rectangle and the Q box by the hatched rectangle. PRTF is shown as a dimer, as suggested by Jarvis *et al.* (1989). [Modified from Bender and Sprague (1987) and Jarvis *et al.* (1988) and reprinted with the permission of Cell Press and the American Society for Microbiology.]

tion as α-specific UAS elements. The only sequence those UAS elements share is a 26-bp segment containing the Q and P boxes (Fig. 2). *STE3* contains one copy of this sequence, and *MFα1* contains two copies. In addition, a third α-specific gene, *MFα2,* also contains a version of the QP sequence, and a 44-bp segment containing QP from *MFα2* acts as an α-specific UAS (Jarvis *et al.,* 1988). That the QP sequence indeed confers α-specific transcription has been corroborated by synthesis of an oligonucleotide of precisely the *STE3* QP sequence and demonstration that the synthetic oligonucleotide serves as an α-specific UAS (Fig. 4; J.-J. Hwang-Shum and G. Sprague, unpublished). Moreover, deletion of just the P segment from the chromosomal *STE3* locus abolishes transcription of that locus (Jarvis *et al.,* 1988). Thus, the QP sequence is necessary and sufficient for α-specific transcription.

To investigate the role of the P-box portion of the QP sequence, we tested the UAS activity of two other synthetic oligonucleotides, the P sequence from *STE3* [P(*STE3*)] and a perfectly symmetric P sequence

[P(PAL)]. P(*STE3*) was inactive as a UAS element. P(PAL), however, was active as a UAS, not only in α cells but also in **a** and **a**/α cells (Fig. 4; Jarvis *et al.*, 1988). Taken together, these UAS experiments imply that a transcription factor present in all three cell types can interact with symmetric P sequences and activate transcription. However, the transcription factor can activate transcription from the P sequences actually found in α-specific UAS elements if and only if the UAS also contain the adjacent Q sequence and α1 protein is present in the cell.

The results of experiments to investigate the binding of proteins to α-specific UAS elements mirror the UAS activity results. α1 binds cooperatively with a second protein to these UAS, and neither protein binds efficiently in the absence of other (Bender and Sprague, 1987; Tan *et al.*, 1988). In particular, extracts of α cells contain a protein activity, which turns out to be α1 and PRTF, that can bind to α-specific UAS elements. However, α1 produced in *Escherichia coli* or purified from yeast does not bind to the UAS. Likewise, crude extracts prepared from **a** or **a**/α cells do not contain a protein activity that can bind efficiently to the UAS. A mixture of α1 (again, produced in *E. coli* or purified from yeast) and an **a** or **a**/α cell extract, however, results in the formation of an activity that readily binds to the UAS. Thus, there is synergism in the binding of α1 and a second protein to α-specific UAS elements. This second protein binds poorly to P(*STE3*) and efficiently to P(PAL) whether or not α1 is present, and thus has the properties expected of PRTF. The number of distinct protein species that actually makes up

UAS Activity

	α	**a**	**a**/α
QP (*STE3*)	+	−	−
Q	−	−	−
P (*STE3*)	−	−	−
P (PAL)	+	+	+

FIG. 4. Summary of UAS activities of QP (*STE3*) and related DNA fragments. The indicated DNA fragments were inserted adjacent to a plasmid-borne *CYC1–lacZ* hybrid gene deleted for its native UAS. The plasmids were introduced into isogenic α, **a**, and **a**/α strains by transformation, and the β-galactosidase activity of the resultant transformants was assayed (Jarvis *et al.*, 1988). Plus indicates significant levels of β-galactosidase activity; minus indicates a level of β-galactosidase no greater than that found in transformants carrying the *CYC1–lacZ* gene without any UAS element. [Modified from Bender and Sprague (1987) and reprinted with the permission of Cell Press.]

PRTF is not known; for simplicity, I assume that PRTF is a single species.

In conclusion, it appears α-specific transcription is largely the result of $\alpha 1$ protein recruiting a general transcription factor to bind to α-specific UAS elements and thereby activating transcription. It is possible that $\alpha 1$ contributes to α-specific transcription in other ways as well. For example, $\alpha 1$ may directly participate in transcription activation when it is bound to UAS elements, or it may induce conformational changes in PRTF that improve the capacity of PRTF to activate transcription. These latter possibilities are suggested by the observation that PRTF can bind *in vitro* to some P sequences that do not serve as efficient UAS elements *in vivo* (Bender and Sprague, 1987; Tan *et al.*, 1988). However, the biological significance of this *in vitro* finding is unclear.

B. MAT$\alpha 2$ AND CONTROL OF **a**-SPECIFIC GENES

PRTF is present in all three cell types and therefore is likely involved in the transcription of genes other than α-specific genes. As one approach to identify such genes, the upstream control regions of many yeast genes have been examined for the presence of P-box sequences. Unexpectedly, the clearest examples of these sequences are found within the UAS elements of **a**-specific genes (Fig. 2; Johnson and Herskowitz, 1985; Bender and Sprague, 1987; Jarvis *et al.*, 1988).

Two lines of evidence indicate that these P boxes contribute to the expression of **a**-specific genes. First, it has been shown that deletion of P-box DNA from the **a**-specific *BAR1* gene reduces its transcription (Kronstad *et al.*, 1987). In addition, a synthetic oligonucleotide that contains the *STE6* P-box sequence serves as a UAS, as measured by its ability to activate transcription of a reporter gene when substituted for the native UAS of the reporter gene (Johnson and Herskowitz, 1985; Jarvis *et al.*, 1988; Keleher *et al.*, 1988). Second, *in vitro* DNA-binding experiments reveal that PRTF can bind to the P boxes of **a**-specific genes (Bender and Sprague, 1987; Keleher *et al.*, 1988; Tan *et al.*, 1988). The versions of this sequence found at **a**-specific genes are more nearly symmetric than their counterparts at α-specific genes. As a consequence, PRTF appears capable of binding alone to the **a**-specific P boxes; no proteins analogous to $\alpha 1$ are required for binding. Thus, PRTF and its binding site contribute to transcription of **a**-specific genes in **a** cells (Fig. 5). An **a**-specific UAS is more than just a P box, however. Other sequences within the UAS contribute to expression and confer sensitivity to other regulators (Kronstad *et al.*, 1987; see Section III,E below).

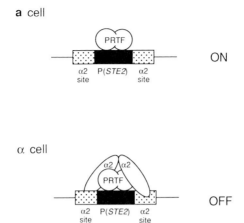

FIG. 5. Model for regulation of **a**-specific genes. The control regions of **a**-specific genes contain the two sequence elements depicted here: a nearly symmetric P box (solid rectangle) and flanking α2 binding sites (stippled rectangles). In **a** cells, the transcription factor PRTF can bind to the P-box sequence and thereby contributes to transcription activation of this gene set (ON). In α cells, PRTF and α2 bind coopera-tively and impose repression on the gene set (OFF). The simultaneous binding of these two proteins to the indicated sequences is based on the work of Keleher *et al.* (1988). [Modified from Bender and Sprague (1987) and Jarvis *et al.* (1988) and reprinted with the permission of Cell Press and the American Society for Microbiology.]

Given that PRTF is present in all three cell types and can activate transcription of **a**-specific P boxes, how is transcription of these genes limited to **a** cells? What is the role of α2? The simple answer to this question is that α2 is a sequence-specific DNA-binding protein; on binding to its target sequence α2 achieves repression. The unexpected finding is that binding of PRTF to the P box, which is flanked by α2 binding sites, is also essential to bring about α2-mediated repression. The experimental support for these statements is presented below.

Johnson and Herskowitz (1985) defined a 32-bp sequence from the **a**-specific *STE6* gene that could function as an operator *in vivo* to bring about repression of a reporter gene. When this operator was placed between the UAS and TATA elements of the reporter gene, transcription was repressed in cells that contained α2 protein. In addition, α2 protein was shown to bind to the operator *in vitro*. The P boxes of **a**-specific genes are located within the 32-bp α2 operator; in fact, they constitute the central core of the operator. Thus, a simple model for α2-mediated repression proposes that binding of α2 to its operator sterically excludes the transcription activator PRTF from the center of the operator. This simple model does not hold, however. Rather, Keleher *et al.* (1988) demonstrated that a second protein, termed GRM (general regulator of mating type), can also bind to the $\alpha2$ operator. Either $\alpha2$ or GRM can bind independently to the operator, but together the two proteins bind cooperatively. A dimer of $\alpha2$ occupies the two ends of the operator, and GRM binds to the center, that is, to the P box (Keleher *et al.*, 1988; Sauer *et al.*, 1988). Because both PRTF and GRM bind to P boxes, it seems probable that they are the same protein species. Indeed, both proteins are encoded by *MCM1* (see Section III,C), so for simplicity I use the PRTF terminology. The ability of the operator to confer repression *in vivo* requires that the binding sites for both $\alpha2$ and PRTF be intact. It appears, therefore, that repression requires the cooperative binding of both $\alpha2$ and PRTF to the operator. The role of PRTF might be to assist the binding of $\alpha2$ so that it can bind efficiently to the operator *in vivo*. $\alpha2$ in turn may mask the portion of PRTF that enables it to stimulate transcription.

As indicated by the foregoing discussion, $\alpha2$ protein has several biological activities. How are these activities distributed in the protein? Deletion analysis has divided the $\alpha2$ protein into two domains, an amino-terminal domain that is required for dimerization, cooperative binding with PRTF, and interaction with **a**1 (see Section III,D) and a carboxy-terminal domain that binds to DNA (Hall and Johnson, 1987; Goutte and Johnson, 1988; Keleher *et al.*, 1988; Sauer *et al.*, 1988). This latter domain contains a helix–turn–helix DNA-binding motif (Anderson *et al.*, 1981; Sauer *et al.*, 1982) and shows extensive homology with the homeo domains of *Drosophila* proteins involved in developmental decisions (Laughon and Scott, 1984; Shepherd *et al.*, 1984).

In summary, PRTF has two roles in the expression of **a**-specific genes (Fig. 5). In **a** cells, PRTF binds to the P boxes of this gene set and contributes to their transcription activation. In α and **a**/α cells PRTF binds cooperatively with $\alpha2$ to bring about repression of the gene set. A remarkable and unexpected feature of this repressor–operator system is its ability to act over a distance. The operator need not overlap a UAS in order to permit repression by $\alpha2$: at least when tested at reporter genes, it brings about repression when placed upstream of a UAS or between a UAS and the transcription start point (Johnson and Herskowitz, 1985). In principle, then, expression of any gene could be made **a** specific simply by insertion of an $\alpha2$ operator somewhere within its promoter. How is repression at a distance achieved? One possiblity, suggested by Keleher *et al.* (1988), is that the operator-bound PRTF–$\alpha2$ complex contacts the transcription machinery but does so too strongly to allow transcription to proceed. In effect, $\alpha2$ and PRTF lock the transcription machinery in place. The model has precedent in bacterial gene regulation. Straney and Crothers (1987) have shown that lac repressor and RNA polymerase do not occlude one another's binding, but rather appear to interact so tightly that RNA polymerase is not released to elongate a transcript.

C. PHEROMONE AND RECEPTOR TRANSCRIPTION FACTOR IS ENCODED BY THE *MCM1* GENE AND IS HOMOLOGOUS TO A MAMMALIAN TRANSCRIPTION FACTOR

PRTF has been defined solely on the basis of its DNA binding properties. In an effort to identify the structural gene for PRTF, we considered whether mutations in any previously identified gene conferred a phenotype consistent with that expected for a cell harboring mutant PRTF molecules. The simple expectation for a PRTF mutant cell would be a failure to express α- and **a**-specific genes, leading to a nonmating phenotype. However, if PRTF were required for expression of essential genes as well as for α- and **a**-specific genes (a reasonable possibility given that PRTF is found in **a**/α cells), then complete loss of PRTF function would be lethal. Therefore, viable mutant alleles should be defective for only some PRTF activities. For example, such alleles could encode a form of PRTF unable to interact with α1 but still able to bind to symmetric P boxes. In this case, transcription of α-specific genes would be depressed, but transcription of other genes would be largely unaffected.

The properties of *MCM1* mutant strains fit these expectations. The *mcm1-1* allele was isolated as a mutation that affected maintenance of certain minichromosomes (Maine *et al.*, 1984). It was subsequently observed that *mcm1-1* caused a mating defect, but only in α cells, and that transcription of α-specific genes, but not **a**-specific genes, was severely depressed. Moreover, deletion of *MCM1* was lethal (Passmore *et al.*, 1988). *MCM1* is therefore a candidate for the PRTF structural gene. Support for this relationshp comes from both *in vitro* and *in vivo* experiments (Jarvis *et al.*, 1989). Electrophoresis mobility shift assays were used to show that truncated MCM1 proteins encoded by deletion derivatives of *MCM1* formed protein–DNA complexes of novel mobility on P-box-containing DNA. That is, changes in the molecular weight of MCM1 had a corresponding affect on the migration of the protein–DNA complex. Moreover, antibodies raised to an MCM1 peptide bound specifically to the complexes formed on the α-specific *STE3* UAS. Together, these observations imply that MCM1 binds to P-box-containing DNA *in vitro*.

That MCM1 also acts at P-box DNA *in vivo* is supported by two observations. First, overexpression of MCM1 stimulates the transcription conferred by *STE3* UAS elements. Moreover, overexpression of MCM1 allows P (*STE3*), which normally is inactive as a UAS, to function as a UAS. High concentrations of MCM1 presumably favor the formation of protein–DNA complexes, according to the laws of mass action, and thereby stimulate transcription (Jarvis *et al.*, 1989). Second, as already noted, transcription of α-specific genes is severely depressed in *mcm1-1* mutant strains (Passmore *et al.*, 1988). All told, these results strongly suggest that *MCM1* encodes PRTF, at least in part. Further support for this conclusion has been provided by several other research groups (Hayes *et al.*, 1988; Passmore *et al.*, 1989; G. Ammerer, personal communciation; C. Keleher and A. Johnson, personal communication). For example, Keleher and Johnson have shown that purified α2 and MCM1 synthesized in *E. coli* bind cooperatively to the α2 operator, as expected if MCM1 encodes PRTF/GRM. The number of distinct protein species that actually makes up PRTF is not known, so other genes may also encode this protein activity.

A system homologous to the yeast P box–PRTF system operates to achieve transcription activation of some mammalian genes. This possibility was first suggested by the observation that a sequence similar to the P box is found upstream of genes, such as c-*fos* and actin (Fig. 6; Gilman *et al.*, 1986; Treisman, 1986; Greenberg *et al.*, 1987; Hayes *et al.*, 1988; Norman *et al.*, 1988). These genes are poorly transcribed in quiescent cells, but their transcription is rapidly induced when cells are exposed to serum growth factors. This transcription induction is mediated by the serum response element (SRE), a relative of the P box, and likely involves the serum response factor (SRF), a protein that binds to SRE (Treisman, 1987). A portion of the SRF amino acid sequence, deduced from its gene sequence, shows a high degree of similarity to a segment of MCM1: in a stretch of 81 amino acids, 55 are identical between the two proteins (Dubois *et al.*, 1987; Norman *et al.*, 1988; Passmore *et al.*, 1988; Fig. 7). This segment of SRF is required for binding to SRE. The properties of a small set of deletion derivatives of

```
P (PAL)      CTTCCTAATTAGGAAG

c-fos        TGTCCATATTAGGACA

ACTIN        TGCCCATATTTGGCGA
```

FIG. 6. Homology between the P box and the mammalian serum response element (SRE). The P(PAL) sequence is compared with the SRE sequences from the c-*fos* and actin genes. Regions of strong similarity are indicated in boldface.

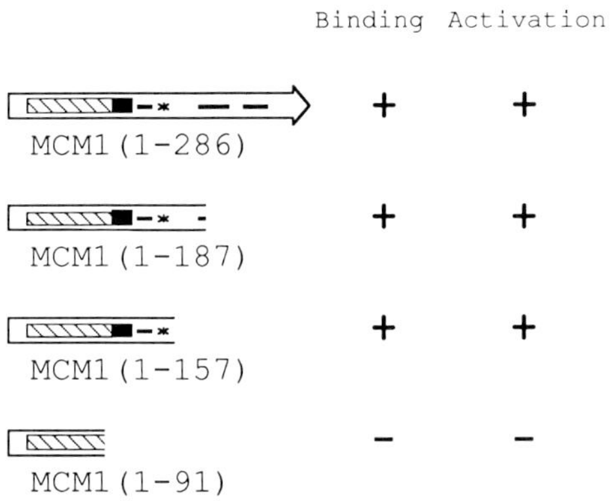

FIG. 7. Homology between MCM1 and the mammalian serum response factor (SRF). The open reading frame corresponding to MCM1 is depicted as an open arrow. Within the arrow, the hatched rectangle represents the region of homology between MCM1 and SRF. Other features of MCM1 are also indicated. Solid boxes indicate a stretch of acidic amino acids, lines indicate polyglutamine tracts, and asterisks indicate the position of the oligopeptide used to raise antibodies to MCM1. The extent of the open reading frame that remains in *MCM1* deletions is indicated below the full-length MCM1 open reading frame. The ability of wild-type and mutant MCM1 proteins to bind to DNA *in vitro* and activate transcription *in vivo* is indicated. [Reproduced from Jarvis *et al.* (1989) with the permission of Cold Spring Harbor Laboratory.]

MCM1 are consistent with the idea that the homologous segment is also the DNA binding domain of MCM1 (Fig. 7; Jarvis *et al.*, 1989).

The known biochemical properties of SRF are insufficient to explain the complex regulation of serum-inducible genes (Norman *et al.*, 1988). Thus, the SRF–SRE system may employ modulators of SRF function analogous to α1 and α2 to achieve this regulation. In this regard, recent experiments have identified an additional protein that forms a ternary complex with SRF at the SRE (Shaw *et al.*, 1989). The presence of this protein in the complex appears to be required for serum responsiveness.

D. MATa1, MATα2, AND CONTROL OF HAPLOID-SPECIFIC GENES

The a1 and α2 proteins act together as a negative regulator of haploid-specific genes. These genes include *MATα1*, *STE5*, *RME1*, and *HO*, a gene required for mating-type interconversion. Control is exerted at the level of transcription as RNA from these genes is not present in *MATa*/ *MATα* diploids but is present in *mata1*/*MATα* or *MATa*/*matα2* mutant diploids (Klar *et al.*, 1981; Nasmyth *et al.*, 1981; Jensen *et al.*, 1983; Mitchell and Herskowitz, 1986; J. Thorner, personal communication).

The first insight into the molecular mechanism of this negative regulation emerged from the study of the *MATα1* promoter. Deletion of a segment of the promoter renders transcription constitutive, that is, insensitive to control by **a**1–α2 (Siliciano and Tatchell, 1984). This segment contains a 20-bp sequence that is present 2 times in the upstream regions of both *STE5* and *RME1* (J. Thorner, personal communication; A. Mitchell, personal communication) and 10 times in the upstream control region of *HO* (Miller *et al.*, 1985; Russell *et al.*, 1986). Synthetic versions of this sequence serve as operators when placed between the UAS and TATA elements of a reporter gene, bringing that gene under **a**1–α2 control (Miller *et al.*, 1985; Goutte and Johnson, 1988). Thus, the **a**1–α2 operator is sufficient to achieve repression in an **a**/α cell, thereby conferring haploid-specific expression on a gene.

Goutte and Johnson (1988) have shown that this operator serves as the binding site for a protein activity present in extracts of cells that contain both the *MAT***a**1 and *MAT*α2 genes. They have demonstrated directly that α2 is bound to the operator, and it is likely that **a**1 is as well. This latter supposition is based on genetic evidence that α2 and **a**1 interact physically and on the observation that the sequence of **a**1 predicts a helix–turn–helix motif found in many DNA-binding proteins (Anderson *et al.*, 1981; Astell *et al.*, 1981; Sauer *et al.*, 1982; Laughon and Scott, 1984; Shepherd *et al.*, 1984). In any event, it is clear that **a**1 alters the binding specificity of α2: α2 alone cannot bind to the **a**1–α2 operator, whereas, in the presence of **a**1, α2 now binds to this operator. Binding of α2 to the α2 operator is not competed by the **a**1–α2 site, and conversely, binding of **a**1–α2 to the **a**1–α2 site is not competed by the α2 operator. Therefore, α2 must exhibit two different DNA binding modes, one that recognizes the α2 operator and one that recognizes the **a**1–α2 operator. Because both **a**-specific and haploid-specific genes must be repressed in **a**/α cells, the level of expression of **a**1 and α2 must be set to ensure that both α2 and **a**1–α2 repressor activities are present. Indeed, extracts of normal **a**/α cells do contain both binding activities.

Remarkably, α-specific genes also contain versions of the **a**1–α2 operator between their UAS and TATA elements (Miller *et al.*, 1985; Jarvis *et al.*, 1988). That these versions of the operator actually function *in vivo* was confirmed by the demonstration that a synthetic version of the sequence from *STE3* brought a reporter gene under **a**1–α2 control (H. Dunstan and G. Sprague, unpublished). In accord with this observation we have found that provision of α1 in **a**/α diploids (by joining the α1 coding sequence to a heterologous promoter) is not sufficient to allow transcription of α-specific genes (Ammerer *et al.*, 1985). Thus, α-specific genes are under redundant control in a/α cells. Not only is a required activator of their transcription (α1) not synthesized, but in addition they are under direct negative control by $a1-\alpha2$. Why a/α cells go to such lengths to ensure that α-specific genes are not expressed is not clear.

E. OTHER REGULATORS OF α-SPECIFIC, a-SPECIFIC, AND HAPLOID-SPECIFIC GENES

Many of the genes involved in mating are subject to a regulatory process that is layered onto the cell-type-specific transcriptional regulation, namely, transcription induction in response to treatment of mating-competent cells with pheromone. In some cases the effect is modest. For example, transcription of the α-specific *STE3* gene and of the a-specific *BAR1* and *STE2* genes increases 5- to 10-fold after pheromone treatment (Manney, 1983; Hagen and Sprague, 1984; Hartig *et al.*, 1986; Kronstad *et al.*, 1987). In contrast, transcription of the haploid-specific *FUS1* is highly inducible (50-fold or more; McCaffrey *et al.*, 1987; Trueheart *et al.*, 1987). Even the basal level of transcription of these genes requires that the signal transduction pathway be intact. This conclusion follows from the observation that transcript levels are reduced (in some cases modestly, in other cases severely) in *ste4, ste5, ste7, ste11,* and *ste12* mutants. Because these five genes are required for pheromone response, the implication is that unstimulated cells contain a low level of the intracellular pheromone signal, which contributes to the transcription activation of pheromone responsive genes (Hartig *et al.*, 1986; McCaffrey *et al.*, 1987; Fields *et al.*, 1988).

Transcription induction by pheromone appears to be mediated by the short sequence TGAAACA, the pheromone response element (PRE) (Kronstad *et al.*, 1987; Van Arsdell *et al.*, 1987). The upstream region of *BAR1* contains two copies of the PRE; *FUS1* contains four copies. Deletion of the two copies of the PRE sequence present at the *BAR1* gene has two consequences. First, the basal level of expression is reduced, and, second, pheromone-mediated induction is abolished. Thus, the UAS of a-specific genes contain at least two elements that contribute to their activity, the P box and the PRE. Further evidence that the PRE confers pheromone responsiveness is the finding that multiple copies of the sequence can serve as a UAS element for a reporter gene and cause that gene to be highly inducible by pheromone (D. Hagen and G. Sprague, unpublished; J. Thorner, personal communciation).

Recently, Dolan *et al.* (1989) and Errede and Ammerer (1989) have shown that the STE12 protein, the product of one of the genes required for pheromone response, is part of the protein–DNA complexes that form on the PRE of a-specific genes. Thus, it is attractive to imagine that the STE12 protein may be sensitive to the pheromone-generated intracellular signal. For example, STE12 activity may change in response to alterations in the level of the signal, leading to increased transcription of genes that contain a PRE. Because both STE12 and PRTF can be bound to the same UAS, it is possible that they interact physically. Thus far, however, STE12 has not been detected in protein–DNA complexes on α-specific UAS elements. One explanation is that the interaction of STE12 with these DNAs or with proteins bound to them is not stable to the *in vitro* manipulations used to examine protein–DNA complex formation. Alternatively, STE12 may never join complexes formed on α-specific UAS elements, in which case transcription induction from these UAS would occur by a mechanism that is independent of STE12 action. This seems unlikely, however, because the basal level of transcription of α-specific genes requires wild-type STE12.

V. Prospectives and Summary

Considerable progress has been made in understanding the molecular basis for the generation of yeast cell types, in particular the mechanism by which differential transcription of cell-type-specific gene sets is achieved. At the same time, a new set of issues is raised that awaits molecular explanation. (1) PRTF apparently interacts with at least two proteins, α1 and α2. How does it do so? What segments of PRTF are involved in these interactions? PRTF does not appear to have the ability to interact with other proteins via a so-called leucine zipper motif (Landschultz *et al.*, 1988). Perhaps whatever mechanism is used will also apply to interactions between other regulatory proteins. (2) Does α1 contact DNA and recruit PRTF to bind to α-specific UAS elements via those DNA contacts and via protein–protein interaction with PRTF? Or does α1 act as an allosteric effector and cause a conformational change in PRTF that permits PRTF to contact both P and Q DNA? These questions are especially pertinent since α1 appears to have little or no intrinsic capacity to bind to DNA. (3) How do PRTF and α1 contact DNA (assuming, of course, that α1 does)? The predicted amino sequences of both proteins (Astell *et al.*, 1981) appear to lack the features that characterize the two well-studied DNA binding motifs, the helix–turn–helix motif (Anderson *et al.*, 1981; Sauer *et al.*, 1982) and the zinc finger (Miller *et al.*, 1985; Klug and Rhodes, 1987; Evans and Hollenberg, 1988). (4) How do $\alpha2$ and $\mathbf{a}1-\alpha2$ achieve repression at a distance?

The strategies that are used by yeast to generate distinct cellular phenotypes are simple but elegant. Two features of yeast cell type determination are of special interest and appear to have the potential to operate in other organisms. First, gene expression can be modulated by transcription repressors as well as by activators. Most commonly, transcriptional regulation of eukaryotic genes is thought to involve sequence-specific DNA-binding proteins that function as transcription activators. Regulation is achieved by controlling the availability or activity of the activator. The finding that $\alpha2$ and $\mathbf{a}1-\alpha2$ are negative regulators of the transcription of cell-type-specific genes serves as a reminder that repressors may also be important regulators in eukaryotic cells. Indeed, the glucucorticoid receptor appears to function as a repressor in some settings (Oro *et al.*, 1988; Sakai *et al.*, 1988).

Second, combinatorial association of proteins is a powerful and versatile method to generate regulatory species with distinct activities. Two examples emerge from the study of yeast cell type. (1) $\alpha2$ acts in concert with two other proteins to create repressors with different DNA binding specificities. $\alpha2$–PRTF serves to repress **a**-specific genes, whereas $\mathbf{a}1-\alpha2$ serves to repress haploid-specific genes. (2) The PRTF–P-box system is unusually flexible (Fig. 8). A symmetric PRTF binding site,

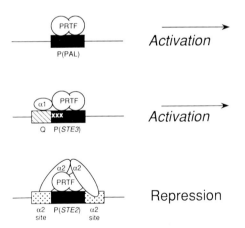

FIG. 8. Summary of the activities of PRTF. PRTF can act in a constitutive transcription activator by binding to symmetric versions of the P box such as P(PAL). Combinatorial association of PRTF with $\alpha1$ or $\alpha2$ generates complexes with distinct biological activities. With $\alpha1$, PRTF serves as a coactivator of transcription; with $\alpha2$, it serves a corepressor of transcription.

P(PAL), acts as an unregulated UAS element. By a slight modification of the P box and addition of a Q box, the UAS becomes an $\alpha1$–PRTF site, rendering gene expression α specific. By flanking the P box with $\alpha2$ sites, an $\alpha2$–PRTF operator is created, conferring a-specific gene expression. Thus, a single protein can function as a generic transcription activator, as a cell-type-specific activator (with $\alpha1$ as a coactivator), and as a cell-type-specific repressor (with $\alpha2$ as a corepressor). It seems likely that homologs to the PRTF–P-box system that are found in other species will show this same versatility.

More generally, the theme of combinatorial control will likely recur as other regulatory systems are explored and understood. Indeed, the mammalian transcription activators FOS and JUN appear to offer the opportunity for combinatorial control of gene expression. The JUN protein has different DNA binding affinities when dimerized with itself than when dimerized with FOS (Curran and Franza, 1988; Halazonetis *et al.*, 1988; Kouzarides and Ziff, 1988; Nakabeppu *et al.*, 1988; Sassone-Corsi *et al.*, 1988; Schuermann *et al.*, 1989). Moreover, the existence of families of FOS- and JUN-related proteins leads to the possibility that heteromeric complexes with subtly different DNA binding properties could be formed, which would magnify the combinatorial possibilities.

REFERENCES

Alberts, B., Bray, D., Lewis, J., Raff, M., Roberts, K., and Watson, J. D. (1983). "Molecular Biology of the Cell." Garland, New York.

Ammerer, G., Sprague, G. F., Jr., and Bender, A. (1985). Control of yeast α-specific genes: Evidence for two blocks to expression in *MAT*a/*MAT*α diploids. *Proc. Natl. Acad. Sci. U.S.A.* **82**, 5855–5859.

Anderson, W. F., Ohlendorf, D. H., Takeda, Y., and Matthews, B. W. (1981). Structure of the cro repressor from bacteriophage λ and its interaction with DNA. *Nature (London)* **290**, 754–758.

Astell, C. R., Ahlstrom-Sonasson, L., Smith, M., Tatchell, K., Nasmyth, K. A., and Hall, B. D. (1981). The sequence of the DNAs coding for the mating-type loci of *Saccharomyces cerevisiae*. *Cell* **27**, 15–23.

Bender, A., and Sprague, G. F., Jr. (1986). Yeast peptide pheromones, a-factor and α-factor, activate a common response mechanism in their target cells. *Cell* **47**, 929–937.

Bender, A., and Sprague, G. F., Jr. (1987). MATα1 protein, a yeast transcription activator, binds synergistically with a second protein to a set of cell-type-specific genes. *Cell* **50**, 681–691.

Bender, A., and Sprague, G. F., Jr. (1989). Pheromones and pheromone receptors are the primary determinants of mating specificity in the yeast *Saccharomyces cerevisiae*. *Genetics* **121**, 463–476.

Betz, R., Duntze, W., and Manney, T. R. (1978). Mating-factor-mediated sexual agglutination in *Saccharomyces cerevisiae*. *FEMS Microbiol. Lett.* **4**, 107–110.

Blumer, K. J., Reneke, J. E., and Thorner, J. (1988). The STE2 gene product is the ligand binding component of the α-factor receptor of *Saccharomyces cerevisiae*. *J. Biol. Chem.* **263**, 10836–10842.

Bucking-Throm, E., Duntze, W., Hartwell, L. H., and Manney, T. R. (1973). Reversible arrest of haploid yeast cells at the initiation of DNA synthesis by a diffusible sex factor. *Exp. Cell Res.* **76**, 99–110.

Burkholder, A. C., and Hartwell, L. H. (1985). The yeast α-factor receptor: Structural properties deduced from the sequence of the *STE2* gene. *Nucleic Acids Res.* **13**, 8463–8475.

Cohen, D. R., and Curran, T. (1988). *fra-1*: A serum-inducible, cellular immediate-early gene that encodes a fos-related antigen. *Mol. Cell. Biol.* **8**, 2063–2069.

Cross, F., Hartwell, L. H., Jackson, C., and Konopka, J. B. (1988). Conjugation in *Saccharomyces cerevisiae*. *Annu. Rev. Cell Biol.* **4**, 429–457.

Curran, T., and Franza, B. R., Jr. (1988). Fos and jun: The AP-1 connection. *Cell* **55**, 395–397.

Dietzel, C., and Kurjan, J. (1987). The yeast *SCG1* gene: A Gα-like protein implicated in the **a**- and α-factor response pathway. *Cell* **50**, 1001–1010.

Dixon, R. A. F., Kibilka, B. K., Strader, D. J., Benovic, J. L., Dohlman, H. G., Frielle, T., Bolanowski, M. A., Bennet, C. D., Rands, E., Diehl, R. E., Mumford, R. A., Slater, E. E., Sigal, I. S., Caron, M. S., Lefkowitz, R. J., and Strader, C. D. (1986). Cloning of the gene and cDNA for mammalian β-adrenergic receptor and homology with rhodopsin. *Nature (London)* **321**, 75–79.

Dolan, J. W., Kirkman, C., and Fields, S. (1989). The yeast STE12 protein binds to the DNA sequence mediating pheromone induction. *Proc. Natl. Acad. Sci. U.S.A.* **86**, 5703–5707.

Dubois, E., Bercy, J., Descamps, F., and Messenguy, F. (1987). Characterization of two new genes essential for vegetative growth in *Saccharomyces cerevisiae*: Nucleotide sequence determination and chromosome mapping. *Gene* **55**, 265–275.

Errede, B., and Ammerer, G. (1989). STE12, a protein involved in cell-type-specific transcription and signal transduction in yeast, is part of protein–DNA complexes. *Genes Dev.* **3**, 1349–1361.

Evans, R. M., and Hollenberg, S. M. (1988). Zinc fingers: Gilt by association. *Cell* **52**, 1–3.

Fehrenbacher, G., Perry, K., and Thorner, J. (1978). Cell–cell recognition in *Saccharomyces cerevisiae*: Regulation of mating-specific adhesion. *J. Bacteriol.* **134**, 893–901.

Fields, S., and Herskowitz, I. (1985). The yeast *STE12* product is required for expression of two sets of cell-type specific genes. *Cell* **42**, 923–930.

Fields, S., and Herskowitz, I. (1987). Regulation by the yeast mating type locus of *STE12*, a gene required for expression of two sets of cell-type-specific genes. *Mol. Cell. Biol.* **7**, 3818–3821.

Fields, S., Chaleff, D. T., and Spargue, G. F., Jr. (1988). Yeast *STE7*, *STE11*, and *STE12* genes are required for expression of cell-type-specific genes. *Mol. Cell. Biol.* **8**, 551–556.

Flessel, M. C., Brake, A. J., and Thorner, J. (1989). The *MFα1* gene of *Saccharomyces cerevisiae*: Genetic mapping and mutational analysis of promoter elements. *Genetics* **121**, 223–236.

Gilman, M. Z., Wilson, R. N., and Weinberg, R. A. (1986). Multiple protein binding sites in the 5' flanking region regulate c-*fos* expression. *Mol. Cell. Biol.* **6**, 4305–4314.

Goutte, C., and Johnson, A. D. (1988). **a**1 protein alters the DNA binding specificity of α2 repressor. *Cell* **52**, 875–882.

Greenberg, M. E., Siegfried, Z., and Ziff, E. B. (1987). Mutation of the c-*fos* dyad symmetry element inhibits inducibility *in vivo* and the nuclear regulatory factor binding *in vitro*. *Mol. Cell. Biol.* **7**, 1217–1225.

Hagen, D. C., and Sprague, G. F., Jr. (1984). Induction of the yeast α-specific *STE3* gene by the peptide pheromone **a**-factor. *J. Mol. Biol.* **178**, 835–852.

Hagen, D. C., McCaffrey, G., and Sprague, G. F., Jr. (1986). Evidence the yeast *STE3* gene encodes a receptor for the peptide pheromone **a** factor: Gene sequence and implications for the structure of the presumed receptor. *Proc. Natl. Acad. Sci. U.S.A.* **83**, 1418–1422.

Halazonetis, T. D., Georgopoulos, K., Greenberg, M. E., and Leder, P. (1988). c-Jun dimerizes with itself and with c-fos, forming complexes of different DNA binding affinities. *Cell* **55**, 917–924.

Hall, M. N., and Johnson, A. D. (1987). Homeo domain of the yeast repressor α2 is a sequence-specific DNA-binding domain but is not sufficient for repression. *Science* **237**, 1007–1012.

Hartig, A., Holly, J., Saari, G., and MacKay, V. L. (1986). Multiple regulation of *STE2*, a mating-type-specific gene of *Saccharomyces cerevisiae*. *Mol. Cell. Biol.* **6**, 2106–2114.

Hartwell, L. H. (1980). Mutants of *Saccharomyces cerevisiae* unresponsive to cell division control by polypeptide mating hormone. *J. Cell Biol.* **85**, 811–822.

Hayes, T. E., Sengupta, P., and Cochran, B. H. (1988). The human c-*fos* serum response factor and the yeast factors GRM/PRTF have related DNA-binding specificities. *Genes Dev.* **2**, 1713–1722.

Herskowitz, I. (1988). Life cycle of the budding yeast *Saccharomyces cerevisiae*. *Microbiol. Rev.* **52**, 536–553.

Inokuchi, K., Nakayama, A., and Hishinuma, F. (1987). Identification of sequence elements that confer cell-type-specific control of *MFα1* expression in *Saccharomyces cerevisiae*. *Mol. Cell. Biol.* **7**, 3185–3193.

Jahng, K.-Y., Ferguson, J., and Reed, S. I. (1988). Mutations in a gene encoding the α subunit of a *Saccharomyces cerevisiae* G protein indicate a role in mating pheromone signaling. *Mol. Cell. Biol.* **8**, 2484–2493.

Jarvis, E. E., Hagen, D. C., and Sprague, G. F., Jr. (1988). Identification of a DNA segment that is necessary and sufficient for α-specific gene control in *Saccharomyces cerevisiae*: Implications for regulation of α-specific and **a**-specific genes. *Mol. Cell. Biol.* **8**, 309–320.

Jarvis, E. E., Clark, K. L., and Sprague, G. F., Jr. (1989). The yeast transcription activator PRTF, a homolog of the mammalian serum response factor, is encoded by the *MCM1* gene. *Genes Dev.* **3**, 936–945.

Jenness, D. D., Burkholder, A. C., and Hartwell, L. H. (1983). Binding of α-factor pheromone to yeast **a** cells: Chemical and genetic evidence for an α-factor receptor. *Cell* **35**, 521–529.

Jensen, R. E., Sprague, G. F., Jr., and Herskowitz, I. (1983). Regulation of yeast mating-type interconversion: Feedback control of *HO* gene expression by the mating-type locus. *Proc. Natl. Acad. Sci. U.S.A.* **80**, 3035–3039.

Johnson, A. D., and Herskowitz, I. (1985). A repressor (*MATα2* product) and its operator control expression of a set of cell type specific genes in yeast. *Cell* **42**, 237–247.

Kassir, Y., and Simchen, G. (1976). Regulation of mating and meiosis in yeast by the mating type locus. *Genetics* **82**, 187–206.

Kassir, Y., Granst, D., and Simchen, G. (1988). *IME1*, a positive regulator of meiosis in *S. cerevisiae*. *Cell* **52**, 853–862.

Keleher, C. A., Goutte, C., and Johnson, A. D. (1988). The yeast cell-type-specific repressor α2 acts cooperatively with a non-cell-type-specific protein. *Cell* **53**, 927–936.

Klar, A. J. S., Strathern, J. N., Broach, J. R., and Hicks, J. B. (1981). Regulation of transcription in expressed and unexpressed mating type cassettes of yeast. *Nature (London)* **289**, 239–244.

Klug, A., and Rhodes, D. (1987). "Zinc fingers": A novel protein for nucleic acid recognition. *Trends Biochem. Sci.* **12**, 464–469.

Konopka, J. B., Jenness, D. D., and Hartwell, L. H. (1988). The C-terminus of the *S. cerevisiae* α-pheromone receptor mediates an adaptive response to pheromone. *Cell* **54**, 609–620.

Kouzarides, T., and Ziff, E. (1988). The role of the leucine zipper in the fos–jun interaction. *Nature (London)* **336**, 646–651.

Kronstad, J. W., Holly, J. A., and MacKay, V. L. (1987). A yeast operator overlaps an upstream activation site. *Cell* **50**, 369–377.

Kubo, T., Fukuda, K., Mikami, A., Maeda, A., Takahashi, H., Mishina, M., Haga, T., Haga, K., Hirose, T., and Numa, S. (1986). Cloning, sequencing, and expression of complementary DNA encoding the muscarinic acetylcholine receptor. *Nature (London)* **323**, 411–416.

Kurjan, J. (1985). α-Factor structural gene mutations in yeast: Effect on α-factor production and mating. *Mol. Cell. Biol.* **5**, 787–796.

Kurjan, J., and Herskowitz, I. (1982). Structure of a yeast pheromone gene (*MFα*): A putative α-factor precursor contains four tandem copies of mature α-factor. *Cell* **30**, 933–943.

Landschultz, W. H., Johnson, P. F., and McKnight, S. L. (1988). The leucine zipper: A hypothetical structure common to a new class of DNA binding proteins. *Science* **240**, 1759–1764.

Laughon, A., and Scott, M. P. (1984). Sequence of a *Drosophila* segmentation gene: Protein structure homology with DNA-binding proteins. *Nature (London)* **310**, 25–31.

Lipke, P. N., Wojciechowicz, D., and Kurjan, J. (1989). *AGα1* is the structural gene for the *Saccharomyces cerevisiae* α-agglutinin, a cell surface glycoprotein involved in cell–cell interactions during mating. *Mol. Cell. Biol.* **9**, 3155–3165.

McCaffrey, G., Clay, F. J., Kelsay, K., and Sprague, G. F., Jr. (1987). Identification and regulation of a gene required for cell fusion during mating of the yeast *Saccharomyces cerevisiae*. *Mol. Cell. Biol.* **7**, 2680–2690.

MacKay, V., and Manney, T. R. (1974). Mutations affecting sexual conjugation and related processes in *Saccharomyces cerevisiae*. II. Genetic analysis of nonmating mutants. *Genetics* **76**, 273–288.

MacKay, V. L., Welch, S. K., Insley, M. Y., Manney, T. R., Holly, J., Saari, G. C., and Parker, M. L. (1988). The *Saccharomyces cerevisiae BAR1* gene encodes an exported protein with homology to pepsin. *Proc. Natl. Acad. Sci. U.S.A.* **85**, 55–59.

Maine, G. T., Sinha, P., and Tye, B.-K. (1984). Mutants of *S. cerevisiae* defective in the maintenance of minichromosomes. *Genetics* **106**, 365–385.

Manney, T. R. (1983). Expression of the *BAR1* gene in *Saccharomyces cerevisiae*: Induction by the α mating pheromone of an activity associated with a secreted protein. *J. Bacteriol.* **155**, 291–301.

Marsh, L., and Herskowitz, I. (1988). The STE2 protein of *Saccharomyces kluyveri* is a member of the rhodopsin/β-adrenergic receptor family and is responsible for recognition of the pdptide ligand α factor. *Proc. Natl. Acad. Sci. U.S.A.* **85**, 3855–3859.

Michaelis, S., and Herskowitz, I. (1988). The **a**-factor pheromone of *Saccharomyces cerevisiae* is essential for mating. *Mol. Cell. Biol.* **8**, 1309–1318.

Miller, J., McLauchlan, A. D., and Klug, A. D. (1985). Repetitive zinc-binding domains in the protein transcription factor IIIA from *Xenopus* oocytes. *EMBO J.* **4**, 1609–1614.

Mitchell, A. P., and Herskowitz, I. (1986). Activation of meiosis and sporulation by repression of the *RME1* product in yeast. *Nature (London)* **319**, 738–742.

Miyajima, I., Nakafuku, M., Nakayama, N., Brenner, C., Miyajima, A., Kaibuchi, K., Arai, K.-I., Kaziro, Y., and Matsumoto, K. (1987). *GPA1*, a haploid-specific essential gene, encodes a yeast homolog of mammalian G protein which may be involved in the mating factor-mediated signal transduction pathway. *Cell* **50**, 1011–1019.

Nakabeppu, Y., Ryder, K., and Nathans, D. (1988). DNA binding activities of three murine jun proteins: Stimulation by fos. *Cell* **55**, 907–915.

Nakafuku, M., Itoh, H., Nakamura, S., and Kaziro, Y. (1987). Occurrence in *Saccharomyces cerevisiae* of a gene homologous to the cDNA coding for the α subunit of mammalian G proteins. *Proc. Natl. Acad. Sci. U.S.A.* **84**, 2140–2144.

Nakayama, N., Miyajima, A., and Arai, K. (1985). Nucleotide sequence of *STE2* and *STE3*, cell type-specific sterile genes from *Saccharomyces cerevisiae*. *EMBO J.* **4**, 2643–2648.

Nakayama, N., Miyajima, A., and Arai, K. (1987). Common signal transduction system shared by *STE2* and *STE3* in haploid cells of *Saccharomyces cerevisiae*: Autocrine cell-cycle arrest results from forced expression of *STE2*. *EMBO J.* **6**, 249–254.

Nasmyth, K., and Shore, D. (1987). Transcription regulation in the yeast life cycle. *Science* **237**, 1162–1170.

Nasmyth, K. A., Tatchell, K., Hall, B. D., Astell, C., and Smith, M. (1981). A position effect in the control of transcription at yeast mating type loci. *Nature (London)* **289**, 244–250.

Nathans, J., and Hogness, D. S. (1984). Isolation and nucleotide sequence of the gene encoding human rhodopsin. *Proc. Natl. Acad. Sci. U.S.A.* **81**, 4851–4855.

Norman, C., Runswick, M., Pollock, R., and Triesman, R. (1988). Isolation and properties of cDNA clones encoding SRF, a transcription factor that binds to the c-*fos* serum response element. *Cell* **55**, 989–1003.

Oro, A. E., Hollenberg, S. M., and Evans, R. M. (1988). Transcriptional inhibition by a glucocorticoid receptor–β-galactosidase fusion protein. *Cell* **55**, 1109–1114.

Passmore, S., Maine, G. T., Elble, R., Christ, C., and Tye, B. (1988). *Saccharomyces cerevisiae* protein involved in plasmid maintenance is necessary for mating of *MATα* cells. *J. Mol. Biol.* **204**, 593–606.

Passmore, S., Elble, R., and Tye, B.-K. (1989). A protein involved in minichromosome maintenance in yeast binds a transcription enhancer conserved in eukaryotes. *Genes Dev.* **3**, 921–935.

Reneke, J. E., Blumer, K. J., Courchesne, W. E., and Thorner, J. (1988). The carboxy-terminal segment of the yeast α-factor receptor is a regulatory domain. *Cell* **55**, 221–234.

Rine, J. D. (1979). Regulation and transposition of cryptic mating type genes in *Saccharomyces cerevisiae*. Ph.D. dissertation, University of Oregon, Eugene, Oregon.

Rine, J. D., Sprague, G. F., Jr., and Herskowitz, I. (1981). The *rme1* mutation of *Saccharomyces cerevisiae*: Map position and bypass of mating type locus control of sporulation. *Mol. Cell. Biol.* **1**, 958–960.

Rose, M. D., Price, B. R., and Fink, G. R. (1986). *Saccharomyces cerevisiae* nuclear fusion requires prior activation by α factor. *Mol. Cell. Biol.* **6**, 3490–3497.

Russell, D. W., Jensen, R., Zoller, M. J., Burke, J., Errede, B., Smith, M., and Herskowitz, I. (1986). Structure of the *Saccharomyces cerevisiae HO* gene and analysis of its upstream regulatory region. *Mol. Cell. Biol.* **6**, 4281–4294.

Sakai, D. D., Helms, S., Carlstedt-Duke, J., Gustafsson, J.-A., Rottman, F. M., and Yamamoto, K. R. (1988). Hormone-mediated repression: A negative glucocorticoid response element from the bovine prolactin gene. *Genes Dev.* **2**, 1144–1154.

Sassone-Corsi, P., Ransone, L. J., Lamph, W. W., and Verma, I. M. (1988). Direct interaction between fos and jun nuclear oncoproteins: Role of the "leucine zipper" domain. *Nature (London)* **336**, 692–695.

Sauer, R. T., Yocum, R. R., Doolittle, R. F., Lewis, M., and Pabo, C. O. (1982). Homology among DNA-binding proteins suggests use of a conserved super-secondary structure. *Nature (London)* **298**, 447–451.

Sauer, R. T., Smith, D. L., and Johnson, A. D. (1988). Flexibility of the yeast $\alpha 2$ repressor enables it to occupy the ends of its operator, leaving the center free. *Genes Dev.* **2**, 807–816.

Schuermann, M., Neuberg, M., Hunter, J. B., Jenuwein, T., Ryseck, R.-P., Bravo, R., and Müller, R. (1989). The leucine repeat motif in fos protein mediates complex formation with jun/AP-1 and is required for transformation. *Cell* **56**, 507–516.

Shaw, P. E., Schröter, H., and Nordheim, A. (1989). The ability of a ternary complex to form over the serum response element correlates with serum inducibility of the human c-*fos* promoter. *Cell* **56**, 563–572.

Shepherd, J. C. W., McGinnis, W., Carrasco, A. E., DeRobertis, E. M., and Gehring, W. J. (1984). Fly and frog homeo domains show homologies with yeast mating type regulatory proteins. *Nature (London)* **310**, 70–71.

Siliciano, P., and Tatchell, K. (1984). Transcription and regulatory signals at the mating type locus in yeast. *Cell* **37**, 969–978.

Singh, A., Chen, E. Y., Lugovoy, J. M., Chung, C. N., Hitzeman, R. A., and Seeburg, P. H. (1983). *Saccharomyces cerevisiae* contains two distinct genes coding for α-factor pheromone. *Nucleic Acids Res.* **11**, 4049–4063.

Sprague, G. F., Jr., and Herskowitz, I. (1981). Control of yeast cell type by the mating type locus. I. Identification and control of expression of the **a**-specific gene, *BAR1*. *J. Mol. Biol.* **153**, 305–321.

Sprague, G. F., Jr., Blair, L. C., and Thorner, J. (1983a). Cell interactions and regulation of cell type in the yeast *Saccharomyces cerevisiae*. *Annu. Rev. Microbiol.* **37**, 623–660.

Sprague, G. F., Jr., Jensen, R., and Herskowitz, I. (1983b). Control of yeast cell type by the mating type locus: Positive regulation of the α-specific *STE3* gene by the *MATα1* product. *Cell* **32**, 409–415.

Straney, S. B., and Crothers, D. M. (1987). Lac repressor is a transient gene-activating protein. *Cell* **51**, 699–707.

Strathern, J., Hicks, J., and Herskowitz, I. (1981). Control of cell type in yeast by the mating type locus: The $\alpha 1-\alpha 2$ hypothesis. *J. Mol. Biol.* **147**, 357–372.

Stryer, L., and Bourne, H. R. (1986). G proteins, a family of signal transducers. *Annu. Rev. Cell Biol.* **2**, 391–419.

Suzuki, K., and Yanagishima, N. (1985). An α-mating-type-specific mutation causing a specific defect in sexual agglutinability of the yeast *Saccharomyces cerevisiae*. *Curr. Genet.* **9**, 185–189.

Tan, S., Ammerer, G., and Richmond, T. J. (1988). Interactions of purified transcription factors: Binding of yeast *MATα1* and PRTF to cell type-specific, upstream activating sequences. *EMBO J.* **7**, 4255–4264.

Teague, M. A., Chaleff, D. T., and Errede, B. (1986). Nucleotide sequence of the yeast regulatory gene *STE7* predicts a protein homologous to protein kinases. *Proc. Natl. Acad. Sci. U.S.A.* **83**, 7371–7375.

Treisman, R. H. (1986). Identification of a protein-binding site that mediates transcriptional response of the c-*fos* gene to serum factors. *Cell* **46**, 567–574.

Treisman, R. H. (1987). Identification and purification of a polypeptide that binds the c-*fos* serum response element. *EMBO J.* **6**, 2711–2717.

Trueheart, J., Boeke, J. D., and Fink, G. R. (1987). Two genes required for cell fusion during yeast conjugation: Evidence for a pheromone-induced surface protein. *Mol. Cell. Biol.* **7**, 2316 2328.

Van Arsdell, S. W., Stetler, G. L., and Thorner, J. (1987). The yeast repeated element sigma contains a hormone-inducible promoter. *Mol. Cell. Biol.* **7**, 749–759.

Whiteway, M., Hougan, L., Dignard, D., Thomas, D. Y., Bell, L., Saari, G. C., Grant, F. J., O'Hara, P., and MacKay, V. L. (1989). The *STE4* and *STE18* genes of yeast encode potential β and γ subunits of the mating factor receptor-coupled G protein. *Cell* **56**, 467–477.

Wilkinson, L. E., and Pringle, J. R. (1974). Transient G1 arrest of *S. cerevisiae* cells of mating type α by a factor produced by cells of mating type **a**. *Exp. Cell Res.* **89**, 175–187.

Wilson, K. L., and Herskowitz, I. (1984). Negative regulation of *STE6* gene expression by the α2 product of yeast. *Mol. Cell. Biol.* **4**, 2420–2427.

GENETIC CONTROL OF CELL TYPE AND PATTERN FORMATION IN *Caenorhabditis elegans*

Paul W. Sternberg

Howard Hughes Medical Institute, Division of Biology,
California Institute of Technology, Pasadena, California 91125

I. Introduction

From the perspective of a developmental geneticist, the single most striking feature of *Caenorhabditis elegans* as an experimental organism is our knowledge of its complete cell lineage. This description of the invariant pattern of cell divisions from the single-cell zygote to the 959 somatic nuclei of the adult hermaphrodite or the 1031 somatic nuclei of the adult male provides a background against which any experimental perturbation stands out. Essentially every cell division and cell speci-

63

ADVANCES IN GENETICS, Vol. 27

fication event is thus amenable to genetic analysis, and in principle we can hope to identify genes acting at each cell division. Because the cell lineage implies an *order* of events controlling cell-type specification, a first approximation to the genetic regulatory hierarchy controlling the cell lineage falls out naturally from simply knowing the phenotypes of loss-of-function mutations in genes affecting the cell lineage. If the cell lineage merely consisted of a series of differentiative events, the genetic pathway specifying the lineage would simply branch. However, the cell lineage involves multiply used "sublineages," and includes multipotent cells that may interact via feeback circuits. We consider such complexities here in addition to reviewing more straightforward genetic pathways.

Because of the staggering amount of detailed information concerning development in *C. elegans,* typically in the form of cell names such as "Abprapaapa," in this review we concentrate on only a few aspects of the cell lineage, notably those cell divisions producing the vulva and a few neuronal lineages. These examples serve to illustrate the approaches used to study *C. elegans* development as well as the general conclusions drawn from these studies.

II. Molecular Genetics of Development

Essentially all aspects of *C. elegans* biology have been reviewed recently (Kenyon, 1988; Wood, 1988). We provide in this section a brief review of the methodologies and technologies for the study of *C. elegans* development, genetics, and molecular biology as is relevant to the following discussion, for both the kinds of experimental approaches taken as well as those likely in the near future.

A. DEVELOPMENT

The *C. elegans* life cycle consists of a period of embryogenesis followed by four larval stages (L1, L2, L3, and L4) and then adulthood. Each larval stage ends with a 1- to 2-hour quiescent period (lethargus), followed by shedding of the cuticle (molting). The entire process takes approximately 55 hours at 20°C. Newly hatched L1 larvae have about 550 cells; about 50 of these cells divide during postembryonic (larval) development to generate the adult complement of cells. Living animals developing on a microscope slide may be observed with Nomarski optics at 1250× magnification, and individual cells may be followed as they divide, migrate, differentiate, or die. Such direct observation has elucidated the complete pattern of cell divisions from the single-celled zygote to the adult (the cell lineage) (Sulston and Horvitz, 1977; Kimble and Hirsh, 1979; Sulston *et al.*, 1980, 1983).

The invariance of *C. elegans* development arises not only from the rigidly determined cell lineage, but also from highly reproducible cell–cell interactions. If a cell is always formed in the same position within the developing animal, it will always be subject to the same signals from neighboring cells and consequently might have the same fate. This invariance is more a useful experimental tool than a limit to the generality of conclusions drawn from studies of *C. elegans* development: comparative studies of nematode development indicate that autonomously specified invariant cell lineages are not far removed from the more plastic, variable lineages observed in vertebrate development (Sternberg and Horvitz, 1981, 1982). In particular, variability in cell lineage is observed as the extent of cell proliferation in a species increases. Moreover, fates specified by cell interaction in one species can be specified by asymmetric cell division in another species.

To analyze a cell lineage, one must be able to assay the state of any cell, often prior to its differentiation. In the absence of molecular markers for cell type, the "type" of a cell is defined by its differentiated properties or the cell lineage it generates. Certain precursor cells each generate the stereotypical patterns of cell divisions with identical sets of progeny cell types. Such repeated lineages are known as "sublineages" (Chalfie *et al.*, 1983; Sternberg and Horvitz, 1981, 1982). Since the state of a precursor cell is inferred from its progeny, precursor cells that generate identical sublineages are thought to be identical.

As is the case in other organisms, some nematode cells become specified to be distinct types either by cell–cell interaction or by asymmetric cell division, the generation of two sister cells of different types when a cell divides. Although much of *C. elegans* development appears to be cell autonomous, the types of certain cells are specified by cell–cell interactions (reviewed in Sternberg and Horvitz, 1984). The best studied cases involve small groups of cells ("equivalence groups") that share several potential fates and whose developmental fates depend on cell–cell interactions. In all such cases there is a strict correlation of the position of a cell and its type; the specification of the types of the cells in equivalence groups is thus a simple example of pattern formation. The autnomous mechanisms by which two sister cells come to differ in their fate are not understood in any organism. For some asymmetric cell divisions in *C. elegans,* genes necessary for the asymmetry of a cell division have been identified; some of these are discussed below (see Sections III,B,3 and IV,E).

Cell-type specification in response to intercellular signals involves two general classes of phenomena: signaling between tissues (induction) and signaling within a tissue ("lateral signaling" or "lateral inhibition"). Nematode development involves both types of interactions. The major examples of induction are induction of the vulva by the gonad, stimulation of mitosis or inhibition of meiosis in the germ line nuclei by the "distal tip cell" of the somatic gonad, and the interaction of descendants of the P1 blastomere with AB descendants (Sulson and White, 1980; Kimble, 1981; Kimble and White, 1981; Priess and Thomson, 1987). Regulative interactions involve equivalence groups, sets of equipotent cells whose fates are specified, at least in part, by lateral interactions among the set (Kimble et al., 1979; for review, see Sternberg, 1988b). Examples of regulative interactions include the anchor cell/ventural uterine precursor, the vulval precursor cells, the male pre-anal ganglion precursor cells, sensory ray precursors in the lateral hypodermis, and AB.a and AB.p blastomeres or their descendants.

Pairs of bipotent cells, such as the two so-called AC/VU cells of the hermaphrodite ventral uterus, provide the simplest examples of equivalence groups. In a given animal, either of two homologous cells (Z1.ppp or Z4.aaa) becomes the anchor cell (AC) while the other cell becomes a ventral uterine precursor (VU) cell. This "decision" occurs at random (Kimble and Hirsh, 1979). After destruction of one of the two cells, the remaining cell becomes the AC (Kimble, 1981). The AC fate is said to be primary and the UV fate secondary (Kimble et al., 1979). Similarly, in the male, two bilateral homologs, B.alaa and B.araa, each have the potential to adopt the α fate, which is primary, or the β fate, which is secondary. How do the two cells in each group become different? Possibly, one of the two cells receives a signal from a third cell, or the two cells may compete for a site, for example, on extracellular matrix. In these types of models, the information to become different is imparted from outside the group. In another general class of models for such bipotent cells, no information from outside the group is required; such models can be illustrated by a simple example (Fig. 1) (Sternberg, 1988b; Seydoux and Greenwald, 1989). Consider two cells, each expressing a membrane-bound signal molecule and a receptor for that signal, in which occupied receptor both activates primary-specific functions including the receptor and inhibits secondary-specific functions including production of the signal. Such a coupled system of positive and negative feedback results in a highly unstable situation. Random fluctuations in the various components will tend to drive the system to either of two stable states: one cell expressing signal (primary functions) and the other expressing receptor (secondary functions), or vice

initial

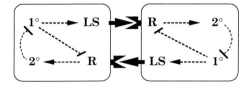

final

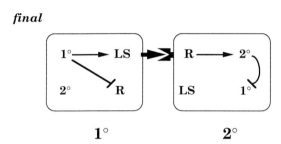

1°   2°

FIG. 1. Bistable cell states and equivalence groups. In this simple general model, each bipotent cell initially produces a lateral signal (LS) and a receptor (R) for that signal. Occupied receptor leads to stimulation of primary-specific functions (1°) including the receptor (arrow) and inhibition of the secondary-specific functions (2°) including LS (bar). Such a coupled system is unstable and, owing to random fluctuations in any component, will evolve to either of two stable states, with one cell (1°) producing LS and the other cell (2°) producing receptor. Such bistable cells can come in pairs [e.g., AC/VU in *C. elegans* hermaphrodites (Seydoux and Greenwald, 1989)], a linear array [e.g., VPCs in a *C. elegans lin-15* mutant (Sternberg, 1988a)], or a two-dimensional array [e.g., the neurogenic region in insect embryos (Doe and Goodman, 1985)].

versa. A number of formally equivalent models can be constructed, involving, for example, regulation at various levels, with extra steps, or with different biochemistries.

An important feature of this class of models is that the initial meta-stable state can be very susceptible to a slight push: the final state can easily be biased from outside (induction) or inside a cell. For example, in a *lin-15* mutant, as described below (Section III,A), the inductive signal from the anchor cell pushes the cell closest to it to become a primary (1°) cell (Sternberg, 1988a).

B. GENETICS

In this section we briefly review several basic methods of genetic analysis as applied to *C. elegans*: mutant isolation, dosage analysis including definition of null phenotype, epistasis and gene interaction studies, and mosaic analysis. These methods play an important role in later sections, each being applied to some genes or genetic pathways.

A description of *C. elegans* genetic nomenclature (see Table 1) is in order (Horvitz *et al.*, 1979). All genetic loci are given names consisting of three italicized letters describing a general class of genes and a number distinguishing complementation groups within each class. For example, *lin-7* is the seventh complementation group of mutations with defective cell lineage. Specific alleles are designated by an italicized lowercase laboratory designation and a unique number. For example, *e1413* is the 1413th mutation identified at the Medical Research Council (MRC) Laboratory in England. Phenotypes are designated by a three-letter abbreviation, capitalized and in roman type. For example, Lin refers to a mutant with altered cell lineage, Lin-7 to the specific phenotype associated with a *lin-7* mutation.

1. Mutant Isolation

A vast majority of *C. elegans* mutations have been induced with ethyl methanesulfonate (EMS), although transposable elements, γ- or X-rays, formaldehyde, diepoxybutane, and ultraviolet radiation (UV),

TABLE 1
Relevant *Caenorhabditis elegans* Terminology

Symbol	Meaning
P6.p	Posterior daughter of the blast cell P6
P6.pa	Anterior daughter of P6.p
lin-3	Genetic locus
Lin-3	Phenotype caused by a mutation in the *lin-3* locus
lin-3(e1417)	Particular allele of *lin-3*
Vul	The vulvaless phenotype, or a gene defined by mutations that cause a Vul phenotype
Muv	The multivulva phenotype, or a gene defined by mutations that cause a Muv phenotype
Df	Genetic deficiency, presumably a deletion
Dp	Genetic duplication, either attached to a chromosome or segregating as a minichromosome ("free" duplication)
VPC	One of six vulval precursor cells, namely, P3.p, P4.p, P5.p, P6.p, P7.p, or P8.p
AC	Anchor cell
Pn	Any or all of the 12 cells, P1, P2, P3, P4, P5, P6, P7, P8, P9, P10, P11, P12
Pn.p	Any or all of the 12 cells, P1.p, P2.p, P3.p, P4.p, P5.p, P6.p, P7.p, P8.p, P9.p, P10.p, P11.p, P12.p
VU	Ventral uterine precursor cell

among other treatments, have been used effectively. Typically, two types of mutagenesis have been carried out to isolate recessive zygotic mutants: F_2 screens and F_1 noncompelementation screens. In the standard F_2 mutant hunt (Brenner, 1974), virgin P_0 hermaphrodites are mutagenized, hitting both male and female germ lines (Fig. 2A). A number of F_1 hermaphrodites per petri plate are allowed to self, producing F_2 progeny homozygous for recessive mutations. With wild-type parents, approximately 100 F_1 haploid chromosome sets can be screened on each 9-cm petri plate. The parent strain could be wild type (strain N2) or an existing mutant. The number of chromosome sets (haploid genomes) screened is 2 times the number of F_1 hermaphrodites, as each has two potentially mutagenized chromosomes.

In a F_1 noncomplementation screen (Greenwald and Horvitz, 1980), one parent is mutagenized and mated to a strain carrying an existing

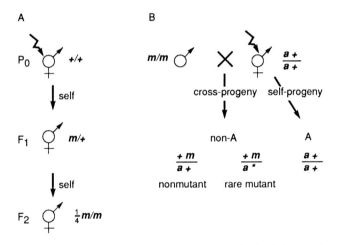

FIG. 2. Generalized mutant screens. (A) F_2 screen. Virgin P_0 hermaphrodites are mutagenized and allowed to self, producing up to 300 F_1 self-progeny. If a sperm or ovum from the P_0 has suffered a mutation, an F_1 hermaphrodite will be heterozygous for that mutation $(m/+)$. The heterozygous F_1 will segregate one-fourth homozygous mutants (m/m) in the F_2. For a typical recessive, F_2 animals are screened for the mutant phenotype. Dominant mutations are identified by F_1 mutant animals; maternal-effect mutations are identified by F_3 mutant animals. (B) F_1 noncomplementation screen. To recover additional alleles at a particular locus, animals heterozygous for an existing mutation and a mutagenized chromosome are constructed and examined for a phenotype. In the example shown here, males homozygous for the starting mutation (m) are mated to mutagenized hermaphrodites homozygous for a closely linked marker (a). The linked marker serves two purposes: to distinguish self- from cross-progeny and to enable the new mutation (*) to be followed during subsequent analyses.

mutation (Fig. 2B). Ideally, both chromosomes are marked so that the three genotypes arising from the cross can be distinguished (m/m, $m/*$, and $*/*$, where m is the original mutation and $*$ is the new mutation). Such a screen can also be done without a cross, by mutagenizing a balanced heterozygote. The number of chromosome sets screened is equal to the number of F_1 animals.

2. Null Phenotypes

Inference of the function of a wild-type gene product requires, at a minimum, understanding the phenotypic consequence of loss of gene function. At present, the gene disruption technology of fungi is not available in *C. elegans*, so other methods are required. While "modern" techniques of using antisense RNA (Melton, 1985; Rosenberg *et al.*, 1985) or dominant-negative products (Herskowitz, 1987) to inactivate function *in vivo* of a cloned gene can in principle be used in *C. elegans*, much of our current knowledge arises from classic genetic studies of gene function.

Several types of evidence have been garnered to define null phenotypes in *C. elegans*. In most cases, only one or a few types of evidence can be obtained. Ultimately, a mutation can be identified as null using molecular probes, for example, demonstrating that no protein product or RNA transcript is detectable.

   a. *Frequency of Mutation.*   The frequency of induction by EMS of knockout mutations is approximately 1/2000 to 1/5000 haploid chromosome sets for typical genes (Brenner, 1974; Greenwald and Horvitz, 1980). If mutations are obtained at such a frequency, and are the most common class, they are arguably null mutations.

   b. *Amber-Suppressible Alleles.*   Suppression by an amber-suppressing tRNA mutation is excellent evidence that a mutation affects a protein product but is not, in general, good evidence for a null mutation. In particular, Ferguson and Horvitz (1985) identified non-null amber mutations in five of seven genes with amber alleles (*lin-2, lin-18, lin-24, lin-34,* and *let-23*). There are numerous examples from other organisms of nonnull amber mutations. For example, Mitchell (1985) showed for *GLN1*, the structural gene for glutamine synthetase (glutamate–ammonia ligase) in yeast, that the typical amber mutant protein has activity and often is semidominant.

   c. *Phenotype Not Enhanced in Trans to a Deficiency.*   The widely used "phenotype enhancement" test for null alleles can rule out that a given mutation is null but does not constitute proof. Specifically, if a homozygous mutation results in a less severe or less penetrant phenotype than the mutation *in trans* to a deficiency, then that mutation is not null.

  d. Noncomplementation Screen. An often conclusive experiment to define a null phenotype is to carry out a genetic screen that can recover true null mutations at a locus and ascertain what is the most frequent class of mutations arising in such a screen. For example, one can screen for new mutations that fail to complement an existing mutation which is itself viable *in trans* to a deficiency (Df) for the locus and, moreover, that confer an observable phenotype (Greenwald and Horvitz, 1980). Such a screen could, in principle, recover null mutations for the locus in question. For example, *lin-3(e1417)/Df* is viable and confers a completely penetrant vulvaless phenotype. Therefore, null alleles of *lin-3* could be recoverd in such a screen with 100% efficiency. In several such screens, a majority of mutations identified by their failure to complement *e1417* for the Vul phenotype confer a lethal phenotype when homozygous, indicating that the null phenotype is lethality (Ferguson and Horvitz, 1985; R. Hill and P. Sternberg, unpublished). One caveat is that deletion of a linked locus might suppress dominantly the lethality of mutant/null, so, even though mutant/Df is viable, mutant/null is not, and null alleles would not be recovered in a noncomplementation screen.

  e. Allelic Series. Some loci are defined by a set of mutations of varying severity (either in the proportion of individuals displaying a phenotype or in the severity of the phenotype displayed). In addition, animals heterozygous for two mutations of different severity often display a phenotype intermediate between those of the relevant homozygotes. In such an allelic series, the argument can be made that the loss-of-function phenotype is at least the most severe of the series.

3. Epistasis and Interactions

  Gene interactions have been investigated in two ways, by construction of multiply mutant strains and by isolation of extragenic suppressor mutations. A test of epistasis between two mutations (e.g., *A* and *B*) that confer distinct phenotypes is performed by comparison of the phenotype of the double-mutant strain to each of the single-mutant strains. if the *A–B* double mutant has the same phenotype as mutant *A*, then the *A* mutation is said to be *epistatic* to the *B* mutation. Epistasis indicates that the two mutations affect a common pathway. Interpretation of the order of gene action requires additional information (see Section V,B). if the *A–B* double mutant has a phenotype different than either of the single mutants, the mutations are *coexpressed*. If the double mutant is more like wild type than either of the single mutants, one mutation *suppresses* the other.

Double-mutant strains are constructed as segregants from parents heterozygous at several loci. The segregation of the individual mutations under study can be followed by scoring tightly linked markers *in cis* or *in trans,* or simply by scoring the phenotype of each mutation. For example, a strain carrying both a vulvaless and a multivulva mutation can be constructed by picking vulvaless F_1 progeny of a doubly heterozygous mother. If such vulvaless F_1 animals segregate multivulva F_2 progeny, then the multivulva mutation is epistatic to the vulvaless mutation, and the double-mutant stain has be constructed. In parallel, F_1 multivulva animals are picked and their progeny are screened for vulvaless animals. By carrying out both methods of construction, the double-mutant strain is generated regardless of which mutation is epistatic (Ferguson *et al.,* 1987).

The isolation of extragenic suppressor mutations by mutagenesis defines new genes that interact with the gene defined by the starting mutation. While in general the logic of interpreting extragenic suppression is similar to that of interpreting epistasis tests, extragenic suppressor mutations might not confer a phenotype in an otherwise wild-type background and would thus only be identified by their ability to suppress an existing mutation. Suppression in *C. elegans* has been reviewed by Hodgkin *et al.* (1987). Striking examples of epistasis pathways have come from studies of sex determination mutants (Hodgkin, 1987a) and dauer larvae formation (Riddle, 1988).

4. Mosaics

For a variety of developmental genetic questions, genetic mosaic analysis is crucial. In cases of cell interaction, mosaics provide a means to ascertain whether a gene product is required in the signaling cell, in the responding cell, or in both. For a gene with general effects (as evidenced by pleiotropic effects of mutations or general pattern of expression), mosaics provide a means of identifying the focus of action. For example, mosaic analysis can indicate whether a gene is essential, because it is necessary for the viability of all cells or acts in a few key cells. Mosaic analysis has been applied in only a few cases, but it will clearly become a genetic tool of increasing importance (Herman, 1984, 1987; Herman and Kari, 1985; Kenyon, 1986; Park and Horvitz, 1986; Austin and Kimble, 1987; Seydoux and Greenwald, 1989).

Genetic mosaics are created by mitotic loss of a "free duplication," essentially a minichromosome, carrying a dominant allele (typically the wild-type allele) of the gene of interest. Loss of the dominant allele uncovers the recessive mutations on the chromosomes. Formally, there is little difference between mosaic analysis in *Drosophila* and *Caenorhabiditis*. In the latter, however, the cell lineage is known, and loss of a duplication in a particular lineage can be accurately determined. Major limitations to mosaic analyses have been the incompleteness of the existing collection of free duplications (Dps) and the lack of well-characterized cell-specific markers. Approximately 50% of the genetic map is covered by free duplications (Edgely and Riddle, 1988); however, not all of these duplications are suitable for mosaic analysis.

C. MOLECULAR BIOLOGY

The facility of *C. elegans* molecular genetics has been greatly increased by several recent advances described in this section: transposon mutagenesis, mapping with restriction fragment length polymorphisms (RFLPs), DNA-mediated transformation, *in situ* hybridization to meiotic chromosomes, and genome mapping. Taken together, these advances allow the cloning of genes identified only by mutations, the creation of transgenic strains harboring genes mutated *in vitro,* and the identification of genetic loci corresponding to cloned genes.

First, several transposons (Tc1, Tc3, etc.) have been identified (Emmons *et al.,* 1983; Liao *et al.,* 1983; Collins *et al.,* 1989a). These elements will transpose in particular genetic backgrounds (Eide and Anderson, 1985; Emmons *et al.,* 1986; Moerman *et al.,* 1986). "Mutator" strains that display a high frequency of transposition (as high as 10^{-3}) have been derived (Collins *et al.,* 1989b). Some mutators have been genetically mapped, with the efficiency of transposition monitored by an insertion or an excision assay (Herman and Shaw, 1987). Cloning genes by "transposon tagging" has been successful in a number of cases (e.g., Greenwald, 1985; Moerman *et al.,* 1986; Way and Chalfie, 1988).

Second, efficient methods for identifying RFLPs have been developed (Baillie *et al.,* 1985; Cox *et al.,* 1985). The commonly used Bristol strain of *C. elegans* has about 50 copies of the Tc1 element (Emmons *et al.,* 1979). In contrast, the Bergerac strain has about 500 copies of Tc1. Thus, the Tc1 elements provide a rich source of RFLPs. RFLPs linked to a gene of interest can be obtained by successive backcrosses of a segment of a Bergerac chromosome into the Bristol background. An RFLP can then be mapped to as high a resolution as necessary by multifactor crosses using both visible markers and other RFLPs (Ruvkun *et al.,* 1989). Sequences flanking the desired Tc1 element insertion site can then be cloned.

Third, DNA-mediated transformation is now standard practice

(Stinchcomb *et al.*, 1985; Fire, 1986; McCoubrey *et al.*, 1988; Spieth *et al.*, 1988; Way and Chalfie, 1988). Transformation is accomplished by microinjection of DNA into germ line nuclei or the syncytial germ line cytoplasm of adult hermaphrodites. Exogenous DNA integrates at random sites in the genome in a few copies, is stably maintained, and can be expressed (Fire, 1986). Transgenic animals can be selected by co-transformation with a dominant selectable marker, e.g., the *sup-7* amber suppresor tRNA (Bolten *et al.*, 1984), or with other cloned genes.

Fourth, clones can now be mapped to portions (~5 map units) of chromosomes by *in situ* hybridization of biotinylated probes to meiotic chromosomes (Albertson, 1985).

Lastly, an effort to obtain a set of cosmid clones that span the entire genome has been undertaken (Coulson *et al.*, 1986, 1988; A. Coulson, J. Sulston, R. Waterston, Y. Kohara, D. Albertson, and R. Fishpool, personal communication). The first step is to "fingerprint" random cosmid clones (35–45 kb in length) and match fingerprints to identify overlapping cosmids. A second step involves hybridization of cosmid clones to yeast artificial chromosome (YAC) clones and vice versa, resulting in 95% of the genome being contained in about 200 "contigs," sets of overlapping clones. Approximately 65% of the genome is present as clones that have been mapped genetically.

A genomic map in the form of an ordered set of clones can be used to localize cloned DNAs to genetic map positions by hybridization in an analogous manner to *in situ* hybridization to *Drosophila* polytene chromosomes. The resolution of such a procedure would be, on average, approximately 40 kb, or 0.13 map units (mu). The metric correlating physical and genetic map distances in *C. elegans* should be 330 kb/mu given a haploid genome of 100,000 kb and a total genetic map distance of 300 mu. This metric varies with position, however, and is considerably higher in the clusters on each linkage group. For example, a metric of 1300 kb/mu holds in the chromosome III cluster (Greenwald *et al.*, 1987).

III. Vulval Development

Development of the hermaphrodite vulva is one of the most intensively studied aspects of *C. elegans* development and, as such, constitutes a major focus of this article. This section first describes the salient features of wild-type vulval development that form the basis for understanding not only the phenotypes associated with mutants defective in this process but also the inferred roles of the wild-type gene products.

Then an overview of the genetic pathway is provided before delving into the details of the genes acting in the pathway and the interactions of genes that determine vulval precursor cell (VPC) fate.

A. DEVELOPMENT OF THE VULVA

Induction of the *C. elegans* hermaphrodite vulva provides a unique opportunity to study the control of cell type by cell–cell interactions. According to current hypothesis, vulval induction involves only seven cells: six multipotent vulval precursor cells (the VPCs) of the hypodermis and the anchor cell (AC) of the gonad. Each VPC becomes specified to be one of three cell types (1°, 2°, or 3°). It is possible that the progeny of the VPCs have some plasticity; however, to a first approximation, the hypothesis that the relevant cell–cell interactions occur on or among the VPCs and not their daughters allows interpretation of a vast majority of the data. Once specified, each VPC generates a cell lineage that appears to be specified cell autonomously; therefore, the type of a VPC can be inferred from the lineage it generates (Sternberg and Horvitz, 1986). The vulval cell lineages occur over a period of 5 hours and consist of only three rounds of cell division (Sulston and Horvitz, 1977). Thus, it is straightforward to observe these lineages. The cell–cell interactions involved in vulval induction have been investigated by destroying particular cells with radiation from a laser microbeam and examining the fates of the remaining cells. During vulval development a pattern of three cell types is established by two intercellular signals (Fig. 3) (Sternberg and Horvitz, 1986; Sternberg, 1988a). Several of the key results from such experiments are summarized in Fig. 4.

In the absence of interactions with other cells (the anchor cell or VPCs), each VPC will remain in the "ground state" (3°) and generate nonspecialized epidermal cells rather than vulval cells. An inductive signal from the anchor cell is thus necessary for a VPC to become a 1° or 2° cell type and generate vulval cells.

The inductive signal appears to be spatially graded; the potency of the signal received by each VPC specifies which type of cell it becomes. The inductive signal does not have to act via other VPCs because a single VPC responds to the signal appropriately. An isolated VPC can be a 1°, 2°, or 3° cell, depending on its distance from the anchor cell (Sternberg and Horvitz, 1986). Thus, the type of each VPC depends primarily on its distance from the anchor cell rather than on its interactions with other VPCs. The VPC closest to the anchor cell, P6.p, becomes a 1° cell; the two adjacent cells, P5.p and P7.p, become 2° cells; and the most distal cells, P3.p, P4.p, and P8.p, become 3° cells. The

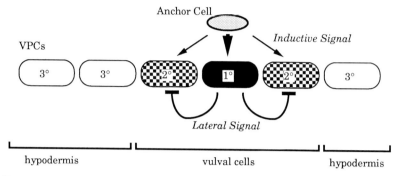

FIG. 3. Vulval induction and pattern formation. Each of six vulval precursor cells (the VPCs) has the potential to generate three distinct lineages (1°, 2°, or 3°). The 1° and 2° lineages generate vulval cells; the 3° lineage generates nonspecialized epidermis. 3° is the ground state. The anchor cell of the gonad, in particular of the ventral uterus, induces three VPCs to generate vulval tissue. Moreover, the pattern of VPC fates is established primarily by this inductive signal. In addition to the anchor cell signal, at least some VPCs signal their neighbor: the 1° VPC prevents its immediate neighbors from also becoming 1° by a "lateral signal." [Adapted from Sternberg (1988a).]

inductive signal can act at a distance: in mutants which have a displaced gonad and the anchor cell located dorsally within an animal, VPCs can be induced, although the pattern of cell types is not necessarily normal (Thomas and Horvitz, 1990; E. Hedgecock, personal communciation).

If a VPC proximal to the anchor cell is destroyed by irradiation with a laser microbeam, a more distal precursor takes its place (Sulston and White, 1980). The names of the three cell types (1°, 2°, and 3°) derive from the hierarchy of changes in cell type after ablation: a primary cell type is the one that is replaced by either a secondary or a tertiary cell after it is destroyed; a secondary cell is one that will replace a primary cell or will be replaced by a tertiary cell (Kimble *et al.*, 1979). In some previous publications these cells were referred to as type 1, type 2, and type 3 (Sternberg, 1988a,b). These "regulative" interactions simply may indicate that one VPC normally prevents a more distal VPC from gaining access to the signal. Alternatively, such regulation may reflect the release of distal VPCs from a lateral inhibitory signal. Indeed, studies of VPC interactions in a *lin-15* multivulva mutant indicate that a 1° VPC prevents its immediate neighbors from also becoming 1° (Sternberg, 1988a). In particular, in a *lin-15* mutant the ground state of a VPC is 1°. If there are two VPCs present and close together, typically one VPC is 1°, the other 2°. If, however, the anchor cell is present, the

Experiment	Result	Conclusion
A Intact wild-type animal	3 3 2 $\overset{AC}{1}$ 2 3	Invariant pattern of cell types
B Ablate P6.p	3 2 1 $\overset{AC}{X}$ 2 3 3 3 2 $\overset{AC}{X}$ 1 2	VPCs are multipotent
C Ablate anchor cell	3 3 3 $\overset{x}{3}$ 3 3	The anchor cell induces the vulva
D Isolate P7.p	X X X $\overset{x}{X}$ 3 X	The ground state is 3°
E Isolate P7.p in *lin-15*	X X X $\overset{x}{X}$ 1 X	The ground state in a *lin-15* mutant is 1°
F Ablate anchor cell in *lin-15*	2 1 2 $\overset{x}{1}$ 2 1 1 2 1 $\overset{x}{2}$ 2 1	1° VPCs inhibit their neighbors from also becoming 1°
G Intact *lin-15*	2 1 2 $\overset{AC}{1}$ 2 1	The anchor cell influences the pattern of VPCs in a *lin-15* mutant
H wild-type isolated P8.p	X X X X $\overset{AC}{X}$ 1 X X X X $\overset{AC}{X}$ 2 X X X X $\overset{AC}{X}$ 3	The anchor cell can induce a 2° fate directly

FIG. 4. Summary of cell–cell interaction experiments. AC, Anchor cell; X, ablated cells. References: (A) Sulston and Horvitz (1977); (B) Sulston and White (1980), Sternberg and Horvitz (1986); (C) Kimble (1981); (D–G) Sternberg (1988a); (H) Sternberg and Horvitz (1986), P. Sternberg (unpublished).

VPC closest to the anchor cell becomes 1°. Therefore, VPCs in a *lin-15* mutant still respond to inductive signal from the anchor cell as well as to the lateral inhibitory signal. Moreover, the inductive signal biases the competition for 1° such that the cell closest to the anchor cell becomes 1°. Therefore, in the wild type, the inductive signal may have two effects: first, to allow VPCs to compete to be 1° and, second, to bias the competition such that P6.p always wins and becomes 1°.

The 1° and two 2° lineages generate a total of 12 cells, 10 of which are binucleate (at 25°C, all 12 can be binulceate) (White, 1988). These vulval cells plus the anchor cell ultimately form the vulva (Sulston and Horvitz, 1977; Kimble, 1981; Sternberg and Horvitz, 1986). Smaller groups of vulval cells also form vulvallike structures. This fact allows the types of VPCs to be inferred from morphological features visible in the dissecting microscope at 25× magnification (P. Sternberg, unpublished). For example, cells produced by a 1° lineage make a *vulva* (in the presence of the anchor cell) or a *bump* (a "pseudovulva") on the ventral surface of the animal (in the absence of the anchor cell) (Fig. 6). Cells produced by a 2° lineage can join a vulva or pseudovulva formed by the cells produced by a 1° lineage, or they can form a pseudovulva themselves. Kimble (1981) observed that the anchor cell is not required just for induction: if the anchor cell is destroyed after induction has taken place, the 12 vulval cells do not undergo normal morphogenesis. The role of the anchor cell in morphogenesis may simply involve a required attachment of the vulva to the uterus.

B. Pathway of Gene Action

The model for vulval development described above suggests that vulval development consists of four major steps in a branched pathway (Fig. 5A) (Sternberg and Horvitz, 1986): (1) generation of multipotent VPCs and the anchor cell; (2) specification of VPC type involving the production of and response to intercellular signals; (3) execution of the 1°, 2°, and 3° lineages resulting in production of the characteristic numbers and types of progeny cells; and (4) morphogenesis of the vulva by interactions among the anchor cell and the vulva cells produced by the 1° and 2° lineages.

Mutants with abnormal vulval cell lineages have been readily obtained because they can be recognized with a dissecting microscope and are fertile (Horvitz and Sulston, 1980; Ferguson and Horvitz, 1985). Mutants defective in vulval development have either of two basic phenotypes (Fig. 6): vulvaless (Vul) mutants lack a vulva and are egglaying defective; multivulva (Muv) mutants have extra vulval cells that form bumps on the ventral surface of the animal (see below) (Horvitz and Sulston, 1980; Sulston and Horvitz, 1981; Ferguson and Horvitz, 1985). A set of mutations that disrupt vulval development has been characterized and assigned to individual steps based on their phenotypes, epistatic interactions, and temperature-sensitive periods

(Ferguson *et al.*, 1987; Sternberg and Horvitz, 1989). The genes acting at each of these steps are described below in Section III,C. The specification step is discussed in greater detail in Section III,D.

1. Generation of the Vulval Precursor Cells and the Anchor Cell

The generation of the VPCs is controlled by at least five genes. These genes have been identified by mutations that alter the fates of VPCs and that are epistatic to mutations which act at later steps in the pathway (Ferguson *et al.*, 1987). *lin-26* is necessary for Pn.p cells, including the presumptive VPCs, to follow a hypodermal as opposed to a neuronal pathway of development. In a *lin-26* mutant all Pn.p cells (including the presumptive VPCs) become neurons or neuroblasts (Fixsen *et al.*, 1985). *lin-24* and *lin-33* are involved early in VPC development. Dominant mutations at either of these loci cause some VPCs to die. Surviving VPCs can generate vulval lineages, become 3°, or generate neuronal cells. A number of mutations, exemplified by *n300* (which is associated with a reciprocal translocation and has not been assigned a gene name), cause VPCs to fuse with hyp7, the large hypodermal syncytium enveloping the animal (Ferguson *et al.*, 1987; S. Clark, K. Edwards, and R. Horvitz, personal communication). *lin-25* mutations cause a few Pn.p cells to divide during the L1 stage. As discussed below, these cells (and the undivided Pn.p cells) are not normal VPCs. Two genes, *unc-83* and *unc-84*, are necessary for Pn nuclear migration. In mutants defective in either of these genes, Pn cells fail to produce posterior daughters (Pn.p cells), and thus no VPCs are formed (Sulston and Horvitz, 1981). These mutations provide a genetic means to ablate the VPCs (e.g., Sternberg and Horvitz, 1986).

Another class of genes, the so-called heterochronic genes, control stage-specific events including vulval formation. For example, in a dominant *lin-14* mutant, Pn.p cells divide during the L3 stage but generate neuronal-like nuclei rather than vulval tissue, based on morphological criteria (Ambros and Horvitz, 1984). In hermaphrodites carrying a loss-of-function *lin-14* mutation, the vulva is induced one stage early. In general, each of the above mentioned genes is represented by only a few alleles (Ferguson and Horvitz, 1985), suggesting that there are many genes with roles in generation of VPCs yet to be identified.

The production of the anchor cell is controlled by the *lin-12* locus (Greenwald *et al.*, 1983). While only a single anchor cell is present in wild-type hermaphrodites (Kimble and Hirsh, 1979), *lin-12* mutants have between zero and four anchor cells, depending on the type of allele (Greenwald *et al.*, 1983; P. Sternberg, unpublished). Dominant gain-of-function [*lin-12(d)*] mutations prevent formation of an anchor cell,

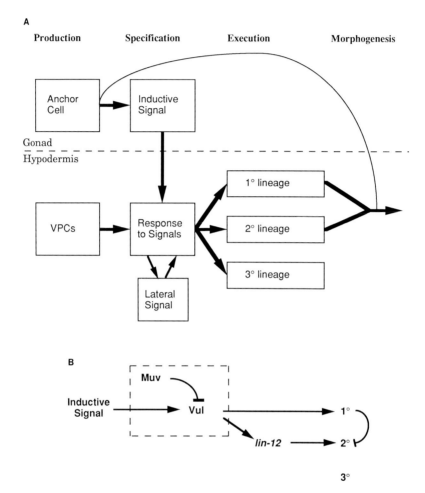

FIG. 5. Pathway of vulval formation. (A) Developmental pathway. Vulval formation involves four tissues: gonad, hypodermis, neurons, and muscle, only the first two
of which we consider here. Development of these tissues follows four major steps.
During the **Production** step, the anchor cell and the VPCs are generated from the Z
and P lineages, respectively. During the **Specification** step, the anchor cell produces
inductive signal and the VPCs produce a lateral inhibitory signal. These signals are
interpreted by the VPCs, leading to the specification of their fates as 1°, 2°, or 3°.
During the **Execution** step, the VPCs divide and produce particular sets of progeny
cells. During the **Morphogenesis** step, the 12 total progeny of the 1° and 2° lineages,
together with the anchor cell, undergo cooperative morphogenesis to form the adult
vulva. [Modified from Sternberg and Horvitz (1986).] (B) Genetic pathway of VPC
specification. The bare outlines of the genetic pathway are shown. The output of the
pathway is in the form of cell-type-specific functions. The input is the inductive signal.
Activation of the 2° lineage requires inductive signal, the Vul genes, and *lin-12*, in

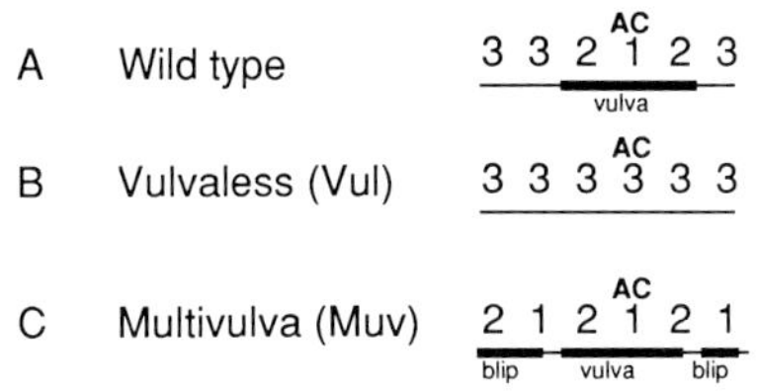

A   Wild type
$$\underline{3\ 3\ 2\ \overset{\text{AC}}{1}\ 2\ 3}$$
vulva

B   Vulvaless (Vul)
$$3\ 3\ 3\ \overset{\text{AC}}{3}\ 3\ 3$$

C   Multivulva (Muv)
$$\underline{2\ 1\ 2\ \overset{\text{AC}}{1}\ 2\ 1}$$
blip   vulva   blip

FIG. 6. Schematic of vulva specification mutant phenotypes. (A) Wild type (Sulston and Horvitz, 1977), (B) vulvaless (Sulston and Horvitz, 1981), and (C) multivulva (Ferguson *et al.*, 1987; Sternberg, 1988a).

while recessive loss-of-function [*lin-12(0)*] mutations cause the production of four anchor cells owing to the misspecification of particular gonadal cells. In particular, whereas wild-type hermaphrodites have three ventral uterine precursor (VU) cells and a single anchor cell, *lin-12(0)* mutants have four anchor cells and no UV cells. During wild-type development, the level of *lin-12* activity specifies the types of particular cells, including the anchor cell. An elegant genetic mosaic analysis has clearly demonstrated that for the AC/VU decision *lin-12* acts in a cell-autonomous manner (Seydoux and Greenwald, 1989). In particular, *lin-12* expression is required for a cell to become a VU precursor.

2. Specification

Specification of the types of vulval VPCs depends on over 15 known genes (Ferguson *et al.*, 1987; Ferguson and Horvitz, 1989; Sternberg and Horvitz, 1989). These genes fall into three major classes: (1) Vul, (2) Muv, and (3) *lin-12*. Minor classes of genes are discussed in Section III,D,3.

1. Five Vul genes (*let-23, lin-2, lin-3, lin-7,* and *lin-10*) are necessary for vulval induction. Loss of function of any of the Vul genes causes VPCs to remain in the ground state, 3°. In principle, these genes could act either in the anchor cell to control production of the inductive signal or in the VPCs to control the response to the signal. A genetic argument

---

that order. Activation of the 1° lineage requires inductive signal and the Vul genes but not *lin-12*. Production of the lateral inhibitory signal requires at least some Vul products. The Muv products act antagonistically to Vul production, although the details of this antagonism are not known; the Muv and Vul genes are enclosed by a dashed box to emphasize this uncertainty. 1° functions must somehow inhibit 2° functions, as a strong inductive signal, acting via the Vul genes, can override the 2° fate specified by hyper-*lin-12* activity.

described below (Section III,B,2,a) suggests that at least four of these Vul genes act in the VPCs and hence in response to the signal. These Vul genes are thus excellent candidates for encoding components of the signal transduction apparatus that senses the inductive signal and specifies the specialized properties of the 1° and 2° cells.

2. The Muv genes (*lin-15* and eight other related genes) are a second class of genes involved in specification of VPC type (Ferguson *et al.*, 1987; Ferguson and Horvitz, 1989; Sternberg and Horvitz, 1989). Mutations in these genes cause all six VPCs to be of the 1° or 2° types even in the absence of the anchor cell. Thus, the Muv mutations act in the VPCs. The wild-type Muv gene products may encode components of the signal transduction apparatus that act antagonistically to those encoded by the Vul genes, or they may encode negative regulators of the Vul genes.

3. *lin-12* specifies the 2° cell type (Greenwald *et al.*, 1983; Sternberg and Horvitz, 1989). The roles of *lin-12* in specifying the fates of the anchor cell and the VPCs are distinct (Sternberg and Horvitz, 1989). Recessive mutations that eliminate *lin-12* activity [*lin-12(0)*] prevent VPCs from becoming 2° cells. Semidominant mutations that increase *lin-12* activity [*lin-12(d)*] allow VPCs to become 2° cells in the absence of the inductive signal. In the absence of the anchor cell, a high level of *lin-12* activity causes that cell to be 2°, while a low level of *lin-12* activity causes it to be 3°. In the presence of the anchor cell, the VPCs near the anchor cell become 1° regardless of the level of *lin-12* activity; therefore, *lin-12* does not specify whether a cell is 1°. The *lin-12* product is similar to a set of proteins that includes the precursor to epidermal growth factor (EGF), the low-density lipoprotein (LDL) receptor, and various yeast cell cycle control proteins (Greenwald, 1985; Yochem *et al.*, 1988). This observation suggests that *lin-12* is an extracellular protein and might be involved in cell–cell interactions (see Section IV,D,1 for discussion).

a. Tissue of Action. As mentioned above, Muv mutations act in the VPCs. This fact has been used to examine whether the known Vul genes act in the anchor cell or in the VPCs (Ferguson *et al.*, 1987). Specifically, since a Muv mutation acts in the VPCs any Vul mutation that suppresses the phenotype of a Muv mutation must also act in the VPCs. Mutations in four Vul genes (*lin-2, lin-7, lin-10,* and *let-23*) suppress the phenotype of at least one Muv mutation. Therefore, each of these Vul genes may have a site of action in the VPCs as opposed to in the anchor cell. There are two complications to this interpretation of the data. First, a Vul gene could act in the VPCs as well as in the anchor cell. Second, it is conceivable that a Muv mutation causes VPCs to be a 1° or a 2° cell in the absence of the anchor cell because that Muv mutation allows the inductive signal to be produced by the VPCs, thus a Vul mutation could act in the AC in the wild type but in the VPCs in a Muv mutant. In summary, it remains possible that each Vul gene normally acts in both tissues, or normally acts in the anchor cell, but is derepressed in a Muv strain. This important question can be addressed by genetic mosaic analysis. The interpretation of interactions between *lin-3* and Muv is more complicated and is left until Section III,C,2.

   b. *Production of the Inductive Signal.*   *lin-12(d)* mutations eliminate the inductive signal simply by removing the cell that produces the signal (see Section III,C,1). There are no genes identified that are known to control production of the inductive signal.

   There are several possible reasons why mutations perturbing the inductive signal have not been identified. One possibility is that the relevant genes are required earlier in development, and thus mutants defective in such genes would not survive long enough for the vulval defect to be observed. However, rare viable alleles of essential genes have been identified, e.g., for *let-23* and *lin-3* vulvaless alleles that are viable have been isolated even though the null phenotype for each locus is larval lethality, with an arrest prior to vulval induction. A second possibility is that the signal is encoded by redundant functions, and thus eliminating only one gene might not eliminate the signals. For example, the genes encoding the yeast mating pheromones, the peptides α-factor and **a**-factor, are each duplicated and both must be knocked out to cause a signaling defect (e.g., Kurjan, 1985); these genes were identified by molecular techniques and not by conventional mutagenesis. Another possibility is that the criterion for localizing the action of vulvaless mutations is spurious. For example, if Muv mutations allow the production of inductive signal by the VPCs, a Vul mutation affecting production of the inductive signal nonetheless would be epistatic to the Muv mutation. By the current criterion, that Vul mutation would be inferred to act in the VPCs.

   c. *Specification versus Execution.*   Candidates for mutations affecting specification rather than execution are those that result in misspecification of VPC types. In many cases, however, such misspecification results in the absence of a particular type of VPC. For example, *lin-7* mutants do not have 1° or 2° VPCs. Could *lin-7* be required for the *execution* of 1° and 2°? The evidence against this possibility is that *lin-7;lin-15* double mutants have 1° and 2° VPCs and *lin-7;lin-12(d)* double mutants have 2° VPCs; thus, *lin-7* is not required for execution of 1° and 2° lineages. Similarly, a *lin-15* mutation abolishes 3° VPCs, but in a *let-23;lin-15* mutant the VPCs can be (but are not always) 3°, and thus *lin-15* is not absolutely necessary for 3°. A different argument can be made for *lin-12* and 2°. *lin-12* is necessary for 2° VPCs, and no other mutation restores 2° VPCs. Thus, *lin-12* could be necessary for the execution of 2°. However, three other genes are known to be involved in 2° lineages. Therefore, *lin-12* might regulate the activity of those 2°-specific genes, but then we would say that *lin-12* is involved in *specification*, as it is a "master regulator" of the 2° cell type.

  d. Interactions of the Vul, Muv, and lin-12 Genes. Sternberg and Horvitz (1989) have provided evidence for the hypothesis that the combined actions of the Vul, Muv, and *lin-12* genes could be sufficient to specify the fates of the VPCs. These observations can be summarized as follows.

1. The Vul and Muv products control the decision between 3° and non-3° (i.e., 1° or 2°) cell types. *lin-12* controls the decision between the 2° and non-2° (i.e., 1° or 3°) cell types.
2. In the absence of *lin-12*, VPCs can be 1° or 3°, depending on whether they receive the inductive signal. VPCs near the anchor cell(s) become 1°, those more distal become 3°. The wild-type Vul products promote the 1° cell type. The wild-type Muv products promote the 3° cell type. In general, Muv and Vul mutations suppress each other: the phenotype of a Muv–Vul double mutant is more wild type than either of the single mutants (Ferguson *et al.*, 1987; Sternberg and Horvitz, 1989; M. Han, R. Aroian, G. Jongeward, and P. Sternberg, unpublished).
3. In the absence of an inductive signal, the level of *lin-12* activity controls the decision between 2° and 3°. A high level of *lin-12* activity promotes the 2° cell type; a low level of *lin-12* activity promotes the 3° cell type.
4. In particular double mutants, all six VPCs are of one type regardless of their positions or the presence of the anchor cell. All VPCs are 1° in a *lin-12(0);lin-15* mutant, all VPCs are 2° in a *lin-12(d)*–Vul, mutant, and all VPCs are 3° in a *lin-12(0)*–Vul mutant.

3. Execution

  a. Execution of the Vulval Cell Lineages. The type of a VPC is defined by the lineage it generates and the morphogenetic behavior of the progeny cells (Sternberg and Horvitz, 1986). Once specified, each VPC generates a particular sublineage, characterized by the extent and symmetry of the divisions and the types of progeny cells produced. The functions responsible for the differences among the 1°, 2°, and 3° lineages are interesting in two respects. These functions are likely to be regulated by the specification pathways described above. In addition, these functions involve fascinating problems in cell biology: the control of cell proliferation, the independence of mitosis and cytokinesis, cell attachment, and cell polarity. The differences among the three types of vulval lineages might come about by any of a number of combinations of specific functions: functions specific to 1°, functions specific to 2°, functions specific to 1° and 2° but not 3°, functions specific to 2° and 3° but not 1°, etc. Cell-type-specific gene expression in *Saccharomyces cerevisiae* is an excellent paradigm (see Sprague, this volume).

Candidates for genes whose actions are specific for one of the three vulval precursor cell types have been identified (Ferguson *et al.*, 1987; Sternberg and Horvitz, 1988). Mutations in three genes (*lin-11, lin-17,* and *lin-18*) affect the 2° but not the 1° or 3° lineages; hence, they are referred to as 2°-specific genes. For example, *lin-11* mutations disrupt the 2° lineages: in the wild type, P5.p generates three types of vulval cells, while in *lin-11* mutants P5.p generates only a single type of vulval cell. Mutations affecting the execution of the lineages are coexpressed with mutations affecting specification of VPC type. For example, in double mutants of genotype *lin-11;lin-12(d)* all six VPCs generate abnormal lineages. These 2°-specific genes may be responsible for the difference between the 2° and 1° vulval lineages. Genes whose actions are specific for the 1° or 3° cell types have not yet been identified.

b. Execution of the Primary Vulval Lineage.   As mentioned above, no genes specific to the 1° lineage have been identified. We expect that such genes exist because 1° cells differ from 2° cells. The unique feature of 1° progeny is that two of the four progeny cells attach to the anchor cell (White, 1988), suggesting that 1°-specific functions exist. Progeny of the 2° sublineage cannot attach to the anchor cell (Sternberg and Horvitz, 1989).

c. Execution of the Secondary Vulval Lineage.   Three genes, *lin-11, lin-17,* and *lin-18,* necessary for the execution of the 2° lineage have been identified by mutations that disrupt this lineage (Ferguson *et al.,* 1987). While mutations in each gene disrupt the asymmetry of the 2° lineage, the particular phenotypes differ. *lin-11* mutations result in the production of only "L" progeny by both daughters of the 2° VPC. *lin-17* and *lin-18* mutations result in symmetric lineages, or in lineages with reversed polarity. *lin-18* mutations appear specific for the 2° lineage, while *lin-17* mutations affect many asymmetric lineages (Sternberg and Horvitz, 1988). A *lin-11* mutation is epistatic to either *lin-17* or *lin-18* mutations (P. Sternberg, unpublished). *lin-17;lin-18* double mutants are inviable (D. Roberts and R. Horvitz, personal communication; P. Sternberg, unpublished).

These observations suggest that *lin-17* and *lin-18* affect a common step, while *lin-11* acts at a separate step in the same pathway. As discussed below, *lin-17* has been argued to act, in general, at or prior to the affected cell division (Sternberg and Horvitz, 1988). These observations lead to two plausible hypotheses for the specification of the 2° lineage by *lin-11, lin-17,* and *lin-18. lin-17* and *lin-18* could establish the asymmetry and correct polarity of the 2° VPC division, and *lin-11* could regulate cell type in response to that asymmetry. (A variation would be that *lin-11* is segregated to one daughter cell.) *lin-11* then promotes the "TN" phenotype or inhibits the "L" phenotype. This hypothesis predicts that *lin-11* is specific to one of two 2° VPC daughters. A second hypothesis is that *lin-11* activates *lin-17* and/or *lin-18*, thereby allowing the VPCs to generate asymmetry. This hypothesis predicts that *lin-11* acts in the 2° VPC. Note that these hypotheses are based on specific expression of the gene products to the 2° lineage. Other models can be constructed in which the gene products are active in all VPCs.

lin-17 has been proposed to control the generation of the asymmetry in particular precursor cells that leads to the differential activities of fate-determination factors in the two daughter cells (Sternberg and Horvitz, 1988). The two observations leading to this contention are, first, that *lin-17* mutations are required for the asymmetric division of precursor cells in a variety of tissues and, second, that, for certain precursor cells, *lin-17* mutations result in sister cells that are equal in size as well as identical in fate. The existing *lin-17* mutations result in a variable phenotype in the 2° VPC linage, which can be either normal, symmetric, or asymmetric but of reversed polarity in *lin-17* mutants. Reversed polarity means that the fates of two sister cells are interchanged. Since the null phenotype of *lin-17* is not known, either of two hypotheses for the role of this gene remain viable (Fig. 7). Loss of *lin-17* function might result in a symmetric division. However, partial loss of *lin-17* function could be imagined to result in the observed variable phenotype because, for example, random localization of a small number of fate "determinants" might sometimes result in reversed polarity, sometimes in symmetry, or sometimes in normal asymmetry. A second possibility is that loss of *lin-17* activity results in a cell division of

A   B

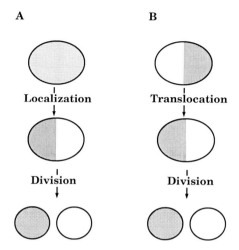

FIG. 7. Determinant models for asymmetric cell division. (A) Localization model. Determinants (stippled) are synthesized and localized to one part of the cell. At division, the determinants are segregated to one daughter cell. Loss of localization would lead to a symmetric cell division, which generates two daughter cells with identical, although possibly abnormal, fates. (B) Polarity reversal model. Determinants are normally localized to one part of the cell, then translocated to the opposite end of the cell. Segregation of division proceeds as in A. Failure to translocate would lead to an asymmetric cell division of reversed polarity.

reversed polarity, with the asymmetry of the cell division established independent of *lin-17*. Variable phenotypes, in particular a symmetric cell division, might arise if, for example, localized "determinants" underly the asymmetry of a cell division and if *lin-17* were required to translocate those determinants. Partial loss of *lin-17* function might result in partial translocation of the determinants, in essence homogenizing the cell, which would then divide symmetrically.

4. Morphogenesis

Morphogenesis of the vulva involves the 12 cells generated by the two 2° and the single 1° VPC, as well as the anchor cell. In addition, 8 muscle cells attach to the vulva, and 8 neurons innervate those muscles (White, 1988). While a number of mutants display abnormal vulval morphogenesis, these have not been studied in detail. For example, several of the egg-laying defective (Egl) mutants identified by Trent *et al.* (1983) have normal numbers of vulval cells but have abnormal vulval morphogenesis and dysfunctional vulvae. Several male abnormal (Mab) mutants, isolated for abnormal male tails, have protruding vulvae (Hodg-kin, 1983).

C. GENES CONTROLLING CELL TYPE

This section reviews the current status of the genetic analysis of several key loci involved in VPC specification.

1. lin-12

Loss-of-function *lin-12* mutations affect both the gonad and the VPCs: the gonad has from two to four anchor cells owing to the transfor-mation of VUs to the anchor cell fate, and VPCs are incapable of becoming 2° (Greenwald *et al.*, 1983; P. Sternberg, unpublished). Semi-dominant mutations of *lin-12* can affect either or both tissues (i.e., the anchor cells or the VPCs). Some *lin-12(d)* mutations, such as *n302*, result in the transformation of the anchor cell to the VU fate but do not affect the VPCs except indirectly. Other *lin-12(d)* mutations, such as *n137*, result in transformations of the anchor cell to the VU fate and of the VPCs to the 2° fate. Still other *lin-12(d)* mutants are defective in the VPCs but not in the anchor cell (Greenwald *et al.*, 1983; P. Sternberg and I. Greenwald, unpublished). These two defects of *lin-12* mutants can thus be separated genetically (Table 2). In addition, the tem-perature-sensitive periods (TSPs) for these defects are distinct: the TSP for the AC defect is during the L2 stage, while the TSP for the VPC defect is during the L3 stage.

The results of gene dosage analysis (e.g., Table 3) for *lin-12* indicate that *lin-12(d)* mutations are hypermorphic, that is, they result in an increase in *lin-12(+)* activity, as the defect in a *lin-12(d)/+* animal is more severe than in a *lin-12(d)/lin-12(0)* animal. In addition, this dosage analysis indicates that a 2-fold increase in *lin-12* activity, from *lin-12(d)/lin-12(0)* to *lin-12(d)/lin-12(d)*, can result in almost a com-plete change in cell fate: one copy of *n379* results in 1% Egl⁻, while two copies of *n379* results in 98% Egl⁻. (In these experiments, an anchor cell-deficient animal is Egl⁻, an animal with an anchor cell is Egl⁺.) Thus, the AC/VU fate decision can be extremely sensitive to the level of *lin-12* activity.

The role of *lin-12* in these two developmental decisions may be simi-lar. In the AC/VU pair, *lin-12* is necessary for inhibition of a super-numerary AC. In a *lin-15* mutant background, *lin-12* is necessary for the inhibition of supernumerary 1° VPCs (Sternberg and Horvitz, 1989).

TABLE 2
Independent Mutability of Gonadal and
Hypodermal Functions of *lin-12*

lin-12(d) genotype	Gonad AC?	Hypodermis VPC fate	Plate phenotype
n137/n137	−	2°	Egl⁻ Muv
n137/n137n720	−	2°	Egl⁻ Muv
	+	1°, 2°	Egl⁺ Muv
n302/n302	−	3°	Egl⁻ Vul
+/+	+	1°, 2°, 3°	Egl⁺ Wt

2. *lin-3*

The *lin-3* locus was defined initially by two Vul alleles (Horvitz and Sulston, 1980; Ferguson and Horvitz, 1985). Additional alleles have been identified by F_1 noncomplementation screens (Ferguson and Horvitz, 1985; R. Hill and P. Sternberg, unpublished). All but one of these alleles confer a lethal phenotype, suggesting that the null phenotype of *lin-3* is larval lethality. A Vul allele *in trans* to a deficiency for the locus is viable and completely penetrant for the Vul phenotype, indicating that null alleles can be obtained by an F_1 noncomplementation screen. Other lethal alleles had been obtained previously in screens for lethals balanced by the translocation *nT1* (Clark *et al.,* 1988; D. Clark, personal communication).

TABLE 3
Dosage Analysis of *lin-12*[a]

	Percentage Egl⁻	
Genotype	*n379*	*n302*
+/0	0	0
+/+	0	0
d/0	1	22
d/+	7	56
d/+/+	33	81
d/d	98	100

[a] +, Wild-type allele; 0, null allele; *d*, dominant allele, either *n379* or *n302;* Egl⁻, egg-laying defective, an indication of the AC to VU transformation. Data from Greenwald *et al.* (1983) and Ferguson and Horvitz (1985).

lin-3 is the only "specification" Vul that may not act in the VPCs. Ferguson *et al.* (1987) reported that a strong heteroallelic combination of *lin-3* mutations, *n378/n1059,* could partially suppress the Muv phenotype a temperature-sensitive allele of *lin-15* at an intermediate temperature. However, we now know that this Muv phenotype is partially dependent on the anchor cell (P. Sternberg, unpublished); thus, *lin-3* may suppress only the anchor cell-dependent vulval cells and, hence, might not act in the VPCs.

3. *lin-2, lin-7, and lin-10*

lin-2, lin-7, and *lin-10* are defined by recessive loss-of-function mutations that result in a vulvaless phenotype (Ferguson and Horvitz, 1985; Ferguson *et al.,* 1987; Kim and Horvitz, 1990). The only phenotypes of mutants defective in any of these genes, or in a triple mutant, are in the vulva. However, these mutations suppress certain male defects of multivulva mutations (P. Sternberg, E. Ferguson, and R. Horvitz, unpublished), suggesting that the products of these genes may also act in the male cells. The products of the three genes appear to act in concert, as complex, or in a common pathway.

4. *let-23*

The *let-23* locus was originally defined by three larval lethal alleles (Herman, 1978; Sigurdson *et al.,* 1984) and one cold-sensitive, semilethal vulvaless allele, *n1045* (Ferguson and Horvitz, 1985). Additional *let-23* alleles have been isolated in two screens (R. Aroian and P. Sternberg, unpublished). Two recessive extragenic suppressors of *lin-15* are *let-23* alleles. One of these, *sy1,* is completely penetrant yet viable *in trans* to a deficiency for the *let-23* locus. New EMS-induced mutations that fail to complement *sy1* have been obtained at high frequency. A majority of such alleles are homozygous lethal, the remainder semilethal. These observations indicate that *let-23* has a lethal null phenotype. While the null phenotype of this locus is lethality, functions necessary for viability and for vulval induction are to some extent independently mutable (R. Aroian and P. Sternberg, unpublished). Certain *let-23* genotypes result in a male mating defect, including at least a spicule defect (R. Aroian, H. Chamberlin, and P. Sternberg, unpublished). What can be said about the genetic organization of this locus? The *n1045* mutation is amber suppressible (Ferguson and Horvitz, 1985). Since *n1045* affects, to various extents, vulval induction, viability, and male mating, these phenotypes most likely reflect a defect in a shared protein domain. Thus, even if there are multiple transcripts encoded by *let-23* corresponding to the genetically defined functions, *n1045* affects a common exon.

5. *lin-15*

In strains partially deficient in *lin-15* activity, all VPCs are 1° or 2°. Thus, the *lin-15* locus encodes a negative regulator of the response to the inductive signal or of the cell-type specific functions that endow a 1° or 2° cell with their specialized properties (Ferguson *et al.*, 1987; Sternberg, 1988a).

The null phenotype of *lin-15* is not known. Existing mutations *in trans* to a deficiency cause animals to be small and sterile (Ferguson and Horvitz, 1985; L. Huang and P. Sternberg, unpublished). This phenotype is more severe that that of existing mutants, suggesting that the null phenotype of *lin-15* is sterility, or worse. An F_1 noncomplementation screen for new *lin-15* alleles that should have recovered null mutations at high efficiency failed to turn up *lin-15* alleles, suggesting that *lin-15* nulls are relatively rare with EMS as a mutagen (L. Huang and P. Sternberg, unpublished).

Genetic studies have suggested that the *lin-15* locus encodes two separable functions, *lin-15A* and *lin-15B* (Ferguson and Horvitz, 1985, 1989). Both functions must be defective to result in a multivulva (Muv) phenotype. In addition to recessive Muv alleles $(A^- B^-)$, recessive alleles that are A^- or B^- exist. While an A^- mutant is wild type by itself, in combination with a mutation in another, unlinked gene (e.g., *lin-9,* a recessive B^- mutation) it confers a Muv phenotype (see Section III,D,2). Similarly, *lin-15B* mutations interact with unlinked recessive *lin-8* mutations (A^-) to confer a Muv phenotype. Evidence that these are two functions of the same gene is as follows. First, a heat-sensitive allele, *n765,* is $A^- B^-$ at the restrictive temperature of 25°C, but is B^- at the permissive temperature of 15°C. Second, $A^- B^-$ alleles such as *n309,* as well as deficiencies, fail to complement both A^- and B^- alleles.

D. SPECIFICATION OF VULVAL PRECURSOR CELL TYPE

This section considers the specification of VPC type in detail, first examining the two branches of the genetic pathway, *lin-12* and Muv/Vul, and then the regulatory circuitry as a whole.

1. *lin-12 Pathway*

The *lin-12* pathway is defined by *lin-12* and interacting genes. Genes that interact with *lin-12* have been identified by extragenic suppressor mutations that suppress *lin-12* dominant or *lin-12* hypomorphic mutations (Thomas *et al.*, 1990; M. Sundaram and I. Greenwald, personal communication). One such locus, *sup-17*, may encode a product that acts in the same pathway as *lin-12*. For example, *sup-17* could be necessary for *lin-12* expression or activity, could be the ligand for *lin-12*, or could act downstream of *lin-12*. A viable *sup-17* mutation, *n316*, suppresses various phenotypes of *lin-12(d)* mutants and confers a vulval defect in a *lin-12($^+$)* background: 2° and occasionally 1° lineages fail to occur (P. Sternberg and R. Horvitz, unpublished).

There are several possible roles for *lin-12* in vulval induction (Sternberg and Horvitz, 1989).

1. In the lateral inhibition pathway. Since *lin-12* is necessary for lateral inhibition in a *lin-15* multivulva mutant, *lin-12* is strongly implicated as having a role in this process, either as signal, receptor, or another component.

2. In an autocrine mode. *lin-12* might be produced by a 2° VPC in response to the inductive signal and act on that 2° cell, as either signal or receptor. The argument for an autocrine role is that *lin-12* activity is both necessary and sufficient to specify 2°, and a single VPC can become 2° (thus, unless *lin-12* is produced by the anchor cell, it must act in or on the cell that produces it).

3. A receptor for one of two inductive signals. The attraction of this hypothesis stems from the fact that *lin-12* is also necessary for a second signaling event involving the anchor cell, namely, inhibition by the anchor cell of its homolog (the presumptive VU cell) from also becoming an anchor cell. Assuming that *lin-12* acts cell autonomously in the VPCs as it does in the AC/VU equivalence group (Seydoux and Greenwald, 1989), *lin-12* might be a receptor for an inductive signal. However, *lin-12* cannot be the only receptor for an inductive signal since induction occurs in the absence of *lin-12* activity (Sternberg and Horvitz, 1989). The observation that *lin-12* is necesasry for 2° fates in a *lin-15* mutant lacking an anchor cell supports the hypothesis that *lin-12* acts in the VPCs.

Overall, the simplest hypothesis is that *lin-12* acts in the VPCs as a receptor for a signal produced by VPCs. Its role in lateral inhibition and in the specification of 2° cell type suggests that *lin-12* could be receptor for the lateral signal, if the lateral signal acts by stimulating 2° (see below). Another possibility is that *lin-12* encodes the lateral signal, which can also act on the cell that produces it to specify 2°.

2. *Muv/Vul Pathway*

Vul genes and Muv genes control an important aspect of VPC specification (Ferguson and Horvitz, 1985, 1989; Ferguson *et al.*, 1987;

Sternberg and Horvitz, 1989). In general, the Vul genes are necessary for 1° and 2° fates, while the Muv genes are necessary for 3° fates. As described below, a complex pattern of interactions is observed among the Vul and Muv genes.

F_2 screens identified five Vul genes (*lin-2, lin-3, lin-7, lin-10,* and *let-23*) and five Muv genes (*lin-13, lin-15, lin-34,* and the *lin-8;lin-9* double mutant) (Horvitz and Sulston, 1980; Ferguson and Horvitz, 1985). Additional Muv and Vul genes have been identified by extragenic suppressor mutations. Suppressors of the Muv phenotype caused by *lin-15* mutations define at least five loci (M. Han, R. Aroian, and P. Sternberg, unpublished; S. Clark and R. Horvitz, personal communication). A few of these loci were previously identified by Vul or lethal (Let) mutations. For example, at least two *lin-15* suppressors are *let-23* alleles. Additional loci have been identified by extragenic suppressors of the Vul phenotype of *lin-10* (S. Kim, D. Parry, and R. Horvitz, personal communication) or *let-23* (G. Jongeward and P. Sternberg, unpublished).

New loci that interact with existing mutations in Muv loci have been identified as synergistic mutations. The Muv phenotype of the strain CB1322 proved to result from mutations at two loci, *lin-8* and *lin-9* (Horvitz and Sulston, 1980). Either of the single mutants has no phenotype (Ferguson *et al.,* 1987). Starting with a *lin-8* mutation, Ferguson and Horvitz (1989) identified additional mutations that, in combination with the *lin-8* mutation, confer a Muv phenotype. These mutations define *lin-35, lin-36, lin-37* as well as *lin-9* and *lin-15.* Similarly, starting with a *lin-9* mutation, mutations defining *lin-38* as well as *lin-15* were identified. By constructing various double-mutant combinations, Ferguson and Horvitz (1989) demonstrated that mutations fell into two classes, A^- and B^-, and combination of A^- plus B^- mutations is necessary to obtain a Muv phenotype. Moreover, they showed that *lin-8* and *lin-38* are mutable to A^-; that *lin-9, lin-35, lin-36,* and *lin-37* are mutable to B^-; and that *lin-15* is mutable to A^- or B^-. As few of the loci have been analyzed in detail, it remains possible that each locus can mutate to A^-, B^-, or A^-B^-. The simplest hypothesis is that *lin-8* is only mutable to A^-, and *lin-9* only to B^-. These genetic data suggest that either the A or the B pathway is sufficient to prevent inappropriate 1° and 2° fates.

The order of action of the Muv and Vul genes is unresolved. Since null mutations in Muv genes have not been identified, a clean test of epistasis is not possible. Given the existing not-necessarily-null mutations, one of three outcomes is observed for each combination of Muv and Vul mutations: (1) the Muv/Vul strain is Muv; (2) the strain is Vul; or (3) the strain has a variable phenotype, ranging from Muv to wild type to Vul. On face value, outcome (1) suggests that the Muv gene acts downstream of the Vul gene because the pathway acts like a regulatory or signal processing pathway as opposed to a biochemical or assembly pathway (see Section V,B). Similarly, outcome (2) suggests that the Muv gene acts upstream of the Vul gene. Outcome (3) can be interpreted in several ways. The mutations used might be hypomorphs, and partial suppression is being observed. Or, the mutations inactivate antagonistic functions. Given the number of genes (at least 18), each of these possibilities could be correct for different pairs of Muv and Vul genes. Multiple genes that are required at one apparent step in a pathway may act in concert (as a multimeric protein) or may act in a subpathway. A large number of specific hypotheses remain to be tested. For example, is *lin-2* necessary for transcription of other Vul genes? Is *lin-2* a negative regulator of *lin-15*?

3. Other Genes

Several of the genes (*lin-25*, *lin-1*, and *lin-31*) analyzed by Ferguson *et al.* (1987) could not be assigned to a single step in the pathway. Such genes might act at multiple steps or might be required for some general aspect of cell type that, if defective, interferes with several subsequent steps.

In a *lin-25* mutant, Pn.p cells can divide during the L1 stage. The additional cells look like VPCs and in some respects act like VPCs. For example, this phenotype of *lin-25* is coexpressed with the *lin-12(d)* multivulva phenotype: as many as eight 2° lineages have been observed in a single *lin-25;lin-12(d)* double-mutant animal (P. Sternberg, unpublished). *lin-25* mutants are defective in VPC specification. Typically, fewer than one VPC proliferates and no vulval cells are generated. This defect is not wholly due to an execution defect, since, as mentioned above, normal 2° lineages can be generated if a *lin-12(d)* mutation is present, apparently by passing the *lin-25* specification defect. A third possible defect in *lin-25* mutants is in execution or morphogenesis: the temperature-sensitive period for *lin-25* extends past the vulval cell divisions (Ferguson *et al.*, 1987). This result is one example in which a TSP is most informative: the TSP extends *past* the time at which phenotypic defects are observed. *lin-25* vulvaless mutations are the only specification Vul mutations that are epistatic to *lin-1* but not epistatic to *lin-12(d)*. These observations suggest that *lin-25* may act downstream of *lin-1* but upstream of *lin-12* (Fig. 8).

The *lin-1* locus, although defined by Muv mutations, is considered distinct from the Muv genes such as *lin-15*. *lin-1* mutations are epistatic to the Vul phenotype of *let-23* (Ferguson *et al.*, 1987). Also, the

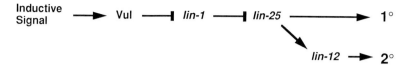

FIG. 8. Proposed roles of *lin-1* and *lin-25*. Epistasis studies suggest the genetic pathway shown. *lin-25* is proposed to be an activator of 1°-specific functions and of the *lin-12* pathway. *lin-1* is proposed to be a negative regulator of *lin-25*. Arrows represent positive regulation, bars represent negative regulation. Vul indicates the action of the Vul genes *lin-3, lin-2, lin-7, lin-10,* and *let-23*. The pathway is derived as follows: *lin-12(d)* mutations are epistatic to *lin-25* mutations; thus, *lin-25* is proposed to act upstream of *lin-12*. *lin-25* is necessary for the specification and/or execution of 1° and 2° fates, and it is epistatic to *lin-1* multivulva mutations. Thus, *lin-25* is proposed to act downstream of *lin-1*. The coexpression of the *lin-1* and *lin-12(d)* phenotypes in a double-mutant strain is explained by *lin-1* mutations allowing 1°-specific functions to be expressed in a 2° VPC. Note that this interpretation of *lin-25* is clouded by its possible roles in VPC generation as described in the text.

lin-1 phenotype is coexpressed with the *lin-12* phenotype in double mutants. A *lin-1* mutation makes the 2° lineages abnormal in a *lin-12(d)* mutant and makes the 1° lineages abnormal in a *lin-12(0)* mutant (P. Sternberg, unpublished). *lin-1* may thus be necessary for both specification and execution of VPC fates. A simple hypothesis is that *lin-1* acts downstream of Vul, Muv, and *lin-12* as a repressor of 1°- and 2°-specific functions (Fig. 8), and, much as an *matα2⁻* mutant of yeast results in simultaneous expression of two sets of cell-type-specific genes (Bender and Sprague, 1989), a *lin-1* mutation might result in the simultaneous expression of 1° and 2° functions. The epistasis of *lin-25* over *lin-1* suggests that pathway shown in Fig. 8.

The *lin-31* locus is defined by mutations conferring a variable multivulva phenotype. Pn.p cells can divide during the L2 stage, and their progeny divide during the L3 stage. Ectopic divisions produce either extra vulval tissue or neuronlike cells.

4. Regulatory Circuitry

The regulatory circuitry underlying the specification of VPC type must account for the following observations (Sternberg and Horvitz, 1986, 1989; Sternberg, 1988a): (1) The Vul genes (*lin-2, lin-3, lin-7, lin-10,* and *let-23*) are necessary for 1° and 2° cell types. The Vul genes may thus mediate the response to inductive signal. (2) *lin-12* is necessary to specify 2°. (3) The Vul genes specify 2° via *lin-12*, since a *lin-12(d)* mutation is epistatic to a Vul mutation. (4) *lin-12* is necessary for the lateral signal or its action; *lin-12* could activate the lateral signal, or the lateral signal could activate *lin-12*. (5) The lateral signal inhibits differentiation at 1°. (6) 2° functions are inactivated in a 1° cell, since a high level of inductive signal forces P6.p to become 1° even in a *lin-12(d)* mutant. (7) The Muv genes promote 3°. (8) The Muv genes are negative regulators of *lin-12*. (9) The anchor cell stimulates 1° via Vul genes but not via *lin-12*. (10) An isolated VPC can become 2°.

The scheme shown in Fig. 9 is one simple model taking these observations into account; a variety of other models also account for the data. The antagonism between 1° and 2° cell types might arise from an *intracellular* antagonism between 1°- and 2°-specific functions, with the *intercellular* lateral signal coupling adjacent cells (Sternberg and Horvitz, 1989). The intracellular antagonism might arise as follows. Lateral signal activates 2°-specific functions via its receptor. 2°-specific functions inactivate 1°-specific functions. 1°-specific functions inactivate the lateral signal receptor and thus 2°-specific functions (other possibilities include 1°-specific functions inactivating 2°-specific functions directly). Thus, a cell can either have 1°-specific functions ON and lateral signal receptor and 2°-specific functions OFF, or *lin-12* and 2°-specific functions ON (Table 4). The intercellular antagonism might occur by the lateral signal acting via its receptor. The lateral signal might be *lin-12* or might be received by *lin-12*. In either case, reception of lateral signal would promote 2° and inhibit 1° cell types, because 2° functions are activated and thus 1°-specific functions are inhibited. A 1° cell might be refractory to the lateral signal because the response to lateral signal is shut off, for example, by inactivating a *lin-12*-encoded receptor. Such a model easily explains the pattern of cell types in a *lin-15* mutant. This scheme is one specific version of the general type of model described in Fig. 1, with the anchor cell biasing the competition between 1° and 2°.

The inductive signal has, in essence, two effects on VPCs. One effect is to allow the VPCs to compete via the circuitry described above, for the 1° fate. The second effect is to bias the competition such that the cell VPC receiving a higher level of inductive signal, or receiving inductive signal sooner, wins the competition and differentiates as 1°.

lin-12 must be able to work in both autocrine and paracrine signaling pathways (Sternberg and Horvitz, 1989). The paracrine role is implied by the requirement of *lin-12* for lateral inhibition. The autocrine role is implied by the observations that *lin-12* is necessary and sufficient to specify 2°, and that an isolated VPC can become 2°. Thus, the isolated VPC must produce and respond to *lin-12* product. A simple hypothesis to account for the dual role of *lin-12* is that *lin-12* might encode a receptor for lateral signal and that lateral signal can act on the cell that

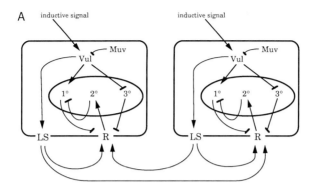

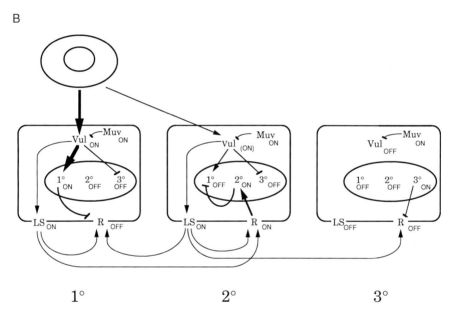

1°   2°   3°

FIG. 9. Model for the regulatory circuitry underlying VPC specification. (A) Circuitry. The possible circuitry (simplified) acting within and between two VPCs is shown. Arrows represent positive regulation and bars represent negative regulation. *lin-12* could be either the lateral signal (LS) or its receptor (R). (B) State of circuits after inductive pattern formation. Three of six VPCs are shown, with the anchor cell above the three VPCs. ON, Function is active; (ON), function is partially active; OFF, function is inactive. The thickness of each arrow indicates the relative strength of the regulatory interaction. In a 3° cell, 3° functions are ON and Muv inhibits Vul; thus, neither 1° nor 2° functions are ON. In a 2° cell, the LS receptor is activated by lateral signal (LS) (a possible autocrine stimulation); thus, 2° functions are ON and 1° functions are OFF. In a 1° cell, Vul activates 1° functions, which (1) inhibit the response to LS, (2) inhibit 2° (shown here by inhibition of the LS receptor), and (3) activate the LS receptor in neighboring cells. [Adapted from Sternberg and Horvitz (1989).]

TABLE 4
Proposed Activity States of Functions in
Vulval Precursor Cells

Cell type	Function				
	1°	2°	3°	LS	LS receptor
1°	ON	OFF	OFF	ON	OFF
2°	OFF	ON	OFF	ON	ON
3°	OFF	OFF	ON	OFF	OFF

produces it. For example, the receptor for a single inductive signal could directly activate a *lin-12*-encoded receptor, or the signal transduction pathway for the induction (Muv/Vul) pathway could feed into that of the *lin-12* pathway. Other more complicated possibilities are also consistent with existing data.

IV. Other Pathways of Cell-Type Specification

The vulva is only one of the genetic pathways of cell-type specification actively being studied in *C. elegans*. This section briefly discusses several other pathways notable because they are being intensively studied at the molecular level.

A. HETEROCHRONIC GENES

Ambros (1989) has elucidated a pathway of cell fate specification involving several genes that are necessary for appropriate stage-specific developmental events, the so-called heterochronic genes (Ambros and Horvitz, 1984, 1987). *lin-14* and *lin-28* act upstream of more specific genes such as *lin-29*, which controls a particular state-specific decision: to divide or to differentiate and secrete a specialized cuticle (the lateral alae). The *lin-4* gene is proposed to act as a negative regulator of *lin-14* because a recessive *lin-4* mutation has the same phenotype as a dominant gain-of-function *lin-14* mutation. Moreover, a loss-of-function *lin-14* mutation is epistatic to a *lin-4* mutation. Stage-specific cuticle genes are candidates for targets of *lin-29* action. Since other stage-specific events are not controlled by *lin-29*, Ambros proposed that *lin-14* and *lin-28* regulate the activities of other, yet to be identified, regulators of stage-specific cell type. *lin-14* has been molecularly cloned

(Ruvkun *et al.*, 1989), and Ruvkun and Giusto (1989) have examined by immunofluorescence the distribution of *lin-14* protein. They found that *lin-14* protein is localized to the nuclei of all cells affected in a *lin-14* mutant as well as in some additional cells. Their results strikingly support the inferences from the genetic studies: *lin-14* protein occurs at a high level early in postembryonic development and at a low level later in development. Moreover, a high level of *lin-14* protein persists in a mutant carrying a gain-of-function mutation in *lin-14*. Thus, *lin-14* activity may be regulated at the level of amount of protein.

B. Programmed Cell Death

The genetic and morphological pathway of programmed cell death in nematodes has been reviewed by Horvitz *et al.* (1982) and Horvitz (1988), and it is included here as it represents one of the best understood cell differentiation pathways. The current view is that particular cell-type regulatory genes specify which cells will initiate the cell death program. Some genes that control which cells will die have been identified. *egl-1* is necessary to prevent the hermaphrodite-specific neuron (HSN) from dying in the hermaphrodite (Trent *et al.*, 1983; Ellis and Horvitz, 1986; Desai *et al.*, 1988; Desai and Horvitz, 1989). *lin-39* might be necessary to prevent the VC motor neurons from dying since a recessive *lin-39* mutation results in the death of the six VC neurons (Fixsen *et al.*, 1985). Initiation of cell death requires the cell death defective loci *ced-3* and *ced-4* (Ellis and Horvitz, 1986). Engulfment of the dead cell requires *ced-1* and *ced-2* (Hedgecock *et al.*, 1983). Degradation of DNA requires the *nuc-1* gene, which encodes a major deoxyribonuclease (Albertson *et al.*, 1978).

C. *mab-5*

The *mab-5* gene is involved in cell fate decisions in several tissues (Kenyon, 1986). In general, the misspecified cells in a *mab-5* loss-of-function mutant act as do more anteriorly located cells. These observations suggest that *mab-5* is necessary for cells to distinguish in some general sense anterior versus posterior. Another type of *mab-5* defect is that a cell which would normally migrate posteriorly instead migrates anteriorly, suggesting that anterior migration is a default state and that *mab-5* is necessary to reverse the direction of migration. Costa *et al.* (1988) inferred that *mab-5* encodes a homeobox-containing protein, and proposed that *mab-5* is a transcriptional regulator. Moreover, Costa *et al.* examined expression of *mab-5* mRNA by hybridization *in situ* to whole mounts of *C. elegans* larvae and found that *mab-5* is preferentially expressed in the posterior region of the animal. As Kenyon (1986) demonstrated by genetic mosaic analysis that the action of *mab-5* is cell autonomous, *mab-5* can be thought to act in either of two ways. *mab-5* might regulate the expression of genes required for cells to respond to an external signal to be posterior. Alternatively, *mab-5* may be involved in a process by which cells inherit a state corresponding to "posteriorness." Identification of an extracellular posterior signal would rule out the latter possibility. In at least one of these tissues (the lateral hypoderm), *mab-5* interacts in an antagonistic manner with *lin-22*. Specifically, *mab-5* promotes the production of rays in the male lateral hypodermis whereas *lin-22* prevents the production of rays, promoting alae formation.

D. *glp-1*

The *glp-1* locus is necessary for at least two aspects of *C. elegans* development. *glp-1* is required *maternally* for pharynx development, possibly involving a cell–cell interaction at the four-cell blastomere stage (Priess and Thomson, 1987; Priess *et al.*, 1987). *glp-1* is also required *zygotically* for proliferation of germ-line nuclei (Austin and Kimble, 1987). Mitotic proliferation of the germ-line nuclei requires the presence of a regulatory cell, the distal tip cell (Kimble and White, 1981). Thus, *glp-1* could be imagined to be required either in the distal tip cell or in the germ line. Mosaic analysis indicates that *glp-1* acts in the germ line rather than in the distal tip cell, and thus it may encode a product necessary for a response to the inductive signal from the distal tip cell.

E. HIERARCHY OF HOMEOBOX GENES

In *Drosophila,* a number of regulatory hierarchies involving homeobox-containing genes have been proposed (see Steinmann-Zwicky *et al.,* Gaul and Jäckle, Rushlow and Levine, Kaufman *et al.,* McGinnis *et al.,* and Campos-Ortega, this volume). The DNA sequences of three *C. elegans* genes defined by mutations which perturb various aspects of cell fate or cell lineage indicates that these loci encode homeobox-containing proteins. What can be said about the interrelationships of these genes during development?

The *mec-3* gene was identified by mutations causing a specific defect in the six microtubule cells mediating the response to light touch (Chalfie and Sulston, 1981). The *mec-3* product encodes an inferred product with similarity in general to the various classes of *Drosophila* homeobox proteins (Way and Chalfie, 1988). In *mec-3* mutants, the presumptive microtubule-rich (touch) cells are present, but they differentiate a nontouch neurons. *mec-3* mutants are the only mutants defective in any of the 18 *mec* genes that produce the cells which would normally differentiate as touch cells, but which do not express any characteristics of touch cells. Chalfie and co-workers (e.g., Chalfie, 1984) argue that *mec-3* specifies the *mec-3* cell type. Epistatic interactions of *mec-3* mutations with other *mec* mutations support this hypothesis. Dominant mutations of the *mec-4* gene, necessary for touch cell functioning, cause an atypical death of the touch cells. Expression of this killing effect is cell autonomous (Herman, 1987). The alellism of *mec-4* with the dominant "*mec-13*" mutation suggested that *mec-4* encodes a differentiation product and that the dominant mutations result in formation of a cytotoxic product. Since *mec-3* mutations are epistatic to dominant *mec-4* mutations, *mec-3* likely acts upstream of *mec-4*. In this view, in the absence of *mec-3*, no toxic product is expressed and the cells differentiate, but not as touch cells. Moreover, ectopic expression of *mec-3* in a *mec-4(dominant)* background leads to novel cells dying (Way and Chalfie, 1988). The *mec-7* gene is necessary for the production of a touch cell-specific microtubule and is another excellent candidate for a gene activated by *mec-3* (Chalfie *et al.*, 1986; Chalfie and Au, 1989). In summary, *mec-3* may be "master regulator" of touch cell type that activates transcription of touch cell-specific genes. How does *mec-3* come to be active in only certain cells? Part of the answer to this question involves the *unc-86* gene.

The *unc-86* gene was identified by mutations that cause a defect in particular neural lineages (Horvitz and Sulston, 1980; Chalfie *et al.*, 1983). Like *mec-3*, the *unc-86* product contains a homeobox domain. This product has additional similarity, in the so-called POU domain, to a family of mammalian transcription factors, including OCT-1, OCT-2, and PIT-1 (reviewed by Levine and Hoey, 1988). How does this proposed role of *unc-86* jibe with the nature of cell lineage defects observed in *unc-86* mutants? *unc-86* mutants display reiterative cell lineages. Consider the QL lineage (Fig. 10). In the wild type, QL divides asymmetrically; the anterior daughter (QL.a) generates a migrating neuron and a cell that dies. The posterior daughter (QL.p) generates a touch cell, another neuron, and a cell that dies. In an *unc-86* mutant, QL.p behaves like QL, generating an anterior daughter that behaves like QL.a and a posterior daughter that behaves like QL. In other words, the QL.p lineage is not generated. Given the hypothesis that *unc-86* encodes a transcriptional regulator, one can imagine two simple possibilities for

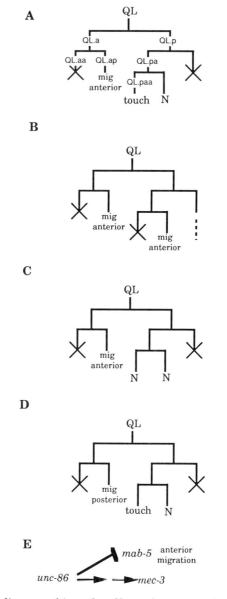

FIG. 10. The QL lineage: a hierarchy of homeobox gene action. (A) The wild-type QL neuroblast lineage produces two cells that die (X), one microtubule-rich touch receptor (touch), one posteriorly migrating neuron (mig), and another neuron (N). (B) In an *unc-86* mutant, QL.p acts like QL, producing an anterior daughter that generates a migrating neuron and a cell destined to die. QL fails to generate a touch cell (Chalfie *et al.*, 1983; Finney *et al.*, 1988). (C) In a *mec-3* mutant, QL.paa differenti-

the role of this gene product in the QL lineage. *unc-86* could activate transcription of genes necessary in QL to establish a switch to the QL.p fate at that cell division. In the absence of gene products (in an *unc-86* mutant) the posterior daughter would fail to switch cell types. A second possibility is that *unc-86* activates genes necessary for QL.p-specific functions. These possibilities might be distinguished when it is known in which cells the *unc-86* product is expressed. While elucidation of the whole story may well require identification of other genes that control the remaining aspects of this lineage; study of *unc-86* should provide an excellent starting point for such analysis.

In the example of the QL neuroblast lineage, *mec-3* is specific to the touch cell type (QL.paa). *unc-86* is necessary for the QL.p branch of the QL lineage. No touch cell is produced in an *unc-86* mutant. Therefore, *unc-86* may be necessary for *mec-3* to act. Since at least one cell division intervenes between the postulated cell in which *unc-86* acts and the cell in which *mec-3* acts, the effect of *unc-86* on *mec-3* is likely to be indirect. However, by identifying genes downstream of *unc-86*, and upstream of *mec-3*, in a regulatory hierarchy, one might hope to link the regulation of these two homeobox genes. In addition, the pattern of programmed cell death in the QL lineage is altered. Since, as described above, the cell death pathway is controlled by *ced-3* and *ced-4*, *unc-86* must, directly or indirectly, regulate the activity of *ced-3* and *ced-4*.

Finally, QL.ap is a neuron that migrates posteriorly to a position in the tail. This migration requires the activity of the *mab-5* gene, a third example of a *C. elegans* gene encoding a homeobox-containing product. In general, *mab-5* acts cell autonomously (Kenyon, 1986). If *mab-5* acts in QL.ap, *unc-86* may be a negative regulator of *mab-5* activity. The role of *mab-5* might be to activate genes required to migrate posteriorly, such as *mig-1* (cell *mig*ration defective; Hedgecock *et al.*, 1987).

F. Sex Determination and Sexual Differentiation Genes

Hodgkin (1987b) has argued that the major sex determination pathway controls the sexual identity of somatic tissues via the regulation of "downstream genes" by *tra-1*, the output of the somatic pathway. As pointed out by Hodgkin, *tra-1* may control sex-specific functions

---

ates as a neuron distinct from a touch cell (Chalfie and Sulston, 1981; Way and Chalfie, 1988). (D) In a *mab-5* mutant, QL.ap migrates anteriorly rather than posteriorly (Kenyon, 1986; Costa *et al.*, 1988). (E) In a proposed regulatory hierarchy for *unc-86*, *mec-3*, and *mab-5*, *unc-86* might positively regulate *mec-3* and negatively regulate *mab-5*.

through an elaborate regulatory hierarchy, or it may act directly on each sex-specific gene. Candidates for downstream genes include *mab-3*, necessary in males to repress vitellogenin synthesis and to promote aspects of male tail development (Shen and Hodgkin, 1988; see Villeneuve and Meyer, this volume), and *egl-1*, necessary to prevent the male-specific programmed death of the HSN neuron (Trent *et al.*, 1983; Desai and Horvitz, 1989).

V. Genetic Control of Cell Lineage

A. DEVELOPMENTAL CONTROL GENES

A major emphasis in developmental genetics has been the identification of "control genes," genes that act as components of switches specifying various aspects of cell type. Criteria for such control genes are several. A gene defined by two classes of mutations with opposite effects on gene activity corresponding to opposite effects on a developmental decision is an excellent candidate for a gene controlling that aspect of development. If one class of mutations inactivates a given gene and another class of mutations of the same locus increases the activity of that gene, the level of activity of that gene can be inferred to specify the relevant developmental decision. Examples from *C. elegans* include *lin-12*, specifying alternative cell types (Greenwald *et al.*, 1983), *lin-14*, specifying early versus late developmental events (Ambros and Horvitz, 1987), and *tra-1*, controlling male versus hermaphrodite development (Hodgkin, 1987b). Such genes are argued to encode an activity not just *necessary* for the developmental process, but to *control* that process. In no sense is the veracity of this conclusion guaranteed, but such control genes identify *possible* points of control over development.

A second criterion is that mutations have a related set of pleiotropic effects. For example, most phenotypes caused by *lin-12* mutations can be interpreted as due to defects in choices between alternative cell fates (Greenwald *et al.*, 1983). The various phenotypes of *lin-17* mutants can be interpreted as due to defects in asymmetric cell division (Sternberg and Horvitz, 1988). A *mab-5* mutation causes a variety of cell lineage defects which have the common feature that cells fail to recognize their posterior positions (Kenyon, 1986). Genes such as *lin-12, lin-17,* and *mab-5* are likely to be involved in general mechanisms of cell-type determination rather than in the idiosyncrasies of particular cell types, especially since these genes act in a diverse set of lineages including mesodermal, neuronal, and epidermal lineages.

Another, more general criterion is that the pathway in which the gene acts leads to a developmental decision. For example, the decision between the P11 and P12 ectoblast fates depends on two genes: *let-23* is necessary for the P12 cell fate, and *lin-15* is necessary for the P11 cell fate (Fixsen *et al.*, 1985; R. Aroian and P. Sternberg, unpublished). These loci together may control the decision between the P11 and P12 fates. Similarly, during male lateral hypodermal development *mab-5* promotes the production of rays, while *lin-22* promotes the production of alae, a cuticular specialization, suggesting that the *mab-5/lin-22* pathway specifies the ray versus alae decision (Kenyon, 1986).

B. PATHWAYS AND EPISTASIS

A genetic pathway can be constructed by examining the phenotypes of single and multiple mutants to ascertain epistasis and interactions, as well as by identifying and analyzing extragenic suppressors. While the *order* of action is not a direct conclusion of such experiments, one can deduce whether genes act in a common pathway, or at a common step within a pathway. Additional information can suggest the order of action. In general, in a pathway defined by intermediate blocks, e.g., biochemical, assembly, or developmental pathway, the earlier block will be epistatic. For example, a Vul mutation resulting in loss of an anchor cell [*lin-12(n302)*] is likely to act prior to a gene defined by a mutation affecting execution of 2° lineages. Indeed, the *lin-12(n302)* mutation is epistatic to a *lin-11* mutation causing abnormal execution of the 2° lineage (E. Ferguson and P. Sternberg, unpublished). However, in a control or signal processing pathway, a bypass or constitutive defect will be epistatic to an earlier block. In the VPC specification pathway, the epistasis of a *lin-12(d)* mutation that results in the signal-independent specification of 2° fates over a Vul mutation that blocks induction of 2° fates implies that *lin-12* acts after the Vul gene.

The order of gene action in one lineage is not necessarily the same as the order in another lineage. For example, based on their phenotypes, *lin-12* acts prior to *lin-17* in the vulval lineages but after *lin-17* in the male B lineage (Fig. 11). During vulval development, *lin-12* specifies the 2° lineage, and *lin-17* is necessary for the execution of that lineage. Similarly, during male pre-anal ganglion development, *lin-12* acts prior to *lin-17*. In this case, *lin-17* also acts during the primary lineage, which is independent of *lin-12*. In contrast, during male tail development, *lin-17* acts at the first B division, and *lin-12* acts at two later steps, in the Bγ/Bδ equivalence group, and in specifying B.pa/B.pp. Similarly, during somatic gonad development in both the male and

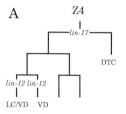

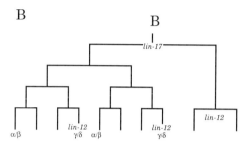

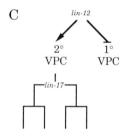

FIG. 11. Orders of action of pleiotropic developmental control genes. (A) During the male somatic gonad (Z4) lineage, *lin-17* acts at the first division, while *lin-12* acts in two of the five progeny cells. (B) During the male ectoblast (B) lineage, *lin-17* acts at the first division, while *lin-12* acts at two subsequent steps: the γ/δ equivalence group and in the B.pa/B.pp fate decision. (C) During the vulval lineages, *lin-12* specifies the 2° lineage, while *lin-17* acts at the first division of the 2° VPC. In the VPC and Z4 cases, *lin-17* and *lin-12* mutations are coexpressed. *lin-17;lin-12* double mutants have six abnormal 2° lineages, and males have up to eight linker cells (LCs). [Data summarized from Sulston and Horvitz (1977), Sternberg and Horvitz (1986, 1988), and Greenwald *et al.* (1983).]

hermaphrodite, *lin 17* acts prior to *lin-12*. These observations indicate that the genes act as part of distinct mechanisms, each of which is used multiple times during development. The epistasis in each case supports the proposed order of action (Sternberg and Horvitz, 1988; P. Sternberg, unpublished).

If a developmental decision has more than two outcomes, circular epistasis relationships can be observed. For example, during VPC spec-ification, *lin-12(d)* (all 2°) is epistatic to a Vul mutation (all 3°), which is epistatic to a Muv mutation (1° and 2), which in turn is epistatic to *lin-12(d)*. This epistasis is a consequence of the branching of the under-lying genetic pathway. It could be argued that *lin-12(d)* mutations are gain of function and lead to this anomalous result. However, the same set of epistasis relations would hold if a mutation eliminating a hypo-thetical negative regulator of *lin-12* were used instead of a *lin-12(d)* mutation. Another interpretation of the epistasis data is that the path-way involves feedback, with 1° inhibiting 2°. Thus, a Muv mutation causes activation of 1° functions in a cell independently of *lin-12*, and 1° functions inactivate 2°. The pathway is a loop, and circular epistasis results.

C. SUBLINEAGES AND GENETIC SUBPROGRAMS

The observation that multiple precursor cells generate identical pat-terns of cell divisions producing identical or nearly identical sets of progeny cells suggests that such precursor cells are in an identical state of determination, and each executes what can be thought of as a devel-opmental "subprogram." For example, 18 cells in the *C. elegans* male generate by a stereotypical division pattern two neurons, one structural cell, and a cell that dies (Sulston *et al.,* 1980). The specification of such sublineages is hypothesized to result from the same molecular events occurring in each precursor cell and its progeny.

Evidence in favor of this hypothesis comes in several forms. First, mutations altering the execution of one example of a sublineage by and large will alter all examples of that sublineage. A second type of evi-dence is that deviant sublineages can easily be converted to a consensus sublineage by simple physical or genetic perturbation. A third type of evidence is that the sublineage is modular, that is, a precursor cell can be moved in time, space, or sex and still generate the correct sublineage. The following describes in detail one example of a sublineage. Other examples of apparent subprograms include the programmed cell death by "suicide" (described in Section IV,B), the Q lineage (Section IV,E and Fig. 10), execution of the 2° vulval lineage (Section III,B,3), the "postdeirid" lineage, and the "ray" sublineage.

The Pn lineages provide an example of variations on a sublineage (Fig. 12). Each of twelve Pn cells divides to generate an anterior daugh-ter that is a neuroblast (Pn.a) and a posterior daughter (Pn.p) that is hypodermal. Each Pn.a neuroblast generates five progeny over a period of 5 hours. If we choose one Pn.a lineage, for example, P6.a, as a

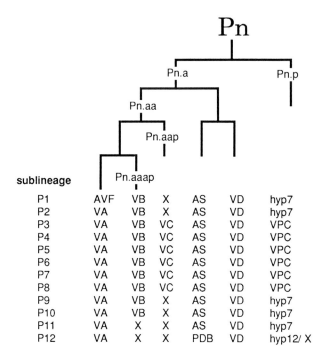

FIG. 12. Pn lineages. Possible fates of each progeny cell are shown underneath the lineage tree for each particular Pn lineage. [Adapted from Sulston and Horvitz (1977).]

standard, then the other Pn.a neuroblast lineages can be considered as variations. P6.aap differentiates as a VC motoneuron, as does P3.aap, P4.aap, P5.aap, P7.aap, and P8.aap. However, P1.aap, P2.aap, P9.aap, P10.aap, P11.aap, and P12.aap die. The *lin-39* mutation *n709* causes P(3,4,5,6,7,8).aap to die as well (Fixsen *et al.,* 1985; R. Ellis and R. Horvitz, personal communication). Thus, the *lin-39* gene may be responsible, in part, for the difference between the P(1,2,9,10,11,12) and P(3–8) sublineages. A second difference among the Pn sublineages is in the fate of the Pn.aaap cell. For P1–P10, this progeny cell differentiates as a VB motoneuron, while P11.aaap and P12.aaap die. The *mab-5* gene is required for this death (Kenyon, 1986).

The posterior hypodermal daughters, Pn.p, have three general fates. P(1,2,9,10,11).p fuse with hyp7, the large syncytium enveloping most of the adult organism. P(3,4,5,6,7,8).p are in the vulval equivalence group; they are the six VPCs. P12.p divides, generating a unique hypodermal cell (P12.pa) and a posterior daughter that dies. Mutations that transform the fates of the Pn.p cells include the following. *let-23* mutations cause P12.p to fuse with hyp7, acting like P(1,2,9,10,11).p do. The *n300* mutation causes P(3–8).p, the VPCs, to join hyp7 (Ferguson *et al.*, 1987). Thus all 12 Pn.p cells are one mutational step away from each other. In addition, a *lin-15* mutation causes P11.p to act like P12.p, dividing to generate a unique hypodermal cell and a posterior daughter that dies (Fixsen *et al.*, 1985; P. Sternberg, unpublished). The coordinate change in the Pn sublineages are evident from the mutations just described.

D. COMPLEXITY

How many gene products are necessary to specify the fate of a cell, that is, how *complex* is the process? Are the same gene products involved, namely, how *interconnected* are the decisions made in distinct cells? These are some of the global questions of developmental genetics, which will be answerable only in the long run. However, from the existing data we can observe clear trends.

A measure of the complexity of cell-type specification is the number of genes involved in the process, which can be ascertained by a saturation analysis (an attempt to identify all the genes acting at that step in development). The most intensively studied cell-type decisions are those involved in VPC specification (Ferguson *et al.*, 1987), HSN differentiation (Desai *et al.*, 1988; Desai and Horvitz, 1989), and touch cell differentiation (Chalfie and Au, 1989). Is the number of genes involved in specifying VPC fates surprising? It is not, given what we know about cellular regulatory mechanisms. For example, a signal transduction cascade might be expected to involve a multicomponent receptor, a three-subunit G protein, a second-messenger producing enzyme, protein kinases, and nuclear transcriptional regulators, not to mention the transcriptional regulators of the signal transduction components. For comparison, the complex regulation of transcription of the yeast *HO* gene involves at least two genes for cell-type control (*MATa1*, *MATα2*), two for mother–daughter control (*SW15, SIN3/SDI1*), two for cell cycle control (*SW14, SW16*), and five others (*SWI1, SWI2, SWI3, SIN1, SIN2*) (reviewed by Nasmyth and Shore, 1987).

The extent of interconnectedness of cell-type specification can be deduced by examining the pleiotropic effects of mutations affecting cell-type specification. Other indications come from examination of the spatial localization of such gene products. There are several clear examples of one gene being involved in distinct cell-type decisions. *lin-12, lin-14, lin-15, lin-17, lin-22, let-23, mab-5,* and *unc-86* provide examples. Other pleiotropies are not well understood. For example, *lin-3* is an essential gene as well as being required for VPC specification (Ferguson and Horvitz, 1985; R. Hill and P. Sternberg, unpublished). In contrast, other genes appear to be required only for particular cell types. For example, *mec-3* may be necessary only for microtubule cell differentiation (Chalfie and Au, 1989).

VI. Conclusions

The inferred products of the four *C. elegans* developmental control genes whose sequences have been published include three with homeobox domains (*mec-3, mab-5,* and *unc-86*; Costa *et al.,* 1988; Finney *et al.,* 1988; Way and Chalfie, 1988) and one with similarity to growth factors or receptors (*lin-12*; Yochem *et al.,* 1988). This small sample, taken together with the larger number of *Drosophila* developmental control genes which have been sequenced (described elsewhere in this volume), suggests that many such genes encode products with familiar functions: transcriptional regulators, growth factors or receptors, protein kinases, or components of second message systems. While we may hope that some of the many unidentified gene products may involve novel biochemical activities or intriguing cell structures, it is clear that a majority of developmental decisions involve the basic components of the cellular regulatory machinery. If developmental genetics continues its success streak, many problems will be reduced to understanding the detailed mechanisms of this machinery, in the context of developing multicellular developing organisms. The other remaining problems might involve heretofore unthinkable mechanisms. The utility of developmental genetics as a tool for understanding cellular regulatory mechanisms is in embedding a particular gene product solidly into the context of a pathway. Genetics is at its most powerful in identifying interacting components, and study of such interacting components is necessary to complete our understanding of any mechanism.

An open question is the extent to which similar proteins act in analogous mechanisms in different organisms. The role of homeobox-containing transcriptional regulators in multifactor control of gene expression is one excellent example, as is emerging from studies of yeast cell type specification (Sprague, this volume) and fruit fly segmentation (Steinmann-Zwicky *et al.,* Gaul and Jäckle, Rushlow and Levine, Kaufman *et al.,* McGinnis *et al.,* and Campos-Ortega, this volume). The parallel roles of *lin-12* and *Notch* in lateral inhibition in nematodes and fruit flies, respectively, constitute another example. Often, the peculiarities of a particular organism or aspect of development allow the study of a particular gene product in more detail with less effort. If similarities in structure are indeed a reasonably reliable indicator of analogous function, then principles gleaned from one organism can be applied to the similar product in another organism, suggesting fruitful lines of inquiry. By elucidating detailed molecular mechanisms of cell-type determination and pattern formation either in *C. elegans* or in other organisms, and by understanding, in parallel, the genetic pathways controlling development in each of several organisms, we may hope someday to be able to unravel the genetic circuitry that underlies animal development.

ACKNOWLEDGMENTS

I am grateful to Ira Herskowitz and Bob Horvitz for teaching me how to think about the big picture while enjoying work in the trenches. I thank Raffi Aroian, Helen Chamberlin, Denise Clark, Scott Clark, Alan Coulson, Kaye Edwards, Ron Ellis, Chip Ferguson, Min Han, Russell Hill, Ed Hedgecock, Bob Horvitz, Linda Huang, Gregg Jongeward, Stuart Kim, Iva Greenwald, Diane Parry, Denise Roberts, John Sulston, Meera Sundaram, and Jim Thomas for communicating results prior to publication. Renee Thorf provided expert and timely handling of the manuscript.

Research in my laboratory has been supported by the U.S. Public Health Service, the Howard Hughes Medical Institute, the National Science Foundation, the Searle Scholars Program, the March of Dimes Birth Defects Foundation, and the Lucille P. Markey Charitable Trust. The Caenorhabditis Genetics Center, supported by contract N01-AG-9-2113 between the National Institutes of Health and University of Missouri, provided many strains.

REFERENCES

Albertson, D. G. (1985). Mapping muscle protein genes by *in situ* hybridization using biotin-labeled probes. *EMBO J.* **4**, 2493–2498.

Albertson, D. G., Sulston, J. E., and White, J. G. (1978). Cell cycling and DNA replication in a mutant blocked in cell division in the nematode *Caenorhabditis elegans. Dev. Biol.* **63**, 165–178.

Ambros, V. (1989). A hierarchy of regulatory genes controls a larval to adult developmental switch in *C. elegans. Cell* **57**, 49–57.

Ambros, V., and Horvitz, H. R. (1984). Heterochronic mutants of the nematode. *Science* **226**, 409–416.

Ambros, V., and Horvitz, H. R. (1987). The *lin-14* locus of *C. elegans* controls the time of expression of specific postembryonic developmental events. *Genes Dev.* **1**, 398–414.

Austin, J., and Kimble, J. (1987). *glp-1* is required in the germ line for regulation of the decision between mitosis and meiosis in *Caenorhabditis elegans. Cell* **51**, 589–599.

Baillie, D. L., Beckenbach, K. A., and Rose, A. M. (1985). Cloning within the *unc-43* to *unc-31* interval (linkage group *IV*) of the *Caenorhabditis elegans* gene using Tc1 linkage. *Can. J. Genet. Cytol.* **27**, 457–466.

Bender, A., and Sprague, G. F. (1989). Pheromones and pheromone receptors are the primary determinants of mating specificity in the yeast *Saccharomyces cerevisiae. Genetics* **121**, 463–476.

Bolten, S. L., Powell-Abel, P., Fischhoff, D. A., and Waterston, R. W. (1984). The *sup-7(st5)* X gene of *C. elegans* encodes a transfer RNA-Trp-UAG amber suppressor. *Proc. Natl. Acad. Sci. U.S.A.* **81,** 6784–6788.

Brenner, S. (1974). The genetics of *Caenorhabditis elegans*. *Genetics* **77,** 71–94.

Chalfie, M. (1984). Genetic analysis of nematode nerve-cell differentiation. *Bioscience* **34,** 295–299.

Chalfie, M., and Au, M. (1989). Genetic control of differentiation of the *Caenorhabditis elegans* touch receptor neurons. *Science* **243,** 1027–1033.

Chalfie, M., and Sulston, J. (1981). Developmental genetics of the mechanosensory neurons of *Caenorhabditis elegans*. *Dev. Biol.* **82,** 358–370.

Chalfie, M., Horvitz, H. R., and Sulston, J. E. (1983). Mutations that lead to reiterations in the cell lineages of *Caenorhabditis elegans*. *Cell* **24,** 59–69.

Chalfie, M., Dean, E., Reilly, E., Buck, K., and Thomson, J. W. (1986). Mutations affecting microtubule structure in *Caenorhabditis elegans*. *J. Cell Sci. Suppl.* **5,** 257–271.

Clark, D. V., Rogalski, T. M., Donati, L. M., and Baillie, D. L. (1988). The *unc-22(IV)* region of *Caenorhabditis elegans:* Genetic analysis of lethal mutations. *Genetics* **119,** 345–353.

Collins, J., Saari, B., and Anderson, P. (1989a). Activation of a transposable element in the germ line but not the soma of *Caenorhabditis elegans*. *Nature (London)* **328,** 726–728.

Collins, J., Forbes, E., and Anderson, P. (1989b). The Tc3 family of transposable genetic elements in *Caenorhabditis elegans*. *Genetics* **121,** 47–55.

Costa, M., Weir, M., Coulson, A., Sulston, J., and Kenyon, C. (1988). Posterior pattern formation in *C. elegans* involves position-specific expression of a gene containing a homeobox. *Cell* **55,** 747–756.

Coulson, A. R., Sulston, J., Brenner, S., and Karn, J. (1986). Toward a physical map of the genome of the nematode *Caenorhabditis elegans*. *Proc. Natl. Acad. Sci. U.S.A.* **83,** 7821–7825.

Coulson, A., Waterston, R., Kiff, J., Sulston, J., and Kohara, Y. (1988). Genome linking with yeast artificial chromosomes. *Nature (London)* **335,** 184–186.

Cox, G. N., Carr, S., Kramer, J. M., and Hirsh, D. (1985). Genetic mapping of *Caenorhabditis elegans* collagen genes using DNA polymorphisms as phenotypic markers. *Genetics* **109,** 513–528.

Desai, C., and Horvitz, H. R. (1989). *Caenorhabditis elegans* mutants defective in the functioning of the motor neurons responsible for egg laying. *Genetics* **121,** 703–721.

Desai, C., Garriga, G., McIntire, S. L., and Horvitz, H. R. (1988). A genetic pathway for the development of the *Caenorhabditis elegans* HSN motor neurons. *Nature (London)* **336,** 638–646.

Doe, C. Q., and Goddman, C. S. (1985). Early events in insect neurogenesis. II. The role of cell interactions and cell lineage in the determination of neuronal precursor cells. *Dev. Biol.* **111,** 206–219.

Edgely, M., and Riddle, D. (1988). "*C. elegans* Genetic Map." The Caenorhabditis Genetics Center, Columbia, Missouri.

Eide, D., and Anderson, P. (1985). The gene structure of spontaneous mutations affecting a *C. elegans* myosin heavy chain gene. *Genetics* **109,** 67–79.

Ellis, H. M., and Horvitz, H. R. (1986). Genetic control of programmed cell death in the nematode *Caenorhabditis elegans*. *Cell* **44,** 817–829.

Emmons, S., Klass, M. R., and Hirsh, D. (1979). Analysis of the constany of DNA sequences during development and evolution of the nematode *C. elegans*. *Proc. Natl. Acad. Sci. U.S.A.* **76,** 1333–1337.

Emmons, S., Yesner, L., Ruan, K. S., and Katzenberg, D. (1983). Evidence for a transposon in *Caenorhabditis elegans*. *Cell* **32,** 55.

Emmons, S. W., Roberts, S., and Ruan, K.-S. (1986). Evidence in a nematode for regulation of transposon excision by tissue-specific factors. *Mol. Gen. Genet.* **202**, 410–415.

Ferguson, E. L., and Horvitz, H. R. (1985). Identification and characterization of 22 genes that affect the vulval cell lineages of *Caenorhabditis elegans. Genetics* **110**, 17–72.

Ferguson, E. L., and Horvitz, H. R. (1989). The multivulva phenotype of certain *C. elegans* mutants results from defects in two functionally redundant pathways. *Genetics* **123**, 109–121.

Ferguson, E. L., Sternberg, P. W., and Horvitz, H. R. (1987). A genetic pathway for the specification of the vulval cell lineages of *Caenorhabditis elegans. Nature (London)* **326**, 259–267.

Finney, M., Ruvkun, G., and Horvitz, H. R. (1988). The *C. elegans* cell lineage and differentiation gene *unc-86* encodes a protein containing a homeo domain and extended sequence similarity to mammalian transcription factors. *Cell* **55**, 757–769.

Fire, A. (1986). Integrative transformation of *Caenorhabditis elegans. EMBO J.* **5**, 2673–2680.

Fixsen, W., Sternberg, P., Ellis, H., and Horvitz, R. (1985). Genes that affect cell fates during development of *Caenorhabditis elegans. Cold Spring Harbor Symp. Quant. Biol.* **50**, 99–104.

Greenwald, I. (1985). *lin-12,* a nematode homeotic gene, is homologous to a set of mammalian proteins that includes epidermal growth factor. *Cell* **43**, 583–590.

Greenwald, I. S., and Horvitz, H. R. (1980). *unc-93(e1500):* A behavioral mutant of *Caenorhabditis elegans* that defines a gene with a wild-type null phenotype. *Genetics* **96**, 147–164.

Greenwald, I. S., Sternberg, P. W., and Horvitz, H. R. (1983). *lin-12* specifies cell fates in *C. elegans. Cell* **34**, 435–444.

Greenwald, I. S., Coulson, A., Sulston, J., and Preiss, J. (1987). Correlation of the physical and genetic maps in the *lin-12* region of *Caenorhabditis elegans. Nucleic Acids Res.* **15**, 2295–2307.

Hedgecock, E. M., Sulston, J. E., and Thomson, J. N. (1983). Mutations affecting programmed cell death in the nematode *Caenorhabditis elegans. Science* **220**, 1277–1279.

Hedgecock, E. M., Culotti, J. G., Hall, D. H., and Stern, B. D. (1987). Genetics of cell and axon migrations in *Caenorhabditis elegans. Development* **100**, 365–382.

Herman, R. K. (1978). Crossover suppressors and balanced recessive lethals in *Caenorhabditis elegans. Genetics* **88**, 49–65.

Herman, R. K. (1984). Analysis of genetic mosaics of the nematode *Caenorhabditis elegans. Genetics* **108**, 165–180.

Herman, R. K. (1987). Mosaic analysis of two genes that affect nervous system structure in *Caenorhabditis elegans. Genetics* **116**, 377–388.

Herman, R. K., and Kari, C. K. (1985). Muscle-specific expression of a gene affecting acetylcholinesterase in the nematode *Caenorhabditis elegans. Cell* **40**, 509–514.

Herman, R. K., and Shaw, J. E. (1987). The transposable element Tc1 in the nematode *Caenorhabditis elegans. Trends Genet.* **3**, 222–225.

Herskowitz, I. (1987). Functional inactivation of genes by dominant negative mutations. *Nature (London)* **329**, 219–222.

Hodgkin, J. (1983). Male phenotypes and mating efficiency in *Caenorhabditis elegans. Genetics* **103**, 43–64.

Hodgkin, J. (1987a). Sex determination and dosage compensation in *Caenorhabditis elegans. Annu. Rev. Genet.* **21**, 133–154.

Hodgkin, J. (1987b). A genetic analysis of the sex-determining gene, *tra-1,* in the nematode *Caenorhabditis elegans. Genes Dev.* **1**, 731–745.

Hodgkin, J., Kondo, K., and Waterson, R. H. (1987). Suppression in the nematode *Caenorhabditis elegans. Trends Genet.* **3**, 325–329.

Horvitz, H. R. (1988). Genetics of cell lineage. *In* "The Nematode *Caenorhabditis elegans*" (W. B. Wood, ed.), pp. 157–190. Cold Spring Harbor Laboratory, Cold Spring Harbor, New York.

Horvitz, H. R., and Sulston, J. E. (1980). Isolation and genetic characterization of cell lineage mutants of the nematode *Caenorhabditis elegans. Genetics* **96**, 435–454.

Horvitz, H. R., Brenner, S., Hodgkin, J., and Herman, R. (1979). A uniform genetic nomenclature for the nematode *Caenorhabditis elegans. Mol. Gen. Genet.* **175**, 129–133.

Horvitz, H. R., Ellis, H. M., and Sternberg, P. W. (1982). Programmed cell death in nematode development. *Neurosci. Comment.* **1**, 56–65.

Kenyon, C. (1986). A gene involved in the development of the posterior body region of *C. elegans. Cell* **46**, 477–487.

Kenyon, C. (1988). The nematode *Caenorhabditis elegans. Science* **240**, 1448–1453.

Kim, S., and Horvitz, R. (1990). In preparation.

Kimble, J. (1981). Lineage alterations after ablation of cells in the somatic gonad of *Caenorhabditis elegans. Dev. Biol.* **87**, 286–300.

Kimble, J., and Hirsh, D. (1979). Postembryonic cell lineages of the hermaphrodite and male gonads in *Caenorhabditis elegans. Dev. Biol.* **70**, 396–417.

Kimble, J., and White, J. G. (1981). On the control of germ cell development in *Caenorhabditis elegans. Dev. Biol.* **81**, 208–219.

Kimble, J., Sulston, J., and White, J. (1979). Regulative development in the postembryonic lineages of *Caenorhabditis elegans. In* "Cell Lineages, Stem Cells and Cell Differentiation" (N. Le Duoarin, ed.), INSERM Symposium No. 10, pp. 59–68. Elsevier, Amsterdam.

Kurjan, J. (1985). α-factor structural gene mutations in *Saccharomyces cerevisiae*: Effects on α-factor production and mating. *Mol. Cell. Biol.* **5**, 787–796.

Levine, M., and Hoey, T. (1988). Homeobox proteins as sequence-specific transcription factors. *Cell* **55**, 537–540.

Liao, L. W., Rosenzweig, B., and Hirsh, D. (1983). Analysis of a transposable element in *C. elegans. Proc. Natl. Acad. Sci. U.S.A.* **80**, 3585–3589.

McCoubrey, W. K., Nordstrom, K. D., and Meneely, P. M. (1988). Microinjected DNA from the X chromosome affects sex determination in *Caenorhabditis elegans. Science* **242**, 1146–1151.

Melton, D. A. (1985). Injected antisense RNAs specifically block messenger RNA translation *in vivo. Proc. Natl. Acad. Sci. U.S.A.* **82**, 144–148.

Mitchell, A. P. (1985). The *GLN1* locus of *Saccharomyces cerevisiae* encodes glutamine synthetase. *Genetics* **111**, 243–258.

Moerman, D. G., Benian, G. M., and Waterston, R. H. (1986). Molecular cloning of the muscle gene *unc-22* in *C. elegans* by Tc1 transposon tagging. *Proc. Natl. Acad. Sci. U.S.A.* **83**, 2579–2583.

Nasmyth, K., and Shore, D. (1987). Transcriptional regulation in the yeast life cycle. *Science* **237**, 1162–1170.

Park, E.-C., and Horvitz, H. R. (1986). *C. elegans unc-105* mutations affect muscle and are suppressed by other mutations that affect muscle. *Genetics* **115**, 853–867.

Priess, J. R., and Thomson, J. W. (1987). Cellular interactions in early *Caenorhabditis elegans* embryos. *Cell* **48**, 241–250.

Priess, J. R., Schnabel, H., and Schnabel, R. (1987). The *glp-1* locus and cellular interactions in early *Caenorhabditis elegans* embryos. *Cell* **51**, 601–611.

Riddle, D. R. (1988). The dauer larvae. In "The Nematode Caenorhabditis elegans" (W. B. Wood, ed.), pp. 393–412. Cold Spring Harbor Laboratory, Cold Spring Harbor, New York.

Rosenberg, U. B., Priess, A., Seifert, E., Jäckle, H., and Knipple, D. C. (1985). Production of phenocopies by Krüpple antisense RNA injections into Drosophila embryos. Nature (London) 313, 703–706.

Ruvkun, G., and Giusto, J. (1989). The Caenorhabditis elegans heterochronic gene lin-14 encodes a nuclear protein that forms a temporal developmental switch. Nature (London) 338, 313–320.

Ruvkun, G., Ambros, V., Coulson, A., Waterston, R., Sulston, J., and Horvitz, H. R. (1989). Molecular genetics of the Caenorhabditis elegans heterochronic gene in lin-14. Genetics 121, 501–516.

Seydoux, G., and Greenwald, I. (1989). Cell autonomy of lin-12 function in a cell fate decision in C. elegans. Cell 57, 1237–1245.

Shen, M. M., and Hodgkin, J. (1988). mab-3, a gene required for sex-specific yolk protein expression and a male-specific lineage in C. elegans. Cell 54, 1019–1031.

Sigurdson, D. C., Spanier, G. J., and Herman, R. K. (1984). Caenorhabditis elegans deficiency mapping. Genetics 108, 333–345.

Spieth, J., MacMorris, M., Broverman, S., Greenspoon, S., and Blumenthal, T. (1988). Regulated expression of a vitellogenin fusion gene in transgenic nematodes. Dev. Biol. 130, 285–293.

Sternberg, P. W. (1988a). Lateral inhibition during vulval induction in Caenorhabditis elegans. Nature (London) 335, 551–554.

Sternberg, P. W. (1988b). Control of cell fates in equivalence groups in C. elegans. Trends Neurosci. 11, 259–264.

Sternberg, P. W., and Horvitz, H. R. (1981). Gonadal cell lineags of the nematode Panagrellus redivivus and implications for evolution by the modification of cell lineage. Dev. Biol. 88, 147–166.

Sternberg, P. W., and Horvitz, H. R. (1982). Postembryonic cell lineages of the nematode Panagrellus redivivus: Description and comparison with those of Caenorhabditis elegans. Dev. Biol. 93, 181–205.

Sternberg, P. W., and Horvitz, H. R. (1984). The genetic control of cell lineage during nematode development. Annu. Rev. Genet. 18, 489–524.

Sternberg, P. W., and Horvitz, H. R. (1986). Pattern formation during vulval development in Caenorhabditis elegans. Cell 44, 761–772.

Sternberg, P. W., and Horvitz, H. R. (1988). lin-17 mutations of C. elegans disrupt asymmetric cell divisions. Dev. Biol. 130, 67–73.

Sternberg, P. W., and Horvitz, H. R. (1989). The combined action of two intercellular signalling pathways specifies three cell fates during vulval induction in C. elegans. Cell 58, 679–693.

Stinchcomb, D. T., Shaw, J. E., Carr, S. H., and Hirsh, D. (1985). Extrachromosomal DNA transformation of C. elegans. Mol. Cell. Biol. 5, 3484–3496.

Sulston, J., and Horvitz, H. R. (1977). Abnormal cell lineages in mutants of the nematode Caenorhabditis elegans. Dev. Biol. 82, 41–55.

Sulston, J., and Horvitz, H. R. (1981). Postmebryonic cell lineages of the nematode Caenorhabditis elegans. Dev. Biol. 56, 110–156.

Sulston, J. E., and White, J. B. (1980). Regulation and cell autonomy during postembryonic development of Caenorhabditis elegans. Dev. Biol. 78, 577–597.

Sulston, J. E., Albertson, D. G., and Thomson, J. N. (1980). The Caenorhabditis elegans male: Postembryonic development on nongonadal structures. Dev. Biol. 78, 542–576.

Sulston, J. E., Schierenberg, E., White, J. G., and Thomson, J. N. (1983). The embryonic cell lineage of the nematode *Caenorhabditis elegans. Dev. Biol.* **100,** 64–119.

Thomas, J., and Horvitz, R. (1990). In preparation.

Thomas, J., Ferguson, E., and Horvitz, R. (1990). In preparation.

Trent, C., Tsung, N., and Horvitz, H. R. (1983). Egg-laying defective mutants of the nematode *Caenorhabditis elegans. Genetics* **104,** 619–647.

Way, J., and Chalfie, M. (1988). *mec-3,* a homeobox-containing gene that specifies differentiation of the touch receptor neurons in *C. elegans. Cell* **54,** 5–16.

White, J. (1988). The anatomy. *In* "The Nematode *Caenorhabditis elegans*" (W. B. Wood, ed.), pp. 81–122. Cold Spring Harbor Laboratory, Cold Spring Harbor, New York.

Wood, W. B., ed. (1988). "The Nematode *Caenorhabditis elegans*." Cold Spring Harbor Laboratory, Cold Spring Harbor, New York.

Yochem, J., Weston, K., and Greenwald, I. (1988). The *Caenorhabditis elegans lin-12* encodes a transmembrane protein with overall similarity to *Drosophila Notch. Nature (London)* **335,** 547–550.

THE REGULATORY HIERARCHY CONTROLLING SEX DETERMINATION AND DOSAGE COMPENSATION IN
Caenorhabditis elegans

Anne M. Villeneuve and Barbara J. Meyer

Department of Biology, Massachusetts Institute of Technology,
Cambridge, Massachusetts 02139

ADVANCES IN GENETICS, Vol. 27

I. Introduction

The decision of an organism to develop as either a male or a female (or as a hermaphrodite) poses a variety of intriguing problems, both for the organism confronting this decision and its consequences and for the investigator attempting to understand how such a decision is made and executed. How do cells choose between alternative cell fates? Once an initial decision is made, how do cells subsequently become restricted in developmental potential, committed to a particular fate prior to the time when differentiation takes place? After differentiation occurs, how is the differentiated state maintained? Further, how does the organism ensure that all cells or tissues make the same choice? What aspects of this process involve intercellular communication, and what functions are cell autonomous? What are the mechanisms by which the pertinent intracellular and/or intercellular signals are transduced? What role does the maternally supplied oocyte cytoplasm play in the decisions directed by the zygotic genome? These questions are easily recognized as encompassing many of the central issues common to all fundamental developmental processes.

In *Caenorhabditis elegans,* the manner in which sexual dimorphism arises during development adds a further dimension to this already substantial problem. Sexual dimorphism in *C. elegans* is pervasive, occurring in all tissue types and arising in almost all major branches of the cell lineage. Moreover, the strategies for generating dimorphism are diverse. These include sex-specific programmed cell death, generation and division of sex-specific blast cells, adoption of alternative lineages by common primordia, and initiation of sex-specific gene expression in tissues that have identical cell lineages. Finally, sexual differentiation occurs at many different times during development in different tissues in the organism. Thus, the final outcome of the sex determination decision (i.e., male versus hermaphrodite development) depends on the well-coordinated orchestration of a large number of distinct developmental events that are spatially, mechanistically, and temporally separated.

The approach that has been used to study sex determination in *C. elegans* and the related phenomenon of X-chromosome dosage compensation (see below) is that of developmental and molecular genetics. In this article, we discuss what is known about the large number of genes that have been identified as playing important roles in these two processes and present a model for how these genes act together in a regulatory hierarchy to coordinately control both the choice of sexual fate and the level of X-chromosome gene expression. Armed with this information, we examine in the last several sections what this genetic and molecular analysis has begun to teach us and what we can hope to learn in the future about various aspects of the fundamental developmental issues posed in the above questions.

II. Sexual Dimorphism in *Caenorhabditis elegans*

Caenorhabditis elegans naturally exists as two sexes, a self-fertilizing hermaphrodite, which has two X chromosomes (XX), and a male, which has a single X chromosome (XO). Hermaphrodites produce both sperm and oocytes and can reproduce either by internal self-fertilization, giving rise to self-progeny broods consisting almost entirely of XX hermaphrodites, or by cross-fertilization with males, giving rise to broods composed of equal numbers of XX hermaphrodite and XO male cross-progeny. Additionally, XO males arise spontaneously at a low frequency (0.2%) in self-fertilizing hermaphrodite populations as a result of meiotic X-chromosome nondisjunction (Hodgkin *et al.*, 1979).

Many nematode species related to *C. elegans* have separate male and female sexes and must reproduce by cross-fertilization. The female sex of the close relative *Caenorhabditis remanei* (Sudhaus, 1974) is anatomically very similar to the *C. elegans* hermaphrodite, suggesting that the *C. elegans* hermaphrodite may be viewed as a modified female that has acquired the ability to produce a small number of sperm in the germ line before switching to oogenesis. This view is strengthened by the fact that single gene mutations can eliminate spermatogenesis in the hermaphrodite while leaving male spermatogenesis intact, resulting in male/female *C. elegans* strains (Doniach, 1986b; Schedl and Kimble, 1988; see Sections IV and VI).

The degree of sexual dimorphism in *C. elegans* is extensive; of 959 somatic nuclei in the hermaphrodite, at least 30% exhibit overt sexual specialization, while at least 40% of the 1031 somatic nuclei in the male are sexually specialized. The developmental and anatomical differences between the two sexes have been described in detail (reviewed in Hodgkin, 1988; Kimble and Ward, 1988; Sulston, 1988; White, 1988), so only a brief outline is provided here. Figure 1 illustrates some of the most obvious sexually dimorphic features. The hermaphrodite has a two-armed gonad in which approximately 320 sperm (160 per gonad arm) are produced during the L4 larval stage, followed by a switch to oogenesis in adulthood. Sperm are stored in a compartment known as the spermatheca, and oocytes are fertilized as they pass through this compartment and move into the uterus. Eggs are laid through the

A

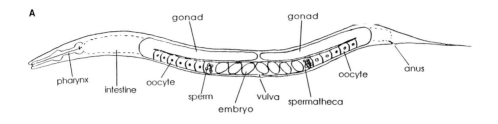

B

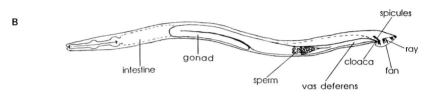

Fig. 1. Schematic diagrams of (A) an adult hermaphrodite (XX) and (B) an adult male (XO) highlighting many of the major anatomical differences. [For simplicity, only part of the intestines (drawn with dashed lines) are shown.] See the text for description.

vulva, an opening in the ventral hypodermis through which in-semination by males also occurs. The hermaphrodite tail is relatively unspecialized, and tapers to a thin point referred to as the "spike" or "whip."

The male anatomy is dramatically different. The male is both shorter and thinner than the hermaphrodite and is highly specialized for mat-ing. It has a single armed gonad (in which approximately 3000 sperm are produced) that is connected via the vas deferens to the cloaca, through which sperm are released. The male tail is equipped with various specialized sensory and copulatory structures (e.g., rays, hook, spicules) that enable the male to locate the vulva and successfully inseminate the hermaphrodite (Fig. 2).

In addition to the features shown in Fig. 1, there is extensive dimor-phism in both the musculature and the nervous system. Sex-specific muscles and sexually specialized and sex-specific neurons are involved in both the control and execution of egg-laying behavior in hermaphro-dites (Trent *et al.,* 1983; Desai *et al.,* 1988) and mating behavior, copula-tion, and male-specific locomotory behavior in males (White, 1988; J. Sulston, personal communication; G. Garriga, personal commu-nication). Finally, the intestine can also be regarded as sexually dimor-phic in that vitellogenins (yolk proteins) are produced only in the hermaphrodite intestine (Kimble and Sharrock, 1983), and intestinal

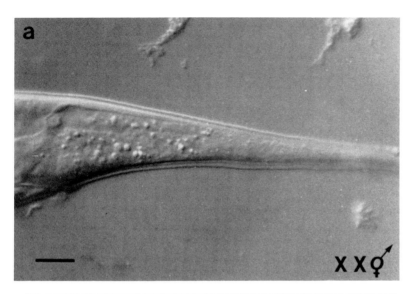

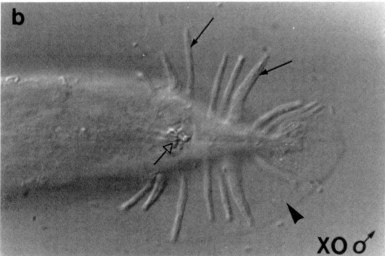

FIG. 2. Nomarski photomicrographs showing the dramatic sexual dimorphism of the adult *C. elegans* tail. The hermaphrodite tail [(a) lateral view] is relatively unspecialized, tapering to a thin whiplike structure. The male tail (or copulatory bursa) is highly specialized for copulation [(b) ventral view]. The large arrowhead points to the fan, an acellular structure formed from the adult cuticle matrix; the solid arrows indicate 2 of the 18 sensory rays, which are used to search for the hermaphrodite vulva; the open arrow indicates the position of the hook and the spicules (both in different planes of focus), structures used in copulation. Bar, 10 μm.

cells are larger in adult hermaphrodites than in adult males (Schedin, 1988).

III. The Primary Sex Determination Signal: The X/A Ratio

In organisms where sex is determined chromosomally, a variety of primary sex-determining mechanisms are possible. In mammals, for example, the Y chromosome acts as a dominant male determinant (Ford *et al.*, 1959; Jacobs and Strong, 1959; Welshons and Russell, 1959); XY, XXY, or XXXY diploid individuals are male, while XO, XX, or XXX diploid individuals are female. In *Drosophila*, on the other hand, sex is determined by the ratio of the number of X chromosomes to the number of sets of autosomes (the X/A ratio) (Bridges, 1925). Diploid flies with two X chromosomes (XX or XXY) develop as females, while those with a single X chromosome (XO or XY) develop as males, regardless of the presence or absence of a Y chromosome.

In *C. elegans* there is no Y chromosome, and the primary determinant of sex is the X/A ratio, as in *Drosophila*. That the ratio of X chromosomes to sets of autosomes is critical in determining sex, rather than the absolute number of X chromosomes per se, was first recognized by Nigon (1949a, 1949b, 1951). Nigon had originally shown that diploid animals with two X chromosomes (2X/2A = 1) are hermaphrodite, while those with a single X chromosome (1X/2A = 0.5) are male. In subsequent work he showed that tetraploid animals with four X chromosomes (4X/4A = 1) or three X chromosomes (3X/4A = 0.75) are hermaphrodite, while those with two X chromosomes (2X/4A = 0.5) are male. Madl and Herman (1979) further showed that triploid animals with three X chromosomes (3X/3A = 1) are hermaphrodite, while those with two X chromosomes (2X/3A = 0.67) are (usually) male. Thus, a ratio of 0.75 or greater elicits hermaphrodite development, while a ratio of 0.67 or less results in male development.

Madl and Herman (1979) also showed that the addition of partial X-chromosome duplications (Dps) to 2X/3A animals can cause a shift toward hermaphrodite development. 2X/3A animals carrying the large duplication *mnDp10* are completely shifted to the hermaphrodite fate, whereas 2X/3A animals carrying either of two smaller duplications, *mnDp9* or *mnDp25*, can develop as *either* hermaphrodites, males, or intersexes (see Fig. 3). Two other duplications, *mnDp8* and *mnDp27*, had no effect in these experiments. It was concluded that the X chromosome contains at least three (and perhaps many more) dose-sensitive sites that cumulatively comprise the numerator of the X/A ratio. As

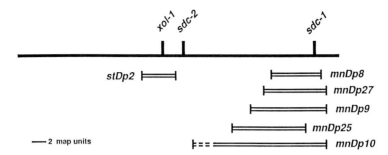

FIG. 3. Genetic map of the X chromosome indicating the approximate genetic extents of the X-chromosome duplications discussed in the text, as well as the locations of *xol-1*, *sdc-1*, and *sdc-2*, three X-linked genes involved in the coordinate control of the sex determination and dosage compensation processes.

yet, however, no specific "numerator loci" have been further localized genetically. Essentially nothing is known about the autosomal portion (or denominator) of the signal. Recent experiments and observations that may provide clues about the nature of the X/A signal are discussed later (Section XI).

As a consequence of the primary sex-determining mechanism, males (XO) and hermaphrodites (XX) possess different numbers of X chromosomes and, hence, different doses of all X-linked genes. *Caenorhabditis elegans* compensates for this difference in X-linked gene dosage by equalizing X-specific transcript levels between the two sexes (Meyer and Casson, 1986; Donahue *et al.*, 1987). The level of X-chromosome expression, like the choice of sexual fate, appears to be set in response to the primary sex-determining signal, the X/A ratio (Villeneuve and Meyer, 1987; Miller *et al.*, 1988; Nusbaum and Meyer, 1989).

Since the sole genetic difference between males and hermaphrodites is the relative dose of X chromosomes, it follows that *C. elegans* must possess some mechanism for "counting" chromosomes. The organism must assess the value of the X/A ratio and then subsequently transmit this information to the genes that control sex determination and the genes that control X-chromosome dosage compensation. A large number of genes that act in a branched pathway connecting the initial signal with its ultimate consequences (i.e., male or hermaphrodite sexual differentiation and an appropriate level of X-linked gene expression) has been identified (Fig. 6, Table 1). Several genes appear to act at early steps in the regulatory hierarchy to coordinately control both sex determination and dosage compensation in response to the X/A signal; these have been identified on the basis of mutations that

TABLE 1

Phenotypes of Sex Determination and Dosage Compensation Mutants[a]

Gene	Number of alleles	XX phenotype	XO phenotype
Genes required for both sex determination and dosage compensation			
Wild type		**Hermaphrodite**	**Male**
sdc-1(lf)	15	Small, Egl hermaphrodites, small pseudomales, and intersexes (maternal rescue)	**Male**
sdc-2(lf) (weak)	8	Small pseudomales, rare intersexes, and Egl hermaphrodites	**Male**
sdc-2(lf) (strong)	>20	Dead, masculinized	**Male**
xol-1(lf)	5	**Hermaphrodite**	Dead, feminized
Genes required for proper sex determination			
her-1(lf)	>25	**Hermaphrodite**	Hermaphrodite
her-1(gf)	2	Masculinized Egl hermaphrodites, intersexes, and pseudomales	**Male**
tra-1(lf)	>50	Low fertility male	Low fertility male
tra-1(gf)	>20	Female	Female
tra-2(lf)	>50	Incomplete, nonmating male	**Male**
tra-2(gf)	12	Female	**Male**
tra-3(lf)	4	Incomplete, nonmating male (complete maternal rescue)	**Male**
fem-1(lf)	>25	Female (some maternal rescue)	Female (some maternal rescue)
fem-2(lf)	11	Female (complete maternal rescue)	Female (strong maternal rescue)
fem-3(lf)	>35	Female (weak maternal absence effect)	Female (both maternal rescue and maternal absence effects)
fem-3(gf)	9	Male germ line, female soma	**Male**
fog-1	>40	Female	Female germ line, male soma
fog-2(lf)	16	Female	**Male**
Genes required for dosage compensation			
dpy-21(lf)	>20	Dpy, Egl hermaphrodite	**Male**
dpy-26(lf)	4	Dead; few Dpy, Egl hermaphrodites (maternal rescue)	**Male**
dpy-27(lf)	13	Dead; few Dpy, Egl hermaphrodites (maternal rescue)	**Male**

TABLE 1 (*Continued*)

Gene	Number of alleles	XX phenotype	XO phenotype
dpy-28(lf)	2	Dead; few Dpy, Egl hermaphrodites (maternal rescue)	**Male**
dpy-29(lf)[b]	11	Dead; few Dpy, Egl hermaphrodites (maternal rescue)	**Male**

[a] Boldface type is used to indicate wild-type phenotypes. *sdc*, Sex determination and dosage compensation; *xol*, XO lethal; *tra*, transformer; *her*, hermaphroditization; *fem*, feminization; *fog*, feminization of germ line; *dpy*, dumpy; *lf*, loss of function; *gf*, gain of function; Egl, egg-laying defective. For a more complete description of the genes and their mutant phenotypes, see the text.

[b] Additional data suggest that *dpy-29* may be a gene of the *sdc* class (see Section IX).

simultaneously disrupt both processes. Other genes act at steps in the hierarchy after the control of these two processes have diverged. Some of these genes are required only for the proper choice of sexual fate, while others appear to be involved only in controlling X-chromosome expression.

In the next several sections, we discuss these genes and the gene interactions involved in the regulatory hierarchy controlling sex determination and dosage compensation in *C. elegans*. In order to develop the rationale for the placement of individual genes at particular steps in this hierarchy, we have chosen to discuss the genes in the following order: (1) genes required for proper sex determination but not for dosage compensation, (2) genes required specifically for dosage compensation, (3) genes involved in the coordinate control of the sex determination and dosage compensation processes.

IV. Genes That Control Sex Determination but Not Dosage Compensation

Seven autosomal genes that control the choice of sexual fate in *C. elegans* but are not required for X-chromosome dosage compensation have been described in detail; these genes appear to be involved in determining sexual fate in all tissues where sexual dimorphism is observed. Hermaphrodite sexual development requires the activities of the *tra* genes (transformer); loss-of-function mutations in these genes transform XX animals into fertile males or pseudomales. Male development, on the other hand, requires the activities of the gene *her-1* (hermaphroditization) and the *fem* genes (feminization). Loss-of-function mutations in *her-1* transform XO animals into fertile hermaphrodites, and loss-of-function mutations in the *fem* genes transform XO animals into spermless (but cross-fertile) females. The *fem* genes are also required for spermatogenesis in hermaphrodites, so that *fem* XX animals are females as well. In this section, we describe the genetic characterization of these sex-determining loci, and, where applicable, we also discuss recent progress in the molecular analysis of several of the genes.

A. TRANSFORMER (*tra*) GENES

1. *tra-1*

The gene *tra-1* is represented by more than 50 recessive loss-of-function alleles and more than 20 dominant gain-of-function alleles. The most common class of *tra-1* loss-of-function alleles (which includes many amber mutations) results in the transformation of XX animals into superficially normal, sometimes fertile mating males (Hodgkin and Brenner, 1977; Hodgkin, 1987c). These *tra-1* XX males are less fertile than wild-type XO males, however. Close examination reveals that while nongonadal tissues develop with normal male morphology, for some alleles (e.g., *e1099*) the male gonads are sometimes small, morphologically abnormal, and contain reduced numbers of sperm. For other alleles (e.g., *e1871am*), male somatic gonad morphology is normal but only a small amount of spermatogenesis occurs, after which a few oocytes are produced (Hodgkin, 1987c; Schedl *et al.*, 1989). Whether either class of alleles represents the complete lack of *tra-1* function is a question that awaits molecular resolution.

In addition to strong alleles that can cause complete masculinization of XX animals, several weaker loss-of-function alleles resulting in less complete masculinization have been identified. Many of these alleles fall into an allelic series of decreasing severity, producing phenotypes ranging from XX pseudomales with nearly normal male morphology but that fail to sire cross-progeny, to animals with reduced and abnormal male tail structure and both sperm and oocytes in a male gonad, to animals that have male gonads but are otherwise completely hermaphrodite. Several other recessive *tra-1* alleles do not fit easily into this allelic series; the most notable is *tra-1(e1488)*, which produces XX animals with wild-type male tail anatomy and mating behavior but hermaphrodite gonad morphology and a hermaphrodite germ line.

Although the primary role of *tra-1* is to promote hermaphrodite development in XX animals, *tra-1* also appears to play a minor role in XO males. Specifically, strong loss-of-function alleles can cause abnormal gonad development, reduced spermatogenesis, and some oogenesis in XO as well as XX males (Hodgkin, 1987c; Schedl *et al.*, 1989).

Dominant gain-of-function *tra-1* alleles have effects that are opposite to the effects of loss-of-function mutations (Hodgkin, 1980, 1987c). For some alleles, *tra-1(gf)/+* XO animals develop as fertile females that are completely feminized in both soma and germ line; for other alleles feminization of XO animals is incomplete, leading to intersexual development. All alleles also feminize the germ lines of *tra-1(gf)/+* XX animals [and *tra-1(gf)/Df* XX animals where tested]. Gene dosage experiments with the allele *tra-1(e1575gf)* suggest that these dominant feminizing mutations affect the sex-specific and germ line-specific regulation of *tra-1* (see below) rather than simply elevating the level of *tra-1* activity. Fine-structure genetic mapping with several *tra-1* alleles indicates that the dominant and recessive mutations map to two distinct regions of the *tra-1* gene, and that the gene itself is quite large (~0.2 map units, based on recombinational distances).

The fact that loss-of-function and gain-of-function *tra-1* mutations cause reciprocal transformations in sexual fate indicates that *tra-1* acts as a genetic switch (Hodgkin, 1983b). According to this model, the presence of *tra-1* activity specifies female development, while the absence of *tra-1* activity specifies male development. It is therefore possible to use *tra-1(lf)* and *tra-1(gf)* mutations to build male/female *C. elegans* strains in which sex is determined not by X-chromosome dosage but by the alleles present at the *tra-1* locus. The fact that such strains are comprised of males and females rather than males and hermaphrodites indicates that in wild-type *C. elegans* strains, additional regulation of *tra-1* in the hermaphrodite germ line must occur in order to permit spermatogenesis during the L4 larval stage.

Analysis of the intersexual phenotypes produced in over 200 XX animals that were genetically mosaic for *tra-1* has led to the conclusion that *tra-1* functions cell (or lineage) autonomously in the determination of sexual fates (C. Hunter and W. Wood, personal communication). That is, the presence or absence of the wild-type *tra-1* gene in a particular cell (or cell lineage) determines whether that cell will adopt the hermaphrodite or male fate. Specifically, cells adopting the male [*tra-1(−)*] fate were always clonally related in these experiments, indicating that loss of *tra-1(+)* in a given cell lineage affects those cells, and only those cells, that are derived from that lineage. [The only exception to this generalization is that induction of vulval development in cells of the ventral hypodermis depends on the presence of *tra-1(+)* in the lineally unrelated anchor cell of the gonad: this result is expected based on previous experiments reviewed by Sternberg (1990) in this volume.] That the presence (or absence) of *tra-1(+)* in a particular cell is critical (rather than in an ancestor cell at some earlier point in the cell lineage) has been shown directly only for the intestinal cell lineage, in which the sexually dimorphic fates of sister cells can be readily examined. Multiple examples of mosaic intestines have been identified in which a single cell fails to express yolk proteins (as assayed by *in situ* hybridization to yolk protein mRNA in partially dissected intestines), indicating that sister cells can express different differentiated sexual fates (C. Hunter and W. Wood, personal communication).

Hodgkin has recently identified a cosmid clone that detects DNA rearrangements in (at least) three independent *tra-1* mutant alleles (J. Hodgkin, personal communication). Interestingly, one of these is the atypical recessive allele *e1488* (see above), and two are complex spontaneous *tra-1(gf)* alleles.

2. tra-2

Recessive loss-of-function mutations in *tra-2* also cause masculinization of XX animals. Strong loss-of-function mutations (putative null alleles) transform XX animals into nonmating pseudomales. These *tra-2* XX pseudomales have male gonads in which extensive spermatogenesis occurs, male tail and copulatory structures, at least some male-specific muscles, and a partially developed male nervous system (Hodgkin and Brenner, 1977). Moreover, unlike some strong *tra-1(lf)* mutants, the *tra-2(0)* XX pseudomales never produce oocytes. The male tail structures are morphologically abnormal, however, and the animals never exhibit male mating behavior, presumably because of nervous system defects.

The absolute level of *tra-2* activity appears to be very important for proper *tra-2* function in promoting hermaphrodite development. The strongest indication of this is the fact that the *tra-2* locus is haplo-insufficient: XX animals with only a single dose of the wild-type gene [*tra2(0)/ +* or *Df/ +*] exhibit an incompletely penetrant egg-laying defective (Egl) phenotype resulting from the lack of a pair of hermaphrodite-specific motor neurons, the HSNs (Trent *et al.*, 1983; Desai and Horvitz, 1989). The HSNs normally die during embryogenesis in male embryos (Sulston *et al.*, 1983), and their absence in these XX animals is indicative of mild sexual transformation. A weak loss-of-function *tra-2* allele, *n1106*, was in fact identified based on an HSN⁻ Egl phenotype in homozygous *n1106/n1106* hermaphrodites (Desai and Horvitz, 1989).

Temperature-shift experiments using the allele *tra-2(b202ts)* were performed to determine the temperature-sensitive period(s) (TSPs) for *tra-2* in the development of a variety of sexually dimorphic features (Klass *et al.*, 1976). The TSP for gonad sexual phenotype determined in these experiments extends from mid-embryogenesis (~8 hours before hatching at 20°C) to the L1 larval molt; this period ends shortly after the gonad cell lineages diverge between males and hermaphrodites but long before significant morphogenesis of the gonad has taken place (Fig. 4). For male tail development, the TSP extends from mid-embryogenesis until midway through the L1 larval stage. This TSP ends after several male-specific blast cells have already taken on a characteristic male-specific appearance but before they have begun to divide, and also before the divergence of many other lineages giving rise to male tail structures. The TSP for the ability to produce oocytes extends from just before hatching until the L4 larval molt, ending several hours prior to the onset of oogenesis in the adult.

In addition to the more than 50 *tra-2* loss-of-function alleles, there are also 12 dominant gain-of-function *tra-2* alleles (Doniach, 1986a; Schedl and Kimble, 1988; J. Kimble, personal communication). Eleven of these have no obvious effects in XO animals but feminize the germ lines of XX animals, causing them to be spermless females. These

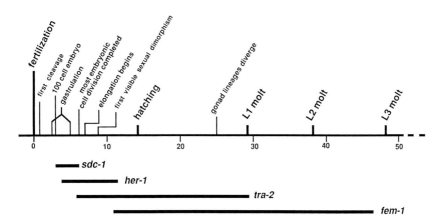

FIG. 4. Temperature-sensitive periods for somatic gonad sexual phenotype. The time line indicates various landmarks during the embryonic and larval development of *C. elegans*; the numbers below the line represent hours after fertilization at 20°C. The bars below the line compare the TSPs for sexual phenotype of the somatic gonad determined for the sex-determining genes *sdc-1*, *her-1*, *tra-2*, and *fem-1*. The temporal order of the TSPs for these genes is entirely consistent with the order of gene function inferred from genetic epistasis (see the text).

tra-2(gf) mutants appear to be defective in the transient down-regulation of *tra-2* activity that is required to permit spermatogenesis in the hermaphrodite germ line (Doniach, 1986a; a more complete discussion of germ line sex determination is deferred to Section VI). Six of these 11 alleles are also apparently partial loss-of-function mutations, since they cause partial masculinization of the hermaphrodite soma when placed *in trans* to a *tra-2* null allele. The twelfth *tra-2(gf)* allele, *e2020,* not only feminizes the germ line of XX animals but also causes partial feminization of both germ line and soma in some homozygous XO animals, suggesting that this allele partially overcomes the normal negative regulation of *tra-2* activity in XO animals (Doniach, 1986a).

The *tra-2* locus, which has been cloned by transposon tagging (P. Okkema and J. Kimble, personal communication), spans a region of at least 10 kb (defined by insertions and deletions in various mutant alleles). Probes spanning this region detect two putative *tra-2* transcripts in RNA from adult hermaphrodites. Six of the 12 *tra-2(gf)* mutations have DNA alterations near the 3′ end of the transcription unit, 2 of which are insertions of the transposon Tc1. Partial sequence of a *tra-2* cDNA shows that the 3′ untranslated region (3′ UTR) contains two identical copies of a 28-bp direct repeat separated by a 4-bp spacer; 3 of the *tra-2(gf)* mutations contain identical deletions of a single copy of the repeat plus the spacer, and a fourth, the exceptional allele *e2020,* contains a deletion of 108 bp that entirely removes both copies of the repeat as well as a point mutation in the coding region that does not alter the protein sequence (P. Okkema, P. Kuwabara, and J. Kimble, personal communication). These data suggest a regulatory role for the direct repeats in the 3′ UTR of *tra-2*; however, the level of regulation of *tra-2* activity in either germ line or soma is not yet known. Interestingly, the 6 other *tra-2(gf)* alleles are all of the class that cause a partial reduction in *tra-2* activity and all contain point mutations in the coding region (P. Okkema and J. Kimble, personal communication).

3. tra-3

The gene *tra-3* is also required for hermaphrodite sexual differentiation and is dispensable in XO males (Hodgkin and Brenner, 1977; Hodgkin, 1985). Like *tra-2,* loss-of-function mutations in *tra-3* transform XX animals into nonmating pseudomales. Unlike *tra-2,* however, *tra-3* mutations exhibit a strong maternal rescue such that homozygous *tra-3* progeny of heterozygous *tra-3/ +* mothers develop as hermaphrodites; these hermaphrodites produce broods of self-progeny comprised entirely of masculinized XX animals. Zygotic expression of *tra-3(+)* is also sufficient to provide wild-type function, since *tra-3/ +* cross-progeny of *tra-3/tra-3* mothers are wild-type hermaphrodites. The masculinization observed in *tra-3* XX animals is somewhat less complete than in *tra-2(0)* mutants. In particular, a small number of *tra-3* animals exhibit intersexual development and a few are self-fertile, especially at low temperatures.

Only four *tra-3* alleles have been identified to date, presumably due in part to the fact that the maternal rescue effect precludes their isolation in the F_2 generation. Three of the four alleles are amber mutations. These amber alleles can be completely suppressed by extremely weak amber-suppressor mutations, suggesting that the *tra-3* gene product may be required only in small quantities to promote hermaphrodite development (Hodgkin, 1985).

B. Hermaphroditization (*her*) Gene

The *her-1* gene, like *tra-1*, acts as a genetic switch. *her-1* is required for male but not hermaphrodite development (Hodgkin, 1980); nearly all of the more than 25 recessive loss-of-function mutations in *her-1* [including at least two putative null alleles (Trent *et al.*, 1988)] result in the complete transformation of XO animals into fertile, anatomically normal hermaphrodites. These *her-1* XO hermaphrodites are reduced in fertility compared to wild-type XX hermaphrodites, however, producing large numbers of inviable nullo-X zygotes. The unexpectedly high frequency of nullo-X gametes (~70%) produced by these animals indicates that an unpaired X chromosome sometimes tends to be lost during hermaphrodite meiosis; XO males reliably produce 50% nullo-X gametes, suggesting that the ability to properly segregate unpaired X chromosomes is a male-specific function (Hodgkin, 1980).

The temperature-sensitive period (TSP) for the role of *her-1* in determining somatic gonad sexual phenotype has been determined using the temperature-sensitive allele *her-1(e1561ts)* (Hodgkin, 1984). This TSP begins during gastrulation, extends through the period of embryonic elongation, and ends about 3 hours before hatching (at 20°C; Fig. 4). This period is complete more than 12 hours before the gonad becomes visibly dimorphic near the end of the first larval stage, indicating that the *her-1* gene product is made, and probably functions, long before overt sexual differentiation of this tissue takes place.

While *her-1(lf)* mutations cause feminization of XO animals, two semidominant gain-of-function *her-1* mutations cause variable masculinization of XX animals; XO animals are unaffected (Trent *et al.*, 1988; A. Villeneuve, B. Meyer, and S. Kim, unpublished). Homozygous *her-1(gf)* XX animals range in phenotype from Egl hermaphrodites that lack HSNs (a phenotype identical to the haplo-insufficient phenotype of the *tra 2* locus; see above), to self-fertile Egl intersexes with partially masculinized tails, to apparent pseudomales that have male gonads, male body shape, and morphologically abnormal male tails. In heterozygous animals, the penetrance of the mutant phenotypes is lower, but the range of phenotypes observed is similar. Is *her-1* activity both necessary *and* sufficient to elicit wild-type male sexual differentiation? If so, then full activation of *her-1* in XX animals should lead to their complete transformation into mating males. Clearly the *her-1(gf)* mutations currently in hand do not cause such a complete sexual transformation; however, both alleles were isolated because of their dominant egg-laying defective phenotypes (Trent *et al.*, 1983; S. Kim, personal communication), and the extent to which they activate *her-1* in XX animals is unknown. The answer to this question will most likely be resolved molecularly.

By mapping cDNAs in the region of a transposon insertion present in a spontaneous intragenic revertant of *her-1(n695gf)*, C. Trent and W. Wood (personal communication) have identified a putative *her-1* transcription unit. A portion of this transcribed region was subsequently shown to be deleted in DNA from a single *her-1(lf)* allele (C. Trent, personal communication). Two transcripts from this region, an abundant 0.9-kb transcript and a less abundant 1.3-kb transcript, are present in populations containing both XX and XO embryos, but they are absent or greatly reduced in pure XX populations. Moreover, two of four *her-1(lf)* alleles tested result in disappearance of the larger of the two transcripts from XX/XO populations. The fact that two of the mutants still have the transcripts is equally informative, since it argues that the transcripts are XO-specific and therefore likely to be products of the *her-1* gene itself, rather than male-specific and simply under control of *her-1*. Both transcripts are also present in *her-1(n695gf)* XX animals, consistent with the prediction that the masculinization caused by this mutation results from the inappropriate activation of the *her-1* gene in XX animals; the relative levels of the transcripts in *n695* XX animals versus wild-type XO animals have not yet been determined quantitatively.

C. FEMINIZATION *(fem)* GENES

Loss-of-function mutations at any of three independent loci, *fem-1*, *fem-2*, or *fem-3*, cause both XX and XO animals to develop as spermless (but cross-fertile) females. In these females, oogenesis begins at about the time that spermatogenesis would normally begin in hermaphrodites, implying that the germ cells as well as the XO soma have undergone a transformation in sexual fate. While the three genes are similar in that they are all required for male somatic sexual differentiation and spermatogenesis in both males and hermaphrodites, there are important differences among them that are described below.

1. fem-1

More than 25 *fem-1(lf)* mutations, including 8 putative null alleles, have been identified in a variety of screens. While null mutations transform both XX and XO animals into females (Doniach and Hodgkin, 1984), the XO females are greatly reduced in fertility, producing only a small number of gametes. *fem-1* mutations exhibit a strong maternal rescue such that complete feminization occurs only in homozygous *fem-1* progeny from homozygous *fem-1* mothers (m^-z^-; m = mother, z = zygote). *fem-1* XO progeny from *fem-1/+* mothers (m^+z^-) develop as incomplete males, having reduced male tails with stunted or missing rays and male-shaped gonads that contain some oocytes in addition to sperm. These phenotypes indicate that maternal *fem-1(+)* product is sufficient to direct some male development. *fem-1* XX progeny from *fem-1/+* mothers (m^+z^-) are usually females, but about 20% produce a limited amount of sperm and are therefore self-fertile hermaphrodites. Maternal *fem-1(+)* product is not required by the zygote for either spermatogenesis or male somatic development, however, since *fem-1/+* XX or XO cross-progeny from *fem-1* mothers (m^-z^+) develop as wild-type hermaphrodites or wild-type males, respectively.

A large number of *fem-1* alleles are temperature sensitive, producing a broad range of intersexual phenotypes in XO animals. The original allele *fem-1(hc17ts)* (Nelson *et al.*, 1978; previously called *isx-1*) feminizes only the gonad and germ line of XO animals, yielding intersexes with male tails, male musculature, male behavior, and hermaphroditic or abnormal intersexual gonads containing oocytes and no sperm. The temperature-sensitve period for gonad development determined for this allele extends from just before hatching until late in the third larval (L3) stage, suggesting that *fem-1* activity is required when sexual differentiation of the gonad actually occurs, that is, throughout the period when the male and hermaphrodite gonad lineages diverge and the basic morphology of the gonad develops (Fig. 4). In contrast, the TSP for the ability of XO animals to produce sperm is short, beginning late in L2 and ending just before the L3 molt, before spermatogenesis takes place. (Since oocyte production in the XO germ line was not assayed in this experiment, it is not known when *fem-1* is required to prevent oogenesis in the male.) The TSP for self-progeny production in XX animals (which is a measure of the number of sperm produced) is much longer, extending from shortly before hatching to mid-L4, ending just at the time that hermaphrodite spermatogenesis normally begins.

The *fem-1* gene has been cloned by chromosomal walking from linked DNA polymorphisms (A. Spence, personal communication), an enterprise that was greatly facilitated by an ongoing project aimed at constructing a physical map of overlapping cosmid clones spanning the entire *C. elegans* genome (Coulson *et al.*, 1986, 1988). DNA polymorphisms have been identified in two *fem-1* alleles, and the gene has been localized to a genomic region of 6 kb by means of DNA-mediated transformation rescue of a *fem-1(0)* mutant. A 2.5-kb RNA that is present in both males and hermaphrodites is transcribed from sequences located within this region. Experiments are underway to further delimit the location of the *fem-1* gene within this region and to determine the primary structure of the product it encodes.

2. fem-2

The *fem-2* locus is unique among the *fem* genes in that even the strongest alleles (11 alleles total) cause complete feminization of XO animals only at 25°C, and only when both mother and zygote are homozygous mutant (m⁻z⁻) (Hodgkin, 1986). At 25°C XO m⁻z⁻ *fem-2* animals develop as females, at 20°C as sterile intersexes, and at 15°C as incomplete males. The presence of a *fem-2(+)* allele in the mother (m⁺z⁻) is sufficient to direct male somatic development in XO animals at both 20 and 25°C, although the 25°C males fail to produce cross-progeny and the 20°C males are reduced in fertility, presumably due to reduced numbers of sperm. The fact that all alleles isolated to date show similar temperature sensitivity in XO animals suggests a more stringent requirement for *fem-2(+)* activity at high temperatures.

XX animals also exhibit a strong maternal rescue. While m⁻z⁻ *fem-2* XX animals are spermless females at all temperatures, all m⁺z⁻ *fem-2* XX animals are self-fertile hermaphrodites (although brood size, which is limited by the number of sperm produced, is somewhat reduced). Both XX and XO phenotypes are completely zygotically rescued; *fem-2/+* cross-progeny of *fem-2* mothers (m⁻z⁺) develop as wild-type hermaphrodites or wild-type males, respectively.

The TSP for self-progeny production in *fem-2* XX animals (for the weak allele *b245ts*) extends from mid-L1 until the L2 molt (Kimble *et al.*, 1984); this period is somewhat shorter than that for *fem-1*, but contained within it. A separate temperature-shift experiment with this allele suggests that, in XO animals, *fem-2(+)* is required starting in L2 and continuing through adulthood to prevent oogenesis in the XO germ line; no direct comparison can be made with *fem-1*, since different parameters were scored in the *fem-1* experiment. A strong *fem-2* allele could be used to determine the TSP(s) for various aspects of male somatic sexual differentiation, but no experiments of this type have been reported.

3. fem-3

The *fem-3* gene is represented by 35 loss-of-function and 9 gain-of-function alleles. Putative *fem-3* null mutations exhibit partial maternal rescue in XO animals, such that m^+z^- XO animals develop as intersexes and can occasionally be self-fertile, while m^-z^- XO animals are invariably completely female (Hodgkin, 1986). There is no maternal rescue of *fem-3* XX animals, which are female regardless of maternal genotype. Moreover, the *fem-3* locus is weakly haplo-insufficient in XX animals, causing 5–10% of *fem-3/+* XX animals to develop as females.

Unlike *fem-1* and *fem-2*, *fem-3(lf)* mutations also exhibit a maternal *absence* effect, implying a requirement for the maternal endowment of *fem-3(+)*, particularly for male gonad development (Hodgkin, 1986). Fully one-third of XO *fem-3/+*(m^-z^+) cross-progeny of *fem-3* mothers have either abnormal gonads or hermaphrodite gonads and partial vulval development, although all have apparently normal male tails. Additionally, the incidence of females is increased in m^-z^+ *fem-3/+* XX animals compared to *fem-3/+* XX self-progeny of heterozygous mothers.

The temperature-sensitive period for self-progeny production in XX animals was determined using the temperature-sensitive weak *fem-3* allele *e2006* (Hodgkin, 1986). This TSP is later and longer than those determined for *fem-1(hc17)* and *fem-2(b245)*, extending from mid-L2 through early adulthood, suggesting that *fem-3* activity is required continuously throughout the period of hermaphrodite spermatogenesis.

In contrast to *fem-3(lf)* mutations, the existing *fem-3(gf)* mutations result in a semidominant Mog (masculinization of germ line) phenotype: homozygous XX animals are somatically hermaphrodite, but the germ line produces a vast excess of sperm and no oocytes (XO animals are wild-type males) (Barton *et al.*, 1987). Gene dosage experiments indicate that these mutations do not simply result in an increase in *fem-3* activity, but rather must affect some aspect of the hermaphrodite germ line-specific regulation of *fem-3* (see Section VI). All of the *fem-3(gf)* alleles are temperature sensitive, with 100% of XX animals (for most alleles) displaying the Mog phenotype at 25°C and self-fertility at

15°C. The TSP for the ability of a *fem-3(gf)* mutation to prevent the switch to oogenesis extends from mid-L4, when spermatogenesis begins, to approximately the time of the onset of oogenesis in early adulthood in wild-type hermaphrodites. Interestingly, following a shift from permissive to restrictive temperature in late L4, some animals can successfully initiate oogenesis and then shift back to spermatogenesis, suggesting that the switch to oogenesis that normally occurs in wild-type hermaphrodites is not irreversible and must be actively maintained.

All *fem-3(gf)* mutations isolated to date masculinize only the germ line and not the soma of XX animals. These mutations, however, were all selected as suppressors that restore self-fertility to *fem-1(hc17ts)* or *fem-2(b245ts)* XX animals at the restrictive temperature (Barton *et al.*, 1987). The fact that all the *fem-3(gf)* alleles are germ line specific and temperature sensitive might in fact be due to the specific methods used to select them.

fem-3 has been cloned by transposon tagging (Rosenquist and Kimble, 1988). Six *fem-3* mutant alleles contain insertions of the transposon Tc1, and three other *fem-3(lf)* mutations contain DNA rearrangements within a 4.5-kb region. At least three different mRNA species are produced by the *fem-3* transcription unit, one of which is present only in embryos and one of which is present only in L4 larvae and adults. While no XO-specific transcripts have been detected at any stage in development, *fem-3* RNA is approximately 6-fold more abundant in XO than in XX embryos. This difference in embryonic *fem-3* message levels must be achieved by posttranscriptional regulation of maternal message, presumably at the level of message stability, since experiments using mutants that produce transcripts of altered mobility demonstrate that the zygotic *fem-3* genes do not become activated until after the end of embryonic development (T. Rosenquist and J. Kimble, personal communication). Mutants that lack a germ line have been used to demonstrate that *fem-3* RNA is restricted to the germ line in adult XX hermaphrodites but is present in both germ line and somatic tissues in adult XO males (Rosenquist and Kimble, 1988).

Sequencing of *fem-3* cDNAs did not reveal any sequence similarity to genes in standard databases, nor any striking structural motifs (T. Rosenquist and J. Kimble, personal communication). The complete or partial sequences of the *fem-3(gf)* alleles reveal that 8/9 contain a single base-pair substitution in the 3' untranslated region of the gene, whereas the one other allele contains a deletion of 114 bp centered around the region where the point mutations occur (J. Ahringer and J. Kimble, personal communication). These data suggest that this part of the message is important for the negative regulation of *fem-3* which permits the switch from spermatogenesis to oogenesis in the hermaphrodite germ line. This type of regulation probably does not occur at the level of message stability, however, since similar amounts of message are detected in wild-type and *fem-3(gf)* XX adults (J. Ahringer and J. Kimble, personal communication).

D. OTHER GENES

Up to this point, we have considered only those genes for which loss of function is known to have a profound effect on the sex determination decision (i.e., complete or near complete reversal of sexual fate). Mutations in several genes that appear to play less central roles in the sex determination process have been isolated on the basis of more subtle phenotypes. For example, mutations in the gene *egl-41 V* (egg-laying defective) cause an egg-laying defect due to the death of the hermaphrodite-specific HSN neurons (Doniach, 1986b; Desai and Horvitz, 1989). Closer examination of *egl-41* hermaphrodites, particularly at 15°C, reveals further subtle masculinization, including survival of the male-specific CEM neurons, which normally die in hermaphrodites, and enlargement or division of several male-specific blast cells. Additionally, a few cells that normally divide only in the hermaphrodite have been observed to divide in *egl-41* XO males, possibly indicative of a slight feminization. Three of the four *egl-41* alleles were isolated on the basis of the HSN⁻ Egl phenotype, and one allele was isolated as a suppressor of a weak dominant feminizing *tra-2(gf)* allele. Further evidence that *egl-41* mutations affect the global sex determination process comes from the fact that these mutations enhance the masculinization caused by weakly masculinizing mutations in other sex determination genes (Doniach, 1986b; Villeneuve and Meyer, 1987; Desai and Horvitz, 1989). Because all alleles are semidominant and it is not known how the mutations affect gene function, however, the wild-type role of *egl-41(+)* in sex determination is unclear.

Another class of mutations that have minor effects on sex determination (six alleles defining four autosomal genes) have been isolated as suppressors of the incompletely penetrant masculinization caused by *her-1(n695sd)* (J. Plenefisch and B. Meyer, unpublished; J. Manser, C. Trent, and W. Wood, personal communication). While feminizing effects of these mutations can be observed in combination with other weak sex-transforming mutations, mutations in three of the four genes have no obvious phenotypes alone in either XX or XO animals; a mutation in the fourth has no phenotype in XX animals, and only a small fraction of XO males show slight feminization of copulatory structures. Again, since it is not known how these mutations affect gene function, the wild-type roles of these genes in sex determination cannot be deduced from the available data.

V. A Pathway for Sex Determination in Somatic Tissues: Genetic and Molecular Epistasis among the Autosomal Sex-Determining Genes

How do the genes described in Section IV collectively act to govern the choice of somatic sexual fate dictated by the primary sex-determining signal, the X/A ratio? A hierarchy of epistasis among these genes has been deduced from the somatic sexual phenotypes of double mutants containing mutations that produce opposite sexual transformations. (Genetic interactions for germ line sex determination are discussed in Section VI.) For example, a double mutant containing a *tra-1* (masculinizing) mutation and a *her-1* (feminizing) mutation develops as a male irrespective of the X/A signal; thus *tra-1* is said to be epistatic to *her-1*. This genetic formalism for the relationship between these genes is depicted as *her-1* < *tra-1*. Compilation of the data from a large number of double mutants yields the following order: *her-1* < *tra-2, tra-3* < *fem-1, fem-2, fem-3* < *tra-1* (Hodgkin, 1980, 1986; Doniach and Hodgkin, 1984).

This analysis forms the basis for the regulatory pathway of gene action proposed by Hodgkin (1984, 1987a, 1987b; also, Hodgkin *et al.*, 1985) and depicted in Fig. 5. The epistasis among these sex-determining genes has been interpreted in terms of a series of negative regulatory interactions between genes that control the male fate (*her-1* and the *fem* genes) and genes that control the hermaphrodite fate (the *tra* genes). The logic for interpreting the genetic interactions in terms of negative regulation is based on several considerations. For somatic sex determination, the *tra* genes are required mainly or exclusively in XX animals to promote hermaphrodite somatic development, while the *her* and *fem* genes appear to be required exclusively in XO animals to promote male somatic development. That is, genes with opposite loss-of-function mutant phenotypes appear to carry out their functions in mutually exclusive places (XX versus XO animals). Moreover, it bears restating that mutations in these two classes of genes produce clearly opposite outcomes, male versus hermaphrodite sexual differentiation. Neither of these two outcomes can reasonably be considered a precursor for the other, and therefore interpreting the data in terms of a biosynthetic pathway or a pathway of morphogenesis seems inappropri-

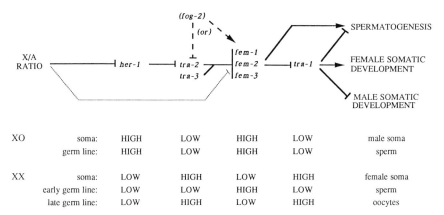

XO	soma:	HIGH	LOW	HIGH	LOW	male soma
	germ line:	HIGH	LOW	HIGH	LOW	sperm
XX	soma:	LOW	HIGH	LOW	HIGH	female soma
	early germ line:	LOW	LOW	HIGH	LOW	sperm
	late germ line:	LOW	HIGH	LOW	HIGH	oocytes

FIG. 5. Modified version of the regulatory pathway proposed by Hodgkin (1987a) for genes controlling sex determination but not dosage compensation. Negative regulatory interactions are indicated by bars, whereas positive regulatory interactions are indicated by arrows. Shown below the pathway are the proposed activity states of the regulatory genes in the soma and germ line of XO and XX animals. The complete rationale for this model is discussed in the text (Sections V and VI).

ate. In further support of a regulatory interpretation is the fact that several of the genes in the pathway are defined by both recessive loss-of-function alleles and dominant gain-of-function alleles that produce opposite sexual transformations; these fulfill the classic criteria for developmental "switch" genes that are involved in regulating a choice between alternative developmental fates (e.g., Lewis, 1978; Cline, 1978, 1979a, 1984; Greenwald *et al.*, 1983; Anderson *et al.*, 1985; reviewed in Hodgkin, 1984).

According to the model, the cascade of negative regulatory gene interactions ultimately controls the activity state of the gene *tra-1:* when X/A is high, *tra-1* is active and hermaphrodite development ensues; when X/A is low, *tra-1* is inactive and male development ensues. In this model, the term "regulation" is used in the broadest possible sense; no specific mechanisms are implied. Possible regulatory mechanisms might include, for example, regulation at the level of transcription, RNA processing, message stability, translation, covalent modification of a gene product, or protein stability. Additionally some gene products might not be regulated at all in the traditional sense of the word; that is, the gene product might be present in both sexes, but active only in combination with the products of other, sex-specific genes. [An experiment showing that *tra-3(+)* activity can be supplied by the germ lines of *fem-1(lf)* or *tra-1(gf)/ +* XO females suggests that this may be the case for *tra-3* (J. Hodgkin, personal communication).] It is important to point out that this pathway may also include positive regulatory steps. For example, where several genes have been assigned to the same step in the regulatory hierarchy, one of the genes may act as a positive regulator of the other gene(s). The fact that mutations in the genes result in similar phenotypes, however, precludes the possibility of deducing an order of gene function from the phenotypes of double mutants.

In the model of Hodgkin, an additional influence of the X/A ratio on the *fem* genes and/or *tra-1* is postulated (indicated by a dotted line in Fig. 5). This is based in part on the fact that while *tra-2* XX animals are pseudomales with abnormal tails and an incompletely male nervous system, *tra-1* XX animals (and *tra-2;tra-1* XX animals) have wild-type male tail morphology and can mate, suggesting that there may be residual *tra-1(+)* activity in *tra-2* XX animals. Moreover, while *tra-2;her-1* XX animals exhibit the abnormal male phenotype of *tra-2* XX pseudomales, *tra-2;her-1* XO males have complete male tails, and many are capable of mating and siring cross-progeny, suggesting that the lower X/A signal in XO animals can be perceived even in the absence of *her-1* activity (Hodgkin, 1980).

The temporal order of gene action inferred from the TSPs of several of the genes in the pathway is entirely consistent with their order in the proposed regulatory hierarchy. Figure 4 summarizes the results of independent temperature-shift experiments in which the TSPs for somatic gonad sexual phenotype were determined for *ts* alleles of *her-1*, *tra-2,* and *fem-1* (see Section IV). (Although other aspects of sexual differentiation were examined in most of the experiments, variations among assays thwart attempts to make direct comparisons.) In the gonad, the TSP for *her-1* precedes that for *tra-2*, and similarly the TSP for *tra-2* precedes that for *fem-1*. Note that the TSP for *her-1* is complete by about 14 hours prior to the first visible indication of sexual dimorphism in gonad development, while *fem-1* seems to be required during the time when sexual differentiation of the gonad is actually taking place.

The recent availability of molecular probes for several of the sex-determining genes provides a direct means for testing some of the predictions made by the Hodgkin model. For example, the gene *her-1* appears to be regulated at the transcript level; putative *her-1* transcripts at present in XO males and absent or reduced in XX hermaphrodites (C. Trent and W. Wood, personal communication). These putative *her-1* transcripts are also absent in *tra-1* XX males, as predicted by the placement of *tra-1* downstream of *her-1* in the sex determination pathway (C. Trent and W. Wood, personal communication). The gene *fem-3* is also regulated at the mRNA level (at least some of the regulation occurs at the level of message stability); *fem-3* RNA is present in the somatic tissues of adult XO males but not adult XX hermaphrodites (Rosenquist and Kimble, 1988). Preliminary experiments suggest that *fem-3* RNA is present in somatic tissues of *tra-3* (m$^-$z$^-$) XX pseudomales (T. Rosenquist and J. Kimble, personal communication). In other words, the absence of *tra-3* activity in XX animals apparently leads to the induction of *fem-3* activity, and thus male development, as predicted if *tra-3* is a negative regulator of *fem-3*.

These few examples clearly vindicate the use of a genetic approach to study a developmental pathway, and they illustrate the utility of a genetic model in suggesting experiments that will be informative about the molecular mechanisms involved in controlling sexual fate. Further molecular analysis will extend our understanding of the gene interactions involved in the sex determination decision beyond the limits of resolution of classic genetic analysis. In particular, the relationship between genes that have been assigned to the same step in the hierarchy (i.e., *tra-2* and *tra-3* or *fem-1, fem-2,* and *fem-3*), which cannot be deduced from epistatic relationships since mutations in them result in similar phenotypes, can be resolved using a combined molecular and genetic approach.

Such a combined approach has been exploited elegantly to show that the genes controlling somatic sexual differentiation in *Drosophila* act in a linear pathway (McKeown *et al.*, 1988). Previous genetic experiments had placed the genes *transformer (tra)* and *transformer-2 (tra-2)* (required for female somatic sexual differentiation) at the same step in the regulatory hierarchy, downstream of *Sex-lethal (Sxl)*, which controls both sex determination and dosage compensation, and upstream of *doublesex (dsx)*, a gene that can be expressed in either of two alternative modes to elicit either male or female development (Cline, 1979a, 1983; Baker and Ridge, 1980). Many aspects of this model were confirmed molecularly by examining the expression of *tra* and *dsx* [both of which produce sex-specific transcripts (Butler *et al.*, 1986; Baker *et al.*, 1987; McKeown *et al.*, 1987)] in various mutant backgrounds (Nagoshi *et al.*, 1988). In their experiments, McKeown *et al.* (1988) constructed a gain-of-function *tra* allele by expressing the female-specific *tra* product under the control of a non-sex-specific heterologous promoter. This construct was used to show that (1) expression of the *tra* female-specific product is sufficient to induce female development in an XY soma, and

(2) the ability of the *tra* female-specific product to elicit female development is dependent on a wild-type *tra-2* gene. These data indicate either that *tra-2* acts downstream of *tra* in the regulatory hierarchy or that *tra-2* is not regulated in a sex-specific manner but is active only in conjunction with active *tra* product.

VI. Germ Line Sex Determination

A. Control of the Sperm/Oocyte Decision

All of the genes described in Sections IV and V are involved in the control of sex determination in both somatic tissues and the germ line. When only the germ line roles of these genes (i.e., the specification of either spermatogenesis or oogenesis) are considered, the hierarchy of epistasis which emerges from the analysis of double mutants is different from that for the soma. The order in the germ line is *her-1* < *tra-2*, *tra-3* < *tra-1* <*fem-1, fem-2, fem-3*. That is, *fem;tra-1* double mutants (either XX or XO) have male somatic tissues but produce only oocytes and no sperm in a male-shaped gonad. Hodgkin *et al.* (1985) suggest that this indicates a dual role for the *fem* genes: (1) to promote male somatic development by preventing the activity of *tra-1* in XO somatic tissues and (2) to promote spermatogenesis in both males and hermaphrodites independently of *tra-1*. This germ line-specific role for the *fem* genes is indicated by an arrow in Fig. 5.

Unlike the *fem* genes, the gene *fog-1* (feminization of germ line) appears to be involved in controlling the choice of sexual fate only in the germ line (Doniach, 1986b; M. K. Barton and J. Kimble, personal communication). *fog-1* mutations prevent spermatogenesis in both sexes, causing germ cells that would normally develop into sperm to instead become oocytes. XX animals homozygous for a *fog-1* mutation develop as females, while XO animals are somatically completely male but produce only oocytes in the germ line. The fact that *fog-1* mutations affect only the germ line suggests that *fog-1* may act downstream of the *fem* genes in controlling the sperm/oocyte decision. While *fog-1* alleles arise at a frequency typical for loss-of-function mutations, they all show a dominant phenotype: *fog-1/+* males produce only a limited number of sperm and then switch to oogenesis (M. K. Barton and J. Kimble, personal communication). This dominant phenotype cannot be explained by haplo-insufficiency since *Df/+* males have wild-type germ lines, and thus the *fog-1* alleles do not appear to be simple loss-of-function mutations. Further genetic analysis is therefore required to determine the wild-type role of *fog-1(+)*.

B. HERMAPHRODITE-SPECIFIC CONTROL OF SPERMATOGENESIS

As mentioned previously, the *C. elegans* hermaphrodite may be considered to be a somatic female with a germ line that switches from an initially male program (spermatogenesis) to a female program (oogenesis) of gametogenesis. How is the male program initiated within the context of the female soma, and how does the switch from the male to the female program take place? First, the phenotype of most *tra-2(gf)* mutations (XX = female, XO = male) suggests that the activity of the hermaphrodite-promoting gene *tra-2* is transiently down-modulated in the wild-type hermaphrodite germ line in order to permit spermatogenesis (Doniach, 1986a). This down-modulation of *tra-2* activity allows a transient activation of the *fem* genes, and the germ line-specific sex determination gene *fog-1*, which specify spermatogenesis. The *tra-2(gf)* mutations presumably render the *tra-2* gene insensitive to down-modulation, and as a consequence only oogenesis can occur. Sequences in the 3′ untranslated region of the *tra-2* gene have been implicated in this germ line-specific regulation (P. Okkema, P. Kuwabara, and J. Kimble, personal communication; see Section IV). Second, the Mog (masculinization of germ line) phenotype in XX animals caused by *fem-3(gf)* mutations suggests that in wild-type XX animals *fem-3* activity must be negatively regulated following hermaphrodite spermatogenesis to allow the switch to oogenesis. Interesting, sequences important for this aspect of *fem-3* regulation have similarly been localized to the 3′ untranslated region of the gene (J. Ahringer and J. Kimble, personal communication; see Section IV).

One gene, *fog-2*, is specifically required for permitting the transient spermatogenesis in hermaphrodites (Schedl and Kimble, 1988). Loss-of-function *fog-2* mutations produce a phenotype identical to that caused by most *tra-2(gf)* mutations, transforming XX animals into females but leaving XO males unaffected. Schedl and Kimble (1988) have proposed that the wild-type *fog-2* activity promotes hermaphrodite spermatogenesis by allowing the transient activation of the *fem* genes and *fog-1* in the hermaphrodite germ line. This positive regulation might be achieved either by direct regulation of the *fem* genes or gene products or indirectly via down-regulation of *tra-2* activity (Fig. 5).

VII. In Pursuit of Downstream Differentiation Functions

Once the functional state of *tra-1* is established in response to the X/A ratio, how does the presence or absence of *tra-1* activity elicit the hermaphrodite- or male-specific programs of sexual differentiation in different tissues and cells in the organism? Clearly, the execution of developmental programs for generating sex-specific structures will involve the function of many genes that act in both sexes [e.g., *lin-12* (Greenwald *et al.*, 1983), *lin-17* (Sternberg and Horvitz, 1988), *sem-4* (M. Stern, personal communication)] as well as genes whose activities may be required primarily, if not exclusively, in a single sex [e.g., *lin-10* (Ferguson and Horvitz, 1985; S. Kim, personal communication) and *egl-1* (Trent *et al.*, 1983; Desai and Horvitz, 1989) in the hermaphrodite; *mab-3* (Shen and Hodgkin, 1988) in the male]. Whether any specific gene or gene product represents a direct and immediate target of *tra-1* regulation is a question not readily resolved by genetic means.

Two major complementary approaches should be instrumental in elucidating the mechanisms by which the choice of sexual fate specified by *tra-1* (and the *fem* genes) results in the final pattern of sex-specific differentiation. Since the molecular characterization of the *tra-1*, *fem-1*, and *fem-3* genes is currently underway, an obvious approach is to look for clues about the biochemical functions of these genes based on the analysis of primary sequence and/or subcellular localization of the gene products. Further, potential targets may be identified biochemically on the basis of physical interactions.

A second strategy is to study the regulation of sex-specific gene products. In *C. elegans,* one manifestation of sexual differentiation is the transcriptional activation of genes encoding two classes of extremely abundant sex-specific proteins. The vitellogenins, or yolk proteins (encoded by the *vit* genes), are expressed exclusively in the intestines of adult hermaphrodites (Kimble and Sharrock, 1983; Blumenthal *et al.*, 1984), while the major sperm proteins, or MSPs (encoded by the *msp* genes, a large multigene family), are expressed exclusively in the male germ line and in the L4 hermaphrodite germ line during spermatogenesis (Klass and Hirsh, 1981; Klass *et al.*, 1982; Ward and Klass, 1982). The *vit* genes and many of the *msp* genes have been cloned (Burke and Ward, 1983; Blumenthal *et al.*, 1984; Klass *et al.*, 1984; Spieth and Blumenthal, 1985); experiments aimed at identifying cis-acting elements, and ultimately trans-acting factors, involved in the sex-specific regulation of the *vit* genes (Spieth *et al.*, 1985, 1988; T. Blumenthal, personal communication) and the spermatogenesis-specific regulation of the *msp* genes (Klass *et al.*, 1988; Ward *et al.*,

1988) are underway. The hope is that these two approaches will eventually converge to link the specification of sexual fate with the execution of sex-specific developmental programs.

One candidate for a gene that may act at an intermediate step between *tra-1* and *vit* gene expression is *mab-3* (male abnormal). Loss-of-function mutations in *mab-3* affect a small subset of sexually dimorphic functions (Shen and Hodgkin, 1988). Specifically, they result in (1) expression of yolk proteins in the male intestine, (2) alterations in the cell lineages of several ray precursor cells that give rise to the sensory rays of the male tail, and (3) abnormal morphogenesis of the male tail. *mab-3* males appear otherwise normal in all respects, and hermaphrodites are apparently completely unaffected. The fact that the ray lineages in *mab-3* are abnormal rather than transformed to the hermaphrodite fate suggests that *mab-3(+)* is probably involved in the execution of male-specific differentiation programs in these tissues rather than in specifying their sexual fates.

VIII. X-Chromosome Dosage Compensation

A. DEMONSTRATION OF DOSAGE COMPENSATION IN
Caenorhabditis elegans

In organisms where the relative number and/or type of sex chromosomes is the primary determinant of sexual fate, males and females (or hermaphrodites in the case of *C. elegans*) differ in the dosage of sex-linked genes. While some organisms such as birds or butterflies appear to tolerate this discrepancy in gene copy number with no attempt to equalize the total amount of sex-linked gene expression (Johnson and Turner, 1979; Baverstock *et al.*, 1982), many other organisms possess mechanisms to compensate for the difference in sex-linked gene dosage. This phenomenon of dosage compensation has been studied extensively in both *Drosophila* and mammals, which have evolved radically different mechanisms to solve the problem (see below). *Caenorhabditis elegans* likewise compensates for the difference in sex-linked gene dosage between males (XO) and hermaphrodites (XX), by equalizing X-specific transcript levels between the sexes (Meyer and Casson, 1986: Donahue *et al.*, 1987).

For a rigorous demonstration of dosage compensation for a particular gene, it is not sufficient to show simply that its expression is equal between XX (hermaphrodite) and XO (male) animals. It is also critical to show that the amount of expression varies with gene copy number in animals of a particular sex; for example, XX animals heterozygous for a deletion of the assayed gene should have one-half the expression of wild-type XX animals. This controls for the possibility that the level of expression of the gene may be strictly controlled via gene-specific feedback mechanisms. Finally, it is important to control for anatomical differences between the sexes, either by examining genes whose expression is limited to a tissue that is anatomically invariant between the two sexes or by using mutations that cause sexual transformations to examine expression in XX and XO animals of the same phenotypic sex. Using these criteria, Meyer and Casson (1986) employed a Northern hybridization assay to provide a direct demonstration of dosage compensation at the level of RNA transcript accumulation for three X-linked genes (uvt-4, uxt-1, and myo-2). Similar results for one of these X-linked transcripts, myo-2 X, were obtained using a somewhat different assay (Donahue et al., 1987).

A genetic approach first conceived by Muller (1950) to demonstrate the phenomenon of dosage compensation in Drosophila has also been used to demonstrate dosage compensation for several C. elegans genes (let-2 and lin-15, Meneely and Wood, 1987; lin-14, DeLong et al., 1987). In this method, the severity of the mutant phenotype produced by an X-linked hypomorphic mutation (which reduces but does not eliminate gene function) is used as an indirect measure of gene activity; if dosage compensation is operating, then the severity of the phenotype in $m/0$ XO animals should be comparable to that in m/m XX animals and less severe than in m/Df XX animals, which carry only a single copy of the gene. The lin-15 and lin-14 hypomorphic mutations can also be used in genetic assays to detect alterations in X-linked gene expression in dosage compensation mutants (DeLong et al., 1987; Meneely and Wood, 1987; see below). In such genetic assays, an increase in X-linked gene expression is reflected by suppression of the mutant phenotype, while a decrease in expression is reflected by enhancement of the mutant phenotype.

While many (perhaps most) X-linked genes in C. elegans are dosage compensated, it is important to point out that not all X-linked genes exhibit compensation. In addition to the three compensated transcripts analyzed, Meyer and Casson (1986) also identified one noncompensated X-linked transcript, uxt-2. Further, genetic experiments suggest that the X-linked amber-suppressor tRNA genes sup-7 and sup-21 (which are presumably transcribed by RNA polymerase III) are not dosage compensated (Hodgkin, 1985).

B. Genes That Implement Dosage Compensation

Five autosomal genes, *dpy-21, dpy-26, dpy-27, dpy-28,* and *dpy-29* (dumpy = short and fat), are known to be required for the proper control of X-chromosome expression in XX animals. (Two X-linked genes, *sdc-1* and *sdc-2,* are important for both sex determination *and* dosage compensation in XX animals, and they are discussed in detail in Section IX,A.) While mutations in all five of these *dpy* genes have been shown to cause overexpression of X-linked genes in XX animals, their involvement in the dosage compensation process was initially inferred from the fact that mutations in the genes result in XX-specific mutant phenotypes (Hodgkin, 1983a; Plenefisch *et al.,* 1989).

Recessive loss-of-function mutations in *dpy-21* (including three amber alleles) result in a Dpy, egg-laying defective (Egl), slightly slow growing phenotype in XX hermaphrodites and a low level of lethality (Hodgkin, 1983a; Plenefisch *et al.,* 1989; L. DeLong, personal communication). *dpy-21* XO males, in contrast, are wild type in length and mature at a normal rate. The Dpy phenotype of *dpy-21* XX animals closely resembles that of triplo-X animals (3X/2A), suggesting that the consequence of loss of *dpy-21* activity is similar to increasing the dose of the entire X chromosome. In keeping with this hypothesis, *dpy-21* 3X/2A animals are apparently inviable, as are *dpy-21(+)* 4X/2A animals (Hodgkin, 1983a).

Loss-of-function mutations in the other four dosage compensation *dpy* genes, *dpy-26, dpy-27, dpy-28,* and *dpy-29,* result in an incompletely penetrant XX-specific L1 larval or embryonic lethality (Hodgkin, 1983a; Plenefisch *et al.,* 1989). The larval arrest phenotype is distinctive and quite similar for all mutant alleles of the four genes, and the percentage lethality at 20°C for the various alleles ranges from 80 to 97% (Plenefisch *et al.,* 1989). XX hermaphrodites that escape this lethality mature very slowly into fertile, extremely Dpy Egl adults; XO animals are fully viable, essentially wild-type males. The extent of the lethality in XX animals homozygous for strong alleles of *dpy-26, dpy-27,* or *dpy-29* is similar to that in animals carrying these mutations *in trans* to a deficiency (Hodgkin, 1983a; Plenefisch *et al.,* 1989), suggesting that the strongest alleles of these genes (which include amber alleles for *dpy-27* and *dpy-29*) may result in a complete loss of function. (No deficiencies are available for the *dpy-28* locus.) Mutations in all four genes exhibit strong maternal rescue, such that while most m^-z^- XX animals, die, all m^+z^- XX animals live. The Dpy phenotype of *dpy-26* and *dpy-28* XX animals, but not *dpy-27* and *dpy-29* XX animals is also rescued by the presence of a wild-type gene in the mother; m^+z^- dpy-26 and dpy-28 XX hermaphrodites are only slightly Dpy and/or Egl and are not reliably distinguishable from wild-type, while m^+z^- dpy-27 and dpy-29 XX hermaphrodites are Dpy. These are not strict maternal-effect mutations, however, since dpy/ + XX cross-progeny of homozygous mutant mothers (m^-z^+) are completely wild type. Curiously, in addition to the XX-specific lethal and Dpy phenotypes, all mutant alleles of dpy-26 and dpy-28 also cause an increased frequency of X-chromosome meiotic nondisjunction, leading to a high incidence of XO males (Him phenotype) among the self-progeny of XX animals.

For all five dosage compensation dpy genes, the phenotypes described above have been shown to be truly XX-specific rather than hermaphrodite-specific (Hodgkin, 1983a; Plenefisch et al., 1989). dpy XO animals transformed into hermaphrodites by a mutation in the sex-determining gene her-1 are wild type in length and fully viable, whereas her-1;dpy XX animals are Dpy and Egl, or dead. Likewise, while dpy XO males are wild type in length and fully viable, tra-1;dpy XX pseudomales are either Dpy, with abnormal male tail morphology, or dead.

The XX-specific Dpy and lethal phenotypes caused by mutations in the maternal-effect dpy genes have been correlated with a disruption of the dosage compensation process. Mutations in dpy-27 and dpy-28 have been shown to cause a 2- to 3-fold elevation of X-specific RNA transcript levels (normalized to autosomal transcripts) specifically in XX animals. By contrast, X-specific transcript levels are normal in dpy-27 and dpy-28 XO animals, consistent with their wild-type phenotype (Meyer and Casson, 1986; L. Casson and B. Meyer, unpublished). Genetic assays corroborate these results for dpy-27 and dpy-28 (DeLong et al., 1987) and further indicate that X-linked gene expression is also elevated in dpy-26 and dpy-29 XX animals but not dpy-29 XO animals (DeLong et al., 1987; Meneely and Wood, 1987; Plenefisch et al., 1989). (The majority of the data suggest that X expression is normal or nearly normal in dpy-26 XO animals as well, although the results from different assays were somewhat equivocal.) Thus, the maternal-effect XX-specific dpy genes appear to be required specifically in XX animals for appropriate levels of X-linked gene expression.

Despite the fact that the dpy-21 Dpy and Egl phenotypes are XX specific, the dpy-21 gene appears to be required for proper X-chromosome expression in both XX and XO animals. Assays of X-specific transcript levels as well as two independent genetic assays indicate an elevation in levels of X-linked gene expression in dpy-21 XX animals similar to that seen in the maternal-effect dpy mutants

(Meyer and Casson, 1986; DeLong *et al.*, 1987; Donahue *et al.*, 1987; Meneely and Wood, 1987). The effects of *dpy-21* mutations on X-chromosome expression in XO animals are less clear, however, since different assays give different results. Measurement of RNA transcript levels and the *lin-15* genetic assay both suggest that *dpy-21* mutations increase X-linked gene expression in XO animals, albeit to a lesser extent than in XX animals (Meyer and Casson, 1986; Meneely and Wood, 1987), while the *lin-14* genetic assay strongly suggests that *dpy-21* mutations decrease *lin-14* expression in XO animals (DeLong *et al.*, 1987). The *lin-14* assay probably reflects the state of gene expression during or prior to the L1 larval stage (Ambros and Horvitz, 1987; G. Ruvkun, personal communication), whereas the *lin-15* assay may reflect expression in later larval stages (Ferguson, 1985); the Northern hybridization assay measures transcript levels in adults. Perhaps the seemingly contradictory results obtained for *dpy-21* XO animals are due to differences in the developmental stages examined using the different assays.

How do the XX-specific *dpy* genes collectively act to implement the dosage compensation process in *C. elegans*? Because mutations in these genes all result in such similar phenotypes, it is not possible to derive an order of gene function from the phenotypes of double mutants. The fact that the penetrance of the XX-specific lethality in XX animals homozygous for mutations in two or more of the *dpy* genes is no greater than the penetrance in any of the single mutants suggests, however, that the *dpy* genes act together in a common process or a single biochemical pathway (Plenefisch *et al.*, 1989). (If they acted in different processes, the lethal effects of two mutations might be expected to be additive.) Although they appear to act in a common process, the *dpy* genes are clearly not functionally equivalent to each other. First, a mutation at any one dosage compensation *dpy* locus fully complements mutations at all other *dpy* loci. Second, additional wild-type copies of one or more *dpy* genes cannot substitute for the loss of a different *dpy* gene (Plenefisch *et al.*, 1989).

While it has been suggested that two additional genes, *dpy-22 X* and *dpy-23 X*, might play a role opposite to that of the XX-specific *dpy* genes in the dosage compensation process, the evidence is not compelling. Mutations in the genes were initially thought to have more severe effects in males than in hermaphrodites, although both males and hermaphrodites are phenotypically Dpy and sickly (Hodgkin, 1974; Hodgkin and Brenner, 1977). Subsequent experiments with a backcrossed *dpy-22* strain showed no bias in viability between *dpy-22* XX hermaphrodites and *dpy-22* XO males, however (DeLong *et al.*, 1987), and *tra-1;dpy-22* XX males appear as sickly and Dpy as *dpy-22* XO males (J. Plenefisch and B. Meyer, unpublished). The *dpy-22* and *dpy-23* mutations do apparently enhance the phenotypes of X-linked hypomorphic mutations in both males and hermaphrodites (DeLong *et al.*, 1987; Meneely and Wood, 1987), as expected for mutations that reduce X-linked gene expression, but control experiments suggest that, particularly for *dpy-23*, this effect may be due to nonspecific consequences of the general ill health of these animals or to effects on autosomal gene expression as well. Further, the *dpy-22* mutation enhances the lethality caused by *dpy-26, dpy-27*, or *dpy-28* mutations in XX animals (Plenefisch *et al.*, 1989), a finding unexpected for mutations that play opposite roles in dosage compensation. All these considerations, together with the fact that *dpy-22* and *dpy-23* are each represented by only a single mutant allele (and it is not known how these mutations affect gene function), make it difficult to draw conclusions regarding the involvement of the genes in the dosage compensation process. For this reason, these genes are not discussed further in this article.

C. Timing of Dosage Compensation

Several lines of evidence suggest that the dosage compensation mechanism in *C. elegans* is functional by mid-embryogenesis. First, the fact that *lin-14* is dosage compensated (DeLong *et al.*, 1987) argues that the dosage compensation mechanism must be operative by the L1 larval stage, since both the TSP for the *lin-14* phenotype assayed and the time of maximum *lin-14* message accumulation occur during L1 (Ambros and Horvitz, 1987; G. Ruvkun, personal communication). Second, the TSP for the XX-specific lethality caused by *dpy-28(y1ts)* extends from approximately 5 to 9 hours after fertilization (at 20°C) and is centered around the time that embryonic cell division ceases and elongation begins (Plenefisch *et al.*, 1989). Because the experiment measured lethality instead of X-chromosome overexpression per se, the embryonic TSP probably reflects a critical time during embryogenesis when the embryo is especially sensitive to overexpression of particular X-linked genes, rather than reflecting the earliest time that overexpression begins. If this is the case, it implies that dosage compensation is already operating even earlier in embryogenesis, prior to the *dpy-28* TSP. Finally, the XX-specific lethal phenotype of all other alleles of the maternal-effect *dpy* genes is cold sensitive, such that over 99% of XX animals die at 15°C (compared to 80–97% at 20°C). This suggests that removal of the dosage compensation genes reveals a cold-sensitive process that is severely affected by the overexpression of certain X-linked genes. The TSP for this cold-sensitive process coincides with the *dpy-28* TSP, further supporting the proposal that the dosage compensation process is functioning by mid-embryogenesis.

D. MECHANISM OF DOSAGE COMPENSATION

The *Drosophila* and mammalian systems clearly illustrate the fact that highly divergent solutions can be used to solve the problem of equalizing the levels of X-linked gene products between the two sexes. In mammals, one of the two X homologs in the cells of female embryos becomes "inactivated" during development (reviewed in Gartler and Riggs, 1983). The inactivated X chromosome becomes condensed into a heterochromatic state and forms the cytologically visible structure known as the Barr body; its DNA replicates out of synchrony with the active chromosomes. Thus, each somatic cell of XY males and of XX females contains only one transcriptionally active X chromosome.

In *Drosophila*, both X chromosomes in the female are transcriptionally active. Dosage compensation is achieved by differential expression of X-linked genes in the two sexes; specifically, transcription of the single X chromosome in the male is hyperactivated to a level twice that of each of the two X chromosomes in the female. Mutations in the class of genes known to implement dosage compensation in *Drosophila*, the *msl* (*male-specific lethal*) and *mle* (*maleless*) genes, result in male (XY)-specific lethality, presumably due to reduced levels of X-linked gene products (reviewed in Lucchesi and Manning, 1987).

A variety of observations suggest that dosage compensation in *C. elegans* probably does not occur by X inactivation. Most X-linked loss-of-function mutations are fully recessive, and animals heterozygous for cell-autonomous X-linked markers are not phenotypic mosaics (Herman, 1984). Moreover, large X-chromosomal deficiencies can be tolerated in heterozygous XX animals. All these observations are inconsistent with models in which only a single X chromosome is active in each individual cell. *Caenorhabditis elegans* more likely employs a mechanism of differential gene expression formally analogous to that used in *Drosophila*, but the fact that the class of genes known to implement dosage compensation in *C. elegans* appear to function mainly or exclusively in hermaphrodites (XX) suggests that the two strategies are probably quite different in detail.

The most obvious type of model for dosage compensation in the worm is that the level of expression of each of the two X chromosomes is *decreased* in XX individuals, via the action of the *dpy* genes, to one-half the basal level of the single X chromosome in XO males. By this model, the products of the *dpy* genes themselves, or of a pathway in which the *dpy* genes act, might effectively act as "repressors" to reduce the rate of X-chromosome expression in XX animals. An alternative possibility that cannot be ruled out is that the XX-specific *dpy* genes might instead act as negative regulators of an as yet undiscovered XO-specific gene (or genes) responsible for hyperactivating the single X chromosome in males. It is interesting, however, that no such genes of this type were identified in a screen of over 22,000 haploid genomes for suppressors of the *dpy-28* XX-specific lethality (Plenefisch *et al.*, 1989).

The observation that some duplications of part of the X chromosome can increase the expression of X-linked genes they do not duplicate (Meneely and Nordstrom, 1988; L. Casson and B. Meyer, unpublished) might tend to lend support to a model for dosage compensation whereby repressors act to reduce X expression in XX animals. The model would suggest that the duplications increase X-chromosome expression because they contain sites that titrate dosage compensation repressor molecules. The fact that some duplications also appear to increase X expression in XO males (Meneely and Nordstrom, 1988), however, is not easily explained by the simplest versions of this model, which predict that repressor molecules should not be present in XO animals.

Ultimately, the question of whether the mechanism of dosage compensation in *C. elegans* involves turning down X expression in 2X animals or turning up X expression in 1X animals, or some combination of the two, will most likely be resolved at the molecular level. One approach, which has yielded the most compelling evidence to date for the hyperactivation mechanism in *Drosophila*, involves comparing the levels of gene expression for an autosomal gene inserted into various locations on the X chromosome to expression levels for the gene at its normal chromosomal location as well as other autosomal insertion sites (Spradling and Rubin, 1983). The availability of a DNA transformation system makes this type of approach feasible in *C. elegans* (Fire, 1986). Preliminary experiments suggest that an autosomal gene inserted onto the *C. elegans* X chromosome can indeed become dosage compensated and can respond to mutations in *dpy-21* (D. Hsu and B. Meyer, unpublished).

IX. A Pathway for the Coordinate Control of Sex Determination and Dosage Compensation

The need for dosage compensation in *C. elegans* arises as a consequence of the primary sex-determining signal, the X/A ratio. Does the dosage compensation mechanism adopt alternative XX or XO modes in response to this same primary signal? That is, are sex determination and dosage compensation coordinately controlled? *A priori*, this need not be the case. In mammals, for example, sex determination and dosage compensation are apparently initiated independently; primary sex determination depends on the presence or absence of a particular region of the Y chromosome (referred to as *Tdy* in mice and *TDF* in humans; reviewed in McLaren, 1988), while X inactivation depends on the number of X chromosomes (reviewed in Gartler and Riggs, 1983). Thus, XXY individuals develop testes and have one inactivated X chromosome. In *Drosophila*, by contrast, the sex determination and dosage compensation processes are coordinately controlled through the activity of the switch gene *Sex-lethal* (*Sxl*) (Lucchesi and Skripsky, 1981; Cline, 1983). In response to a high X/A ratio, *Sxl* becomes activated, resulting in female sexual differentiation and a female level of X-linked gene expression. In flies with a low X/A ratio *Sxl* remains inactive, resulting in male development and hyperactivation of X-chromosome expression.

In *C. elegans*, the existence of single-gene mutations that simultaneously disrupt both sex determination and dosage compensation indicates that these two pathways do indeed share common steps prior to their divergence into independent pathways (Villeneuve and Meyer, 1987; Meyer, 1988; Miller *et al.*, 1988; Nusbaum and Meyer, 1989). That is, both the choice of sexual fate and the appropriate level of X-linked gene expression are established in response to the same primary signal, the X/A ratio. At least three genes, *sdc-1*, *sdc-2*, and *xol-1*, are involved in the coordinate control of these two processes. These genes presumably play a role either in reading the X/A signal and/or in transmitting information about the signal to the two independent sets of genes (or gene products) that control sexual differentiation and the level of X-chromosome expression.

The X-linked genes *sdc-1* and *sdc-2* (sex determination and dosage compensation) are required for the XX, or hermaphrodite, modes of both sex determination and dosage compensation (Villeneuve and Meyer, 1987; Nusbaum and Meyer, 1989), whereas the gene *xol-1* (XO lethal), also X linked, is required for the XO or male modes of both processes (Miller *et al.*, 1988). In this section we discuss the properties of mutations in these genes, as well as the rationale for assigning these genes to early steps in the regulatory hierarchy that is responsible for the coordinate control of sex determination and dosage compensation in *C. elegans*. A model for this regulatory (Miller *et al.*, 1988) is presented in Fig. 6 and is referred to throughout the text of this section.

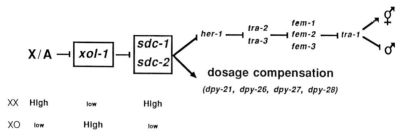

FIG. 6. Proposed gene hierarchy for the coordinate control of sex determination and dosage compensation in *C. elegans* (Miller *et al.*, 1988). Experiments described in Section IX indicate that the genes *xol-1, sdc-1,* and *sdc-2* are involved in the coordinate control of sex determination and dosage compensation, acting at steps that are shared by the two pathways prior to their divergence. A full explanation of the logic for assigning these genes to particular steps in the hierarchy is provided in the text. Negative regulatory interactions are indicated by bars, whereas positive regulatory interactions are indicated by arrows. The descriptions "High" and "low" indicate the activity states of the *xol-1* and *sdc* genes or gene products at different values of the X/A ratio. As discussed in Section V for genes controlling sexual differentiation but not dosage compensation, regulation might occur by any of a variety of molecular mechanisms, including transcriptional control, posttranslational modification, protein–protein interactions, or alternative RNA splicing. Further, some gene products may be present in both sexes but active only in the presence of other, sex-specific, gene products. The remainder of the sex determination branch of the pathway as shown (*her-1* to *tra-1*) is a simplified version of the model of Hodgkin presented in Fig. 5 including only the regulatory interactions for somatic sex determination.

A. SEX DETERMINATION AND DOSAGE COMPENSATION (*sdc*) GENES

1. *sdc-2*

Thirty recessive loss-of-function mutations in *sdc-2* do not affect XO animals, but inappropriately shift both sex determination and dosage compensation in XX animals toward their XO, or male, modes (Nusbaum and Meyer, 1989). The most common class of *sdc-2* alleles are strong loss-of-function (possibly null) mutations that result in XX-specific lethality; rare XX animals that escape this lethality develop as extremely small abnormal pseudomales. Weaker loss-of-function *sdc-2* alleles permit the viability of most XX animals, which are small (Sma), variably Dpy, slow growing, and usually extensively masculinized. The range of sexual phenotypes for these animals includes a few self-fertile, egg-laying defective (Egl) hermaphrodites, a small number of intersexual animals, and a large number of nonmating pseudomales that exhibit nearly complete masculinization in all tissues examined.

The XX specificity of the *sdc-2* Dpy and lethal phenotypes suggested that, in addition to causing sexual transformation, *sdc-2* mutations disrupt dosage compensation. This disruption of dosage compensation has been demonstrated directly by showing that *sdc-2* mutations cause an increase in the levels of several X-linked mRNA transcripts specifically in XX but not in XO animals (Nusbaum and Meyer, 1989). Thus, *sdc-2* has precisely the properties expected for a gene that is required for the control of both sex determination and dosage compensation in XX animals. That is, loss-of-function mutations cause (1) a complete (or nearly complete) reversal of sexual fate, similar to the phenotypes caused by mutations in the *tra* genes, and (2) XX-specific lethality and overexpression of X-linked genes, similar to the phenotypes caused by mutations in the dosage compensation *dpy* genes.

Rapid progress has been made toward the molecular cloning of the *sdc-2* locus. *sdc-2* was mapped using visible genetic markers to within 0.08 map units of *lin-14*, a gene that has been cloned by chromosomal walking from linked restriction fragment length polymorphisms (RFLPs) (Ruvkun *et al.*, 1989). Subsequent mapping of *sdc-2* relative to RFLPs in the region (M. Stern, personal communication; Nusbaum and Meyer, 1989) led to the identification of a cosmid clone from the chromosomal walk that detects DNA alterations in four different *sdc-2* mutant alleles (C. Nusbaum and B. Meyer, unpublished). Studies to further localize the *sdc-2* gene are underway.

2. *sdc-1*

Loss-of-function mutations in *sdc-1*, as in *sdc-2*, cause both masculinization and elevated levels of X-chromosome expression in XX animals, while XO animals are unaffected (Villeneuve and Meyer, 1987). *sdc-1* differs from *sdc-2* in several important respects, however. First, the null phenotype of *sdc-1* is apparently an incompletely penetrant transformation of XX animals toward the male fate; even in the absence of *sdc-1* activity, only some XX animals develop as pseudomales or intersexes, while many develop as hermaphrodites (Villeneuve and Meyer, 1990). Second, unlike *sdc-2*, *sdc-1* mutations do not result in significant XX-specific lethality despite causing increased levels of X-linked gene expression. XX animals are, however, small (Sma) and slow growing, and they are usually egg-laying defective (Egl) and often have an abnormal protruding vulva (Pvul) (if hermaphrodite) or have stunted and abnormal tail morphology (if male). Such XX-specific phenotypes are reminiscent of those caused by mutations in the dosage compensation *dpy* genes, particularly *dpy-21*. Finally, all 15 *sdc-1* mutations exhibit a strong maternal rescue. Homozygous *sdc-1* progeny from *sdc-1*/+ mothers are essentially wild type for most of the phenotypes described above, while homozygous *sdc-1* progeny from homozygous *sdc-1* mothers display the full mutant phenotype. There is no maternal absence effect; *sdc-1/+* cross-progeny from *sdc-1/sdc-1* mothers are fully wild type. Thus, either maternal or zygotic activity can be sufficient for *sdc-1* function. Together the data suggest that while *sdc-1* is clearly important for proper sex determination and dosage compensation in XX animals, it does not function alone in setting the hermaphrodite modes of these two processes.

Temperature-shift experiments with a strongly temperature-dependent *sdc-1* allele define a brief and specific TSP during the first half of embryogenesis for the role of *sdc-1* in most aspects of somatic sex determination (Villeneuve and Meyer, 1990). The TSP extends from approximately the 100-cell stage to the end of embryonic cell proliferation and is complete prior to the onset of embryonic elongation, more than 2 hours before the first visible evidence of sexual dimorphism and nearly a full day before the somatic gonad, one of the tissues scored in the experiment, becomes visibly dimorphic (Fig. 4). The nature of the TSP suggests that *sdc-1* acts early in development to establish the proper choice of sexual fate in cells or cell lineages giving rise to terminally differentiated sexually dimorphic structures. Temperature-shift experiments examining the dosage compensation phenotypes of *sdc-1* likewise suggest that *sdc-1* may be involved in establishment, but not maintenance, of the XX mode of dosage compensation (Villeneuve and Meyer, 1990).

A mosaic analysis of the sex determination phenotype of *sdc-1* yielded an interesting and unusual result. Nearly all mosaic analyses in *C. elegans* reported to date give results consistent with cell- or lineage-autonomous action of the gene assayed; that is, the phenotype of a particular cell depends on the presence or absence of the wild-type gene in that cell. As discussed above, the results of a mosaic analysis of *tra-1* are entirely consistent with a cell-autonomous requirement for the *tra-1* gene in the male–hermaphrodite decision. In the *sdc-1* experiment, a large number of genetic mosaics were generated that had lost the wild-type copy of the *sdc-1* gene from the lineage of either AB or P_1, the two blastomeres derived at the first embryonic cleavage. [In these experiments, *sdc-1* strains with a high (50 to >90%) penetrance of masculinized phenotypes were used, and the wild-type allele was introduced through the father to circumvent the problem of *sdc-1(+)* maternal rescuing activity.] The striking result is that although many genotypically mosaic animals were identified, these animals did not display the *sdc-1* sexual transformation phenotypes (Villeneuve and Meyer, 1990).

Two possible interpretations of this result are that either (1) the *sdc-1* gene is expressed immediately, in the one- or two-celled embryo, and this early expression is sufficient to provide *sdc-1* function, or (2) *sdc-1* acts in a nonautonomous fashion, such that expression in either lineage can compensate for lack of the gene in the other lineage. If the first interpretation is correct, then transcription of the *sdc-1* gene takes place several hours prior to the time that the temperature-shift experiments suggest *sdc-1* functions in the sex determination process. By this model, either *sdc-1* transcripts or product must persist through multiple rounds of embryonic cell division; this type of perdurance, at least for maternally supplied *sdc-1*, is indeed predicted by the fact that *sdc-1* mutants exhibit maternal rescue. It is important to note that this interpretation cannot distinguish whether *sdc-1* functions autonomously or nonautonomously. (A third explanation, discussed in detail in Section XI, is based on the possibility that the X/A signal has been effectively altered in these mosaic animals by the method used to generate them.)

3. The sdc Genes and the Sex Determination Pathway

How do the *sdc* genes fit into the sex determination pathway? The construction of double-mutant strains indicates that the masculinization caused by loss-of-function mutations in either *sdc-1* or *sdc-2* is completely suppressed by loss-of-function mutations in *her-1*, which is normally required for male but not hermaphrodite development (Villeneuve and Meyer, 1987; Nusbaum and Meyer, 1989). This requirement for *her-1(+)* suggests that the masculinization caused by the *sdc* mutations results from inappropriate activation of *her-1* in XX animals. It further suggests that the wild-type products of *sdc-1* and *sdc-2* act, directly or indirectly, as negative regulators of *her-1* activity in XX animals (see Fig. 6); the cloning of *her-1* (C. Trent and W. Wood, personal communication) makes this proposition directly testable. Consistent with this placement of the *sdc* genes upstream of *her-1* in the sex determination pathway is the fact that the TSP for *sdc-1* in controlling somatic sexual phenotype is earlier than that determined for *her-1* for somatic gonad sexual phenotype (Villeneuve and Meyer, 1990; see Fig. 4).

Nusbaum and Meyer (1989) have further shown that the ability of the viable, reduced-function allele *sdc-2(y55)* to masculinize XX animals is dependent not only on the presence of but also on the gene dosage of the *her-1* locus. Specifically, masculinization is almost completely suppressed in *sdc-2(y55)* XX animals that carry only a single copy of the wild-type *her-1* gene (*her-1/ +*). Since *her-1* is not normally haplo-insufficient in XO animals, this result implies that *her-1* can be regulated to an intermediate state in response to partial *sdc-2* activity. Reducing the *her-1(+)* gene dosage by one-half in *sdc-2(y55)* XX animals apparently reduces the amount of *her-1* activity below a level capable of promoting male development.

While a *her-1* mutation suppresses the masculinizing effects of *sdc* mutations, it does not suppress the elevated levels of X-linked gene expression, the XX-specific lethality, or the XX-specific morphological phenotypes (Sma or Dpy, Egl, Pvul) associated with the disruption of dosage compensation (Villeneuve and Meyer, 1987; Nusbaum and Meyer, 1989). This indicates that the effects of *sdc-1* and *sdc-2* on sex determination and dosage compensation are ultimately implemented by independent pathways. Further, the fact that *sdc-1* and *sdc-2* mutations simultaneously disrupt both sex determination and dosage compensation suggests that both processes are responsive to the same primary signal, the X/A ratio. We therefore suggest that *sdc-1* and *sdc-2* act at a step in the regulatory hierarchy prior to the divergence of these two pathways (refer to Fig. 6). Further support for the positioning of *sdc-1* at an early step in the dosage compensation branch of the pathway comes from temperature-shift experiments suggesting that *sdc-1* may function in the establishment, but not maintenance, of the XX mode of dosage compensation.

The fact that *sdc-1* and *sdc-2* act at the same position in the hierarchy suggests that these two genes may act in conjunction to promote hermaphrodite sexual differentiation and a hermaphrodite level of X-chromosome expression in XX animals. Such synergism is suggested by the finding that XX animals homozygous for both an *sdc-1* mutation and an *sdc-2(weak)* mutation are completely inviable, despite the fact that neither mutation alone causes significant lethality (Villeneuve and Meyer, 1990). Genetic analysis does not allow us to distinguish among the multitude of potential models for the *sdc-1–sdc-2* interaction, however. *sdc-1* might act to enhance *sdc-2* activity, for example, by regulating *sdc-2* expression or by stabilizing the *sdc-2* product. Alternatively, *sdc-1* and *sdc-2* may act in parallel in the regulation of sex determination and dosage compensation functions. Ultimately, questions regarding the nature of the interaction between the *sdc* genes will most likely be resolved at the molecular level.

B. XO-SPECIFIC LETHAL (*xol-1*) GENE

1. XO Phenotypes: The Role of xol-1 in the Regulatory Hierarchy

The X-linked gene *xol-1* is required for the XO, or male, modes of both sex determination and dosage compensation (Miller *et al.*, 1988). Five independent recessive loss-of-function mutations in *xol-1* result in embryonic or early L1 larval lethality specifically in XO animals; XX *xol-1* animals are fully viable normal hermaphrodites. The XO-specific lethality presumably results from the disruption of dosage compensation that has been demonstrated for *xol-1* mutants: X-linked transcript levels are reduced to one-half of wild-type levels in XO dying embryos, while transcript levels are normal in XX embryos.

xol-1 mutations also feminize XO animals. Although most *xol-1* XO animals die as embryos, some do hatch and a few are healthy enough to permit scoring of the few sexually dimorphic markers that are visible in newly hatched L1 larvae. In several dying *xol-1* XO larvae, cells have been identified that had clearly adopted hermaphrodite fates. Examples of dying *xol-1* XO larvae in which some cells had clearly adopted male fates have also been observed, however, suggesting that some, but not all, *xol-1* XO animals are transformed toward the hermaphrodite fate. Thus, in *xol-1* XO animals, both sex determination and dosage compensation are apparently shifted toward their XX, or hermaphrodite, modes.

Both the XO-specific lethality and the feminization caused by *xol-1* mutations are completely suppressed by mutations in either *sdc-1* or *sdc-2*, genes that control the XX modes of sex determination and dosage compensation. That is, *xol-1 sdc-1* or *xol-1 sdc-2* XO animals are fully viable wild-type males. (*xol-1* mutations do not suppress the XX-specific phenotypes of *sdc* mutations.) These results suggest that the *xol-1* mutant phenotypes result from inappropriate activation of *sdc-1* and *sdc-2* in XO animals.

The *xol-1* XO-specific lethality, but not the feminization, is also suppressed by mutations in genes required for implementing the hermaphrodite mode of dosage compensation, the XX-specific *dpy* genes *dpy-21*, *dpy-26*, *dpy-27*, and *dpy-28*. For *dpy-21;xol-1* and *dpy-28;xol-1* XO animals, suppression of the lethality has been correlated directly with restoration of X-chromosome expression to nearly wild-type levels. The rescued *dpy;xol-1* XO animals are predominantly hermaphrodites, but the range of sexual phenotypes includes a few intersexual animals as well as some essentially wild-type, fertile males. This suggests that the available *xol-1* mutations alone do not completely feminize all XO animals. Moreover, the percentage of XO animals that develop as hermaphrodites varies depending on the *dpy* mutation used to rescue them, suggesting that the *dpy* mutations themselves may influence the choice of sexual fate. (Indeed, feminizing effects of the *dpy* mutations have been observed previously in animals whose sexual identity is ambiguous, in particular in animals with intermediate X/A ratios.

These observations and their implications are discussed in Section XI.)

The feminization in these rescued *dpy;xol-1* XO animals is completely suppressed by a mutation in the gene *tra-2*, which is required for hermaphrodite sexual differentiation. Thus, complete *xol-1* suppression can be achieved either by a loss-of-function mutation in a single gene that coordinately controls the hermaphrodite modes of both sex determination and dosage compensation (e.g., *sdc-1* or *sdc-2*) or by a combination of mutations in two genes, one that is required for hermaphrodite sex determination (e.g., *tra-2*) and one that is required for hermaphrodite dosage compensation (e.g., *dpy-28*).

These interactions suggest that *xol-1* is the earliest acting gene in the known hierarchy controlling the male–hermaphrodite decision in *C. elegans* (Miller *et al.*, 1988; see Fig. 6). By this model, wild-type *xol-1* activity promotes male development by ensuring that the hermaphrodite modes of sex determination and dosage compensation remain inactive in XO animals. The data suggest that *xol-1* accomplishes this (at least in part) by negative regulation of *sdc-1* and *sdc-2*, genes known to control the hermaphrodite modes of these two processes. The fact that *sdc-1* mutations exhibit maternal rescue indicates that such negative regulation of *sdc-1* by *xol-1* must be achieved by a posttranscriptional mechanism.

xol-1 may not be the only gene acting at this particular step in the hierarchy. While *xol-1* mutations clearly feminize some XO animals, not all *xol-1* XO animals are feminized. Moreover, the *xol-1* lethality and feminization are fully suppressed by *sdc-1* mutations that by themselves cause masculinization in only a small percentage of XX animals and do not result in significant XX-specific lethality. One possible explanation is that the existing *xol-1* mutations do not completely eliminate *xol-1* function. Another possibility is that other genes may act in parallel with *xol-1* to regulate the activities of the *sdc* genes; these could be genes that, like *xol-1*, act to repress the *sdc* activities in XO animals, or they could be an opposite class of genes that are required in XX animals to fully activate the *sdc* genes.

2. xol-1 XX Phenotypes

In some mutant backgrounds, a paradoxical masculinizing effect of *xol-1* mutations on sexual phenotype in XX animals, which is opposite to the feminizing effect of *xol-1* mutations in XO animals, has been observed (Miller *et al.*, 1988). As stated previously, *xol-1* mutations have no obvious effects in XX animals that are otherwise wild type. Unexpectedly, however, *xol-1* mutations have been found to enhance the masculinization of XX animals that are already partially masculinized by a mutation in a sex-determining gene. Specifically, *xol-1* mutations transform some XX animals homozygous for *tra-2(lf)*, *tra-3(lf)*, or *her-1(gf)*, into complete cross-fertile mating males. Such masculinizing effects have been observed for each of the three *xol-1* alleles tested, and gene dosage experiments indicate that the phenotype results from a loss or reduction of gene function. The ability of *xol-1* to play opposite roles in different sexes, particularly when the role in one sex is major and in the other sex minor, is reminiscent of a similar duality exhibited by the gene *tra-1*; *tra-1* plays a crucial role in all aspects of hermaphrodite development but also appears to have a minor role in gonadogenesis and gametogenesis in males (see Section IV).

A wild-type *her-1* gene, normally essential for male development, is not required for the masculinizing effects of *xol-1* mutations in XX animals. Thus, the masculinization is implemented either through a minor pathway that is parallel to the main sex determination pathway or through the main sex determination pathway via a branch that intersects it downstream of *her-1*. The masculinizing effect of *xol-1* mutations in XX animals has not been incorporated into the model presented in Fig. 6 since the basis for the effect is not yet understood. It should be noted, however, that the existence of an activity similar to that of *xol-1* in XX animals was predicted by Hodgkin (1980) based in part on the fact that *her-1;tra-2* XX animals are nonmating pseudomales whereas *her-1;tra-2* XO animals are wild-type males (see Section V).

3. Molecular Cloning of xol-1

The heroic efforts of J. Sulston and A. Coulson in constructing the physical map of the *C. elegans* genome (Coulson *et al.*, 1986, 1988) have been instrumental in the molecular cloning of *xol-1*. A large set of overlapping cosmid and yeast artificial chromosome (YAC) clones completely spans the *xol-1* region of the X chromosome. The *xol-1* locus has been localized within this region by genetic mapping of *xol-1* relative to strain-specific RFLPs, and a cosmid was identified that detects a DNA alteration in one of the *xol-1* mutant alleles (L. Miller and B. Meyer, unpublished). This 40-kb cosmid has been shown to rescue the *xol-1* XO-specific lethal phenotype in DNA-mediated germ line transformation experiments, indicating that the *xol-1* gene is indeed contained within the clone (L. Miller and B. Meyer, unpublished). Experiments to further localize and define the extent of the gene are underway.

C. ANOTHER *sdc* LOCUS?

In Section VIII,B, we described the properties of mutations in *dpy-29*, a gene required for proper dosage compensation in XX animals. All nine *dpy-29* alleles, some of which appear to be null, produce phenotypes similar to those caused by mutations in the other maternal-effect dosage compensation *dpy* genes *dpy-26, dpy-27,* and *dpy-28:* XX-specific lethality and dumpiness, cold sensitivity, maternal rescue, and, for the allele tested, an apparent increase in X-linked gene expression in XX animals (Plenefisch *et al.,* 1989; J. Plenefisch and B. Meyer, unpublished; see Section VIII). Recent evidence suggests, however, that *dpy-29* may not fit into the dosage compensation *dpy* category, but may instead be a gene of the *sdc* class.

Five alleles of a newly identified *tra* locus have been mapped to the *dpy-29* interval on chromosome V (L. DeLong, J. Song, and B. Meyer, unpublished). These *tra* mutations result in a highly penetrant transformation of XX animals into nonmating pseudomales (similar in phenotype to *tra-2* XX pseudomales); a small percentage of animals are less fully masculinized, developing as Egl hermaphrodites or intersexes. The *tra V* mutations exhibit partial maternal rescue, such that the penetrance of the Tra phenotype is reduced in *tra/tra* progeny from *tra/ +* mothers. Failed attempts to obtain recombinants between a *dpy-29* allele and a *tra V* allele suggest that they map less than 0.04 map units apart, on the order of intragenic distances. Moreover, although *dpy-29* mutations alone (or *dpy-29/Df*) show no evidence of masculinization, all alleles of *dpy-29* at least partially fail to complement *tra V* mutations for the Tra phenotype (*dpy-29/tra V* XX animals range from <1 to 90% Tra for the various alleles) (L. DeLong, J. Plenefisch, and B. Meyer, unpublished). (*dpy-29* mutations fully complement mutations in other *tra* genes.) Further, *dpy-29* mutations that exhibit strong failure to complement *tra V* for the Tra phenotype suppress *both* the XO-specific lethality and the feminization caused by *xol-1* mutations, as do mutations in the *sdc* genes (J. Plenefisch and B. Meyer, unpublished).

Is *dpy-29* a cryptic *sdc* gene? Like *sdc-1* and *sdc-2, dpy-29* mutations appear to shift both sex determination and dosage compensation in the same direction. Also, *tra V* appears to act upstream of *her-1* in the sex determination pathway (L. DeLong and B. Meyer, unpublished). Unlike *sdc-1* and *sdc-2,* mutations at the *dpy-29–tra V* locus by themselves appear to affect either only dosage compensation (*dpy-29* alleles) or only sex determination (*tra* alleles), even though most of the *dpy-29* alleles were obtained in screens capable of yielding null alleles at the locus. Whether the *dpy-29* and *tra V* mutations define two neighboring genes or together define a new *sdc* gene with a separately mutable sex determination function is a question awaiting molecular resolution.

X. How Many Genes Are There?

One of the reasons for using a genetic approach to study a developmental process is to identify by mutation many, if not all, of the components involved in the process. As detailed in previous sections, a large number of genes involved in the coordinate control of sex determination and dosage compensation, in sex determination per se, and in the control and/or execution of dosage compensation have been identified in *C. elegans*. What fraction of the genes specifically involved in these processes are actually represented in the current collection?

The branch of the regulatory hierarchy controlling sex determination but not dosage compensation is probably the closest to saturation. Many alleles of the *tra, her-1, fem,* and *fog* genes have been isolated in general screens for mutations causing specific sexual transformations or in screens for revertants of other sex determination mutants [e.g., *tra(lf)* (Hodgkin and Brenner, 1977; Hodgkin, 1987c; Miller *et al.*, 1988), *tra(gf)* (Hodgkin, 1980, 1986; Doniach, 1986a; Schedl and Kimble, 1988), *her-1(lf)* (Hodgkin, 1980; Plenefisch *et al.*, 1989), *fem(lf)* (Hodgkin, 1986), *fem-3(gf)* (Barton *et al.*, 1987), and *fog* (Schedl and Kimble, 1988)]. Alleles of the same genes have been isolated repeatedly using these strategies, suggesting that most of the genes that can be mutated easily to cause complete or near complete reversal of sexual fate have probably been identified. The fact that a recent screen identified three independent alleles of a new *tra* locus on chromosome V (Miller *et al.*, 1988; L. DeLong, J. Song, and B. Meyer, unpublished; see Section IX) that had been missed in earlier screens, however, suggests that some additional loci may yet be discovered. Also, the screens were not entirely general; many were biased against isolating maternal-effect mutations or were designed to select mutations based on specific epistatic interactions with mutations in other sex-determining genes. Further, while the screens may have identified most of the genes that play a central role in this branch of the sex determination pathway, there may be many more genes that play peripheral roles (see Section IV).

The number of genes involved in the coordinate control of sex determination and dosage compensation, or in implementing the dosage compensation process itself, is much less certain. Because mutations that disrupt dosage compensation result in sickness or lethality specifically in either XX or XO animals, general screens aimed at identifying such mutations require more elaborate strategies than those used to identify mutations affecting only sexual phenotype; this reduces the number of genomes that can be screened in a given experiment. A small general screen [~4000 ethyl methanesulfonate (EMS)-mutagenized haploid genomes] for mutations specifically affecting XO animals or maternal-effect mutations specifically affecting XX animals yielded one allele of *dpy-28*, one allele of *sdc-1*, and one allele of *her-1* (Plenefisch *et al.*, 1989). Another type of screen specifically for X-linked mutations that produce either XX- or XO-specific phenotypes yielded one *sdc-2* allele (Nusbaum and Meyer, 1989) and two putative *xol-1* alleles (L. Miller and B. Meyer, unpublished) from a total of 5500 EMS-mutagenized chromosomes. [In this type of screen, the broods of F_1 hermaphrodites of the general genotype *him;m1/m2* (Him hermaphrodites produce both XX and XO progeny; *m1* and *m2* are two closely linked X-chromosome markers in the trans configuration) are screened for the absence of a class of marked animals of a single sex. Such screens should yield not only XX- or XO-specific lethal mutations but also sex-transforming mutations.]

Reversion of the *xol-1* XO-specific lethal phenotype has provided a powerful, efficient, and reasonably general means for isolating many additional alleles of previously known genes required for the XX mode of dosage compensation (or sex determination and dosage compensation), as well as for identifying new loci that are important for these processes (Miller *et al.*, 1988; J. Song, J. Plenefisch, M. Soto, A. Villeneuve, and B. Meyer, unpublished.). The results of this ongoing suppressor screen suggest that most of the genes of the *sdc* class that act downstream of *xol-1*, and many, but probably not all, of the genes required for dosage compensation in hermaphrodites have been identified. Several classes of possible dosage compensation mutations would not be identified in this screen, however; mutations in as yet unidentified genes required specifically for the male (XO) mode of dosage compensation, mutations in other *xol*-like genes, or mutations in *sdc*-like genes that act upstream of or in parallel with *xol-1*.

If the goal of identifying all of the major components in the regulatory hierarchy is to be achieved, exhaustive general screens for mutations specifically affecting either XX or XO animals are clearly warranted. Strategies similar to that used to isolate X-linked mutations of this type can be employed for screening the whole genome on a chromosome-by-chromosome basis; such screens should eventually identify most of the central genes required specifically for sex determination and/or dosage compensation. It is important to point out, however, that there are likely to be additional genes that have important roles in sex determination and dosage compensation but that are also required for other vital processes. Non-sex-specific lethality caused by mutations in such genes would preclude their identification using the kinds of screens described in this section. An example of a gene of this type is the *daughterless* gene of *Drosophila melanogaster*, which plays a crucial role in the initiation of the female sexual pathway in XX embryos (Cline, 1983, 1984) but which has other vital functions (Sandler, 1972; Mange and Sandler, 1973; Cline, 1976). These include essential roles in the generation of the peripheral nervous system in the embryo (Caudy *et al.*, 1988a) and in the development of the presumptive adult epidermal cells during larval stages (Cronmiller and Cline, 1987) in both sexes, and in proper egg membrane formation in the female somatic gonad (Cline, 1976, 1980; Cronmiller and Cline, 1987). As more is understood about the sex determination and dosage compensation genes that have already been identified in *C. elegans*, it will be important to devise new types of mutant screens that will allow identification of additional genes that are important for these processes but have other essential functions and therefore would have been overlooked in previous screens.

XI. The X/A Ratio Revisited

An important current problem in *C. elegans* sex determination is the elucidation of the nature of the primary sex-determining signal, the X/A ratio. What are the signal elements that comprise the numerator (X) and the denominator (A) of the ratio, and how are these counted? How do such seemingly small differences in the value of the X/A ratio (0.67 versus 0.75) result in the dramatically different outcomes of male versus hermaphrodite development? How does the endowment of maternally inherited products present in the oocyte influence the assessment of the X/A ratio?

Loci that act as counted numerator elements of the X/A signal should have several predictable genetic properties. Mutations that remove X-chromosome signal elements (loss of function) should reduce the X/A ratio in XX animals toward the male value, while mutations that duplicate signal elements (gain of function) should increase the X/A ratio in XO animals toward the hermaphrodite value. If the total number of elements that comprise the X/A signal is large, mutating any single numerator locus may not have any detectable consequences, precluding the possibility of identifying such loci genetically. On the other hand, if the total number of numerator elements is small, it is possible to design appropriate experimental conditions under which removal of a single locus results in a shift of both sex determination and dosage compensation toward the male modes, and/or duplication of a single locus leads to a shift of both pathways toward the hermaphrodite modes. Further, mutations in individual numerator elements that act cumulatively in determining sex are expected to have dose-dependent, additive effects; in other words, loss-of-function mutations in two separate numerator loci should fail to complement. Finally, duplication of one element should (at least partially) compensate for loss of another.

In *Drosophila*, two X-linked loci have been identified that fulfill these criteria for numerator signal elements: *sisterless-a* (*sis-a*) and *sisterless-b* (*sis-b*) (Cline, 1988). Removal of two copies of these elements is lethal to XX animals, presumably due to dosage compensation upsets; conversely, addition of two copies of these elements is lethal to XY animals. Further, reducing the dose of either element masculinizes 2X/3A animals (which are normally intersexes), whereas increasing the dose of either element feminizes 2X/3A animals (Cline, 1986, 1988).

No such X-linked numerator signal elements have yet been identified genetically in *C. elegans*, however. None of the three X-linked genes (*xol-1*, *sdc-1*, and *sdc-2*) that are important for both sex determination and dosage compensation is a reasonable candidate for a component of the X/A signal (Miller *et al.*, 1988). Although *xol-1* mutations shift both sex determination and dosage compensation in XO animals to the XX or hermaphrodite modes, they do so as a consequence of *loss* of *xol-1* function rather than *gain* of function. Loss-of-function mutations in both *sdc-1* and *sdc-2* do shift both sex determination and dosage compensation in XX animals in the direction expected for loss of a numerator element (toward the male modes), but several considerations argue that *sdc* genes are also unlikely to be components of the signal. First, both *sdc-1* and *sdc-2* appear to act downstream of *xol-1* in the regulatory hierarchy, while a signal element should act at the earliest step. Second, mutations in *sdc-1* and *sdc-2* fully complement each other. Finally, *sdc-1* mutations exhibit a strong maternal rescue, which is inconsistent with expectations for a numerator element. The phenotypes of *xol-1*, *sdc-1*, and *sdc-2* mutations instead suggest that the wild-type genes are involved in interpreting the signal or in transmitting this information to the downstream regulatory genes of the sex determination and dosage compensation pathways.

Several independent observations demonstrate that in animals with intermediate X/A ratios, mutations that result in elevated levels of X-linked gene expression can produce effects that mimic increasing the X/A ratio. Specifically, whereas one or more mutations in the dosage compensation *dpy* genes *dpy-21*, *dpy-27*, and *dpy-28* have no apparent affect on sex determination in otherwise wild-type diploid animals, such mutations have strong feminizing effects on animals with a 2X/3A karyotype. As discussed in Section III, 2X/3A animals are predominantly male. In striking contrast, 2X/3A progeny produced by mating tetraploid (2X/4A) males with *dpy-21, dpy-27,* or *dpy-28* diploid (2X/2A) mothers are predominantly hermaphrodite or intersex (Hodgkin, 1987b; Plenefisch *et al.*, 1989). For all three genes this feminizing effect is largely due to a maternal absence of the *dpy* gene product, since the 2X/3A progeny produced by the crosses described all have two copies of the wild type *dpy(+)* allele. (For *dpy-27* and *dpy-28* the feminizing effect is somewhat stronger if the tetraploid father is also homozygous for either *dpy-27* or *dpy-28*, respectively.) Additionally, mutations in *dpy-21* shift diploid XO animals carrying a duplication of a large part of the X chromosome from male to intersexual development (Meneely and Wood, 1984). (These experiments were not performed in a way that would distinguish whether it is maternal or zygotic absence of *dpy-21* that is critical for this effect.)

A related phenomenon has been observed in experiments with the gain-of-function mutation *her-1(n695sd)*. XX animals homozygous for the mutation range in sexual phenotype from Egl hermaphrodites to pseudomales (see Section IV), suggesting that the mutation allows partial escape of the *her-1* locus from negative regulation in XX animals. The mutation appears to retain partial sensitivity to regulation by X/A, however, since (1) the masculinization caused by *her-1(n695sd)* is almost completely suppressed in XXX diploid hermaphrodites (3X/2A = 1.5) (Trent *et al.*, 1988); (2) X-chromosome duplications can suppress the *her-1(n695sd)* masculinization phenotype in XX animals (A. Villeneuve and B. Meyer, unpublished); and (3) deficiencies of some regions of the X chromosome can enhance the masculinization (D. Hsu and B. Meyer, unpublished). (Since the *sdc* genes are proposed to act between X/A and *her-1* in XX animals, the fact that X duplications can suppress the *her-1sd* masculinization suggests that it may be possible to increase *sdc* activity beyond the level that normally occurs and is sufficient to promote hermaphrodite development in wild-type XX animals.) The masculinization phenotype of *her-1(n695sd)* is also strongly suppressed by mutations in *dpy-21, dpy-27,* and *dpy-28* (Trent *et al.*, 1988; Plenefisch *et al.*, 1989). (Again, it has not been determined whether it is the maternal or zygotic genotype with respect to the *dpy* gene that is important for this feminizing effect.)

The shift toward the hermaphrodite sexual fate caused by the dosage compensation *dpy* mutations under these unusual circumstances is opposite in direction to the effect of these mutations on the dosage compensation process. This is quite different from the phenotypes produced by mutations in genes involved in the coordinate control of sex determination and dosage compensation; such mutations shift both processes in the same direction, toward either the male modes (*sdc-1* and *sdc-2*) or the hermaphrodite modes (*xol-1*). What is the basis for this seemingly paradoxical feminizing effect of the *dpy* mutations?

Several distinct and speculative models have been suggested to account for this phenomenon; most suggest that the feminizing effects of the dosage compensation *dpy* mutations observed under these unusual circumstances are indirect consequences of the overexpression of X-linked genes documented for these mutations. Most models further assume that relatively constant amounts of maternally supplied *dpy* gene products are normally present in oocytes. For example, one possible interpretation is that the numerator of the X/A ratio is comprised of transcripts rather than chromosomal sites. According to this scenario, the absence of maternally supplied *dpy* gene products in the oocyte leads to a perceived increase in the X/A ratio due to overexpression of these "numerator transcripts" in the zygote (Hodgkin, 1987b; Plenefisch *et al.*, 1989). Alternatively, increased expression of X-linked genes that control the hermaphrodite modes of sex determination and dosage compensation, such as *sdc-1* and *sdc-2*, could alter the *probability* of adopting the hermaphrodite fate in response to an intermediate or ambiguous value of the X/A signal. Since *sdc-1* activity has a strong maternal component, it is quite plausible that overexpression of the maternal *sdc-1* gene could lead to feminization of the zygote. (Note that this type of model postulates that the *sdc* genes may be subject to regulation by maternally provided components of the dosage compensation process that they regulate in the zygote.) A third possibility takes into account the fact that X/A ratio assessment, and indeed all of embryonic development, takes place in the context of a large endowment of maternally supplied gene producs, most of whose functions are not specifically involved in sex determination. Altering the balance of maternal gene products could conceivably affect either assessment and/ or response to the X/A ratio in animals whose sexual identity is ambiguous (Plenefisch *et al.*, 1989).

It is also possible that the feminizing effects are not the result of X-chromosome overexpression per se but are rather the consequences of changes in X-chromosome structure. Since the *dpy* genes encode part of the dosage compensation machinery, some of their products might be intimately associated with X chromatin; in this type of model, the absence of the *dpy* products might alter the accessibility of putative X-chromosome numerator sites and thereby lead to an incorrect assessment of the X/A ratio (Plenefisch *et al.*, 1989).

All of the above models share in common the proposition that the feminizing effects of the *dpy* genes are secondary consequences of disruption of the normal function of these genes in the regulation of X-chromosome expression. Wood *et al.* (1987) proposed a different type of model in which the products of the *dpy* genes, in addition to their role in dosage compensation, might themselves constitute the denominator of the X/A ratio. In this model, the ratio of X chromosomes to *dpy* gene products would in turn determine the level of X-linked gene products (e.g., the *sdc* products) that promote hermaphrodite development. This last model does not take into account the fact that, where tested, the feminizing effects of the *dpy* mutations are strongly dependent on the genotype of the mother. Clearly, the denominator of the X/A ratio cannot be entirely maternally supplied, since 2X/3A and 2X/2A progeny of 2X/2A mothers are male and hermaphrodite, respectively, even though they both have two X chromosomes and both were derived from mothers with two sets of autosomes (noted in Wood *et al.*, 1987). While a role for the *dpy* genes as part of the denominator is possible, the proposal of such a distinct secondary function for these genes is not required by the data.

At this point, it is necessary to reconsider the mechanism(s) by which partial duplications of the X chromosome feminize 2X/3A, or *her-1(n695sd)* XX, animals. In the original experiments by Madl and Herman (1979), it was implicitly assumed that duplications capable of feminizing 2X/3A animals do so because they contain X-chromosome numerator signal elements and therefore increase the value of the X/A ratio. More recent data suggest that such duplications may also cause overexpression of X-linked genes that they do not duplicate (Meneely and Nordstrom, 1988; see Section VIII). This suggests the formal possibility that the feminizing capability of these X-chromosome duplications might be due either to the presence of bona fide numerator signal elements, as was originally proposed, or, by analogy to the effects of the *dpy* mutations, to overexpression of X-linked genes, or to some combination of the two.

While the numerator elements of the X/A ratio have not yet been defined genetically, McCoubrey *et al.* (1988) performed experiments aimed at identifying X-linked numerator elements by assaying the feminizing capability of microinjected DNA sequences. These investigators identified several X-linked DNA clones that can cause partial feminization of the putative *2X;3A;mnDp8* progeny of a cross between injected *4X;4A* hermaphrodites and *1X;2A;mnDp8* males. Such feminizing elements were identified in three different X clones and appear to be less frequently distributed on the autosomes, although additional autosomal controls will be required to show whether such sequences are truly X-chromosome specific.

The assay used in these experiments was a transient one, so the presence and/or copy number of the DNA was not verified in most cases. Additionally, the strains used were not genetically marked in a way that would permit unambiguous identification of progeny with the desired $2X;3A;mnDp8$ karyotype. Moreover, in the few cases tested, there was a lack of correlation between the presence of the DNA and sexual phenotype: not only was the DNA absent from some feminized animals (which might be explained trivially if the copy number of the DNA were below the detection limits), but it was present in high copy number in some animals that were not feminized. The results of these experiments are intriguing but preliminary; more compelling experiments involving the generation of heritable transformants in which the presence and copy number of the DNA can be verified are both feasible and warranted.

What are the feminizing elements identified in these experiments? The authors suggest that these sequences may comprise part of the "X component' (or numerator) of the X/A ratio. Other interpretations should be considered, however, particularly in light of the fact that (1) mutations that increase X-linked gene expression feminize 2X/3A animals and (2) duplications of part of the X chromosome can apparently elevate the expression of X-linked genes that they do not duplicate. An alternative model suggested by these observations is that the X-linked clones in the microinjection experiments might feminize because they contain sites involved in dosage compensation (which might be expected to occur ubiquitously throughout the X chromosome); the introduction of such sites in high copy number could titrate putative negative regulators of X-linked gene expression, thereby elevating the overall level of expression of all dosage compensated X-linked genes.

XII. The Roles of Autonomy, Nonautonomy, and Maternal Influence in the Choice of Sexual Fate

Sex determination in wild-type worms is a high fidelity process. In *C. elegans* diploids, XX animals reliably develop as hermaphrodites, while XO animals reliably develop as males; intersexual animals do not occur. How do all sexually dimorphic cells in the animal make the same decision? Is sex determined independently by each of the cells in a cell-autonomous fashion? Or are some aspects of the sex determination process nonautonomous, involving cell–cell communication? What is the role of the maternally supplied oocyte cytoplasm in the decision directed by the zygotic genome?

In mammals, sex determination and sexual differentiation involve both autonomous and nonautonomous processes. The Y chromosome is the primary determinant of sex (Jacobs and Strong, 1959), and the putative testis-determining gene *Zfy* (formerly, *TDF*) located in the sex-determining region of the human Y chromosome has been cloned (Page *et al.*, 1987). The gene encodes a protein containing multiple putative DNA-binding "zinc finger" domains, suggesting that the earliest step in the cascade controlling sexual differentiation in mammals is cell autonomous. Further, studies with chimeric mice indicate that the presence of a Y chromosome is required cell autonomously for differentiation of the Sertoli cells of the male testis, one of the earliest defined events in male differentiation (Singh *et al.*, 1987; Burgoyne *et al.*, 1988). By contrast, both XX and XY cells can be induced (perhaps by the Sertoli cells) to develop as Leydig cells, the testosterone-producing cells of the testis, indicating a nonautonomous step in the sex determination process (reviewed in McLaren, 1984). Secretion of testosterone by these cells then induces other cells in the animal, both XX and XY, to undergo male sexual differentiation (nonautonomous). Finally, other experiments show that cells must express androgen receptor in the cytoplasm (autonomous) in order to respond to this extracellular signal. Either lack of functional androgen receptor (a condition known as testicular feminization syndrome; Lyon and Hawkes, 1970) or lack of male hormones (e.g., due to castration of fetal mice before gonad differentiation takes place; Jost, 1953) results in differentiation of female secondary sexual characteristics in XY animals.

In *Drosophila*, on the other hand, sex determination appears to take place almost entirely cell autonomously. XX/XO mosaics that have lost an X chromosome at very early nuclear divisions develop as gynandromorphs in which all XX cells of the adult cuticle are female and all XO cells are male (Morgan and Bridges, 1919; Tokunaga, 1962). Further, mosaic analyses have shown that the master switch gene *Sxl*, which controls both sex determination and dosage compensation, as well as *tra, tra-2,* and *dsx*, downstream genes that control sexual differentiation, all function in a cell autonomous fashion in cells giving rise to the adult cuticle (Cline, 1979b; Baker and Ridge, 1980; Sanchez and Nöthiger, 1982). Cline has suggested that the development of the entire fly as either male or female is ensured by the fact that the control of sex determination and the control of dosage compensation are tightly coupled; he postulates that any cells that make the wrong decision will die because of dosage compensation problems and therefore be excluded from the final pattern (Cline, 1984).

While *tra-1*, the last gene in the pathway that controls all aspects of sexual differentiation in *C. elegans*, appears to act in a cell- (or lineage-)

autonomous fashion (C. Hunter and W. Wood, personal communication) it is not yet known whether earlier events in the determination of sexual fate in *C. elegans* occur cell autonomously or nonautonomously. Nor is it known how different cells in the animal are coordinated to ensure that they all make the same decision. It is clear that the type of "editing" suggested by Cline for *Drosophila* (see above) cannot be operating in *C. elegans* due to the essentially invariant nature of the cell lineage which cannot accommodate stochastic error and subsequent cell death. Although these issues are not yet resolved, we describe in this section a variety of experiments that have begun to address the possible roles of autonomy, nonautonomy, and maternal influence in the choice of sexual fate in *C. elegans*.

First, two independent observations suggest that different cells in the organism are not choosing their sexual fates independently. In one case, careful observation of the classes of sexual phenotypes occurring in *sdc-1* mutant strains (in which XX animals range from pseudomales to intersexes to self-fertile hermaphrodites) revealed that there is a strong statistical correlation between tail and gonad sexual phenotype even though these two tissues diverge lineally at the first embryonic cleavage: XX animals with male tails generally have male gonads, animals with hermaphrodite tails generally have hermaphrodite gonads, and animals with tail and gonad of opposite sex are relatively rare (Villeneuve and Meyer, 1990). The correlation holds when individual tissues are examined at a closer level; for example, while intersexual tails with some features of both male and hermaphrodite do occur, tails that are either entirely male or entirely hermaphrodite are much more common.

A similar phenomenon has been observed in experiments examining the classes of intersexuality that occur in animals with an intermediate X/A ratio (Schedin, 1988; C. Hunter, personal communication). Madl and Herman (1979) had previously shown that *2X;3A;mnDp9* or *2X;3A;mnDp25* animals can develop as intersexes in which some cells apparently adopt the hermaphrodite fate and other cells apparently adopt the male fate. (It is not known whether these intersexual phenotypes result from cells differing in the assessment of X/A or from cells differing in the ability to respond to a uniform assessment of X/A at an intermediate value.) While all potential combinations of male and hermaphrodite structures do occur in these intersexes, Schedin and Hunter found that the tendency of intersexes to be either mostly male or mostly hermaphrodite was much greater than would be expected if different cells were choosing their sexual fates independently. In Hunter's experiments, a statistical analysis of the data examining the likelihood that any given pair (or triple) of terminally differentiated sexually dimorphic structures scored would have the same sexual phenotype verified the nonrandom nature of this choice. The analysis failed to reveal any lineage or positional basis for the tendency, however, and no particular cell or structure was identified whose sexual phenotype in an individual animal was a consistent predictor of the general sex tendency of that animal.

How can this phenomenon of nonindependence be explained mechanistically? One possibility is that sex determination in *C. elegans* occurs cell autonomously, but with a strong influence of the maternally provided environment on the choice of sexual fate. It seems likely that the complement of components supplied to the oocyte by the mother is somewhat variable from oocyte to oocyte. This type of model suggests that in the absence of *sdc-1* activity, or with an ambiguous X/A signal, the choice of sexual fate becomes sensitized to this naturally occurring variability. Thus, the maternally provided environment in some oocytes will tend to favor male development, whereas in others hermaphrodite development will be favored.

A second possibility is that sex determination in *C. elegans* is not exclusively cell autonomous, but rather involves cell–cell interactions. This type of model suggests that either (1) there exists an as yet unidentified "master" cell or tissue, in which the X/A ratio is assessed and the decision made, which subsequently communicates the decision to the rest of the cells in the animal, or (2) the X/A ratio is read in many different cells which then produce a diffusible factor, the cumulative level of which acts to coordinate the choice of sexual fate made by all the cells in the body. In either case, the nonautonomous signal is ultimately received and interpreted by individual cells in order to regulate downstream switch genes such as *tra-1* that control sexual identity in a cell-autonomous fashion.

There is additional evidence that the maternal environment does indeed play a role in the choice of sex of the zygote. Specifically, the sex of triploid animals of identical chromosomal constitution can depend on the crosses performed to generate them. Nearly all *2X;3A;mnDp8* or *2X;3A;mnDp27* animals develop as males when produced by crossing *4X;4A* hermaphrodites with *1X;2A;Dp* males, while 40 and 36%, respectively, of *2X;3A;Dp* progeny develop as hermaphrodites or intersexes when produced by crossing *2X;2A;Dp/Dp* hermaphrodites with *2X;4A* males (R. Herman, personal communication). It is not clear whether it is the ploidy of the mother (tetraploid versus diploid) or the presence or absence of the duplication in the mother that is responsible for the striking difference in sex tendency among the $2X;3A;Dp$ progeny of the reciprocal crosses.

In two tissues in the worm, namely, the intestine and the germ line, effects of position on cell fate have been observed. These two tissues are unique in that their ability to express either the male or the hermaphrodite mode of sexual differentiation is not constrained by terminal differentiation of sex-specific structures, and in that the expression of sexually dimorphic functions (yolk protein expression and oogenesis in the hermaphrodite) does not occur until adulthood. In the first case, it was shown that the intestines of $2X;3A;mnDp9$ and $2X;3A;mnDp25$ intersexes can exhibit a mosaic pattern of yolk protein gene expression as assayed by *in situ* hybridization of radiolabeled probes to RNA in partially dissected intestines (Wood *et al.*, 1985; Schedin, 1988). In the mosaic intestines, the pattern of expressing and nonexpressing cells is unrelated to their origins in the cell lineage; cells expressing yolk are always contiguous, and expression invariably occurs in anterior but not in posterior cells. In some cases, there appeared to be a gradient of expression, with cells on the border between expressing and nonexpressing cells showing weak hybridization, raising the possibility that such cells may be exhibiting neither the male nor the hermaphrodite fate but rather an intermediate fate. In a separate experiment, it was found that approximately one-half of *her-1(ts)* XO hermaphrodites raised at 24°C produce first sperm, then oocytes, in the anterior gonad arm (the hermaphrodite pattern) but only sperm in the posterior arm (the male pattern) (P. Schedin, personal communication). In both experiments, cells in the anterior of the animal express the hermaphrodite fate, while cells in the posterior express the male fate. Taken together, these observations suggest that in adult worms the expression of sexually differentiated cell fates may be positionally influenced, although no potential source of positional information has been identified. As stated above, no such positional influences can be demonstrated for the terminally differentiated dimorphic structures that are produced during embryonic and larval development.

The experiments and observations described in this section suggest that in addition to cell-autonomous processes (e.g., *tra-1* function), maternal influence and/or cell–cell communication may play important roles in the sex determination process in *C. elegans*. The cloning and molecular analyses now underway for many of the genes in the sex determination pathway are likely to yield important information about the molecular bases for these classical developmental phenomena.

XIII. Determination and Differentiation

A. Is Sex "Determined" in *Caenorhabditis elegans*?

In *Drosophila*, several lines of evidence have led to the conclusion that an irrevocable sexual pathway commitment is made early in development and subsequently maintained in a heritable fashion. One class of experiments bearing on this issue involves examining the phenotypes that result from changing the X/A signal. XX/XO mosaic animals can be generated in *Drosophila* by a variety of genetic manipulations. X-chromosome loss in early cell divisions gives rise to gynandromorphs, which are mosaics of male (XO) and female (XX) tissues. Attempts to generate XX/XO mosaics at various times later in development have suggested that clones of XO cells induced in XX animals after early embryonic stages either die or are extremely small and unhealthy, presumably due to dosage compensation upsets; XO cells that survive appear female in sexual phenotype (Baker and Belote, 1983; Sanchez and Nöthiger, 1983). These experiments have been interpreted to mean that the X/A signal is assessed early in development (around the blastoderm stage), resulting in an irreversible determination of sexual fate and X-chromosome transcription rate.

A second type of evidence for an early determination of sex in *Drosophila* is the observation that the coarse-grained mosaic intersex phenotypes seen in animals having an ambiguous X/A signal (i.e., 2X/3A = 0.67), or in diploid XX animals in which the ability of cells to activate the key switch gene *Sxl* has been genetically compromised (Cline, 1984), resemble the mosaic phenotype of gynandromorphs. That is, the sizes of clonal patches of male and female tissues in these mosaic intersexes are consistent with commitment occurring at or soon after the blastoderm stage. It is important to point out that the ability to make inferences about the timing of determination based on the analysis of these types of intersexes depends on three critical assumptions, all of which have been shown to be valid for *Drosophila*: (1) Sex determination is cell autonomous (see Section XII). (2) The "mosaic intersex" phenotype, in which animals are composed of a mixture of clones of male cells and female cells, must be clearly distinguishable from a "true intersex" phenotype, in which cells are simultaneously expressing both male and female differentiation functions (discussed in Cline, 1985). (3) At least one key sex-determining gene (in this case *Sxl*) must be capable of being regulated to only two possible stable states (effectively, "ON" or "OFF") in response to the X/A signal (Cline, 1988).

A third type of evidence comes from experiments in which genetically marked mitotic recombinant clones were induced in a mosaic intersex genetic background at various times during the first and second larval instars, long before differentiation of adult sexually dimorphic tissues takes place. All cells in these marked clones were invariably of a single sex, indicating that commitment to a particular sexual fate had taken place prior to the time at which the clones were induced (Cline, 1985).

Together the above experiments suggest that, in *Drosophila*, sex is "determined" in the strictest sense of the word. Do similar determinative events occur in *C. elegans*? Important differences in various aspects of the biology and development of *Drosophila* and *C. elegans* necessitate the use of alternative experimental strategies to address the issue of sex "determination" in the nematode. For example, despite foiled attempts (R. Herman, personal communication), no method for generating mosaics through X-chromosome loss has yet been developed for *C. elegans*. Further, conclusions about the timing of determination cannot reasonably be drawn from analysis of the phenotypes of $2X;3A;Dp$ intersexes because none of the three criteria outlined above is known to be true for *C. elegans*.

Evidence that sex is indeed determined in *C. elegans* has come from temperature-shift experiments involving temperature-sensitive alleles of the sex-determining genes *sdc-1* and *her-1* (see Sections IV and IX; Fig. 4). In particular, the TSPs for *sdc-1* for sexual phenotype of the gonad and tail/copulatory structures are complete nearly a full day before any overt sexual dimorphism occurs in many of the lineages giving rise to these tissues (Villeneuve and Meyer, 1990). The TSP for *her-1* for gonad sexual phenotype is similarly complete more than 12 hours prior to the appearance of sexual dimorphism in the gonad lineage (Hodgkin, 1984). These data indicate that the cells giving rise to these dimorphic structures have become committed to a particular sexual fate by mid-embryogenesis, long before sexual differentiation actually takes place. Other temperature-shift experiments with *sdc-1* additionally suggest that *sdc-1* may be required for proper initiation, but not maintenance, of the hermaphrodite mode of dosage compensation, consistent with the notion that a cell and its descendants can become fixed in a pattern of gene expression (Villeneuve and Meyer, 1990). Together these experiments demonstrate that at least some aspects of sex are indeed "determined" in *C. elegans*.

The experiments do not, of course, directly address the issue of when assessment of the X/A ratio occurs. One experiment described in Section IX, the *sdc-1* mosaic analysis, should be reconsidered here from the perspective of the X/A ratio. Because *sdc-1* is X linked, the free X-chromosome duplication containing the *sdc-1(+)* allele used to generate mosaics may also contain X-chromosome numerator elements, and thus cause a slight increase in the X/A signal. *sdc-1* XX animals are poised precariously between the male and hermaphrodite fates and therefore could be sensitive to such an increase in X dosage. Indeed, subsequent experiments have shown that the attached duplication *mnDp25*, which does not contain *sdc-1(+)*, does suppress the *sdc-1* Tra phenotype (Villeneuve and Meyer, 1990). Thus, it is possible that the failure to observe sexual transformation phenotypes in *sdc-1(−)* cells in the mosaic animals is due to the feminizing effect of an increased X/A ratio rather than to the *sdc-1(+)* gene. If this alternative interpretation of the data is correct, it follows then that either (1) the X/A ratio is assessed immediately and irreversibly in the one- or two-celled embryo, or (2) the X/A signal directs sex determination in a nonautonomous fashion.

B. IRREVERSIBILITY VERSUS PLASTICITY OF SEXUAL DIFFERENTIATION

Many aspects of sexual dimorphism in *C. elegans* are constrained in potential in that terminally differentiated anatomical structures (e.g., the somatic gonad, vulva, sex muscles, and male copulatory organs) are produced via specified cell lineages. Some differences arise as a result of programmed cell death that occurs only in one sex (such as the programmed cell death of the hermaphrodite-specific HSN motor neurons in male embryos), others result from divisions of sex-specific blast cells that divide in one sex but fail to divide in the other sex (such as the B cell that gives rise to the spicules of the male tail or the cells P5.p, P6.p, and P7.p which give rise to the hermaphrodite vulva), and still others result from alternative lineages adopted by common primordia (such as the sex myoblast M which generates the vulval and uterine muscles in the hermaphrodite and the diagonals and other muscles required for copulation in the male) (Sulston and Horvitz, 1977; Sulston *et al.*, 1983). All of these developmental strategies share in common the fact that once sexual differentiation takes place, the ability to adopt the alternative fate is lost or diminished.

There are two notable exceptions to this general rule of constrained developmental potential: the germ line, which naturally switches between male and female programs during hermaphrodite development, and the intestine, which is dimorphic not in anatomy or lineage but in the fact that yolk protein genes are expressed in this tissue only in the hermaphrodite (Kimble and Sharrock, 1983). Only with respect to these tissues that retain the possibility of plasticity in sexually dimorphic function may we reasonably ask which sex determination functions are required to maintain the choice of sexual pathway in adult animals.

The term "sex determination" is perhaps not particularly applicable to the germ line as a whole, since in hermaphrodites primordial germ cells are not set aside as separate sperm or oocyte precursors at an early stage (Hodgkin, 1974). Further, while the TSPs of *fem-1* and *fem-2* for hermaphrodite spermatogenesis suggest that the ability to make sperm is influenced by the presence or absence of these *fem* products during earlier larval stages (Nelson *et al.*, 1978; Kimble *et al.*, 1984), it is not known whether individual germ cell nuclei become committed to the male or female program before differentiation takes place in the L4 or adult. Many individual undifferentiated germ cell nuclei clearly retain the potential to undergo either spermatogenesis or oogenesis. Temperature-shift experiments in XX animals with a *fem-3(lf)* mutation indicate that *fem-3* is needed throughout the entire period of hermaphrodite spermatogenesis; shifting animals from permissive to restrictive temperatures during this time causes germ cells that would otherwise differentiate as sperm to undergo oogenesis instead (Hodgkin, 1986). Also, temperature shifts with a *fem-3(gf)* allele indicate that XX animals which have already switched to oogenesis can switch back to spermatogenesis (Barton *et al.*, 1987). These experiments suggest that the presence of active *fem-3* just prior to or during germ cell differentiation is both necessary and sufficient to cause spermatogenesis in hermaphrodites.

The male germ line differs from the hermaphrodite germ line in that only the male program of germ cell differentiation is normally followed. Temperature-shift experiments suggest that the product of *fem-2* is needed continually through adulthood to maintain spermatogenesis and prevent oogenesis in males (Kimble *et al.*, 1984). Temperature shifts with *her-1(ts)* also suggest a maintenance function for *her-1* in continuing spermatogenesis in the male germ line (Schedin, 1988). This result contrasts with the TSP for *her-1* for gonad sexual phenotype, which is a short and discrete period during embryonic development, suggesting an early determinative role for *her-1* in this tissue (Hodgkin, 1984).

Experiments with *her-1(ts)* further suggest that *her-1* function is required in adult males to maintain the "differentiated" state of the male intestine, that is, to prevent yolk protein expression. Specifically, XO males that have been raised at permissive temperatures to adulthood begin to produce yolk proteins within 36 hours after a shift to restrictive temperatures (Schedin, 1988). Analogous experiments could be performed with *tra-2(ts)* to determine whether *tra-2* function is required in adult hermaphrodites to actively promote and maintain yolk protein expression.

One as yet unexplored area of sexual dimorphism with a potential for plasticity is that of sex-specific behavior. Do sex-specific neurons, once generated, still require the function of sex determination genes in order to maintain the ability to elicit sex-specific behaviors, or once they have differentiated can they continue to function independently of any sex determination functions? In other words, does the sex determination pathway become dispensable to their function? This question can be addressed at both the behavioral and immunohistochemical levels using temperature-sensitive alleles of the sex-determining genes. For example, *tra-2(ts)* egg-laying proficient XX adult animals (raised at the permissive temperature) could be shifted to the restrictive temperature and scored for the ability of the HSN neurons to continue to drive egg-laying behavior and/or to continue production of serotonin, the HSN neurotransmitter (Desai *et al.*, 1988). Similarly, *her-1(ts)* XO mating males (raised at the permissive temperature) could be shifted to the restrictive temperature and scored for the persistence of male mating behavior.

XIV. Concluding Remarks

Throughout this article, we have compared and contrasted various aspects of the sex determination and dosage compensation processes in *C. elegans* with their counterparts in *D. melanogaster* and mammals, the two other systems in which these processes have been most thoroughly studied. We conclude with two final points relevant to these comparisons. The first is that sex determination strategies that initially appear to be quite different may not in fact be so different after all. At face value, the sex-determining mechanism of *C. elegans*, in which the X/A ratio serves as the primary sex-determining signal, seems extremely unlike that found in mammals, where a dominant sex chromosome (the Y chromosome in males) is the primary determinant of sex. Several investigators have shown, however, that *C. elegans* can be readily interconverted from the XX hermaphrodite–XO male state to a ZW female–ZZ male state, in which sex is determined by a dominant sex chromosome (the W chromosome in females), by using a combination of dominant and recessive mutations at a single sex-determining locus (Hodgkin, 1983b; Miller *et al.*, 1988). One can further envision using a temperature-sensitive sex determination mutation to construct a stable *C. elegans* strain in which sex is determined environmentally, depending on the temperature during embryonic and larval development. Thus, it is easy to see how one type of primary sex-determining mechanism might easily and rapidly evolve from another. Evolutionary mechanisms by which this type of change might occur, as well as examples from other organisms in which populations have been changed from one sex-determining system to another under selective pressure, have been discussed extensively by Bull (1983). The demonstrated potential for plasticity in primary sex-determining mechanisms suggests that what appear to be major differences between sex determination systems may in fact result from relatively small modifications of strategies that share many common features.

The opposite side of this argument brings us to the second point, which is that strategies that at first appear quite similar may not in fact be so similar. For example, both *C. elegans* and *Drosophila* use the X/A ratio as the primary sex-determining signal, both dosage compensate using mechanisms involving differential gene expression rather than X inactivation, and both coordinately control their sex determination and dosage compensation processes. These apparent similarities may be purely superficial, however. Clearly, classes of genes have been identified by mutation in *C. elegans* that do not correspond to any identified in *Drosophila*, and vice versa, although this is simply lack of evidence and alone is not a compelling argument for significant differences. Further, while most loss-of-function mutations in *C. elegans* sex-determining genes sexually transform both soma and germ line, loss-of-function mutations in the *Drosophila* genes *tra, tra-2, dsx,* and *ix* transform only soma and not germ line (Schüpbach, 1982; Marsh and Wieschaus, 1987).

As the molecular characterization of the *C. elegans* genes controlling sex determination and dosage compensation proceeds, we can begin to make direct comparisons between the molecules and the regulatory mechanisms involved in the sex determination processes in these two organisms. While the data are far from complete, it is nevertheless informative to examine what is known to date. The *Drosophila* sex-determining genes *da, Sxl, tra, tra-2,* and *dsx* have been molecularly cloned (Maine *et al.*, 1985; Butler *et al.*, 1986; McKeown *et al.*, 1987; Amrein *et al.*, 1988; Baker and Wolfner, 1988; Caudy *et al.*, 1988b; Cronmiller *et al.*, 1988). Much of the molecular analysis of these genes has demonstrated that alternative RNA splicing plays a major role in the male–female decision. Three of the genes, *Sxl, tra,* and *dsx,* give rise to sex-specific transcripts via alternative RNA splicing (Baker *et al.*, 1987; Boggs *et al.*, 1987; Bell *et al.*, 1988), and two of the genes, *Sxl* and *tra-2*, exhibit sequence similarity to RNA-binding proteins, suggesting that they may be directly involved in the splicing machinery responsible for the regulation (Amrein *et al.*, 1988; Bell *et al.*, 1988). Where data are available for nematode sex-determining genes, no sequence similarity to the *Drosophila* genes has been observed (Rosenquist and Kimble, 1988; P. Kuwabara and J. Kimble, personal communication). Further, although (at least) two of the *C. elegans* genes appear to be regulated at the transcript level, this regulation does not appear to occur by sex-specific splicing mechanisms. *her-1* seems likely to be regulated transcriptionally, since the putative *her-1* transcripts are present mainly or exclusively in XO animals (C. Trent and W. Wood, personal communication). The *fem-3* gene produces transcripts that are present in both sexes, but in different amounts and/or in different tissues; at least some of this regulation occurs at the level of message stability (Rosenquist and Kimble, 1988). The molecular data on the nematode sex-determining genes are still rather limited, of course, so similarities with *Drosophila* may yet be detected.

Caenorhabditis elegans and *Drosophila* show an additional striking difference at the level of sex-specific differentiated gene products. Specifically, while the vitellogenins (yolk proteins) of nematodes, sea urchin (*Strongylocentrotus purpuratus*), and two vertebrates, frog (*Xenopus laevis*) and chicken (*Gallus gallus*), are all encoded by members of a single related gene family, the analogous proteins in *Drosophila* are completely unrelated (Spieth *et al.*, 1985; Blumenthal and Zucker-Aprison, 1987).

If sex determination in *C. elegans* and *Drosophila* turns out to be quite different in detail, this should not be at all surprising in light of the fact that nematodes and flies are probably as distantly related from each other by many phylogenetic and taxonomic criteria as either group is from mammals. While biparentalism and gamete dimorphism almost certainly evolved long before the three branches diverged more than 580 million years ago (Margulis and Schwartz, 1982; Margulis and Sagan, 1986), strategies for diversification of the sexes and sex-determining mechanisms have subsequently coevolved along with the obviously different anatomy, physiology, ontogeny, and ecological niches characteristic of the present-day organisms. The large number of genes involved in the male–hermaphrodite decision in *C. elegans* perhaps reflects the evolutionary history that ultimately gave rise to the sex determination system that exists in the species today.

As a final note, we should not be alarmed if the mechanisms by which sexual fate is determined happen to be different in each of the systems in which the process has been best studied. This result would be disturbing only if one considers the study of sex determination from an extremely narrow perspective, that is, from the point of view that one studies the mechanism of sex determination in a model organism in order to learn how sex determination works in general in other organisms. When sex determination is considered in a broader context, as a model for studying fundamental processes in developmental biology, the field is enriched by the finding of such diversity since it increases the number and types of developmental strategies that may be discovered and elucidated.

ACKNOWLEDGMENTS

We gratefully acknowledge our numerous colleagues who allowed us to keep this review as up-to-date as possible by permitting us to cite unpublished data, in particular, members of the laboratories of B. Meyer, J. Hodgkin, J. Kimble, C. Trent, and W. Wood. We also thank T. Cline, S. Kim, and all of the members of our laboratory for challenging discussions, speculations, and flights of fancy that have helped us to formulate some of the arguments and ideas presented in this article. We thank L. Miller, C. Nusbaum, L. DeLong, and M. Basson for comments on portions of the manuscript, and we are especially indebted to R. K. Herman and R. Klein, who graciously took time to review the entire manuscript.

REFERENCES

Ambros, V., and Horvitz, H. R. (1987). The *lin-14* locus of *Caenorhabditis elegans* controls the time of expression of specific postembryonic developmental events. *Genes Dev.* **1**, 398–414.

Amrein, H., Gorman, M., and Nöthiger, R. (1988). The sex-determining gene *tra-2* of Drosophila encodes a putative RNA binding protein. *Cell* **55**, 1025–1035.

Anderson, K. V., Jürgens, G., and Nüsslein-Volhard, C. (1985). Establishment of dorsal–ventral polarity in the Drosophila embryo: Genetic studies on the role of the *Toll* gene product. *Cell* **42**, 779–789.

Baker, B. S., and Belote, J. M. (1983). Sex determination and dosage compensation in *Drosophila melanogaster. Annu. Rev. Genet.* **17**, 345–393.

Baker, B. S., and Ridge, K. A. (1980). Sex and the single cell. I. On the action of major loci affecting sex determination in *Drosophila melanogaster. Genetics* **94**, 383–423.

Baker, B. S., and Wolfner, M. F. (1988). A molecular analysis of *doublesex,* a bifunctional gene that controls both male and female sexual differentiation in *Drosophila melanogaster. Genes Dev.* **2**, 477–489.

Baker, B. S., Nagoshi, R. N., and Burtis, K. C. (1987). Molecular genetic aspects of sex determination in Drosophila. *BioEssays* **6**, 66–77.

Barton, M. K., Schedl, T. B., and Kimble, J. (1987). Gain-of-function mutations of *fem-3,* a sex-determining gene in *Caenorhabditis elegans. Genetics* **115**, 107–119.

Baverstock, P. R., Adams, M., Polkinghorne, R. W., and Gelder, M. (1982). A sex-linked enzyme in birds—Z chromosome conservation but no dosage compensation. *Nature (London)* **296**, 763–766.

Bell, L. R., Maine, E. M., Schedl, P., and Cline, T. W. (1988). *Sex-lethal,* a Drosophila sex determination switch gene, exhibits sex-specific RNA splicing and sequence similarity to RNA binding proteins. *Cell* **55**, 1037–1046.

Blumenthal, T., and Zucker-Aprison, E. (1987). Evolution and regulation of vitellogenin genes. *In* "Molecular Biology of Invertebrate Development" (J. D. O'Connor, ed.), pp. 3–19. Liss, New York.

Blumenthal, T., Squire, M., Kirtland, S., Cane, J., Donegan, M., Spieth, J., and Sharrock, W. J. (1984). Cloning of a yolk protein gene family from *Caenorhabditis elegans*. *J. Mol. Biol.* **174**, 1–18.

Boggs, R. T., Gregor, P., Idriss, S., Belote, J. M., and McKeown, M. (1987). Regulation of sexual differentiation in *D. melanogaster* via alternative splicing of RNA from the *transformer* gene. *Cell* **50**, 739–747.

Bridges, C. B. (1925). Sex in relation to chromosomes. *Am. Nat.* **59**, 127–137.

Bull, J. J. (1983). "The Evolution of Sex Determining Mechanisms." Benjamin/ Cummings, Menlo Park, California.

Burgoyne, P. S., Buehr, M., Koopman, P., Rossant, J., and McLaren, A. (1988). Cell-autonomous action of the testis-determining gene: Sertoli cells are exclusively XY in XX/XY chimaeric mouse testes. *Development* **102**, 443–450.

Burke, D. J., and Ward, S. (1983). Identification of a large multigene family encoding the major sperm protein of *Caenorhabditis elegans*. *J. Mol. Biol.* **171**, 1–29.

Butler, B., Pirrotta, V., Irminger-Finger, I., and Nöthiger, R. (1986). The sex-determining gene *tra* of Drosophila: Molecular cloning and transformation studies. *EMBO J.* **5**, 3607–3613.

Caudy, M., Grell, E. H., Dambly-Chaudiere, C., Ghysen, A., Jan, L. Y., and Jan, Y. N. (1988a). The maternal sex determination gene *daughterless* has zygotic activity necessary for the formation of peripheral neurons in Drosophila. *Genes Dev.* **2**, 843–852.

Caudy, M., Vassin, H., Brand, M., Tuma, R., Jan, L. Y., and Jan, Y. N. (1988b). *daughterless*, a Drosophila gene essential for both neurogenesis and sex determination, has sequence similarities to *myc* and the *achaete–scute* complex. *Cell* **55**, 1061–1067.

Cline, T. W. (1976). A sex-specific temperature-sensitive maternal effect of the *daughterless* mutation of *Drosophila melanogaster*. *Genetics* **84**, 723–742.

Cline, T. W. (1978). Two closely linked mutations in *Drosophila melanogaster* that are lethal to opposite sexes and interact with *daughterless*. *Genetics* **90**, 683–698.

Cline, T. W. (1979a). A male-specific lethal mutation in *Drosophila melanogaster* that transforms sex. *Dev. Biol.* **72**, 266–275.

Cline, T. W. (1979b). A product of the maternally-influenced *Sex-lethal* gene determines sex in *Drosophila melanogaster*. *Genetics* **91**, s22.

Cline, T. W. (1980). Maternal and zygotic sex-specific gene interactions in *Drosophila melanogaster*. *Genetics* **96**, 903–926.

Cline, T. W. (1983). The interaction between *daughterless* and *Sex-lethal* in triploids: A lethal sex-transforming maternal effect linking sex determination and dosage compensation in *Drosophila melanogaster*. *Dev. Biol.* **95**, 260–274.

Cline, T. W. (1984). Autoregulatory functioning of a *Drosophila* gene product that establishes and maintains the sexually determined state. *Genetics* **107**, 231–277.

Cline, T. W. (1985). Primary events in the determination of sex in *Drosophila melanogaster*. *In* "Origin and Evolution of Sex" (H. O. Halvorson and A. Monroy, eds.), pp. 301–327. Liss, New York.

Cline, T. W. (1986). A female-specific lesion in an X-linked positive regulator of the *Drosophila* sex determination gene, *Sex lethal*. *Genetics* **113**, 641–663 (corrigendum: **114**, 345).

Cline, T. W. (1988). Evidence that *sisterless-a* and *sisterless-b* are two of several discrete "numerator elements" of the X/A sex determination signal in *Drosophila* that switch *Sxl* between two alternative stable expression states. *Genetics* **119**, 829–862.

Coulson, A., Sulston, J., Brenner, S., and Karn, J. (1986). Toward a physical map of the *Caenorhabditis elegans* genome. *Proc. Natl. Acad. Sci. U.S.A.* **83,** 7821–7825.

Coulson, A., Waterston, R., Kiff, J., Sulston, J., and Kohara, Y. (1988). Genome linking with artificial chromosomes. *Nature (London)* **335,** 184–186.

Cronmiller, C., and Cline, T. W. (1987). The *Drosophila* sex determination gene *daughterless* has different functions in the germ line versus the soma. *Cell* **48,** 479–487.

Cronmiller, C., Schedl, P., and Cline, T. W. (1988). Molecular characterization of *daughterless,* a *Drosophila* sex determination gene with multiple roles in development. *Genes Dev.* **2,** 1666–1676.

DeLong, L., Casson, L. P., and Meyer, B. J. (1987). Assessment of X chromosome dosage compensation in *Caenorhabditis elegans* by phenotypic analysis of *lin-14*. *Genetics* **117,** 657–670.

Desai, C., and Horvitz, H. R. (1989). *Caenorhabditis elegans* mutants defective in the functioning of the motor neurons responsible for egg laying. *Genetics* **121,** 703–721.

Desai, C., Garriga, G., McIntire, S. L., and Horvitz, H. R. (1988). A genetic pathway for the development of the *Caenorhabditis elegans* HSN motor neurons. *Nature (London)* **336,** 638–646.

Donahue, L. M., Quarantillo, B. A., and Wood, W. B. (1987). Molecular analysis of X chromosome dosage compensation in *Caenorhabditis elegans. Proc. Natl. Acad. Sci. U.S.A.* **84,** 7600–7604.

Doniach, T. (1986a). Activity of the sex-determining gene *tra-2* is modulated to allow spermatogenesis in the *C. elegans* hermaphrodite. *Genetics* **114,** 53–76.

Doniach, T. (1986b). Genetic analysis of sex determination in the nematode *Caenorhabditis elegans*. Ph.D. thesis, Counc. Natl. Acad. Awards, United Kingdom.

Doniach, T., and Hodgkin, J. (1984). A sex-determining gene, *fem-1*, required for both male and hermaphrodite development in *Caenorhabditis elegans. Dev. Biol.* **106,** 223–235.

Ferguson, E. L. (1985). The genetic analysis of the vulval cell lineages of *Caenorhabditis elegans*. Ph.D. thesis, Massachusetts Institute of Technology, Cambridge, Massachusetts.

Ferguson, E. L., and Horvitz, H. R. (1985). Identification and characterization of 22 genes that affect the vulval cell lineages of the nematode *Caenorhabditis elegans. Genetics* **110,** 17–72.

Fire, A. (1986). Integrative transformation of *Caenorhabditis elegans. EMBO J.* **5,** 2673–2680.

Ford, C. E., Miller, O. J., Polani, P. E., de Almeida, J. C., and Briggs, J. H. (1959). A sex-chromosome anomaly in a case of gonadal dysgenesis (Turner's syndrome). *Lancet* **1,** 711.

Gartler, S. M., and Riggs, A. D. (1983). Mammalian X-chromosome inactivation. *Annu. Rev. Genet.* **17,** 155–190.

Greenwald, I. S., Sternberg, P. W., and Horvitz, H. R. (1983). The *lin-12* locus specifies cell fates in *Caenorhabditis elegans. Cell* **34,** 435–444.

Herman, R. K. (1984). Analysis of genetic mosaics of the nematode *Caenorhabditis elegans. Genetics* **108,** 165–180.

Hodgkin, J. (1974). Genetic and anatomical aspects of the *Caenorhabditis elegans* male. Ph.D. thesis, Cambridge University, Cambridge, England.

Hodgkin, J. (1980). More sex-determination mutants of *Caenorhabditis elegans. Genetics* **96,** 649–664.

Hodgkin, J. (1983a). X chromosome dosage and gene expression in *Caenorhabditis elegans*: Two unusual dumpy genes. *Mol. Gen. Genet.* **192,** 452–458.

Hodgkin, J. (1983b). Two types of sex determination in a nematode. *Nature (London)* **304**, 267–268.

Hodgkin, J. (1984). Switch genes and sex determination in the nematode *C. elegans*. *J. Embryol. Exp. Morphol.* **83** (Suppl.), 103–117.

Hodgkin, J. (1985). Novel nematode amber suppressors. *Genetics* **111**, 287–310.

Hodgkin, J. (1986). Sex determination in the nematode *C. elegans*: Analysis of *tra-3* suppressors and characterization of *fem* genes. *Genetics* **114**, 15–52.

Hodgkin, J. (1987a). Sex determination and dosage compensation in *Caenorhabditis elegans*. *Annu. Rev. Genet.* **21**, 133–154.

Hodgkin, J. (1987b). Primary sex determination in the nematode *C. elegans*. *Development* **101** (Suppl.), 5–16.

Hodgkin, J. (1987c). A genetic analysis of the sex-determining gene, *tra-1*, in the nematode *Caenorhabditis elegans*. *Genes Dev.* **1**, 731–745.

Hodgkin, J. (1988). Sexual dimorphism and sex determination. *In* "The Nematode *Caenorhabditis elegans*" (W. B. Wood, ed.), pp. 243–280. Cold Spring Harbor Laboratory, Cold Spring Harbor, New York.

Hodgkin, J., and Brenner, S. (1977). Mutations causing transformation of sexual phenotype in the nematode *Caenorhabditis elegans*. *Genetics* **86**, 275–287.

Hodgkin, J. Horvitz, H. R., and Brenner, S. (1979). Nondisjunction mutants of the nematode *Caenorhabditis elegans*. *Genetics* **91**, 67–94.

Hodgkin, J., Doniach, T., and Shen, M. (1985). The sex determination pathway in the nematode *Caenorhabditis elegans*: Variations on a theme. *Cold Spring Harbor Symp. Quant. Biol.* **50**, 585–593.

Jacobs, P. A., and Strong, J. A. (1959). A case of human intersexuality having a possible XXY sex-determining mechanism. *Nature (London)* **183**, 302–303.

Johnson, M. S., and Turner, J. R. G. (1979). Absence of dosage compensation for a sex-linked gene in butterflies Heliconius. *Heredity* **43**, 71–77.

Jost, A. (1953). Problems of fetal endocrinology: The gonadal and hyphyseal hormones. *Recent Prog. Horm. Res.* **8**, 379–418.

Kimble, J., and Sharrock, W. J. (1983). Tissue-specific synthesis of yolk proteins in *Caenorhabditis elegans*. *Dev. Biol.* **96**, 189–196.

Kimble, J., and Ward, S. (1988). Germ-line development and fertilization. *In* "The Nematode *Caenorhabditis elegans*" (W. B. Wood, ed.), pp. 191–215. Cold Spring Harbor Laboratory, Cold Spring Harbor, New York.

Kimble, J., Edgar, L., and Hirsh, D. (1984). Specification of male development in *Caenorhabditis elegans*: The *fem* genes. *Dev. Biol.* **105**, 234–239.

Klass, M., and Hirsh, D. (1981). Sperm isolation and biochemical analysis of the major sperm protein from *Caenorhabditis elegans*. *Dev. Biol.* **84**, 299–312.

Klass, M., Wolf, N., and Hirsh, D. (1976). Development of the male reproductive system and sexual transformation in the nematode *C. elegans*. *Dev. Biol.* **52**, 1–18.

Klass, M. R., Wolf, N., and Hirsh, D. (1979). Further characterization of a temperature-sensitive transformation mutant in *Caenorhabditis elegans*. *Dev. Biol.* **69**, 329–335.

Klass, M., Dow, B., and Herndon, M. (1982). Cell-specific transcriptional regulation of the major sperm protein from *Caenorhabditis elegans*. *Dev. Biol.* **93**, 152–164.

Klass, M. R., Kinsley, S., and Lopez, L. C. (1984). Isolation and characterization of a sperm-specific gene family in the nematode *Caenorhabditis elegans*. *Mol. Cell. Biol.* **4**, 529–537.

Klass, M., Ammons, D., and Ward, S. (1988). Conservation of the 5'-flanking sequences of transcribed members of the *Caenorhabditis elegans* major sperm protein gene family. *J. Mol. Biol.* **199**, 14–22.

Lewis, E. B. (1978). A gene complex controlling segmentation in *Drosophila*. *Nature* (*London*) **276**, 565–570.

Lucchesi, J. C., and Manning, J. E. (1987). Gene dosage compensation in *Drosophila melanogaster*. *Adv. Genet.* **24**, 371–429.

Lucchesi, J. C., and Skripsky, T. (1981). The link between dosage compensation and sex determination in *Drosophila melanogaster*. *Chromosoma* **82**, 217–227.

Lyon, M. F., and Hawkes, S. G. (1970). An X-linked gene for testicular feminization of the mouse. *Nature* (*London*) **227**, 1217–1219.

McCoubrey, W. K., Nordstrom, K. D., and Meneely, P. M. (1988). Microinjected DNA from the X chromosome affects sex determination in *Caenorhabditis elegans*. *Science* **242**, 1146–1151.

McKeown, M., Belote, J. M., and Baker, B. S. (1987). A molecular analysis of *tra*, a gene in *Drosophila melanogaster* that controls sexual differentiation. *Cell* **48**, 489–499.

McKeown, M., Belote, J. M., and Boggs, R. T. (1988). Ectopic expression of the female transformer gene product leads to female differentiation of chromosomally male Drosophila. *Cell* **53**, 887–895.

McLaren, A. (1984). Chimeras and sexual differentiation. *In* "Chimeras in Developmental Biology" (N. Le Douarin and A. McLaren, eds.), pp. 381–399. Academic Press, New York.

McLaren, A. (1988). Sex determination in mammals. *Trends Genet.* **4**, 153–157.

Madl, J. G., and Herman, R. K. (1979). Polyploids and sex determination in *Caenorhabditis elegans*. *Genetics* **93**, 393–402.

Maine, E. M., Salz, H. K., Cline, T. W., and Schedl, P. (1985). The *Sex-lethal* gene of *Drosophila*: DNA alterations associated with sex-specific lethal mutations. *Cell* **43**, 521–529.

Mange, A. P., and Sandler, L. (1973). A note on the maternal effect mutants daughterless and abnormal oocyte in *Drosophila melanogaster*. *Genetics* **73**, 73–86.

Margulis, L., and Sagan, D. (1986). "Origins of Sex: Three Billion Years of Genetic Recombination." Yale Univ. Press, New Haven, Connecticut.

Margulis, L., and Schwartz, K. V. (1982). "Five Kingdoms." Freeman, New York.

Marsh, J. L., and Wieschaus, E. (1978). Is sex determination in germline and soma controlled by separate genetic mechanisms? *Nature* (*London*) **272**, 249–251.

Meneely, P. M., and Nordstrom, K. D. (1988). X chromosome duplications affect a region of the chromosome they do not duplicate in *Caenorhabditis elegans*. *Genetics* **119**, 365–375.

Meneely, P. M., and Wood, W. B. (1984). An autosomal gene that affects X chromosome expression and sex determination in *Caenorhabditis elegans*. *Genetics* **106**, 29–44.

Meneely, P. M., and Wood, W. B. (1987). Genetic analysis of X-chromosome dosage compensation in *Caenorhabditis elegans*. *Genetics* **117**, 25–41.

Meyer, B. J. (1988). Primary events in *C. elegans* sex determination and dosage compensation. *Trends Genet.* **4**, 337–342.

Meyer, B. J., and Casson, L. P. (1986). *Caenorhabditis elegans* compensates for the difference in X chromosome dosage between the sexes by regulating transcript levels. *Cell* **47**, 871–881.

Miller, L. M., Plenefisch, J. D., Casson, L. P., and Meyer, B. J. (1988). *xol-1*: A gene that controls the male modes of both sex determination and X chromosome dosage compensation in *C. elegans*. *Cell* **55**, 167–183.

Morgan, T. H., and Bridges, C. B. (1919). Contribution to the genetics of *Drosophila melanogaster*. I. The origin of gynandromorphs. *Carnegie Inst. Washington Publ.* **278**, 1–122.

Muller, H. J. (1950). Evidence for the precision of genetic adaptation. *Harvey Lect. Ser.* (Vol. XLIII, 1947–1948), 165–229.

Nagoshi, R. N., McKeown, M., Burtis, K. C., Belote, J. M., and Baker, B. S. (1988). The control of alternative splicing at genes regulating sexual differentiation in *D. melanogaster. Cell* **53**, 229–236.

Nelson, G., Lew, K. K., and Ward, S. (1978). Intersex, a temperature-sensitive mutant in the nematode *Caenorhabditis elegans. Dev. Biol.* **66**, 386–409.

Nigon, V. (1949a). Les modalités de la reproduction et le déterminisme de sexe chez quelques nématodes libres. *Ann. Sci. Nat. Zool.* **2** (Ser. 11), 1–132.

Nigon, V. (1949b). Effets de la polyploidie chez un nématode libre *C. R. Acad. Sci., Paris* **228**, 1161–1162.

Nigon, V. (1951). Polyploidie expérimentale chez un nématode libre, *Rhabditis elegans Maupas. Bull. Biol. Fr. Belg.* **85**, 187–225.

Nusbaum, C., and Meyer, B. J. (1989). The *C. elegans* gene *sdc-2* controls sex determination and dosage compensation in XX animals. *Genetics* **122**, 579–593.

Page, D. C., Mosher, R., Simpson, E. M., Fisher, E. M. C., Mardon, G., Pollack, J., McGillivray, B., de la Chapelle, A., and Brown, L. G. (1987). The sex-determining region of the human Y chromosome encodes a finger protein. *Cell* **51**, 1091–1104.

Plenefisch, J. D., DeLong, L., and Meyer, B. J. (1989). Genes that implement the hermaphrodite mode of dosage compensation in *Caenorhabditis elegans. Genetics* **121**, 57–76.

Rosenquist, T. A., and Kimble, J. (1988). Molecular cloning of *fem-3*, a sex-determining gene in *Caenorhabditis elegans. Genes Dev.* **2**, 606–616.

Ruvkun, G., Ambros, V., Coulson, A., Waterston, R., Sulston, J., and Horvitz, H. R. (1989). Molecular genetics of the *Caenorhabditis elegans* heterochronic gene *lin-14. Genetics* **121**, 501–516.

Sanchez, L., and Nöthiger, R. (1982). Clonal analysis of *Sex-lethal,* a gene needed for sexual development in *Drosophila melanogaster. Wilhelm Roux Arch. Entwicklungsmech. Org.* **191**, 211–214.

Sanchez, L., and Nöthiger, R. (1983). Sex determination and dosage compensation in *Drosophila melanogaster*: Production of male clones in XX females. *EMBO J.* **2**, 485–491.

Sandler, L. (1972). On the genetic control of genes located in the sex-chromosome heterochromatin of *Drosophila melanogaster. Genetics* **70**, 261–274.

Schedin, P. (1988). Tissue-autonomy, timing, and reversibility of sex determination in *Caenorhabditis elegans.* Ph.D. thesis, University of Colorado, Boulder, Colorado.

Schedl, T., and Kimble, J. (1988). *fog-2*, a germ-line-specific sex determination gene required for hermaphrodite spermatogenesis in *Caenorhabditis elegans. Genetics* **119**, 43–61.

Schedl, T., Graham, P., Barton, M. K., and Kimble, J. (1989). Analysis of the role of *tra-1* in germ line sex determination in the nematode *Caenorhabditis elegans. Genetics* **123**, 755–769.

Schüpbach, T. (1982). Autosomal mutations that interfere with sex determination in somatic cells of *Drosophila* have no direct effect on the germline. *Dev. Biol.* **89**, 117–127.

Shen, M., and Hodgkin, J. (1988). *mab-3*, a gene required for sex-specific yolk protein expression and a male-specific lineage in *C. elegans. Cell* **54**, 1019–1031.

Singh, L., Matsukuma, S., and Jones, K. W. (1987). Testis development in a predominantly XX mosaic mouse. *Dev. Biol.* **122**, 287–290.

Spieth, J., and Blumenthal, T. (1985). The *Caenorhabditis elegans* vitellogenin gene family includes a gene encoding a distantly related protein. *Mol. Cell. Biol.* **5**, 2595–2501.

Spieth, J., Denison, K., Zucker, E., and Blumenthal, T. (1985). The *C. elegans* vitellogenin genes: Short sequence repeats in the promoter regions and homology to the vertebrate genes. *Nucleic Acids Res.* **13**, 5283–5295.

Spieth, J., MacMorris, M., Broverman, S., Greenspoon, S., and Blumenthal, T. (1988). Regulated expression of a vitellogenin fusion gene in transgenic nematodes. *Dev. Biol.* **130**, 285–293.

Spradling, A. C., and Rubin, G. M. (1983). The effect of chromosomal position on the expression of the *Drosophila* xanthine dehydrogenase gene. *Cell* **34**, 47–57.

Sternberg, P. (1989). Genetic control of cell type and pattern formation in *C. elegans*. *Adv. Genet.* **27**, 63–115.

Sternberg, P., and Horvitz, H. R. (1988). *lin-17* mutations of *Caenorhabditis elegans* disrupt certain asymmetric cell divisions. *Dev. Biol.* **130**, 67–73.

Sudhaus, W. (1974). Zur Systematik, Verbreitung, Ökologie und Biologie neue und wenig bekannter Rhabditiden (Nematoda). *Zool. Jahrb. Abt. Syst. Oekol. Geogr.* **101**, 417–465.

Sulston, J. (1988). Cell lineage. *In* "The Nematode *Caenorhabditis elegans*" (W. B. Wood, ed.), pp. 123–156. Cold Spring Harbor Laboratory, Cold Spring Harbor, New York.

Sulston, J. E., and Horvitz, H. R. (1977). Post-embryonic cell lineages of the nematode *Caenorhabditis elegans*. *Dev. Biol.* **56**, 110–156.

Sulston, J. E., Schierenberg, E., White, J. G., and Thomson, J. N. (1983). The embryonic cell lineage of the nematode *Caenorhabditis elegans*. *Dev. Biol.* **100**, 64–119.

Tokunaga, C. (1962). Cell lineage and differentiation in the male foreleg of *Drosophila melanogaster*. *Dev. Biol.* **4**, 489–516.

Trent, C., Tsung, N., and Horvitz, H. R. (1983). Egg-laying defective mutants of the nematode *Caenorhabditis elegans*. *Genetics* **104**, 619–647.

Trent, C., Wood, W. B., and Horvitz, H. R. (1988). A novel dominant transformer allele of the sex-determining gene *her-1* of *Caenorhabditis elegans*. *Genetics* **120**, 145–157.

Villeneuve, A. M., and Meyer, B. J. (1987). *sdc-1*: A link between sex determination and dosage compensation in *C. elegans*. *Cell* **48**, 25–37.

Villeneuve, A. M., and Meyer, B. J. (1990). The role of *sdc-1* in the sex determination and dosage compensation decisions in *C. elegans*. *Genetics* **124**, 91–114.

Ward, S., and Klass, M. (1982). The location of the major protein in *Caenorhabditis elegans* sperm and spermatocytes. *Dev. Biol.* **92**, 203–208.

Ward, S., Burke, D. J., Sulston, J. E., Coulson, A. R., Albertson, D. G., Ammons, D., Klass, M., and Hogan, E. (1988). The genomic organization of transcribed major sperm protein genes and pseudogenes in the nematode *Caenorhabditis elegans*. *J. Mol. Biol.* **199**, 1–13.

Welshons, W. J., and Russell, L. B. (1959). The Y chromosome as the bearer of male factors in the mouse. *Proc. Natl. Acad. Sci. U.S.A.* **45**, 560–566.

White, J. (1988). The anatomy. *In* "The Nematode *Caenorhabditis elegans*" (W. B. Wood, ed.), pp. 81–122. Cold Spring Harbor Laboratory, Cold Spring Harbor, New York.

Wood, W. B., Meneely, P., Schedin, P., and Donahue, L. (1985). Aspects of dosage compensation and sex determination in *Caenorhabditis elegans*. *Cold Spring Harbor Symp. Quant. Biol.* **50**, 575–593.

Wood, W. B., Trent, C., Meneely, P., Manser, J., and Burgess, S. (1987). Control of X-chromosome expression and sex determination in embryos of *Caenorhabditis elegans*. *In* "Genetic Regulation of Development: 45th Symposium of the Society of Developmental Biology" (W. Loomis, ed.), pp. 191–199. Liss, New York.

GENETIC CONTROL OF SEX DETERMINATION IN *Drosophila*

Monica Steinmann-Zwicky, Hubert Amrein,* and Rolf Nöthiger

Zoological Institute, University of Zurich,
CH-8057 Zurich, Switzerland

I. Introduction

In 1921, Bridges published the classic model that attempted to explain how sex is determined in *Drosophila*. Observing that genotypes 2X;2A and 3X;3A develop as females, that flies with 1X;2A are males, irrespective of the presence of a Y chromosome, and that flies of genotype 2X;3A are intersexes, he concluded that the ratio of X chromosomes to sets of autosomes (X : A) provides the primary genetic signal for sex determination. Bridges visualized this signal as the result of two sets of counteracting elements, with the X chromosomes carrying female-determining factors and the autosomes male-determining factors.

---

* Present address: Department of Biochemistry and Molecular Biology, Harvard University, Cambridge, Massachusetts 02138.

ADVANCES IN GENETICS, Vol. 27

Males and females differ in many ways. The sexual dimorphism becomes apparent in an animal's behavior, in morphology and function of somatic tissues, and in the germ line. In *Drosophila*, as in many other species, the male has one X chromosome whereas the female has two. Since most X-chromosomal genes are unrelated to sex, a mechanism has evolved to equalize the products of such genes in the two sexes. This process, called dosage compensation, ensures that X-chromosomal genes are transcribed twice as efficiently in males as they are in females.

The primary chromosomal signal of the X : A ratio regulates all aspects of sex (Fig. 1). The genes producing the sexual phenotype, however, are not directly controlled by the X : A ratio. Rather, this signal achieves its effects through a short hierarchy of regulatory genes that transmit the signal to the target genes. In a first step, the X : A ratio leads to differential activity of *Sxl* (*Sex-lethal*), a gene that controls sex determination in the soma and in the germ line as well as dosage compensation. Below *Sxl*, the genetic pathways for the three processes diverge. The second step in the pathway that regulates somatic sex determination is the transmission of the state of activity of *Sxl*, via a short cascade of subordinate control genes, to the bifunctional gene *dsx* (*double-sex*). The third step involves the regulation of the differentiation genes by the products of *dsx*, that is, the realization of the overt sexual phenotype.

In this article, we concentrate on the sex determination pathway in somatic cells, but we also discuss some aspects of dosage compensation and sex determination in the germ line since the three processes are in part controlled by the same genes. For a thorough review of dosage compensation, the readers are referred to Lucchesi and Manning (1987) and for new results on sex determination in the germ line to Schüpbach (1985), Steinmann-Zwicky *et al.* (1989), and Nöthiger *et al.* (1989).

We first deal with the second step from *Sxl* to *dsx*. This part of the regulatory hierarchy has been most extensively and most successfully studied, and consequently has been presented in several recent reviews (Baker and Belote, 1983; Nöthiger and Steinmann-Zwicky, 1985a; Baker *et al.*, 1987; Wolfner, 1988). We then return to the primary signal and discuss how it might implement differential activity of *Sxl*. This first step in the regulatory cascade is still poorly understood and a matter of debate (for review, see Cline, 1985, 1988b). The final section is devoted to the question of how *dsx* achieves differential expression of those genes that actually produce the sexual dimorphism at the phenotypical level (for a short review, see Wolfner, 1988).

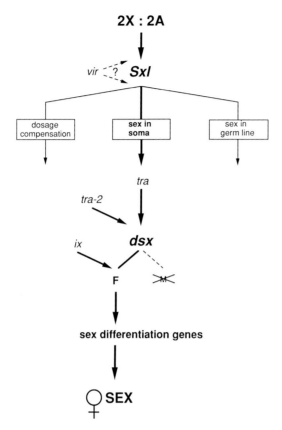

FIG. 1. Summary of the genetic hierarchy that regulates sexual differentiation in *Drosophila*. The central position of the gene *Sxl* (*Sex-lethal*) in the control of the three developmental processes of dosage compensation, sex in soma, and sex in germ line is shown. The role of *vir* (*virilizer*) and its position in the hierarchy are not yet clear, as indicated by the dashed arrows. An X:A ratio of 2:2 is the primary signal that initiates female development. In response to this signal, *Sxl* becomes active around the blastoderm stage and is then maintained in this state throughout development. The state of *Sxl* regulates the activity of subordinate control genes. In the pathway of somatic sex determination, the gene *tra* (*transformer*) is used to transmit the signal further downstream. When *tra* and *tra-2* (*transformer-2*) are active, the bifunctional gene *dsx* (*double-sex*) produces a female-specific transcript which, in conjunction with a functional gene *ix* (*intersex*), forms the female-determining product **F**. This product allows the expression of the female set of differentiation genes, leading to a female sexual phenotype in the soma. With an X:A ratio of 1:2, *Sxl* and then *tra* would remain inactive, and, as a consequence, *dsx* would produce the male-determining product **M**, leading to male somatic development. This alternative is not shown. Similarly, no details are given about the control of dosage compensation or of sex determination in the germ line, since these two processes are less well understood and not discussed in this article.

II. Sex Determination in the Soma: From *Sex-lethal* to *double-sex*

A. SEX-DETERMINING GENES

Spontaneous mutations, detected by chance over the years, led to the discovery of a small group of genes that play an important role in sex determination. A mutation in any one of the genes results in a sexual transformation of either XX or XY animals. The first such mutation was found by Gowen in 1940 (Gowen and Fung, 1957) and was later shown to be a dominant allele of *dsx* and renamed dsx^D (Duncan and Kaufman, 1975). In the meantime, more such genes became known, and a number of new alleles have been generated for all of them. Most of the mutations are recessive and cause partial or complete loss of gene function; a few are dominant, resulting from constitutive gain of function.

1. *Mutant Phenotypes*

The phenotypes of mutations in the known sex-determining genes are listed in Table 1. The evaluation of mutations in *Sxl* and *vir* (*virilizer*) is complicated by the fact that these genes are involved not only in sex determination but also in dosage compensation. Disruption of dosage compensation is lethal. Thus, XX animals lacking *Sxl* product (Sxl^{fl}/Sxl^{fl}) are unviable because both X chromosomes are transcribed at the high rate characteristic of males (Lucchesi and Skripsky, 1981; Gergen, 1987); the gain-of-function allele Sxl^{M1} is lethal to males, probably because the single X chromosome is transcribed at a low rate (Cline, 1978). The sex-transforming effect of such mutations can be visualized only under special circumstances. Whereas whole animals die, clones of mutant cells can survive in mosaic animals which then reveal the transformed male or female character of such cells (Cline, 1979; Sánchez and Nöthiger, 1982). The effect also becomes apparent in animals with two X chromosomes and three sets of autosomes (2X:3A) which normally develop a mixture of phenotypically male and female cells. When such animals carry Sxl^{fl} or Sxl^{M1}, they are transformed into phenotypical males or females, respectively (Cline, 1983b).

Loss of function of *tra* (*transformer*) or *tra-2* (*transformer-2*) results in the transformation of XX animals into phenotypical males, so-called pseudomales. These resemble normal males in every respect except that they are slightly larger and sterile. Somatic cells of XY males are not affected by the mutations, but *XY;tra-2/tra-2* males are sterile because *tra-2* function is required in the male germ line for normal spermatogenesis (Schüpbach, 1982; Belote and Baker, 1983). Loss of

TABLE 1
Sex-Determining Genes, Mutant Phenotypes, and Inferred State of Activity in XX and XY Animals[a]

	Mutation		Phenotype		
Locus	Loss of function	Gain of function	XX	XY	Reference
Sex-lethal (*Sxl*)	Sxl^{f1}		Ψ♂†	♂	Cline (1978)
		Sxl^{M1}	♀	Ψ♀†	
Sxl^{+}			ON	OFF	
virilizer (vir^{H2})	*vir*		Ψ♂†	♂	A. Hilfiker, H. Amrein and R. Nöthiger (unpublished)
vir^{+}			ON	OFF*	
transformer (*tra*)	*tra*		Ψ♂	♂	Sturtevant (1945)
		hs[tra⁺]#	♀	Ψ♀	McKeown *et al.* (1988)
tra^{+}			ON	OFF	
transformer-2 (*tra-2*)	*tra-2*		Ψ♂	♂[s]	Watanabe (1975)
$tra\text{-}2^{+}$			ON	OFF*	
intersex (*ix*)	*ix*		⚥	♂	Morgan *et al.* (1943)
ix^{+}			ON	OFF	
double-sex (*dsx*)	dsx^{mf}		⚥	⚥	Hildreth (1965); Nöthiger *et al.* (1987)
	dsx^{f}		⚥	♂	
	dsx^{m}		♀	⚥	
		dsx^{D}	Ψ♂	♂	
dsx^{f+}			ON	OFF	
dsx^{m+}			OFF	ON	

[a] ON and OFF are used to indicate that functional product of a gene is required (ON) or not required (OFF). ♂, Male; ♀, female; ⚥, intersex; Ψ♂, pseudomale; Ψ♀, pseudofemale; †, lethal in whole animals, but sex may be revealed in clones; *, gene function also required in males, but not for sex determination; [s], sterile; #, *hs-tra-female*, constitutive for *tra* function (McKeown *et al.*, 1988). The *dsx* locus contains a female-specific (dsx^{f+}) and a male-specific (dsx^{m+}) function (see also the text).

function of *ix* (*intersex*) or *dsx* (*double-sex*) produces intersexes in which male and female characters appear simultaneously. The action of *ix* appears to be restricted to chromosomal females, whereas the situation for *dsx* is more complex.

Mutations in *dsx* can be recessive or dominant, and they can affect both chromosomal sexes, only XX, or only XY. The mutations reveal that the locus harbors two genetic functions: dsx^{m} which implements the male program and dsx^f which implements the female program. When males were treated with a mutagen, mutations in only dsx^m or only dsx^f were rare; most mutations (23 of 28) were of the dsx^{mf} type abolishing both functions. Since mutations in dsx^m and dsx^f fully complement each other in trans-heterozygotes, we concluded that the dsx^m and dsx^f functions each have their own and independent domain within the gene but also share a large common domain that is essential for both functions (Nöthiger *et al.*, 1987). This prediction was confirmed by molecular analyses (Baker and Wolfner, 1988; Burtis and Baker, 1989). Three dominant mutations (dsx^D, dsx^{Mas}, dsx^T) appear to express the male-determining function constitutively since $X/X;dsx^D/dsx^{mf}$ animals are not intersexes, but completely transformed pseudomales.

Pseudomales, pseudofemales, and intersexes are sterile. Mutations in *tra, tra-2, dsx,* and *ix*, however, have no sex-transforming effect in the germ line, as shown by transplantation of pole cells which are the progenitors of the germ cells. Depending on the X : A ratio, mutant germ cells form normal eggs when transplanted into a female host, or sperm in a male host (Marsh and Wieschaus, 1978; Schüpbach, 1982). Thus, the reason for the sterility of sexually transformed flies lies in an incompatibility between the sex of the transformed soma and the chromosomal sex of the nontransformed germ line. *Sxl*, however, is also involved in sex determination of the germ line (Steinmann-Zwicky *et al.*, 1989; Nöthiger *et al.*, 1989).

2. Sex Determination Uses a Binary Code of Gene Function

The dramatic effect of a mutation in a single gene on all aspects of somatic sexual differentiation suggests that the quantitative signal of the X : A ratio uses the sex-determining genes to regulate the sexual pathway. This can be demonstrated by increasing the female-determining component of the signal in genotypes that are mutant for *tra* or *dsx* or *ix*. Such flies showed no signs of feminization when one-half of an X chromosome was added so that a ratio of 2.5X:2A was produced. These same X-chromosomal fragments, however, led to an almost complete feminization when added to 2X:3A intersexes that carried wild-type alleles of all sex-determining genes (Steinmann-Zwicky and Nöthiger, 1985a).

Table 1 reveals a complementary pattern of mutant effects: a particular gene function appears to be required either only in the female soma (*Sxl, vir, tra, tra-2, ix, dsx^f*) or only in the male soma (*dsx^m*). From this we can infer the requirement for these genes in wild-type males and females, and this is indicated by "ON" or "OFF" in Table 1. Thus, the X : A ratio implements a complementary pattern of activity of a small group of genes whose task is to direct the cells into the male or female pathway. This pattern of gene activity has characteristics of a binary code in which the ON state as well as the OFF state have informational content, allowing the cells to choose between two alternative developmental pathways.

B. SEX-DETERMINING GENES ACT CELL AUTONOMOUSLY AND ARE PERMANENTLY REQUIRED TO MAINTAIN CELLS IN A GIVEN PATHWAY

Genetic mosaics and temperature-sensitive mutations are useful to study the temporal and spatial requirements of gene function as well as the interaction between cells. Genetic mosaics are animals composed of genetically different cells. In *Drosophila*, such animals can be obtained by chromosome loss, by mitotic recombination, or by transplantation of cells or nuclei.

1. The X : A Ratio Is Read at the Blastoderm Stage

An unstable X chromosome exists that is occasionally lost from a nucleus during the cleavage divisions of an XX (female) zygote (Zalokar *et al.*, 1980). The resulting embryo is then composed of XX and XO nuclei. When the unstable X carries dominant alleles, the adult fly will display the recessive phenotype in those cells that are XO. Whenever such cells come to lie in a sexually dimorphic area of the fly, they form male structures, whereas XX cells form female structures immediately adjacent to them. These mosaic animals, called gynandromorphs, demonstrate that sex is not irreversibly determined at fertilization and that each cell differentiates autonomously according to its own X : A ratio, with no influence from sex hormones.

A change in the X : A ratio from 1.0 to 0.5 during the syncytial nuclear cleavage divisions results in healthy XO cells that build the male part of the gynandromorph. After the blastoderm stage, however, when cells have formed, such a change in the X : A ratio no longer produces mosaics. Only when the remaining X chromosome carries the mutation Sxl^{f1} instead of the wild-type allele do the XO cells survive and form male structures. This led to the conclusion that the XO cells died because their Sxl^{+} gene was irreversibly locked in the active (female) state (Sánchez and Nöthiger, 1983). Gergen (1987) later confirmed that Sxl was already differentially active in male and female cells at the blastoderm stage. He showed that the segmentation gene *runt*, which is expressed at this stage, is dosage compensated and its level of expression in XX embryos dependent on Sxl. Thus, at about 3 hours after fertilization, the sex of an animal is determined in the sense that *Sxl* has become permanently active or inactive, and is now independent of the X : A ratio.

2. Sex-Determining Genes Function throughout Development

In a population of dividing cells, the genotype of a single cell and its clonal descendants can be changed anytime during development by mitotic recombination. The process results in the elimination of the dominant wild-type alleles from a heterozygous cell, rendering one of its daughter cells homozygous for the mutant recessive alleles present on the same chromosome arm (Fig. 2). This technique allows us to change the genotype of a cell from female to male, for example, from *X/X;tra/ +* (female) to *X/X;tra/tra* (male), at progressively later times during development, and we can ask whether the female developmental history or the recently acquired male genotype determines the sexual differentiation of such a cell clone.

The results are clear and informative. XX cells that were made homozygous for *Sxl*[fl], *tra, tra-2,* or *dsx* formed clones whose sexual phenotype corresponded to the cell's new genotype, except when mitotic recombination was induced very late in development (Baker and Ridge, 1980; Sánchez and Nöthiger, 1982; Wieschaus and Nöthiger, 1982). When clones were generated after the second larval instar, male structures still appeared on forelegs, tergites, and analia, but male genital

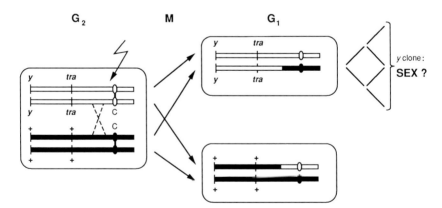

FIG. 2. Mitotic recombination is inducible by X-rays (wavy arrow) in mitotically dividing cells. When the cells are heterozygous for appropriate mutations, mitotic recombination produces a clone of cells labeled with a marker mutation, for example, *y*, and which is homozygous for a mutation whose effects on the sexual phenotype we want to study, for example, *tra*. G_2, M, G_1, Phases of the cell cycle; c, centromere.

structures were no longer found (Wieschaus and Nöthiger, 1982). Male genital clones, however, could again be recovered when the onset of imaginal differentiation was experimentally postponed by *in vivo* culture of the genital disc of irradiated *X/X;tra/ +* female animals, even when mitotic recombination had been induced in the third larval instar (Epper and Nöthiger, 1982). The reason for this is found in the special organization of the genital disc which forms the postabdomen of the adult fly. The genital disc is a bisexual anlage which, in both sexes, harbors separate primordia for the female genitalia, the male genitalia, and the analia, probably corresponding to abdominal segments A8, A9, and A10/11 (Nöthiger *et al.*, 1977; Schüpbach *et al.*, 1978; Janning *et al.*, 1983). In a female disc, the cells of the male genital primordium divide very slowly and do not form any recognizable structures in the adult fly; the same is true for the female genital primordium in the disc of a male (Ehrensperger, 1983). We conclude that, after a genetically male *tra/tra* clone had been generated in the male genital primordium of the irradiated female disc, the extra time granted by *in vivo* culture allowed the cells of the clone to change their pace of cell division from slow to fast. Thus, before terminal differentiation began, they could undergo enough cell divisions to be able to form male genital structures. Clones in the forelegs, tergites, and analia need no extra cell divisions because the same cells form either male or female structures and divide at the same fast pace.

The results of the clonal analyses demonstrate that the products of *Sxl, tra,* and *tra-2* are permanently required to maintain the cells in the female pathway. When mitotic recombination creates a male genotype by removing the wild-type allele from a heterozygous female cell, its daughter cells will "remember" their female history for a few cell divisions, as long as the products of the wild-type allele remain in sufficient number. But then the new male genotype will take command. If this happens before terminal differentiation, the clone will form male structures.

Thus, the sexual pathway remains flexible, and, in this sense, sex is never determined because it can be reversed anytime by changing the state of activity at a sex-determining locus. This flexibility works in both directions. The temperature-sensitive allele *tra-2^{ts}* forces homozygous XX animals to develop as males at 29°C and as females at 16°C. Temperature shifts of mutant animals showed that the initiated sexual pathway could be reprogrammed from male to female, or from female to male, depending on the temperature at which an animal spends a critical phase of development (Belote and Baker, 1982, 1987; Epper and Bryant, 1983).

Sex-determining genes have properties of homeotic genes. They are selector genes (Garcia-Bellido, 1975) whose state of activity specifies one of two alternative pathways. Homeotic and sex-determining genes are permanently required to maintain a given developmental program. Homeotic genes act in spatial domains, so-called compartments (Garcia-Bellido *et al.*, 1979). The sex-determining genes, on the other hand, operate in genetic compartments, namely, XX or XY.

C. Sex-Determining Genes Form a Short Regulatory Cascade with *Sex-lethal* at the Top and *double-sex* at the Bottom

1. Genetic Arguments

The first indications for a hierarchical arrangement of the sex-determining genes as shown in Fig. 1 were provided by double-mutant combinations which showed clear epistatic relations (Baker and Ridge, 1980; Nöthiger *et al.*, 1987). XX or XY animals that were homozygous mutant for dsx^{mf} and a mutation in any of the other genes showed the dsx phenotype. This result assigns a central position to *dsx* and suggests that the other sex-determining genes function as regulators of *dsx*. Mutations in *tra* or *tra-2* are epistatic over mutations in *Sxl*, as demonstrated by the fact that XX animals carrying the female-determining constitutive mutation Sxl^{M1} nevertheless develop as males when they are homozygous mutant for *tra* (Cline, 1979). Conversely, XX animals carrying a viable male-determining mutant allele of *Sxl* (Cline, 1984) are transformed into females by a constitutively expressed *tra* gene (McKeown *et al.*, 1988).

The hierarchical position of *tra* relative to *tra-2* cannot be determined unequivocally on the basis of genetic arguments. If the cascade were strictly linear, *tra-2* would have to be downstream of *tra* because it is epistatic over the constitutively expressed *tra* gene (McKeown *et al.*, 1988).

Similarly, the positions of *ix* and *vir* are unclear. Genetic arguments support the view that the role of *ix* is to render the female-specific product of dsx^f functional, which would place *ix* below *dsx* in the female pathway (Nöthiger *et al.*, 1987). The gene *vir*, on the other hand, must be placed high up in the hierarchy, because it apparently provides functions for both sex determination and dosage compensation. The first *vir* allele was isolated by T. Schüpbach as a temperature-sensitive mutation. At 29°C, it transforms XX flies into sterile intersexes. Males are not affected, nor are females when kept at 25°C. New alleles have now been generated. Certain alleles and allelic combinations lead to sex-specific lethality of XX zygotes, which can be rescued by Sxl^{M1} or *mle* and other male-specific lethals (*msl*) shown to be involved in dosage compensation (for *mle*, see Belote and Lucchesi, 1980; Breen and Lucchesi, 1986). This indicates that *vir* plays a role in dosage compensation, possibly by regulating *Sxl*. The rescued animals, however, are transformed into pseudomales, which reveals the sex-determining function of *vir* and suggests that this gene acts downstream of *Sxl* in the sex-determining pathway, or in parallel to it. Thus, the position of *vir* relative to *Sxl* remains uncertain. The case is further complicated by the fact that certain alleles of *vir* are also lethal in XY males (A. Hilfiker, H. Amrein, and R. Nöthiger, unpublished).

2. Molecular Analysis

The genetic evidence summarized above revealed that sex is primarily determined by the X : A ratio, which directs the key gene *Sxl* into either the active (female) or inactive (male) mode. The differential activity of *Sxl* in turn seems to regulate a small number of downstream genes so that the bifunctional gene *dsx* eventually expresses either the female function or the male function.

Of the six genes listed in table 1, four have been cloned and their structures and transcripts characterized. The four genes are *Sxl* (Maine *et al.*, 1985a; Bell *et al.*, 1988; Salz *et al.*, 1989), *tra* (Butler *et al.*, 1986; McKeown *et al.*, 1987), *tra-2* (Amrein *et al.*, 1988; Goralski *et al.*, 1989), and *dsx* (Baker and Wolfner, 1988; Burtis and Baker, 1989). The genes *tra* and *tra-2* are small; genomic DNA fragments of only 2 and 3.9 kb, introduced into flies by P-mediated transformation, were able to rescue the mutant phenotype of XX animals homozygous for *tra/tra* and *tra-2/tra-2*, respectively. The genes *Sxl* and *dsx*, on the other hand, are large, comprising at least 23 and 45 kb, respectively. All four genes produce two or more different transcripts, and none appears to be primarily regulated at the transcriptional level, although the use of alternative promoters has not been excluded. Thus, the simple ON and OFF states, suggested in Table 1, do not relate to transcriptional activity of these genes.

The gene *Sxl* fulfils various functions that are genetically separable (see Fig. 1 and Section II,C,4). The genetic complexity is paralleled at the molecular level by a pattern of at least 10 distinct but overlapping transcripts that arise from alternative splicing (Bell *et al.*, 1988) and probably also from the use of different transcription start sites (Salz *et al.*, 1989). Some transcripts are found only in very young embryos; others appear later in embryonic development and persist through adult life. The *Sxl* transcripts found in the very early embryos largely overlap those present in later stages and adult flies but differ in the 5′ region, which raises the possibility that the embryonic and the adult transcripts originate from different transcription start sites. It will be important to see whether early transcripts are present in both XX and XY embryos, and, if so, whether they are different in the two sexes and therefore might represent the initial response of *Sxl* to the X : A ratio. The production of large numbers of pure male or pure female embryos by genetic means is now possible (Walker *et al.*, 1989).

In adult females, an abundant *Sxl* transcript of 3.3 kb is found specifically in the germ line. Older embryos, larvae, pupae, and adult flies of both sexes produce three major overlapping transcripts. In females, they are 1.9, 3.3, and 4.2 kb long, owing to differences in their 3′ ends; the three corresponding transcripts in males are about 0.2 kb longer. Sequence analysis of sex-specific cDNAs corresponding to the 1.9-kb class revealed that the size difference results from a small axon of 190 bp present in all male-specific transcripts but absent from all female-specific transcripts; the other seven exons are identical (Fig. 3). The additional exon of 190 bp introduces a stop codon so that all transcripts of the male lack the long open reading frame (ORF) found in the transcripts of females. Thus, only the female is able to make functional products of *Sxl*. Therefore, the ON/OFF states of Table 1 in fact seem to refer to the presence or absence of proteins.

The mechanism of alternative splicing also operates to achieve ON/OFF expression of *tra*. The gene *tra* produces a pre-mRNA that is alternatively spliced (Fig. 3). An unspecific transcript of 1.2 kb is found in both sexes, and a 1.0-kb transcript, largely overlapping the longer transcript, is present only in females. Both transcripts consist of three exons. The size difference between the two transcripts results from the use of two different 3′ splice acceptor sites of the first intron. The additional sequences present in the unspecific transcript (the only transcript of males) introduce a stop codon in the single long ORF present in the female-specific transcript (Boggs *et al.*, 1987). A more detailed picture of *tra* can be found in an article by Belote *et al.* (1989).

The third gene whose sex-specific differential expression is regulated by alternative splicing is *dsx*. This locus produces several transcripts which vary depending on the developmental stage and sex (Baker and Wolfner, 1988). The most interesting transcripts first become detectable in the third larval instar and persist through the pupal stage to adulthood. One has a size of 3.2 kb and occurs only in females; the other is 3.8 kb and occurs only in males. The two transcripts share a common 5′ domain and differ only in their 3′ exons (Fig. 3). In contrast to *Sxl* and *tra*, however, both the female-specific and the male-specific transcripts contain a large ORF (Burtis and Baker, 1989). The presence of transcripts potentially encoding proteins with a large common amino termi-

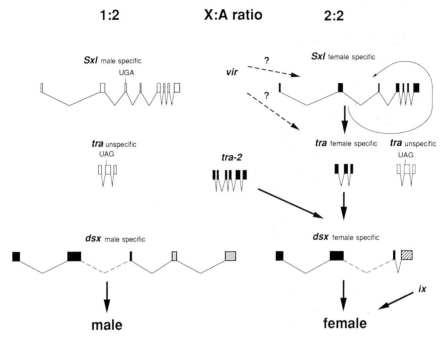

FIG. 3. Regulation of sex determination by alternative splicing. The sex-determining genes (see Table 1) and the relevant transcripts are shown where known. Arrows indicate active regulatory effects. Transcripts potentially encoding functional proteins are drawn with black exons; transcripts with interrupted ORFs are drawn with white exons. The male- and female-specific exons of *dsx* are shown as stippled and hatched boxes, respectively. The dashed line in the *dsx* transcript indicates that this intron is proportionally much larger. The position of *vir* is still unclear (see Fig. 1). Further explanations are given in the text.

nus of 397 amino acids but different sex-specific carboxy termini (30 amino acids in females, 152 in males) is consistent with the genetic data.

So far, alternative splicing emerges as the principle that is used to regulate somatic sex determination in *Drosophila*. The *tra-2* gene, however, appears to be an exception; it is unique among the sex-determining genes inasmuch as its function is required not only in the female soma for sex determination, but also in the male germ line for proper sperm differentiation (Schüpbach, 1982; Belote and Baker, 1983). The regulation of *tra-2* is also different from that of the other molecularly characterized sex-determining genes (Amrein *et al.*, 1988; Goralski *et al.*, 1989). Different transcripts are produced by alternative splicing events in the 5' region of the gene. However, the major transcripts are identical in the soma and germ line of males and females, encoding a putative protein of 264 amino acids (H. Amrein and R. Nöthiger, 1989).

3. Cascade of Alternative Splicing

The fact that the genes *Sxl, tra,* and *dsx* produce sex-specific transcripts allowed Nagoshi *et al.* (1988) to test at the molecular level the hierarchy proposed by geneticists. The data are summarized in Table 2, and the reader can find a comprehensive review in Baker (1989). XX animals that lack the function of *Sxl, tra,* or *tra-2* produce the male-specific instead of the female-specific transcript of *dsx*, indicating that these genes regulate the functional state of the *dsx* locus, as predicted by genetic arguments. Production of the female-specific transcript of *tra* depends on *Sxl* activity but not on the activities of *tra-2* or *dsx*. These observations lead to the gene order *Sxl → tra → dsx*, indicating that *Sxl* activity is required for the female-specific splicing of *tra* pre-mRNA and that *tra* and *tra-2* activities are required for the female-specific splicing of *dsx* pre-mRNA. Mutations in *ix*, although they transform XX animals into intersexes (Table 1), do not alter the female-specific transcript of *dsx*. As we have seen, *tra-2* is placed outside of the linear hierarchy since its expression appears to be independent of *Sxl* or any

TABLE 2

Transcripts and Phenotypical Sex of Mutant Animals[a]

Karyotype	*Sxl*	*tra*	*tra-2*	*dsx*	*ix*	Phenotypical sex	
XX	FT	FT	—	FT	—	Female	⎫ Wild
XY	MT	MT	—	MT	—	Male	⎭ type
XX	*Loss*	MT	—	MT	—	Pseudomale	⎫
XX	FT	*Loss*	—	MT	—	Pseudomale	⎪
XY	MT	*Gain*	—	FT	—	Pseudofemale	⎬ Mutants
XX	FT	FT	*Loss*	MT	—	Pseudomale	⎪
XX	FT	FT	—	*Loss*	—	Intersex	⎪
XX	FT	FT	—	FT	*Loss*	Intersex	⎭

[a] Transcripts (bold capitals) and the phenotypical sex were recorded in wild-type animals and in flies mutant (italics) for one of the sex-determining genes. [Data from Nagoshi *et al.* (1988) and McKeown *et al.* (1988).] **FT,** Female-specific transcripts; **MT,** male-specific transcripts; *Loss,* loss of gene function; *Gain,* constitutive gain of gene function through *hs-tra-female* (McKeown *et al.,* 1988). Transcripts were not monitored for *tra-2* and *ix;* the gene *tra-2* is not sex specifically regulated, and *ix* has not yet been cloned. All genes were present in their wild-type allelic state, except the ones labeled *loss* or *gain.* See also Section II,C,3.

other gene related to sex. It is, however, possible that the product of *tra-2* is subject to translational or posttranslational control by *tra*. The female-specific function of *dsx* may similarly be regulated by the *ix* product.

An elegant and simple test, performed by McKeown *et al.* (1988), confirmed the hierarchy. The authors studied the effects of a female-specific cDNA of *tra* under control of a heat-shock promoter. This construct, called *hs-tra-female*, essentially corresponds to a constitutive gain of *tra* function and was able to transform XY animals into pseudofemales that, as expected, expressed the female-specific transcript of *dsx* (Table 2). It also rescued the sexual phenotype of XX animals lacking proper function of *tra* or *Sxl* but had no effect on flies mutant for *dsx* or *ix*.

In the hierarchy shown in Figs. 1 and 3, the gene interactions occur at the level of alternative splicing. An exciting and simple hypothesis derives from the structure of the putative proteins of *Sxl* and *tra-2*. Both show homology to RNA-binding proteins of yeast and mammals (Amrein *et al.*, 1988; Bell *et al.*, 1988; Goralski *et al.*, 1989; for a concise review on RNA-binding proteins, see Bandziulis *et al.*, 1989). Thus, the *Sxl* protein could interact with the *tra* pre-mRNA and achieve its female-specific splicing. Similarly, the *tra-2* protein, when *tra* protein is present, could interact with the pre-mRNA of *dsx* and impose on it the female-specific splicing. The *tra* protein itself is relatively basic, with no similarity to any known protein (Boggs *et al.*, 1987).

The sex-specific splicing of the *tra* pre-mRNA represents a case of competition between two 3' splice sites. Two different models can account for the choice of either the first (unspecific) or the second (female-specific) site. (1) The *Sxl* protein may bind to and partially block the first site so that efficient splicing can also occur at the second site. (2) The *Sxl* protein could directly interact with the female-specific site and actively make this site available for the splicing machinery. Results obtained with *tra* genes that were manipulated in their 3' splice sites are consistent with the first model. When the unspecific 3' splice site was removed or altered by specific nucleotide changes, the female-specific site was used irrespective of *Sxl* activity. Deletion of the female-specific splice site, on the other hand, led to an accumulation of unspliced RNA, with the amount of unspliced RNA depending on *Sxl* activity. This indicates that the target site for the *Sxl* protein was not affected, and we conclude that in females the protein protects the unspecific acceptor site so that the female-specific site can be efficiently used (Sosnowski *et al.*, 1989). Similar results were obtained in experiments where the galactosidase activity of three different *actin–tra–*

lacZ fusion genes was assayed in flies. In these constructs, either the female-specific or the unspecific, or both, splice acceptor sites of the *tra* gene had been inserted between the *actin* 5' region (promoter–exon 1 –donor splice site and beginning of intron 1) and the *lacZ* reporter gene. Galactosidase activity was independent of *Sxl* activity with the first construct containing only the female-specific splice acceptor site of the *tra* gene but not with the other two constructs (U. Ernst and R. Nöthiger, unpublished).

Nothing is known yet about how *tra-2* and *tra* achieve the female-specific splicing of the *dsx* pre-mRNA, which is their likely target. Molecular data, however, suggest that *tra-2* (and *tra*) protein may perform an active role in the choice of the female-specific exon (see Fig. 6 for the structure of the *dsx* locus). This speculation is based on the following observations: (1) the 5' region of the female-specific exon contains a 13-bp motif that occurs six times (Burtis and Baker, 1989); (2) three *dsx* mutations resulting in constitutive male-specific splicing in XX animals map within this region (Baker and Wolfner, 1988) and apparently prevent the female-specific splice site from being used (Burtis and Baker, 1989); (3) the 3' splice acceptor site of the preceding female-specific intron is an uncommon and poor acceptor site by the criterion that it contains only a very short polypyrimidine stretch, it is, however, highly conserved in *Drosophila melanogaster* and *D. virilis*; and (4) in the absence of *tra-2* (and *tra*) protein, the general splicing machinery does not recognize this splice site, contrary to the "first come–first served" rule (Aebi and Weissman, 1987). These facts support the idea that the *tra-2* protein, perhaps together with the *tra* protein, interacts with this particular region and actively promotes the use of the female-specific splice site. The situation, however, may be more complex since alternative splicing at the *dsx* pre-mRNA is coupled with alternative polyadenylation (Burtis and Baker, 1989), possibly controlled by *tra-2*.

As mentioned before, function of *tra-2* is also required in the male germ line for proper spermatogenesis. In this process, however, its target cannot be the *dsx* pre-mRNA since it has been shown that *dsx* function is not required for normal spermatogenesis (Schüpbach, 1982). The RNA-binding property of the putative protein suggests that *tra-2* may accomplish the processing of a pre-mRNA that is essential in the male germ line.

4. Pivotal Gene Sex-lethal

As shown in Fig. 1, *Sxl* is involved in the regulation of dosage compensation, sex determination in the soma, and sex determination in the germ line. The functions controlling these three processes are genetically separable. XX animals, homozygous for Sxl^{fs1}, are fully viable and develop as phenotypically normal but sterile females, showing that this allele is defective only in the germ-line function (Salz et al., 1987). The allelic combination $Sxl^{fm7,M1}/Sxl^{M1,fm3}$ is reasonably viable; such animals develop as pseudomales (Cline, 1984), but their germ cells form normal eggs when transplanted into normal female hosts (Schüpbach, 1985). Thus, this allelic combination appears defective in the somatic sex-determining function but much less so in the functions providing dosage compensation and not at all in those required for sex determination in the germ line.

Genetic experiments, succinctly summarized in Maine et al. (1985b), reveal that, in the somatic pathway, Sxl performs functions at three regulatory levels. These are initiation, maintenance, and expression of the sexual pathway. Early functions of Sxl, probably provided by the early embryonic transcripts, are required to initiate or implement the female sexual pathway. Later functions are required if the gene is to be maintained in the active state and also to control the expression of subordinate sex-determining genes, such as tra (see Fig. 3). The two alleles Sxl^{f9} and Sxl^{fLS} each appear to be defective in only one of these functions of Sxl. Both alleles are homozygous lethal but complement each other in trans-heterozygous females, which shows that they are deficient in different functions. Further analyses indicate that Sxl^{f9} is defective in initiation, can no longer respond to the X : A ratio, but can maintain and express the female pathway when trans-activated by a wild-type allele present in the same cell. Sxl^{fLS}, on the other hand, is defective in maintenance (and expression) but can initiate the female pathway and trans-activate another allele of Sxl.

Genetic data had suggested that the Sxl gene becomes set in either the active or inactive mode around the blastoderm stage and is then maintained in this state (Sánchez and Nöthiger, 1983). Cline (1984) found evidence for an autoregulatory mechanism that could underly the maintenance phenomenon. The molecular analysis of Sxl provides us with an attractive hypothesis to account for both the maintenance and expression of the initiated pathway. The perpetuation of the sex-specific expression pattern may be achieved if the Sxl protein binds to its own pre-mRNA and directs the female-specific splicing pattern, thus maintaining the female developmental program (Bell et al., 1988). The fact that the same particular sequence $(T)_8C$ is present in the unspecific 3' splice site of tra and in the male-specific 3' splice site of Sxl suggests that the splicing of Sxl pre-mRNA may also occur by blocking this acceptor site analogously to the situation for the tra pre-mRNA (Sosnowski et al., 1989).

As this section has shown, the pathway from Sxl to dsx is well understood, down to the molecular level. The beginning and the end of the entire process of sexual differentiation, above *Sxl* and below *dsx*, however, are far less clear. The next two sections are devoted to a discussion of the primary events that initiate differential expression of *Sxl* and to the question of how the products of *dsx* achieve the differentiation of overt sexual dimorphism.

III. From the Primary Signal to *Sex-lethal*

A. ELEMENTS REQUIRED FOR FEMALE-SPECIFIC EXPRESSION OF *Sex-lethal*

To obtain a functional *Sxl* protein, the *Sxl* gene must be transcribed and the female-specific splice pattern has to be implemented and maintained. Loss-of-function mutations in any gene specifically required for these processes are expected to affect females only, just as mutations in *Sxl* itself. To investigate the hierarchical relation between genes acting in the same genetic pathway, it is useful to combine loss-of-function mutations in one gene with gain-of-function mutations in another. If the mutation Sxl^{M1}, which is female determining, rescues the female-specific phenotype of a loss-of-function mutation in a different gene, we can conclude that the latter gene is required for the female-specific expression of Sxl^+. This criterion makes it possible to test whether genes identified as being involved in sex determination act upstream or downstream of *Sxl*. So far, the genes involved in the regulation of *Sxl* can be subdivided into (1) genes acting maternally, (2) genes with both a maternal and a zygotic component, and (3) genes acting in the zygote.

1. Maternal Genes

The gene *daughterless* (*da*) has several different functions, some zygotic, some maternal, but the function that is required for *Sxl* to make an active product is purely maternal. Embryos lacking zygotic *da* function are lethal (Cronmiller and Cline, 1987) and develop without peripheral neurons and associated sensory structures (Caudy *et al.*, 1988a). A viable hypomorphic allele (*da¹*), however, displays a sex-limited maternal effect. When kept at 25°C, homozygous *da¹/da¹* females have only male progeny; daughters die, but the lethality can be rescued if they carry Sxl^{M1} (Cline, 1978). At 18°C, female progeny survive even in the absence of Sxl^{M1}. The temperature-sensitive period extends through the last 60 hours of oogenesis until 3 hours after fertilization. During this period, a functional da^+ product is required for daughters to activate their Sxl^+ genes (Cline, 1976).

The single transcription unit *da* produces two overlapping transcripts (Caudy *et al.*, 1988b; Cronmiller *et al.*, 1988). In the ovary, a 3.2-kb transcript is present, which indicates that this is the maternal product. In the zygote, both the 3.2- and a 3.7-kb transcript are produced (Caudy *et al.*, 1988b). A putative *da* protein shows sequence similarities with the proteins of *myc, Myo D1*, and three proteins from the *Drosophila achaete–scute* complex (the putative products of T3, T4, and T5), a gene complex which is also required for neurogenesis. All these proteins are somehow involved in a process of cell determination. The presence of specific sequences in *da* and *myc* led to the speculation that *da* may be a DNA-binding protein (Caudy *et al.*, 1988b). The best homology exists between two immunoglobulin enhancer binding proteins and the *da* product. These and the other proteins mentioned share a DNA-binding and dimerization motif. Thus, *da* appears to encode a transcription factor that binds to DNA as a dimer (Murre *et al.*, 1989).

A second gene whose maternal product is also required for the activation of Sxl^+ was named *Daughterkiller* (*Dk*). Females carrying the dominant mutation *Dk* have no female progeny, but again daughters can be rescued by Sxl^{M1}. A deficiency for *Dk* enhances the effect of *da* and Sxl^{f1}, which proves that Dk^+ is required in females and that the three genes act in the same pathway (M. Steinmann-Zwicky, E. Fuhrer-Bernhardsgrütter, D. Franken, and R. Nöthiger, unpublised). The alleles available so far are homozygous lethal, which suggests that *Dk*, like *da*, may have different maternal and zygotic functions. Although the nervous system of mutant embryos is defective, the phenotype displayed is different from that of embryos lacking *da* function (M. Steinmann-Zwicky, unpublished).

2. Genes with Both Maternal and Zygotic Components

The region containing the gene *liz* (Steinmann-Zwicky, 1988), also called *snf* (Oliver *et al.*, 1988) and *fs(1)1621* (Gans *et al.*, 1975), was first identified as a sex-determining region in an analysis of genetic mosaics (Steinmann-Zwicky and Nöthiger, 1985b; see Fig. 4A). The analysis revealed that the duplication of two X-chromosomal regions was crucial in obtaining feminization. One of them, region 3E to 4F, includes *liz*; the other includes *Sxl*. These two regions showed synergistic interactions: genotype *Df 3E to 4F/Sxl^{f1}* is poorly viable, displays patches of tissue with male transformations, and is sterile. The initial results suggested that region 3F to 4B was responsible for these effects. Further studies, however, showed that the mutant phenotypes appear in females that have only one liz^+ allele (located in region 4F) and only one Sxl^+ gene. In addition, *liz* was reported to display a maternal effect so strong that when a mother is heterozygous for the mutation, most daughters carrying only one Sxl^+ allele die (Cline, 1988a; Oliver *et al.*, 1988; Steinmann-Zwicky, 1988).

These interactions show that the gene *liz* is involved in sex determination and dosage compensation, but they do not show its position in the genetic hierarchy. A mutation in a gene that regulates Sxl is expected to be female lethal. The only mutation of *liz* described so far, however, is a viable female-sterile [*fs(1)1621*; Gans *et al.*, 1975]. Combining the mutation *liz* with Sxl^{M1} showed that this constitutive mutation could rescue all female-specific mutant phenotypes associated with *liz*, namely, the sterility of homozygous flies, the zygotic phenotypes resulting from simultaneously deleting one *liz* and one Sxl allele, and the maternal effect (Steinmann-Zwicky, 1988). Thus, in the presence of Sxl^{M1}, liz^+ is dispensable, which shows that the gene participates in the process that achieves the female-specific expression of Sxl^+.

In some experiments, the mutation Sxl^{M1} behaved in an unexpected way. We have seen that XY animals carrying Sxl^{M1} are unviable, presumably because of upsets of dosage compensation and that clones of mutant XO tissue differentiate mostly female structures (Cline, 1979). If, however, XY flies carrying Sxl^{M1} are also mutant for *liz*, they are viable and develop as fertile males. Furthermore, one Sxl^{M1} allele was sufficient to rescue all female-specific zygotic phenotypes of *liz* observed in germ line and soma, but full rescue of the maternal effect was obtained only with two Sxl^{M1} alleles, or with a Sxl^{M1} allele and a wild-type gene of Sxl. Both sets of results suggest that the expression of a single Sxl^{M1} allele is dependent on *liz* product (Steinmann-Zwicky, 1988). A similar observation was made in an experiment involving *sis-b* mutations (Torres and Sánchez, 1989; for *sis-b* see Section III,A,3). Female-specific lethality was fully rescued by a Sxl^{M1} allele only when a wild-type gene was also present. These results can be best explained by assuming that Sxl^{M1} does not always constitutively express the female-specific Sxl functions. In a homozygous stock in which all flies, XY males and XX females, are mutant for both *liz* and Sxl^{M1}, it is obvious that Sxl^{M1} is under control of the X:A ratio (Steinmann-Zwicky, 1988). Other data also support the view that Sxl^{M1} is not really a constitutive mutation. No evidence was found for female expression of Sxl^{M1} in early XX embryos deriving from *da/da* mothers (Gergen, 1987) nor in the XY germ line (Cline, 1983a; Steinmann-Zwicky *et al.*, 1989); in the XO soma, Sxl^{M1} was only partially active (Cline, 1979).

A second gene whose properties might be similar to those of *liz* could lie within region 11D to 12A. Preliminary analysis revealed that daughters from mothers hemizygous for this region are lethal if they are heterozygous for Sxl^{fl}. Females escaping the lethal effect often exhibit patches of male tissue (T. Scott and B. Baker, unpublished; cited in Belote *et al.*, 1985b).

3. Genes Acting in the Zygote

Females homo- or hemizygous for the mutation *sisterless-a* (*sis-a*) are lethal, but they are rescued by Sxl^{M1}. Males are not affected (Cline, 1986). This shows that *sis-a*$^+$ is required for the female-specific expression of Sxl^+. The mutation shows no maternal effect (Cline, 1986; Steinmann-Zwicky, 1988). The only mutation of *sis-a* available so far is probably hypomorphic. Rare escapers of genotype *sis-a/sis-a* occur at low temperature, and these are female; hemizygous escapers are not found. The observation that homozygous clones induced during larval development differentiate female structures was interpreted as pointing out that *sis-a*$^+$ is required only early in development, for setting the initial expression pattern of Sxl^+, and not at later stages (Cline, 1986). These results, however, could also be explained by the hypomorphic nature of the mutation.

Recently, Cline (1988a) discovered a new gene that interacts with both *sis-a* and *Sxl*. This gene, *sisterless-b* (*sis-b*), lies within the *achaete–scute* complex in the *scute–alpha* region. Duplicating both *sis-a* and *sis-b* is lethal to males, but these males are largely rescued if they also carry either the mutation *sis-a* or Sxl^{fl}, or if their mother was mutant for *da*. Thus, males with the double duplication must be lethal because their Sxl^+ gene is at least partly expressing the female-specific product, which leads to a lower transcription of X-linked genes. Females heterozygous for a deficiency of *sis-b* and either the mutation *sis-a* or Sxl^{fl} are unviable, supposedly because they produce too little female-specific *Sxl* product to ensure a low rate of X-chromosomal transcription.

Cline (1988a) found that the mutation sc^{3-1} behaves as a deficiency of *sis-b*. Garcia-Bellido (1979) reported that the viability of females homozygous for sc^{3-1} was poor (10% survived) and that males carrying the mutation were fully viable. Hemizygous females ($sc^{3-1}/Df(1)sc^{19}$) were unviable, which suggests that the mutation is hypomorphic. The *achaete–scute* complex has been thoroughly studied at the molecular level (Campuzano *et al.*, 1985; Villares and Cabrera, 1987). Several deficiencies and mutations that are well characterized were tested further to localize *sis-b* (Torres and Sánchez, 1989). Two mutations, introducing a stop codon or a frameshift in the T4 transcript, displayed poor female viability and female-lethal interactions with *Df(sis-a)*. These effects were rescued by Sxl^{M1}. A different mutation that overexpresses T4 causes male lethality when *sis-a* is also duplicated. These results suggest that *sis-b* function is provided by a product encoded by the T4 transcript. This same transcript, however, has previously been shown to provide the *scute* function (Campuzano *et al.*, 1985; Villares and Cabrera, 1987). The T4 transcript thus might play an important role in both neurogenesis and sex determination. We remind readers that da^+ is also involved in the genetic control of the same two processes and that sequence similarities between *da* and the product of T4 have been described (Caudy *et al.*, 1988a,b).

B. Are the Genes That Are Required for Female-Specific Expression of *Sex-lethal* Elements of the X : A Ratio?

1. Model for the X : A Ratio

Bridges visualized the signal formed by the X : A ratio as the result of two sets of counteracting elements, with the X chromosomes carrying female-determining factors and the autosomes male-determining factors. A search for such factors suggested that feminizing elements are spread over the X chromosome, since various small X-chromosomal duplications were able to feminize triploid intersexes (2X;3A), whereas deletions had opposite masculinizing effects (Dobzhansky and Schultz, 1934). None of the tested autosomal duplications was able to masculinize the phenotype of triploid intersexes (2X;3A), nor did deficiencies feminize them. Therefore, no evidence pointed to the existence of male-determining elements on the autosomes (Pipkin, 1947, 1960).

In light of current knowledge, one might visualize the feminizing elements as stretches of specific DNA sequences located exclusively, or at least predominantly, on the X chromosome. A masculinizing product, encoded by autosomal genes and present in equal amounts in male and female embryos, might be titrated by such DNA sequences, which could be noncoding and serve as a trap. The two X chromosomes of females would bind and neutralize more of the product than the one X chromosome of males. In the absence of free masculinizing product, female development would ensue; in its presence female development would be repressed. Thus, the masculinizing product could be a repressor of *Sxl* (Chandra, 1985). Alternatively, specific X-chromosomal DNA sequences might also serve as a source and code for a feminizing product. According to this model, the autosomal component, which could consist of noncoding DNA sequences, RNA, or protein, could titrate the feminizing product such that none remains in males but sufficient quantities are still present in females, whose two X chromosomes produce twice as much product as the single X of males. Such an X-chromosomal product would then be an activator of female-determining genes, for example, an activator of *Sxl*.

X-chromosomal stretches of DNA that might fulfill the requirements for a trap were recently described in *Caenorhabditis elegans*, a nematode whose primary sex-determining signal is also the X : A ratio. Multiple copies of three different X-chromosomal genes were able to feminize males, whereas none of the autosomal genes tested had such an effect (McCoubrey *et al.*, 1988). Until now, no similar experiment was reported for *Drosophila*. Two sequences, however, were described that might be candidates for either a trap or a source. A middle-repetitive DNA sequence seems to be located specifically on the euchromatic portion of the X chromosome. This 372-bp repeat is distributed among approximately 10 major and 20 minor X-chromosomal sites, is (A + T) rich, and is noncoding (Waring and Pollack, 1987). Another sequence, $(dC\text{-}dA)_n(dG\text{-}dT)_n$, is present on most euchromatic regions, but the density of sites along the X chromosome is at least twice as high as the density on the autosomes. *In situ* hybridization to polytene chromosomes shows conserved chromosomal locations in *Drosophila melanogaster, D. simulans*, and *D. virilis*. Of special interest, however, is the distribution pattern seen in *D. pseudoobscura* and *D. miranda*. In both these species, a chromosome arm that was autosomal in a common *Drosophila* ancestor has been translocated to the X chromosome and, as a new X chromosome arm, has evolved the ability to regulate its genes by dosage compensation. Both these new X chromosomes showed hybridization patterns typical of the X chromosome and different from those of the autosomes (Pardue *et al.*, 1987).

As we have seen, DNA sequences located exclusively or predominantly on the X chromosome could be part of the primary signal for sex determination and dosage compensation. Alternatively, they could also be involved in the process of dosage compensation only and function as cis elements that control the differential level of transcription of X-chromosomal genes in males and females. The sequences would then serve as target sites for a sex-specific product, for example, *Sxl*, and they would not fulfill a function in the process of sex determination. At present it is not yet possible to assign a function to the X-specific sequence nor to the $(dC\text{-}dA)_n(dG\text{-}dT)_n$ sequence.

2. Candidates for the X-Chromosomal Component

Female-specific transcripts of *Sxl* are found in unfertilized eggs (Salz *et al.*, 1989), which will give rise to male and female embryos. Therefore, these transcripts alone cannot provide the discriminating sex-determining function. The choice of sexual pathway is imposed on embryos by genes acting in the zygote. Several of the zygotic genes required for the female-specific expression of *Sxl* are X linked. These genes could form part of the primary signal, that is, the X : A ratio. They could also be genes whose products are required but have no discriminative function in deciding whether male or female *Sxl* transcripts will be produced. A third alternative is that these genes are regulated by the X : A ratio and that they transmit the primary signal to *Sxl*.

According to Cline (1988a), variations in the dose of the X-chromosomal elements that form the primary signal, so-called numerator elements, should have the following effects. (1) Females should be killed by a decrease in the dose of the elements, because they would not express the female-specific functions of *Sxl*. (2) Males should be killed by an increase in the dose of the elements because they would produce feminizing *Sxl* functions. (3) A decrease in the dose of an X : A numerator element should enhance female-lethal effects of a leaky mutation in a different gene, for example, *da*, while an increase in the dose of the same element should suppress such lethal effects.

The first two criteria are certainly useful for a test of putative numerator elements. Furthermore, a decrease in *Sxl* activity in a female, resulting from a missing numerator element, should be rescued by Sxl^{M1}. Improper activation of *Sxl* in males, by a duplication of a numerator, on the other hand, should be rescued by Sxl^{f1} (Cline 1985, 1988a). The third criterion, however, might include many genes involved in the early steps of sex determination. A deficiency of *Dk* enhances the effect of *da* and vice versa, and a duplication of *da* rescues the lethal effect of *Dk*. These genes are not located on the X chromosome, and their maternal, not their zygotic function, is required for sex determination. Therefore, they do not qualify as numerator elements, but other X-chromosomal and zygotic genes might show similar effects.

No single gene alone satisfies the criteria listed, but since the expectations are met by *sis-a* and *sis-b* together (simultaneous hemizygosity for both elements is lethal to females, and duplicating both is lethal to males), Cline considers *sis-a* and *sis-b* to be numerator elements of the X : A ratio whose doses are counted and whose function in the female is to direct the female-specific expression of *Sxl*. Although this hypothesis seems reasonable, it can be questioned whether the criteria described are sufficient to identify X-chromosomal elements that form the primary signal, as opposed to genes that respond to this signal. As Cline's results show, a third element, namely *Sxl* itself, behaves very much the same as *sis-a* and *sis-b*. Hemizygosity for both *Sxl* and either *sis-a* or *sis-b* is lethal to females, and duplicating *Sxl* as well as either *sis-a* or *sis-b* is lethal to males. Why should these males be unviable if *sis-a* and *sis-b* have only a feminizing effect together? Salz *et al.* (1989) consider it likely that the early embryonic transcripts of *Sxl* are involved in the initial response of the gene to the X : A ratio. However, it is also possible that *Sxl* provides part of the primary signal itself.

We have seen that females require *Sxl* activity throughout development, since homozygous Sxl^{f1}/Sxl^{f1} clones generated at any time differentiate male structures (Sánchez and Nöthiger, 1982). We have also seen that Sxl^+ becomes irreversibly set around blastoderm stage, because XX cells that lose an X chromosome after this stage do not form male clones (Sánchez and Nöthiger, 1983). Thus, *Sxl* is the likely target gene responding to the primary signal. A clear prediction for any gene which acts prior to the target gene is that XX clones, homozygous for a loss-of-function mutation and generated before the blastoderm stage, should either form male structures or be lethal because of disrupted dosage compensation. Clones produced after this stage should be female, since *Sxl* no longer depends on the primary signal. Unfortunately, this test cannot distinguish between a gene that provides the primary signal and a gene that transmits this signal to *Sxl*. Neither would it allow us to test whether *Sxl*, which acts as the target, also provides part of the primary signal.

A comparison of some of Cline's observations with results we have published earlier reveals a picture that is complex (Fig. 4). In his experiments, Cline (1988a) tested the viability of whole animals. Lethality was taken to reflect feminization. To circumvent problems of lethality, we assessed the sex and viability of aneuploid genotypes in clones of mosaic flies (Steinmann-Zwicky and Nöthiger, 1985b). This method showed that duplicating *Sxl* or region 3E to 4F, which contains *liz*, has a more dramatic effect than duplicating *sis-b*. In genotype *b* (Fig. 4A) neither *sis-a* nor *sis-b* is duplicated, yet no male clones could be found. This means either that cells of this genotype all choose the female pathway or that male clones are produced but cannot survive because they are aneuploid and because this is more deleterious to male cells than to female cells. Genotype *d*, however, shows that aneuploidy alone does not kill male cells: they appear when *Sxl* is not duplicated. A similar argument can be made for genotype *c*, in which only one copy of region 3E to 4F is present. Whereas these data point to *Sxl* and possibly *liz* as major feminizing elements, it should be mentioned that XY flies carrying a duplication for these two elements alone can survive as males (Steinmann-Zwicky, 1988). Thus, other regions must also be duplicated to obtain feminization. Whereas it is certainly possible that *Sxl* provides part of the primary signal, the role of *liz* is less clear. The

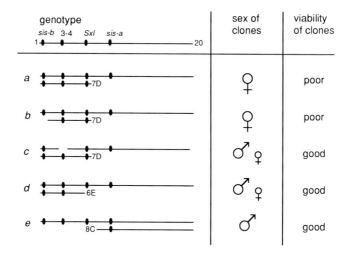

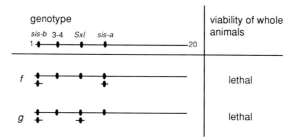

FIG. 4. Viability and sex of aneuploid genotypes. The X chromosome is shown and on it the location of elements known to be involved in the activation of *Sxl*, including *Sxl* itself. (A) Tissue of various aneuploid genotypes was analyzed in mosaic animals. Clones of genotypes *a* and *b* produced only female tissue, and genotype *e* only male tissue. Clones of genotypes *c* and *d* were mostly male, rarely female. The selected examples show that duplications for regions 3E to 4F (named 3–4 on the figure) and 6E to 7D are strongly feminizing. These regions include *liz* and *Sxl*, respectively. Other regions appear to have no strong effect when duplicated. [Data from Steinmann-Zwicky and Nothiger (1985b).] (B) A second experiment tested the viability of aneuploid whole animals. The lethality of genotypes *f* and *g* suggests that duplicating *sis-a* and *sis-b* or *sis-b* and *Sxl* is feminizing and therefore disrupts dosage compensation. A comparison of genotypes *b* and *c* suggests that duplicating *sis-b* and *Sxl* is less feminizing than duplicating region 3E to 4F and *Sxl*, which is in accord with the result observed for genotype *g*. [Data from Cline (1988a).] Numbers 1 and 20 on A and B mark the cytological ends of the X chromosome.

fact that it provides a maternal component makes the gene an unlikely candidate for an element of the X : A ratio.

In our model we have seen that an X-chromosomal element of the X : A ratio could consist either of DNA sequences which serve as a trap or of genes which produce a transcript and perhaps a protein. A transcribed gene that participates in the primary signal has to be expressed in embryos of both sexes and cannot be subject to dosage compensation. This has not yet been tested for the T4 transcript of the *achaete–scute* complex (*sis-b*) and the early transcripts of *Sxl*, which are the only two genes cloned so far. At present we do not yet know how the X : A ratio functions and what elements provide the signal. Several questions await answers. Do the possible candidates, namely, *Sxl, sis-a, sis-b*, and perhaps even the zygotic function of *liz*, provide the primary signal? If so, the expression of these genes should be independent of each other and higher in females than in males. If, on the other hand, the level of transcripts or proteins of some of these genes were to affect the level of expression of other genes of the group, then the latter would not be part of the primary signal. We believe that these questions can be answered only when the functions and interactions of all the genes have been elucidated at the molecular level.

3. Candidates for the Autosomal Component

The number of X chromosomes is measured in reference to the so-called autosomal component. This component may consist of autosomal sequences that could titrate an X-chromosomal product or may be a product of autosomal genes, titrated by X-chromosomal sequences. It is unlikely that the autosomal component is a maternal product. The observation that haploid tissue is female (Santamaria, 1983) whereas cells with one X chromosome and two sets of autosomes are male shows that one set of autosomes, present or absent in a cell, can make the difference between male and female differentiation. Thus, neither *da* nor *Dk* nor the maternal component of *liz* seem to be elements contributing to the autosomal component.

All genetic attempts to localize the putative denominator of the X : A ratio on the autosomes have failed (Pipkin, 1947, 1960; Roehrdanz and Lucchesi, 1981), but it may be that molecular techniques will eventually be more successful. In the housefly *Musca domestica*, sex is determined by a dominant, male-determining factor that is normally localized on the Y chromosome. This factor, however, behaves as a mobile element and in some populations has spread in multiple copies on the autosomes. The effect of this invasion is counteracted by the parallel evolution of a dominant female-determining mutation in a gene on chromosome 4 (Franco *et al.*, 1982). We have proposed (Nöthiger and Steinmann-Zwicky, 1985b) that the epistatic female-determining factor is a constitutive mutation, corresponding to Sxl^{M1} in *Drosophila*, and that the male-determining factor normally represses the wild-type allele of this gene. We see a further parallel in the occurrence of the male-determining factor in multiple copies on the autosomes of *Musca* with the postulated male-determining or denominator elements on the autosomes of *Drosophila*. In the fly *Calliphora* and in the midge *Chironomus thummi thummi*, the male-determining factor is located in a cytologically unique band on one of the polytene chromosomes of males but not females (Ribbert, 1967; Hägele, 1985). This band can now be cloned by microdissection, and it may become possible to look for homologous sequences on the autosomes of *Drosophila*.

There is also an alternative, however, that requires no specific loci, coding or noncoding, to explain the contribution of the autosomes. This model assigns the role of the "autosomal component" to the size of the cells, which is a function of the number of chromosomes per nucleus. In *Drosophila*, cell size is essentially determined by the number of autosomes, which represent around 80% of the chromosome content. Thus, male and female cells have roughly the same size. If the X chromosomes were to produce sex-determining molecules in proportion to their number, the concentration of these molecules would be about twice as high in females as in males, and this could initiate the cascade of sex determination. A 2-fold difference in the concentration of *bicoid* protein has been shown to be sufficient to trigger the activation of the *hunchback* gene in an all or none fashion (Struhl *et al.*, 1989). Similarly, a 2-fold difference in the quantity of a sex-determining product might lead to the activation of *Sxl* in females, but not in males. In this model, no particular chromosomal region would provide the autosomal component, and any search for specific genes or loci would be fruitless.

IV. From *double-sex* to Real Sex

The alternative and mutually exclusive products of the *dsx* locus, **M** (the product of *dsx*^{m}) or **F** (the product of *dsx*^{f}), are the molecular signals for sexual differentiation. Their action leads to the concrete manifestation of sexual phenotype which becomes apparent in morphological, behavioral, and biochemical differences between the sexes. These differences are the result of the activity of male- and female-specific differentiation genes. How do the products of *dsx* achieve this differential, sex-specific expression of differentiation genes?

The same male-determining (**M**) or the same female-determining (**F**) product of *dsx* is probably present and active in every somatic cell of a male or female, respectively. An analogous situation exists in mammals where the sex hormones are ubiquitously distributed, thus reaching every cell and regulating sexual differentiation. In mammals, the signal is produced in an endocrine tissue and reaches the cells via the bloodstream; in *Drosophila*, the signal is produced within each cell.

In both cases, the question is how a single molecule achieves the very different responses displayed by different tissues. The specificity of the sexual response apparently depends on the cell type. For many cells, it seems irrelevant which of the *dsx* products is present, or whether the gene functions at all. In the presence of **M**, or rather in the absence of **F** (see later), however, sex comb bristles are differentiated on the basitarsus of the foreleg and nowhere else; the cells of abdominal segment 8 (A8) remain quiescent, those of A9 form male genitalia, and those of A10/11 male analia; brain cells are programmed for male behavior; and the fat body does not synthesize yolk proteins.

Yolk proteins are subject to time-, tissue-, and sex-specific control of gene expression. They are synthesized in the fat body and ovarian follicle cells of adult females several hours after emergence. How could this regulation be achieved? Specifically, what are the cis- and trans-regulatory signals that are involved in the process? For simplicity, we assume that there are two nonoverlapping sets of sex differentiation genes, one characteristic of males, the other of females, although some aspects of sexual dimorphism may result from genes being expressed at different levels in males and females.

A. ROLE OF *double-sex*

1. *double-sex Acts by Repressing Sex Differentiation Genes*

When flies lack a functional *dsx* gene they develop as intersexes, irrespectively of whether the animal has an XX or XY karyotype. The intersexual phenotype is displayed at the level of individual cells, suggesting that both sets of sex differentiation genes are expressed simultaneously in these animals, and, in fact, within single cells. This type of intersexuality is most clearly seen in the cells that form a sex comb in males or a row of simple bristles in females (Fig. 5a–c). Other sexual characters are also sexually intermediate. Since this intersexuality results from the absence of a functional *dsx* product, we conclude that the sex differentiation genes are constitutively expressed, and that **M** serves to repress the female set, **F** to repress the male set of differentia-

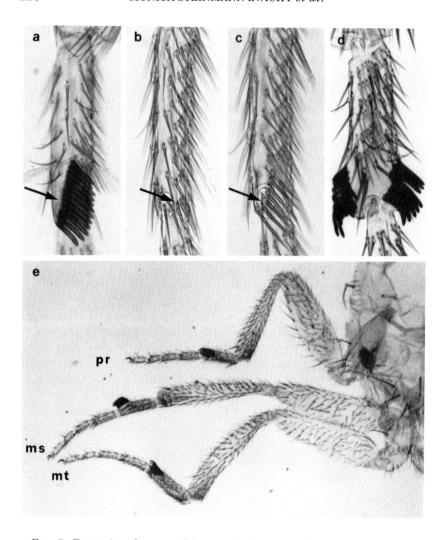

FIG. 5. Formation of a sex comb is controlled by *dsx* and homeotic genes. A normal sex comb is formed by cells located in the distal region of the anterior compartment of the basitarsus of prothoracic legs of males. Arrows point to the row of bristles that form a sex comb in males or intersexes. Basitarsi of (a) normal male, (b) normal female, (c) *dsx* mutant (note the intersexual phenotype at the cellular level displayed by the intermediate shape of the sex comb bristles), (d) *en/en* male (*en* transforms posterior cells into anterior cells), (e) *Pc³/+* male [*Pc³* transforms mesothoracic (ms) and metathoracic (mt) legs into prothoracic (pr) legs]. (Figure 5e courtesy of Ana Busturia and Ginés Morata.)

tion genes. The exclusive presence of either **M** or **F** guarantees a clear alternative expression of sex-specific genes so that a male or a female is formed (Fig. 6).

2. *double-sex Acts in Concert with Homeotic Genes and Positional Information*

We have already stated that the specificity of sexual differentiation, that is, the response of a specific tissue, depends on the cell type. According to our current view, the cell type, or more specifically the segmental identity, is determined by the action of selector genes

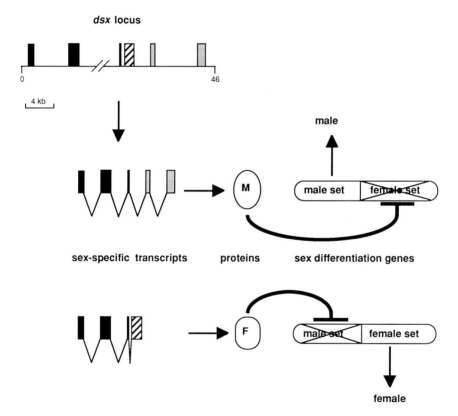

FIG. 6. Structure and function of the *dsx* locus. By alternative splicing, the bifunctional *dsx* locus can produce either a male-specific or a female-specific transcript. The corresponding functional proteins act as male-determining (**M**) or female-determining (**F**) signals that repress genes required for female or male differentiation. The male- and female-specific exons are shown as stippled and hatched boxes, respectively; the common exons are black.

(Garcia-Bellido, 1975), the so-called homeotic genes (for review, see Akam, 1987). The principle is again demonstrated by the cells that form a sex comb. This structure is differentiated in the anterior compartment of the basitarsus on the prothoracic leg of males but not females. In male flies mutant for *en* (*engrailed*), a secondary sex comb is formed on the prothoracic legs as a mirror image of the primary comb, because the posterior cells are transformed into anterior cells by the *en* mutation (Fig. 5d). The homeotic mutations *Pc* (*Polycomb*) or *esc* (*extra sex combs*) transform the meso- and metathoracic legs into prothoracic legs. In the mutants, sex combs may appear on all legs, but again only in males and only in the anterior compartment of the basitarsus (Fig. 5e). The cooperation between the products of *dsx* and those of the homeotic genes can take place even late in development, as shown by homozygous *esc*/*esc* clones. In males, such cells were able to form sex comb bristles on all legs, even when the clones were generated toward the end of the second larval instar (Tokunaga and Stern, 1965).

These examples show that the formation of a sex comb depends on the activity of a set of homeotic genes that specify the segmental and compartmental fate of a cell (Garcia-Bellido *et al.*, 1979). The products of *dsx*, **M** or **F**, are probably present even in tissues in which we cannot detect any sexual dimorphism. Only when such cells are homeotically transformed may the presence of **M** or **F** become effective and apparent.

The sex comb is formed in a unique position along the proximodistal axis of the leg. At present, we must explain this specificity by postulating some sort of positional information that is differentially distributed along the longitudinal axis of the leg. A gene (*Distal-less, Dll*), potentially involved in establishing, or responding to, a gradient of positional information along the proximodistal axis of the limbs, has been identified. Its function is specifically required in the distal portions of the antenna and the legs (Cohen and Jürgens, 1989). The gene was found to contain a homeodomain (Cohen *et al.*, 1989), which suggests that it may regulate the activity of other genes. These results justify some hope that we may be able to find a genetic basis for the proximodistal specification of limbs. In summary, we see that the cell type forms the context in which the products of *dsx*, **M** or **F**, act to bring about a particular type of sexual differentiation.

B. MODES OF REGULATION

The effects of mutations in any of the genes in the cascade from *Sxl* to *dsx* are global, with consequences for all aspects of sex in somatic cells. The question arises whether there are subordinate sex-determining genes with a locally restricted realm of action downstream of *dsx*. Since lack of a signal from *dsx* results in intersexuality, mutations in a tissue-specific sex-determining gene should lead to an intersexual phenotype rather than to a replacement of female structures by male structures. If such genes existed they would have to be bifunctional like *dsx*, but their regulation would have to be different from that of *dsx*. One could imagine that their two functions are constitutive and serve to activate the sex differentiation genes; one of the two functions would normally be repressed by **M** or **F**.

For *Drosophila*, no mutations with a locally restricted *dsx* phenotype are known. Therefore, we believe that such genes do not exist in *Drosophila*, and we favor the simpler hypothesis that the products of *dsx*, **M** or **F**, are the ubiquitous signals that determine the sexual pathway.

What are the target genes of the *dsx* products? A first step toward their identification is to begin at the end and study the terminal differentiation genes. A few sex-specific products are known, such as the yolk proteins (YPs), the chorion proteins (CPs), the sex peptide (SP), and male-specific transcripts (*mst*) of undefined functions. These products are synthesized only at specific times and in specific tissues of either females (YP, CP) or males (SP, *mst*). The regulation occurs at the level of transcription (for review, see Wolfner, 1988). Two different modes of regulation seem to be used to control the expression of terminal differentiation genes. These are discussed below and are schematically summarized in Fig. 7.

1. Sex-Specific Expression of a Gene May Be Under Permanent Control of double-sex

Regulation of gene expression under permanent control of *dsx* occurs in the fat body which synthesizes yolk proteins in adult females. The fat body is a tissue that is present in both sexes, but the genes coding for YPs are transcribed only in adult females. When XX animals are transformed into pseudomales by the mutations *tra*, *tra-2*, or *dsx*^{D}, the *yp* genes in the fat body remain inactive; in XY flies, the genes become active when such animals are feminized by the mutation *dsx*^{m} (Bownes and Nöthiger, 1981) or when they are transformed into pseudofemales by *hs-tra-female*, a transposable construct that constitutively expresses a functional product of *tra* (A. Dübendorfer, M. Steinmann-Zwicky, and R. Nöthiger, unpublished; for details on *hs-tra-female*, see McKeown *et al.*, 1988). These results show that expression of the *yp* genes is controlled by the sex-determining genes, and ultimately by *dsx*.

The transcriptional activity of the *yp* genes in the fat body is under permanent control by *dsx*, as revealed in a classic experiment per-

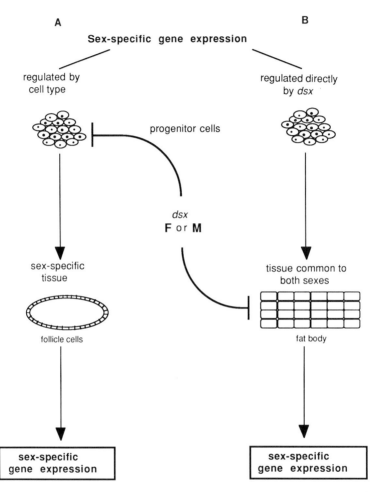

FIG. 7. Regulation of sex-specific gene expression. Two mechanisms by which sex-specific expression of terminal differentiation genes may be achieved are shown. The functional products of *dsx*, **M** or **F**, prevent the expression of a tissue-specific gene. (A) This target gene is itself a regulatory gene that instructs a group of progenitor cells to form or not to form a sex-specific tissue, such as the follicle cells or the paragonia, for example; this cell type then dictates the synthesis of sex-specific products. (B) The target gene of *dsx* is a terminal differentiation gene that is either active or inactive in a tissue common to both sexes. [Modified from Wolfner (1988).]

formed by Belote *et al.* (1985a). When XX flies, homozygous for a temperature-sensitive allele, *tra-2^{ts}*, were raised at 16°C and thus developed as females, the *yp* genes became active in their fat body. But when the animals were raised at 29°C and thus became pseudomales, the *yp* genes remained inactive. When the *X/X;tra-2^{ts}/tra-2^{ts}* adult females were shifted from 16°C to the restrictive temperature of 29°C, the transcription of the *yp* genes gradually decreased and eventually stopped. More surprisingly, the silent *yp* genes of the pseudomales became active within a few hours after the animals were shifted to the permissive temperature of 16°C.

Thus, repression and derepression of the *yp* genes in the fat body of *tra-2^{ts}* mutant animals are reversible, depending on whether the product of *tra-2^{ts}* is inactive (at 29°C) or active (at 16°C). A certain reversibility also exists in the nervous system, as demonstrated by partial reversion of sexual behavior of XX flies mutant for *tra-2^{ts}* upon temperature shifts (Belote and Baker, 1987). Since the gene *tra-2* exerts its effect on the *yp* genes through *dsx*, we conclude that the alternative expression of the *dsx* gene must be equally reversible, even in adult flies.

The sex-specific control of the *yp* genes in the adult fat body can be overthrown by injection of high doses of 20-hydroxyecdysone into adult XY males, which then start to synthesize YPs. Interestingly, the synthesis is restricted to the fat body. Thus, only the sex-specific but not the tissue-specific nor the stage-specific control of the genes is lost (Postlethwait *et al.*, 1980a; Bownes *et al.*, 1983). Since 20-hydroxyecdysone can derepress the *yp* genes in *X/X;dsx^{D}/dsx* and other pseudomales, the hormone does not work through the sex-determining genes (Bownes and Nöthiger, 1981). It could, however, affect their targets, or *dsx* could control the ability of the cells to respond to the hormone.

2. Sex-Specific Expression of a Gene May Be Delegated to a Sex-Specific Cell Type

Regulation by sex-specific cell type is best illustrated by the case of male-specific transcripts (*msts*) which are synthesized in the paragonia (Schäfer, 1986a,b; DiBenedetto *et al.*, 1987). The paragonia are accessory glands of the male copulatory apparatus; their contents are transferred into the female gonoducts during copulation. The biological function of *msts* is largely unknown. One of the transcripts has features of a peptide hormone precursor (Monsma and Wolfner, 1988), and the sex peptide, another product of the paragonia, acts as a pheromone, modifying the behavior and physiology of the mated female (Chen *et al.*, 1988).

The synthesis of these transcripts is under control of the sex-determining genes (Schäfer, 1986b; Chapman and Wolfner, 1988). Unlike the YPs in the fat body, however, this control is indirect. Using again the *tra-2^{ts}* allele and temperature shifts, Wolfner and collaborators (DiBenedetto *et al.*, 1987; Chapman and Wolfner, 1988; Monsma and Wolfner, 1988) found that in *X/X;tra-2^{ts}/tra-2^{ts}* flies, the paragonial transcripts appeared only when morphologically recognizable paragonial structures were formed, that is, when the animals were kept at 29°C at least during the critical period for the determination of paragonia. Shifting these animals to the female-determining temperature of 16°C did not lead to the disappearance of the *mst*s. Similarly, shifting *X/X;tra-2^{ts}/tra-2^{ts}* mutants that had developed into females at the permissive temperature to 29°C did not activate any *mst* genes, indicating that no cell type of a female is capable of transcribing the *mst* genes.

These experiments demonstrate that the functional state of *dsx*, regulated by *tra-2*, determines whether paragonia will develop or not. Once this male-specific tissue is formed, transcription of the *mst* genes is no longer dependent on the activity of sex-determining genes, but on the sex-specific cell type. This second type of regulation seems to be the rule and may operate in many other cases where sex-specific products are synthesized by tissues or cell types that are only formed in one sex. Candidates for this type of regulation are the chorion proteins (Petri *et al.*, 1976) and those yolk proteins that are synthesized in the follicle cells of the ovary (Bownes and Hames, 1978; Brennan *et al.*, 1982).

In some cases, sex-specific gene expression or differentiation depends on cellular interactions. The follicle cells in the ovary transcribe the *yp* genes from stage 8 to 10b (Brennan *et al.*, 1982) and the chorion genes from stage 11 to 14 (Petri *et al.*, 1976). In females that lack a germ line, both the *yp* genes in the ovary and the chorion protein genes are not transcribed even though the somatic components of the ovary are formed (Postlethwait *et al.*, 1980b; DiBenedetto *et al.*, 1987). In the absence of a developing germ line, the follicle cells apparently do not reach the stage where these genes can become active. This situation, then, is not principally different from regulation by cell type. Another example is a muscle that is present in the fifth abdominal segment of males but not females (Lawrence and Johnston, 1984). In chimeric flies composed of male and female cells, this male-specific muscle is formed whenever the innervating neuron is male, irrespective of the sex of the muscle cells. In our view, this phenomenon is again a case of regulation by cell type. We are apparently dealing with an inductive process in which a particular neuron acquires the sex-specific capacity to induce the formation of a muscle.

C. Anatomy of a Sex Specifically Regulated Gene

1. ypl–yp2 Gene Pair Coding for Yolk Proteins 1 and 2 (YP1 and YP2)

There are three genes, *yp1*, *yp2*, and *yp3*, coding for YPs, and they form a cluster on the X chromosome (Barnett *et al.*, 1980). The genes are coordinately regulated and expressed in the fat body and follicle cells of adult females. This time-, tissue-, and sex-specific regulation is somehow modulated by hormones (Jowett and Postlethwait, 1980; Bownes, 1982; for review, see Bownes, 1986).

The *yp1* and *yp2* genes are divergently transcribed, with 1225 nucleotides located between their transcription start sites (Fig. 8A). By generating various deletions, Wensink and collaborators identified cis-acting elements that are necessary for YP synthesis in the fat body or in the follicle cells (Garabedian *et al.*, 1985, 1986; Logan *et al.*, 1989). A short element of 125 nucleotides acts as an enhancer and is able to convey fat body-specific expression to a heterologous gene. Two short elements, 184 and 105 bp long, respectively, specify expression of *yp1* and *yp2* in the follicle cells. In the absence of these three elements the *yp* genes remain inactive, as they are in all other tissues. This indicates that the tissue-specific control is positive, suggesting that the fat body and the follicle cells produce activating factors.

The enhancer specific for fat body expression is necessary and sufficient to confer tissue-, time-, and sex-specific expression on a heterologous promoter (Garabedian *et al.*, 1986). Whether this 3-fold specificity is caused by a single element or whether there are multiple control regions within an element is not known.

2. Genes Coding for Chorion Proteins

The chorion or eggshell of *Drosophila* contains around 20 different proteins that are synthesized in the follicle cells during a short period of 5–6 hours at the end of oogenesis (Mahowald and Kambysellis, 1980). Several of the genes coding for these proteins have been cloned (Parks and Spradling, 1987). We concentrate on the gene *s15-1* which is expressed very late at the end of the choriogenic period.

Mariani *et al.* (1988) found that a DNA segment of 73 nucleotides upstream of the transcription start site of *s15-1* contains sequences that are essential for the tissue-specific and late expression of the gene. Further subdivision of the segment and *in vitro* mutagenesis revealed the existence of three adjacent regulatory elements (Fig. 8B). One element conveys late expression; another element prevents early expression of *s15-1*. The most important sequence, however, is a hexanucleotide, TCACGT, that is essential for the gene to become active at all.

A

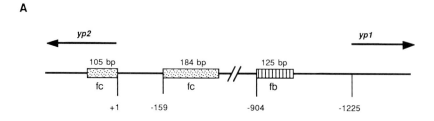

B

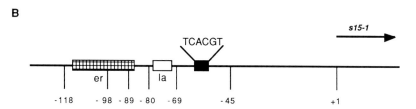

FIG. 8. Structure of the cis-regulatory region of a sex differentiation gene. (A) The closely linked and coordinately expressed yolk protein genes *yp1* and *yp2* are separated by 1225 nucleotides which harbor regulatory sequences with enhancer properties. One of them (fb) is necessary and sufficient for transcription of both genes in the fat body, two others (fc) for transcription in the follicle cells. The elements convey the tissue, stage, and sex specificity of gene expression. (B) Transcription of the chorion protein gene *s15-1* requires the presence of the highly conserved hexanucleotide TCACGT. Two other upstream sequences act as "late activating" (la) and "early repressing" (er) elements that regulate the precise temporal specificity of expression. The elements operate in a sex-specific tissue, the ovarian follicle cells. Numbers give map positions in nucleotides.

When this short sequence is removed or replaced by a different hexanucleotide, *s15-1* cannot be transcribed. Remarkably, the same hexanucleotide is found about 25 nucleotides upstream of the TATA box in all other chorion genes studied so far (Wong *et al.*, 1985), and it is also present in the regulatory DNA of the chorion genes of the silkmoth *Bombyx mori* (Mitsialis and Kafatos, 1985). The chorion proteins, as well as the organization of their genes, differ widely between *B. mori* and *Drosophila*—the two species diverged some 240 million years ago. Nevertheless, when two clustered chorion genes of the silkmoth were introduced into *Drosophila melanogaster*, these genes were properly regulated and transcribed only in the follicle cells, as long as the hexanucleotide TCACGT was present (Mitsialis *et al.*, 1987). The evidence is compelling that this sequence serves as the target for a highly conserved trans-activating factor that is synthesized in the follicle cells and promotes transcription of the chorion genes.

One could expect that TCACGT might also be found in the DNA stretch that is responsible for expression of the *yp* genes, and perhaps of other genes, in the follicle cells. The task of the sex-determining genes, then, would be to dictate the development of a tissue in which a specific trans-acting factor is made which recognizes TCACGT so that transcription can take place. The first analyses with the *yp* genes are disappointing. No sequence motif related to TCACGT is found in the enhancer that is specific for follicle cell expression of the *yp* genes of *Drosophila* (Wensink, personal communication), nor can a clear consensus sequence be detected in the enhancer elements of *yp1–yp2* and those of *yp3* although these genes are coordinately expressed in the fat body and in the follicle cells (Hung and Wensink, 1981, 1983; Garabedian *et al.*, 1987). It thus appears that different genes have different cis-regulatory elements, even when they are expressed in the same tissue at approximately the same time and in the same sex. Nevertheless, it will be interesting to see whether the male-specific transcripts produced in the paragonia form a "family," like the chorion genes, with a common cis-regulatory sequence, and whether such a sequence is also evolutionarily conserved. Seven genes that are transcribed in the male germ line of *Drosophila* share a consensus sequence of 12 nucleotides, located just upstream of the translation start site. In this case, however, the sequence is apparently involved in translational rather than in transcriptional control (M. Schäfer, R. Kuhn, and U. Schäfer, unpublished).

D. Concluding Remarks and a Simple Model

Sex-specific transcription of a gene in a tissue that is common to both sexes, for example, the fat body, requires tissue-specific trans-acting factors (TTAF) that recognize a specific enhancer sequence and thus activate transcription. These factors must be active at the time when the gene is to be transcribed. The product of *dsx*, which is present long before overt sexual differentiation, cannot be involved in the timing. In the case of the *yp* genes, the successful interaction between the postulated TTAF and its cis-regulatory target is somehow prevented by the **M** product of *dsx*. In a simple model, **M** might inactivate TTAF either by forming a nonfunctional complex with it (Fig. 9A) or by directly blocking the cis-regulatory target sequence (Fig. 9B). In the first alternative, in which **M** or **F** would bind to TTAF, we expect that such

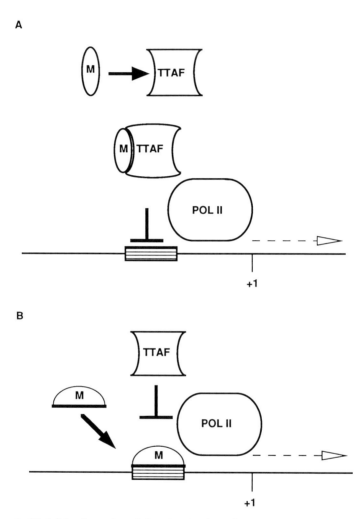

FIG. 9. Model for the action of *dsx*, showing how the products of *dsx* could regulate their target genes. Remember that **M** acts to repress the female-specific genes, **F** to repress the male-specific genes. The simplest model requires three elements: a tissue-specific trans-activating factor (TTAF), the product of *dsx* (**M** or **F**), and a single cis-regulatory sequence to which TTAF can bind and promote transcription by polymerase II (POL II). The product **M** prevents the expression of a female-specific gene, either by forming a nonfunctional complex with TTAF (A) or by blocking the cis-regulatory site (B). A male-specific target gene would be analogously repressed by **F**. This model applies to both mechanisms shown in Fig. 7. Dashed arrows symbolize lack of transcription.

trans-acting factors should have a conserved domain for the interaction with **M** or **F**. The second alternative requires that the putative *dsx* protein should have features of a DNA-binding protein, which, however, does not seem to be the case (Burtis and Baker, 1989).

The search for putative trans-acting factors has barely begun. Mitsis and Wensink (1989) have isolated a DNA-binding protein from a *Drosophila* cell line. This protein, called YPF1, binds with very high affinity and specificity to a sequence of 31 bp in the *yp1* gene. This binding site, however, is within the translated region, 148 bp downstream of the transcription initiation site, whereas the well-defined and only known enhancer element is some 150 bp 5′ upstream of the gene. Furthermore, the protein is found in a cell line and in embryos in which the *yp1* gene is not active, but it is not detectable in the fat body of adult females where the *yp* genes are active. For these reasons, YPF1 is an unlikely candidate for the TTAF of our model.

Sex-specific transcription of a gene in a tissue that is limited to one sex, for example, the follicle cells of the ovary, is analogous to any other tissue-specific gene expression. It is the formation of the tissue which is controlled by *dsx*. In this case, the target gene of *dsx* is not a terminal differentiation gene but probably a regulatory gene whose state of activity directs a group of cells into a specific developmental pathway. For example, whether the precursor cells of abdominal segment 9 undergo rapid cell divisions and form the paragonia plus other parts of the male genitalia, or whether they remain more or less quiescent which is the case in a female, depends on the functional state of *dsx* (Nöthiger *et al.*, 1977, 1987; Schüpbach *et al.*, 1978) and on some trans-acting factors specifically present in these precursor cells. None of these regulatory genes, responding to *dsx* and directing a developmental program, has yet been identified. They could be genes controlling rates and patterns of cell division, with properties of selector genes, and probably themselves controlled by homeotic genes.

The principle of **M** or **F** interacting with a trans-acting factor and a cis-acting sequence of a target gene can account for both modes of regulation depicted in Fig. 7. The developmental level, however, at which *dsx* acts is different for the fat body and the paragonia or the follicle cells. In the fat body, the target gene of *dsx* is likely to be a terminal differentiation gene that synthesizes a sex-specific product; in the paragonia or the follicle cells, it is a regulatory gene that directs a group of cells into a sex-specific developmental pathway.

Sex determination can serve as a paradigm for the regulation of a developmental program. In itself, it is a remarkably simple process: a primary chromosomal signal which is quantitative achieves the result that the key gene *Sxl* makes either an active or an inactive product. This initiates a small cascade which leads to the expression of either the male-specific or female-specific product of *dsx*. Some sort of a bifunctional switch is logically required if two sets of genes are to be regulated such that two mutually exclusive patterns arise. Unfortunately, none of the direct target genes of *dsx* are known so far.

To achieve sexual dimorphism, *dsx* has to cooperate with the homeotic genes. Sex becomes reality through the interaction of the male-determining or female-determining product of *dsx* with tissue-specific trans-activating factors whose synthesis eventually depends on the differential activity of homeotic genes.

ACKNOWLEDGMENTS

We are grateful to many colleagues at home and abroad who helped us with this article by contributing unpublished information, critical comments, technical help, and encouragement. Our own work was supported by the Swiss National Science Foundation, the "Jubiläumsspende" and the "Stiftung für wissenschaftliche Forschung" of the University of Zurich, the Julius Klaus-Stiftung, the Karl Hescheler-Stiftung, and the Georges and Antoine Claraz-Schenkung.

REFERENCES

Aebi, M., and Weissmann, C. (1987). Precision and orderliness in splicing. *Trends Genet.* **3,** 102–107.

Akam, M. (1987). The molecular basis for metameric pattern in the *Drosophila* embryo. *Development* **101,** 1–22.

Amrein, H., and Nöthiger, R. (1989). Erratum. *Cell* **58,** 421.

Amrein, H., Gorman, M., and Nöthiger, R. (1988). The sex-determining gene *tra-2* of *Drosophila* encodes a putative RNA binding protein. *Cell* **55,** 1025–1035.

Baker, B. S. (1989). Sex in flies: The splice of life. *Nature (London)* **340,** 321–324.

Baker, B. S., and Belote, J. M. (1983). Sex determination and dosage compensation in *Drosophila. Annu. Rev. Genet.* **17,** 345–397.

Baker, B. S., and Ridge, K. A. (1980). Sex and the single cell. I. On the action of major loci affecting sex determination in *Drosophila melanogaster. Genetics* **94,** 383–423.

Baker, B. S., and Wolfner, M. F. (1988). A molecular analysis of *double-sex,* a bifunctional gene that controls both male and female sexual differentiation in *Drosophila melanogaster. Genes Dev.* **2,** 477–489.

Baker, B. S., Nagoshi, R. N., and Burtis, K. C. (1987). Molecular genetic aspects of sex determination in *Drosophila. BioEssays* **6,** 66–70.

Bandziulis, R. J., Swanson, M. S., and Dreyfuss, G. (1989). RNA-binding proteins as developmental regulators. *Genes Dev.* **3,** 431–437.

Barnett, T., Pachl, C., Gergen, J. P., and Wensink, P. C. (1980). The isolation and characterisation of *Drosophila* yolk protein genes. *Cell* **21,** 729–738.

Bell, L. R., Maine, E. M., Schedl, P., and Cline, T. W. (1988). *Sex-lethal,* a *Drosophila* sex determination switch gene, exhibits sex-specific RNA splicing and sequence similarity to RNA binding proteins. *Cell* **55,** 1037–1046.

Belote, J. M., and Baker, B. S. (1982). Sex determination in *D. melanogaster:* Analysis of *transformer-2*, a sex determining locus. *Proc. Natl. Acad. Sci. U.S.A.* **79**, 1568–1572.

Belote, J. M., and Baker, B. S. (1983). The dual functions of a sex determination gene in *Drosophila melanogaster*. *Dev. Biol.* **95**, 512–517.

Belote, J. M., and Baker, B. S. (1987). Sexual behavior, its control during development and adulthood in *Drosophila melanogaster*. *Proc. Natl. Acad. Sci. U.S.A.* **84**, 8026–8030.

Belote, J. M., and Lucchesi, J. C. (1980). Control of X chromosome transcription by the maleless gene in *Drosophila*. *Nature (London)* **285**, 573–575.

Belote, J. M., Handler, A. M., Wolfner, M. F., Livak, K. L., and Baker, B. S. (1985a). Sex specific regulation of yolk protein gene expression in *Drosophila*. *Cell* **40**, 339–348.

Belote, J. M., McKeown, M. B., Andrew, D. J., Scott, T. N., Wolfner, M. F., and Baker, B. S. (1985b). Control of sexual differentiation in *Drosophila melanogaster*. *Cold Spring Harbor Symp. Quant. Biol.* **50**, 605–614.

Belote, J. M., McKeown, M., Boggs, R. T., Ohkawa, R., and Sosnowski, B. A. (1989). The molecular genetics of *transformer*, a genetic switch controlling sexual differentiation in *Drosophila*. *Dev. Genet.* **10**, 143–154.

Boggs, R. T., Gregor, P., Idriss, S., Belote, J. M., and McKeown, M. (1987). Regulation of sexual differentiation in *D. melanogaster* via alternative splicing of RNA from the *transformer* gene. *Cell* **50**, 739–747.

Bownes, M. (1982). Hormonal and genetic regulation of vitellogenesis in *Drosophila*. *Q. Rev. Biol.* **57**, 247–274.

Bownes, M. (1986). Expression of the genes coding for vitellogenin (yolk protein) *Annu. Rev. Entomol.* **31**, 507–531.

Bownes, M., and Hames, B. D. (1978). Analysis of yolk proteins in *Drosophila melanogaster*. *FEBS Lett.* **96**, 327–330.

Bownes, M., and Nöthiger, R. (1981). Sex determining genes and vitellogenin synthesis in *Drosophila melanogaster*. *Mol. Gen. Genet.* **182**, 222–228.

Bownes, M., Blair, M., Kozma, R., and Dempster, M. (1983). 20-Hydroxyecdysone stimulates tissue-specific yolk-protein gene transcription in both male and female *Drosophila*. *J. Embryol. Exp. Morphol.* **78**, 249–268.

Breen, T. R., and Lucchesi, J. C. (1986). Analysis of the dosage compensation of a specific transcript in *Drosophila melanogaster*. *Genetics* **112**, 483–491.

Brennan, M. D., Weiner, A. J., Goralski, T. J., and Mahowald, A. P. (1982). The follicle cells are a major site of vitellogenin synthesis in *Drosophila melanogaster*. *Dev. Biol.* **89**, 225–236.

Bridges, C. B. (1921). Triploid intersexes in *Drosophila melanogaster*. *Science* **54**, 252–254.

Burtis, K. C., and Baker, B. S. (1989). *Drosophila doublesex* gene controls somatic sexual differentiation by producing alternatively spliced mRNAs encoding related sex-specific polypeptides. *Cell* **56**, 997–1010.

Butler, B., Pirrotta, V., Irminger-Finger, I., and Nöthiger, R. (1986). The sex-determining gene *tra* of *Drosophila melanogaster*: Molecular cloning and transformation studies. *EMBO J.* **5**, 3607–3613.

Campuzano, S., Carramolino, L., Cabrera, C. V., Ruiz-Gomez, M., Villares, R., Boronat, A., and Modolell, J. (1985). Molecular genetics of the *achaete–scute* gene complex of *D. melanogaster*. *Cell* **40**, 327–338.

Caudy, M., Grell, E. H., Dambly-Chaudière, C., Ghysen, A., Jan, L. Y., and Jan. Y. N. (1988a). The maternal sex determination gene *daughterless* has zygotic activity necessary for the formation of peripheral neurons in *Drosophila*. *Genes Dev.* **2**, 843–852.

Caudy, M., Vässin, H., Brand, M., Tuma, R., Jan, L. Y., and Jan, Y. N. (1988b). *Daughterless*, a *Drosophila* gene essential for both neurogenesis and sex determination, has sequence similarities to *myc* and the *achaete–scute* complex. *Cell* 55, 1061–1067.

Chandra, H. S. (1985). Sex determination: A hypothesis based on noncoding DNA. *Proc. Natl. Acad. Sci. U.S.A.* 82, 1165–1169.

Chapman, K. B., and Wolfner, M. F. (1988). Determination of male-specific gene expression in *Drosophila* accessory glands. *Dev. Biol.* 126, 195–202.

Chen, P. S., Stumm-Zollinger, E., Aigaki, T., Balmer, J., Bienz, M., and Böhlen, P. (1988). A male accessory gland peptide that regulates reproductive behavior of female *D. melanogaster. Cell* 54, 291–298.

Cline, T. W. (1976). A sex-specific, temperature-sensitive maternal effect of the daughterless mutation of *Drosophila melanogaster. Genetics* 84, 723–742.

Cline, T. W. (1978). Two closely linked mutations in *Drosophila melanogaster* that are lethal to opposite sexes and interact with daughterless. *Genetics* 90, 683–698.

Cline, T. W. (1979). A male-specific lethal mutation in *Drosophila melanogaster* that transforms sex. *Dev. Biol.* 72, 266–275.

Cline, T. W. (1983a). Functioning of the genes *daughterless (da)* and *Sex-lethal (Sxl)* in *Drosophila* germ cells. *Genetics* 104, s16–17.

Cline, T. W. (1983b). The interaction between daughterless and Sex-lethal in triploids: A novel sex-transforming maternal effect linking sex determination and dosage compensation in *Drosophila melanogaster. Dev. Biol.* 95, 260–274.

Cline, T. W. (1984). Autoregulatory functioning of a *Drosophila* gene product that establishes and maintains the sexually determined state. *Genetics* 107, 231–277.

Cline, T. W. (1985). Primary events in the determination of sex in *Drosophila melanogaster. In* "Origin and Evolution of Sex" (H. O. Halvorson and A. Monroy, eds.), pp. 301–327. Liss, New York.

Cline, T. W. (1986). A female-specific lethal lesion in an X-linked positive regulator of the *Drosophila* sex determination gene, *Sex-lethal. Genetics* 113, 641–663.

Cline, T. W. (1988a). Evidence that *sisterless-a* and *sisterless-b* are two of several discrete "numerator elements" of the X/A sex determination signal in *Drosophila* that switch *Sxl* between two alternative stable expression states. *Genetics* 119, 829–862.

Cline, T. W. (1988b). Exploring the role of the gene, *Sex-lethal (Sxl)*, in the genetic programming of *Drosophila* sexual dimorphism. *In* "Evolutionary Mechanisms in Sex Determination" (CRC Uniscience Series) (S. S. Wachtel, ed.), pp. 23–36. CRC Press, Boca Raton, Florida.

Cohen, S. M., and Jürgens, G. (1989). Proximal–distal pattern formation in *Drosophila*: Cell-autonomous requirement for *Distal-less* gene activity in limb development. *EMBO J.* 8, 2045–2055.

Cohen, S. M., Brönner, G., Küttner, F., Jürgens, G., and Jäckle, H. (1989). Distal-less encodes a homoeodomain protein required for limb development in *Drosophila. Nature (London)* 338, 432–434.

Cronmiller, C., and Cline, T. W. (1987). The *Drosophila* sex determination gene *daughterless* has different functions in the germ line versus the soma. *Cell* 48, 479–487.

Cronmiller, C., Schedl, P., and Cline, T. W. (1988). Molecular characterization of *daughterless*, a *Drosophila* sex determination gene with multiple roles in development. *Genes Dev.* 2, 1666–1676.

DiBenedetto, A. J., Lakich, D. M., Kruger, W. D., Belote, J. M., Baker, B. S., and Wolfner, M. F. (1987). Sequences expressed sex-specifically in *Drosophila melanogaster* adults. *Dev. Biol.* 119, 242–251.

Dobzhansky, T., and Schultz, J. (1934). The distribution of sex factors in the X-chromosome of *Drosophila melanogaster. J. Genet.* 28, 349–386.

Duncan, F. W., and Kaufman, T. C. (1975). Cytogenetic analysis of chromosome 3 in *Drosophila melanogaster*. Mapping of the proximal portion of the right arm. *Genetics* **80,** 733–752.

Ehrensperger, P. C. (1983). Die Entwicklung der bisexuellen Anlage der Genitalien und Analien, untersucht an verschiedenen Geschlechtsmutanten der Taufliege *Drosophila melanogaster. Mitt. Aarg. Naturforsch. Ges.* **30,** 145–235.

Epper, F., and Bryant, P. J. (1983). Sex-specific control of growth and differentiation in the *Drosophila* genital disc, studied using a temperature-sensitive *transformer-2* mutation. *Dev. Biol.* **100,** 294–307.

Epper, F., and Nöthiger, R. (1982). Genetic and developmental evidence for a repressed genital primordium in *Drosophila melanogaster. Dev. Biol.* **94,** 163–175.

Franco, M. G., Rubini, P. G., and Vecchi, M. (1982). Sex-determinants and their distribution in various populations of *Musca domestica* L. of Western Europe. *Genet. Res.* **40,** 279–293.

Gans, M., Audit, C., and Masson, M. (1975). Isolation and characterization of sex-linked female-sterile mutants in *Drosophila melanogaster. Genetics* **81,** 683–704.

Garabedian, M. J., Hung, M. C., and Wensink, P. C. (1985). Independent control elements that determine yolk protein gene expression in alternative *Drosophila* tissues. *Proc. Natl. Acad. Sci. U.S.A.* **82,** 1396–1400.

Garabedian, M. J., Shepherd, B. M., and Wensink, P. C. (1986). A tissue-specific transcription enhancer from the *Drosophila* yolk protein 1 gene. *Cell* **45,** 859–867.

Garabedian, M. J., Shirras, A. D., Bownes, M., and Wensink, P. C. (1987). The nucleotide sequence of the gene coding for *Drosophila melanogaster* yolk protein 3. *Gene* **55,** 1–8.

Garcia-Bellido, A. (1975). Genetic control of wing disc development in *Drosophila. Ciba Found. Symp., Cell Patterning* **29,** 161–182.

Garcia-Bellido, A. (1979). Genetic analysis of the *achaete–scute* system of *Drosophila melanogaster. Genetics* **91,** 491–520.

Garcia-Bellido, A., Lawrence, P. A., and Morata, G. (1979). Compartments in animal development. *Sci. Am.* **241,** 102–110.

Gergen, J. P. (1987). Dosage compensation in *Drosophila*: Evidence that *daughterless* and *Sex-lethal* control X chromosome activity at the blastoderm stage of embryogenesis. *Genetics* **117,** 477–485.

Goralski, T. J., Edström, J. E., and Baker, B. S. (1989). The sex determination locus *transformer-2* of *Drosophila* encodes a polypeptide with similarity to RNA binding proteins. *Cell* **56,** 1011–1018.

Gowen, J. W., and Fung, S. T. C. (1957). Determination of sex through genes in a major sex locus in *Drosophila melanogaster. Heredity* **11,** 397–402.

Hägele, K. (1985). Identification of a polytene chromosome band containing a male sex determiner of *Chironomus thummi thummi. Chromosoma* **91,** 167–171.

Hildreth, P. E. (1965). Doublesex, a recessive gene that transforms both males and females of *Drosophila melanogaster* into intersexes. *Genetics* **51,** 659–678.

Hung, M.-C., and Wensink, P. C. (1981). The sequence of the *Drosophila melanogaster* gene for yolk protein 1. *Nucleic Acids Res.* **9,** 6407–6420.

Hung, M.-C., and Wensink, P. C. (1983). Sequence and structure conservation in yolk proteins and their genes. *J. Mol. Biol.* **164,** 481–492.

Janning, W., Labhart, C., and Nöthiger, R. (1983). Cell lineage restrictions in the genital disc of *Drosophila* revealed by *Minute* gynandromorphs. *Wilhelm Roux's Arch. Dev. Biol.* **192,** 337–346.

Jowett, T., and Postlethwait, J. (1980). The regulation of yolk polypeptide synthesis in *Drosophila* ovaries and fat bodies by 20-hydroxyecdysone and a juvenile hormone analogue. *Dev. Biol.* **80,** 225–234.

Lawrence, P. A., and Johnston, P. (1984). The genetic specification of pattern in a *Drosophila* muscle. *Cell* **36,** 775–782.

Logan, S. K., Garabedian, M. J., and Wensink, P. C. (1989). DNA regions which regulate the ovarian transcriptional specificity of *Drosophila* yolk protein genes. *Genes Dev.* **3,** 1453–1461.

Lucchesi, J. C., and Manning, E. (1987). Gene dosage compensation in *Drosophila mela-nogaster. Adv. Genet.* **24,** 371–429.

Lucchesi, J. C., and Skripsky, T. (1981). The link between dosage compensation and sex differentiation in *Drosophila melanogaster. Chromosoma* **82,** 217–227.

McCoubrey, W. K., Nordstrom, K. D., and Meneely, P. M. (1988). Microinjected DNA from the X chromosome affects sex determination in *Caenorhabditis elegans. Science* **242,** 1146–1151.

McKeown, M., Belote, J. M., and Baker, B. S. (1987). A molecular analysis of *transformer,* a gene in *Drosophila melanogaster* that controls female sexual differentiation. *Cell* **48,** 489–499.

McKeown, M., Belote, J. M., and Boggs, R. T. (1988). Ectopic expression of the female *transformer* gene product leads to female differentiation of chromosomally male *Drosophila. Cell* **53,** 887–895.

Mahowald, A. P., and Kambysellis, M. P. (1980). Oogenesis. *In* "The Genetics and Biology of *Drosophila*" (M. Ashburner and T. R. F. Wright, eds.), Vol. 2D, pp. 141–224. Academic Press, New York.

Maine, E. M., Salz, H. K., Cline, T. W., and Schedl, P. (1985a). The Sex-lethal gene of *Drosophila*: DNA alterations associated with sex-specific lethal mutations. *Cell* **43,** 521–529.

Maine, E. M., Salz, H. K., Schedl, P., and Cline, T. W. (1985b). Sex-lethal, a link between sex determination and sexual differentiation in *Drosophila melanogaster. Cold Spring Harbor Symp. Quant. Biol.* **50,** 595–604.

Mariani, B. D., Lingappa, J. R., and Kafatos, F. C. (1988). Temporal regulation in development: Negative and positive cis regulators dictate the precise timing of expression of a *Drosophila* chorion gene. *Proc. Natl. Acad. Sci. U.S.A.* **85,** 3029–3033.

Marsh, J. L., and Wieschaus, E. (1978). Is sex determination in germ line and soma controlled by separate genetic mechanisms? *Nature (London)* **272,** 249–251.

Mitsialis, S. A., and Kafatos, F. C. (1985). Regulatory elements controlling chorion gene expression are conserved between flies and moths. *Nature (London)* **317,** 453–456.

Mitsialis, S. A., Spoerel, N., Leviten, M., and Kafatos, F. C. (1987). A short 5'-flanking DNA region is sufficient for developmentally correct expression of moth chorion genes in *Drosophila. Proc. Natl. Acad. Sci. U.S.A.* **84,** 7987–7991.

Mitsis, P. G., and Wensink, P. C. (1989). Identification of yolk protein factor 1, a sequence-specific DNA-binding protein from *Drosophila melanogaster. J. Biol. Chem.* **264,** 5188–5194.

Monsma, S. A., and Wolfner, M. F. (1988). Structure and expression of a *Drosophila* male accessory gland gene whose product resembles a peptide pheromone precursor. *Genes Dev.* **2,** 1063–1073.

Morgan, T. H., Redfield, H., and Morgan, L. V. (1943). Maintenance of a *Drosophila* stockcenter in connection with investigations on the constitution of the germinal material in relation to heredity. *Yearbk. Carnegie Inst.* **42,** 171–175.

Murre, C., McCaw, P. S., and Baltimore, D. (1989). A new DNA binding and dimerization motif in immunoglobulin enhancer binding, *daughterless, MyoD,* and *myc* proteins. *Cell* **56,** 777–783.

Nagoshi, N. N., McKeown, M., Burtis, K. C., Belote, J. M., and Baker, B. S. (1988). The control of alternative splicing at genes regulating sexual differentiation in *Drosophila melanogaster. Cell* **53,** 229–236.

Nöthiger, R., and Steinmann-Zwicky, M. (1985a). Sex determination in *Drosophila. Trends Genet.* **1,** 209–215.

Nöthiger, R., and Steinmann-Zwicky, M. (1985b). A single principle for sex determination in insects. *Cold Spring Harbor Symp. Quant. Biol.* **50,** 615–621.

Nöthiger, R., Dübendorfer, A., and Epper, F. (1977). Gynandromorphs reveal two separate primordia for male and female genitalia in *Drosophila melanogaster. Wilhelm Roux's Arch. Dev. Biol.* **181,** 367–373.

Nöthiger, R., Leuthold, M., Andersen, N., Gerschwiler, P., Grüter, A., Keller, W., Leist, C., Roost, M., and Schmid, H. (1987). Genetic and developmental analysis of the sex-determining gene *double sex (dsx)* of *Drosophila melanogaster. Genet. Res.* **50,** 113–123.

Nöthiger, R., Jonglez, M., Leuthold, M., Meier-Gerschwiler, P., and Weber, T. (1989). Sex determination in the germ line of *Drosophila* depends on genetic signals and inductive somatic factors. *Development* **107,** 505–518.

Oliver, B., Perrimon, N., and Mahowald, A. P. (1988). Genetic evidence that the *sans fille* locus is involved in *Drosophila* sex determination. *Genetics* **120,** 159–171.

Pardue, M. L., Lowenhaupt, K., Rich, A., and Nordheim, A. (1987). $(dC-dA)_n(dG-dT)_n$ sequences have evolutionarily conserved chromosomal locations in *Drosophila* with implications for roles in chromosome structure and function. *EMBO J.* **6,** 1781–1789.

Parks, S., and Spradling, A. (1987). Spatially regulated expression of chorion genes during *Drosophila* oogenesis. *Genes Dev.* **1,** 497–509.

Petri, W. H., Wyman, A. R., and Kafatos, F. C. (1976). Specific protein synthesis in cellular differentiation: III. The eggshell proteins of *Drosophila melanogaster* in their program of synthesis. *Dev. Biol.* **89,** 225–236.

Pipkin, S. B. (1947). A search for sex genes in the second chromosome of *Drosophila melanogaster*, using the triploid method. *Genetics* **32,** 592–607.

Pipkin, S. B. (1960). Sex balance in *Drosophila melanogaster*: Aneuploidy of long regions of chromosome 3, using the triploid method. *Genetics* **45,** 1205–1216.

Postlethwait, J. H., Bownes, M., and Jowett, T. (1980a). Sexual phenotype and vitellogenin synthesis in *Drosophila melanogaster. Dev. Biol.* **79,** 379–387.

Postlethwait, J. H., Laugé, G., and Handler, A. M. (1980b). Yolk protein synthesis in ovariectomized and genetically agametic $[fs(1)X^{87}]$ *Drosophila melanogaster. Gen. Comp. Endocrinol.* **40,** 385–390.

Ribbert, D. (1967). Die Polytänchromosomen der Borstenbildungszellen von *Calliphora erythrocephala. Chromosoma* **21,** 296–344.

Roehrdanz, R. L., and Lucchesi, J. C. (1981). An X chromosome locus in *Drosophila melanogaster* that enhances survival of the triplo-lethal genotype, *Dp-(Tpl). Dev. Genet.* **2,** 147–158.

Salz, H. K., Cline, T. W., and Schedl, P. (1987). Functional changes associated with structural alterations induced by mobilization of a *P* element inserted in the *Sexlethal* gene of *Drosophila. Genetics* **117,** 221–231.

Salz, H. K., Maine, E. M., Keyes, L. N., Samuels, M. E., Cline, T. W., and Schedl, P. (1989). The *Drosophila* female-specific sex-determination gene, *Sex-lethal,* has stage-, tissue-, and sex-specific RNAs suggesting multiple modes of regulation. *Genes Dev.* **3,** 708–719.

Sánchez, L., and Nöthiger, R. (1982). Clonal analysis of Sex-lethal, a gene needed for female sexual development in *Drosophila melanogaster. Wilhelm Roux's Arch. Dev. Biol.* **191,** 211–214.

Sánchez, L., and Nöthiger, R. (1983). Sex determination and dosage compensation in *Drosophila melanogaster*: Production of male clones in XX females. *EMBO J.* **2**, 485–491.

Santamaria, P. (1983). Analysis of haploid mosaics in *Drosophila*. *Dev. Biol.* **96**, 285–295.

Schäfer, U. (1986a). Genes for male-specific transcripts in *Drosophila melanogaster*. *Mol. Gen. Genet.* **202**, 219–225.

Schäfer, U. (1986b). The regulation of male-specific transcripts by sex determining genes in *Drosophila melanogaster*. *EMBO J.* **5**, 3579–3582.

Schüpbach, T. (1982). Autosomal mutations that interfere with sex determination in somatic cells of *Drosophila* have no direct influence on the germ line. *Dev. Biol.* **89**, 117–127.

Schüpbach, T. (1985). Normal female germ cell differentiation requires the female X chromosome to autosome ratio and expression of Sex-lethal in *Drosophila melanogaster*. *Genetics* **109**, 529–548.

Schüpbach, T., Wieschaus, E., and Nöthiger, R. (1978). The embryonic organization of the genital disc studied in genetic mosaics of *Drosophila melanogaster*. *Wilhelm Roux's Arch. Dev. Biol.* **185**, 249–270.

Sosnowski, B. A., Belote, J. M., and McKeown, M. (1989). Sex-specific alternative splicing of RNA from the *transformer* gene results from sequence dependent splice site blockage. *Cell* **58**, 449–459.

Steinmann-Zwicky, M. (1988). Sex determination in *Drosophila*: The X-chromosomal gene *liz* is required for *Sxl* activity. *EMBO J.* **7**, 3889–3898.

Steinmann-Zwicky, M., and Nöthiger, R. (1985a). The hierarchical relation between X-chromosomes and autosomal sex determining genes in *Drosophila*. *EMBO J.* **4**, 163–166.

Steinmann-Zwicky, M., and Nöthiger, R. (1985b). A small region on the X chromosome of *Drosophila* regulates a key gene that controls sex determination and dosage compensation. *Cell* **42**, 877–887.

Steinmann-Zwicky, M., Schmid, H., and Nöthiger, R. (1989). Cell-autonomous and inductive signals can determine the sex of the germ line of *Drosophila* by regulating the gene *Sxl*. *Cell* **57**, 157–166.

Struhl, G., Struhl, K., and Macdonald, P. (1989). The gradient morphogen *bicoid* is a concentration-dependent transcriptional activator. *Cell* **57**, 1259–1273.

Sturtevant, A. H. (1945). A gene in *Drosophila melanogaster* that transforms females into males. *Genetics* **30**, 297–299.

Tokunaga, C., and Stern, C. (1965). The developmental autonomy of extra sex combs in *Drosophila melanogaster*. *Dev. Biol.* **11**, 50–81.

Torres, M., and Sánchez, L. (1989). The *scute* (T4) gene acts as a numerator element of the X : A signal that determines the state of activity of *Sex-lethal* in *Drosophila*. *EMBO J.* **8**, 3079–3086.

Villares, R., and Cabrera, C. V. (1987). The *achaete–scute* gene complex of *D. melanogaster*: Conserved domains in a subset of genes required for neurogenesis and their homology to *myc*. *Cell* **50**, 415–424.

Walker, E. S., Lyttle, T. W., and Lucchesi, J. C. (1989). Transposition of the *Responder* element (*Rsp*) of the *Segregation distorter* system (*SD*) in the X chromosome in *Drosophila melanogaster*. *Genetics* **122**, 81–86.

Waring, G. L., and Pollack, J. C. (1987). Cloning and characterization of a dispersed, multicopy, X chromosome sequence in *Drosophila melanogaster*. *Genetics* **84**, 2843–2847.

Watanabe, T. K. (1975). A new sex-transforming gene on the second chromosome of *Drosophila melanogaster*. *Jpn. J. Genet.* **50,** 269–271.

Wieschaus, E., and Nöthiger, R. (1982). The role of the transformer genes in the development of genitalia and analia of *Drosophila melanogaster*. *Dev. Biol.* **90,** 320–334.

Wolfner, M. F. (1988). Sex-specific gene expression in somatic tissues of *Drosophila melanogaster*. *Trends Genet.* **4,** 333–337.

Wong, Y. C., Pustell, J., Spoerel, N., and Kafatos, F. C. (1985). Coding and potential regulatory sequences of a cluster of chorion genes in *Drosophila melanogaster*. *Chromosoma* **92,** 124–135.

Zalokar, M., Erk, I., and Santamaria, P. (1980). Distribution of ring-X chromosomes in the blastoderm of gynandromorphic *D. melanogaster*. *Cell* **19,** 133–141.

ROLE OF GAP GENES IN EARLY *Drosophila* DEVELOPMENT

Ulrike Gaul*'† and Herbert Jäckle‡

†Max-Planck-Institut für Entwicklungsbiologie,
D-7400 Tübingen, Federal Republic of Germany
‡Institut für Genetik und Mikrobiologie, Universität München,
D-8000 München 19, Federal Republic of Germany

I. Introduction

The organization of the *Drosophila* embryo along its anteroposterior axis is metameric, that is, the embryo is composed of serially repeated units, or body segments, each of which acquires a unique morphology according to its position in the embryo. Owing to the power of genetic analysis it has been possible to dissect the complex processes that generate this metameric organization and to identify the genetic components involved. Analysis of mutant phenotypes led to the definition of classes of genes with similar functions and provided a profound understanding of how metamerization is achieved (Jürgens *et al.*,

* Present address: Howard Hughes Medical Institute and Department of Molecular and Cell Biology, University of California, Berkeley, Berkeley, California 94720.

239

1984; Nüsslein-Volhard *et al.*, 1984, 1987; Wieschaus *et al.*, 1984b; Schüpbach and Wieschaus, 1986).

First, two types of genes can be discriminated: maternal-effect genes which are expressed during oogenesis and zygotic genes which become activated only after fertilization, during the blastoderm stage, and thereafter. The maternal genes seem to furnish the egg with an initial crude organization of the anteroposterior axis, the further elaboration of which is the task of the zygotic genes (for review, see Nüsslein-Volhard *et al.*, 1987). Among the zygotic genes, in turn, two major classes can be distinguished: the segmentation genes act in concert to establish the appropriate number of repeating units, while the homeotic selector genes control the diverse pathways by which each of the segments acquires its specific identity.

The segmentation genes themselves can be grouped into three classes according to their mutant phenotypes: "gap," "pair-rule," and "segment polarity" genes. Mutations in the gap genes cause embryos to develop with large gaps in the array of segments. Mutations in the pair-rule genes cause embryos to develop with only half the normal number of segments, and the characteristics of the remaining segments suggest that elements of each alternate segment have been deleted. Mutations in the segment polarity genes cause deletions, duplications, or reversal of polarity of pattern elements within segments. These findings led to the idea that the three classes of genes act in sequence to subdivide the embryo into progressively smaller subunits (Nüsslein-Volhard and Wieschaus, 1980).

Many of the genes that control the formation of the metameric pattern have been analyzed at the molecular level, and the spatiotemporal patterns of their expression have been visualized by RNA *in situ* hybridization and by immunohistochemical techniques (for review, see Scott and O'Farrell, 1986; Akam, 1987; Ingham, 1988). These studies showed that by the end of blastoderm stage region-specific patterns of gene expression are established which correlate with the regions of phenotypic defects in the corresponding mutants. These findings corroborated the results of transplantation experiments demonstrating that the cells of the late blastoderm have already become determined (Chan and Gehring, 1971; Simcox and Sang, 1983). Moreover, the analysis of expression patterns in mutant embryos revealed the regulatory relationships between genes and confirmed that the three groups of segmentation genes form a regulatory hierarchy.

In this article, we discuss the role of gap genes in establishing the metameric body pattern of the *Drosophila* embryo. Standing highest in the hierarchy of zygotic genes, the gap genes provide a link between the maternally installed information and the subsequent zygotic gene expression. We describe how the domains of gap gene expression become established under the influence of the maternal gene products and discuss the relevance of interactions among the gap genes in the establishment of their patterns. We then examine the function of the gap genes. Gap gene action pertains to both aspects of metamerization: by regulating the expression of the pair rule genes, the gap genes participate in organizing segmentation; at the same time, they control the spatial expression of the homeotic selector genes, thereby guiding the process of diversification. We close with some considerations on the expression of gap genes at postblastoderm stages of development. We primarily summarize and discuss the study of expression patterns of the relevant genes both in the wild type and under various mutant conditions, focusing on the regulatory relationships between them. We include the results available to date on the molecular mechanisms of these gene interactions.

II. Gap Genes: Mutant Phenotypes, Sequence Analysis, and Expression Patterns

A. Phenotypes of Gap Gene Mutants

The gap genes have been so named because mutant embryos lacking the genes exhibit large gaps in the array of segments (Nüsslein-Volhard and Wieschaus, 1980). *hunchback* (*hb*), *Krüppel* (*Kr*), and *knirps* (*kni*) are the classic examples of this group of genes. Their deletion patterns span the entire segmented region of the embryo: mutations in the *hb* gene cause a deletion in the head and thorax, in *Kr* mutants the thorax and anterior abdomen are missing, and *kni* mutant embryos lack the abdomen (Fig. 1). The regions of deletion of the gap gene mutants overlap. The thorax is missing in both *hb* and *Kr* mutants. The anterior abdomen is deleted in both *Kr* and *kni* mutants. This finding led very early to the idea that the gap genes must interact with each other in order to produce a particular part of the segmented region (Meinhardt, 1977).

hb takes a special place among the three genes insofar as it has not only a zygotic but also a maternal component. Embryos mutant for *hb* show a much stronger phenotypic defect if the maternal gene product is also missing. However, embryos lacking the maternal gene product can be rescued if a wild-type copy of the gene is provided paternally, indicating that the zygotic expression alone is sufficient for normal develop-

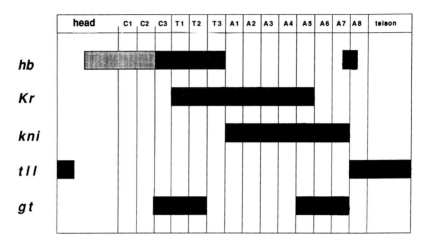

FIG. 1. Segment deletions in gap gene mutants (■). From anterior to posterior the wild-type cuticle pattern elements are acron, head with three gnathal segments (C1–C3), three thoracic segments (T1–T3), eight abdominal segments (A1–A8), and telson. In *hb* mutants, additional lack of the maternal product causes the affected region to be extended (▨).

ment. Moreover, in addition to the gap in the anterior region, *hb* mutant embryos show a defect in the posterior: they lack the posterior part of the seventh and the anterior part of the eighth abdominal segment (Lehmann and Nüsslein-Volhard, 1987a).

More recently, two other genes have been added to the class of gap genes, namely, *tailless* (*tll*) and *giant* (*gt*). Embryos mutant for either of these genes show defects in two separate regions. *gt* embryos have defects in the segmented region, both in the head and in the abdomen (Petschek *et al.*, 1987). *tll* embryos show defects in the terminal regions of the embryo, that is, anteriorly the acron is defective while posteriorly the eighth abdominal segment and the telson are missing (Strecker *et al.*, 1986).

B. SEQUENCE ANALYSIS OF GAP GENES

Three of the gap genes, namely, *hb, Kr,* and *kni,* have been cloned and analyzed at the molecular level (Preiss *et al.*, 1985; Tautz *et al.*, 1987; Nauber *et al.*, 1988). The analyses revealed that all three gap genes share homologies to DNA-binding transcription factors.

Kr and *hb* contain repeat units of the DNA-binding "zinc finger" motif of *Xenopus* transcription factor IIIA (TF IIIA) (Rosenberg *et al.*, 1986; Tautz *et al.*, 1987). This motif has also been found in a large number of other eukaryotic proteins, several of which have been demonstrated to be transcriptional activators (for review, see Klug and Rhodes, 1987; Evans and Hollenberg, 1988). The distinguishing feature of this repeat sequence is the conservation of two cysteine and two histidine residues which, in the case of TF IIIA, have been shown to chelate collectively a zinc ion (zinc fingers) (Klug and Rhodes, 1987). By contrast, the *kni* gene encodes a DNA-binding motif characteristic of ligand-dependent nuclear receptor molecules including hormone receptors and the receptor for retinoic acid (Nauber *et al.*, 1988). The most notable feature of the DNA-binding domain of these factors is a precise spacing of cysteine residues which may fold into zinc finger-like structures (Evans and Hollenberg, 1988). The data suggest that the gap gene products act as transcriptional regulators. Since TF IIIA has been demonstrated to bind to both DNA and RNA, however, a role of gap genes in posttranscriptional regulation cannot be excluded, at least for *hb* and *Kr*.

In this article we focus on *hb* and *Kr,* since most current knowledge about gap genes derives from the molecular analysis of these two genes. The *kni* probe has only recently become available (for reviews of earlier work on these genes, see Gaul and Jäckle, 1987a; Tautz and Jäckle, 1987).

C. EXPRESSION PATTERNS OF *hunchback* AND *Krüppel*

The expression patterns of *hb* and *Kr* have been examined by *in situ* RNA hybridization and by immunohistochemical techniques (Knipple *et al.*, 1985; Gaul *et al.*, 1987; Tautz *et al.*, 1987; Tautz, 1988). Both genes pattern in a dynamic way in a number of locations and tissues throughout embryogenesis. We describe and discuss the postblastoderm patterns in Section VI. For the most part, we concentrate on the early expression of these genes up to early gastrula stages, since by then the segment pattern has been determined.

Owing to maternal transcription of the *hb* gene, *hb* RNA is already present in freshly laid eggs, at which time it is uniformly distributed in the entire embryo. During late cleavage stages, the transcripts exhibit a graded distribution along the anteroposterior axis (Tautz *et al.*, 1987). *hb* protein is first detected just prior to pole cell formation, and its distribution is uneven, with high levels in the anterior one-third and decreasing levels in the posterior two-thirds. Subsequently, the protein expression becomes more and more confined to the anterior half of the embryo.

The zygotic expression of *hb* (see Fig. 2a–d) begins at the syncytial blastoderm stage, with *hb* protein present in the anterior half of the embryo. At midblastoderm, high levels of protein have accumulated in the anterior, and, in addition, a second domain of expression has emerged posteriorly, forming a cap at the posterior pole. At the end of the blastoderm stage, the anterior domain recedes from the anterior tip and becomes inhomogeneous in that three stripes of stronger staining appear within the domain. In the posterior, a stripe replaces the cap (Tautz, 1988). The expression pattern of *hb* in the wild-type embryo corresponds with the various aspects of the *hb* mutant phenotype: the two domains of *hb* expression in the blastoderm stage reflect the defects in the anterior and in the posterior of *hb* mutant embryos. The presence of maternally provided *hb* product in the anterior half may be responsi-

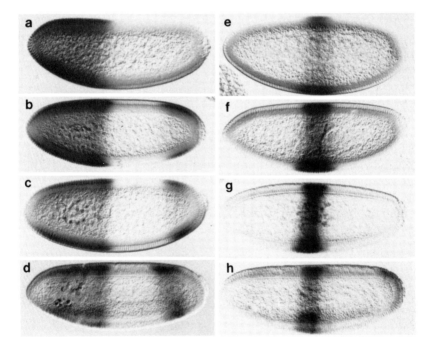

FIG. 2. Protein distribution of *hb* (a–d) and *Kr* (e–h) in wild-type embryos during cellularization of the blastoderm and at the onset of gastrulation, as visualized by antibody staining (the orientation of embryos is anterior to the left and dorsal up).

ble for the fact that one maternal wild-type copy of *hb* is sufficient to reduce the severity of the *hb* phenotype.

To complicate matters, the complex and dynamic expression pattern of *hb* is generated by two different promoters that drive two different transcripts of the *hb* gene (2.9 and 3.2 kb) (see Fig. 3). The two transcripts, however, code for the same protein. The 3.2-kb transcript is expressed both maternally and zygotically, whereas the 2.9-kb transcript is expressed only zygotically. *In situ* hybridizations with transcript-specific probes have revealed that the 2.9-kb transcript forms the anterior domain and the posterior cap, whereas the 3.2-kb message is used for the maternal transcription and for two stripes in the late blastoderm, the posterior stripe and the posteriormost of the three stripes in the anterior region, located at about 50% egg length (EL)

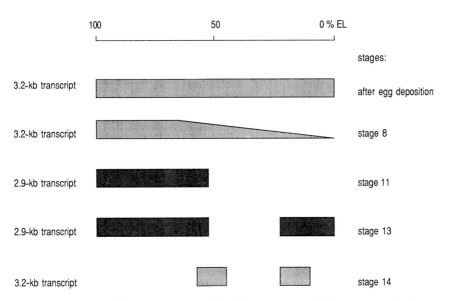

FIG. 3. Pattern of hb transcripts in the wild type, as revealed by *in situ* hybridization with transcript-specific probes. Staging refers to nuclear division cycles as described by Foe and Alberts (1983). After fertilization, the zygotic nuclei divide 13 times before cellularization. The first seven zygotic divisions take place centrally in the yolk. After cycle 8 (i.e., stage 8) a few nuclei are incorporated into the posterior pole plasm to form the polar buds (pole cells). During the course of the next three divisions the somatic nuclei gradually approach the surface of the egg; after having reached the periphery (at stage 11, marking the beginning of the "syncytial blastoderm" stage), these nuclei divide another three times. Thereafter, the somatic nuclei become cellularized (stage 14, mostly referred to as the "blastoderm" stage). The scale at the top indicates the percentage of egg length (EL).

(Schröder *et al.*, 1988). Hence, the complex *hb* pattern can be understood as the superposition of two different patterns generated by the two different promoters of the *hb* gene.

By comparison, the expression of *Kr* before and during the blastoderm stage is fairly simple. *Kr* starts out with a broad band of expression in the central region of the embryo which gradually narrows (see Fig. 2e–h). At midblastoderm, the band of *Kr* staining covers the region between 55 and 39% EL (Gaul *et al.*, 1987).

The domains of *hb* (anterior) and *Kr* distribution overlap to some extent. This can be shown by comparing *hb/ftz* and *Kr/ftz* double-label experiments (see Fig. 9 and Section IV). At midblastoderm, the overlap comprises about four to six cells in width.§

III. Regulation of *hunchback* and *Krüppel* Expression

How does the region-specific expression of *hb* and *Kr* come about, that is, which gene products control the expression of the two genes and cause their patterns to be spatially so well defined? In order to address this question we first have to introduce the maternal-effect genes, whose products provide the egg with the primary organization of the anteroposterior axis during oogenesis.

A. MATERNAL-EFFECT GENES

The maternal genes involved in development of the anteroposterior body pattern have been classified according to their mutant phenotypes into groups that affect (1) the anterior, (2) the posterior, and (3) the terminal regions of the egg (see Nüsslein-Volhard *et al.*, 1987; see also Fig. 4 below). In the following we refer to the genes of the three groups as anterior, posterior, and terminal organizers, respectively.

The anterior organizer mutants all show defects in the anterior head, but they differ with respect to the region of the segment pattern that is affected in addition. While in *bicoid* (*bcd*) embryos the entire anterior portion of the embryo, comprising head and thorax, is defective, *ex-*

---

§ Using a more sensitive antibody staining technique, we recently found that the *Kr* as well as the *hb* domain, and, consequently, the overlap between the two, are actually broader. The *Kr* domain covers the region between about 60 and 33% EL, and the overlap comprises about 8–12 cells, depending on how much the antibody staining has been "pushed." Both *hb* and *Kr* show a gradual decrease of concentration of product toward the margins of their domains, and this means that the overlap is inhomogeneous with respect to protein concentration. High levels of both *hb* and *Kr* protein are present only in the central portion of the overlap (Gaul and Jäckle, 1989).

TABLE 1

List of Genes Referred to in This Article[a]

Gene name (symbol and locus)	Phenotype (aspects of phenotype not discussed here)	Reference(s)[b]
Maternal-effect genes		
bicoid (*bcd*, ANT-C)	Deletion of head and thorax, acron transformed to telson	*1*
exuperantia (*exu*, 2–93)	Weak anterior deletions, weak segmentation defects in posterior abdomen	*2, 3*
swallow (*swa*, 1–14)	Weak anterior deletions, segmentation defects in abdomen (stronger than in *exu*)	*3*
nanos (*nos*, 3–66.2)	Deletion of abdomen (pole plasm present)	*4*
oskar (*osk*, 3–48.5)	Deletion of abdomen (and pole plasm)	*5*
vasa (*vas*, 2–51)	Deletion of abdomen (and pole plasm; null phenotype; sterile eggs)	*2*
staufen (*stau*, 2–83)	Deletion of abdomen (and pole plasm; in addition, anterior head defects)	*2*
torso (*tor*, 2–57)	Deletion of acron and telson	*2*
torsolike (*tsl*, 3–71)	Deletion of acron and telson	*6*
Gap genes		
hunchback (*hb*, 3–48)	See Fig. 1	*7*
Krüppel (*Kr*, 2–107.6)	See Fig. 1, mutants also lack Malpighian tubules	*8*
knirps (*kni*, 3–47)	See Fig. 1	*9*
tailless (*tll*, 3–102)	See Fig. 1	*10*
giant (*gt*, 1–1)	See Fig. 1	*11*
Pair rule and segment polarity genes		
even skipped (*eve*, 2–55)	Null mutants eliminate all segmental periodicity	*12*
fushi tarazu (*ftz*, ANT-C)	Pattern deletions correspond to even numbered parasegments	*9*
engrailed (*en*, 2–62)	Mutants show variable pair rule fusions and segment polarity reversals	*12*
Homeotic genes		
Sex combs reduced (*Scr*, ANT-C)	Segment transformation in parasegments	*9, 13*
Antennapedia (*Antp*, ANT-C)	Segment transformation in parasegments 3–5	*9, 13*
Ultrabithorax (*Ubx*, BX-C)	Segment transformation in parasegments 5–13	*14*

(continued)

TABLE 1 *(Continued)*

Gene name (symbol and locus)	Phenotype (aspects of phenotype not discussed here)	Reference(s)[b]
forkhead (*fkh*, 3–95)	Transformation of nonsegmental terminal regions into segmental derivatives, especially deletion of hindgut and Malpighian tubules	15
caudal (*cad*, 2–38)	Elimination of anal tuft and anal sense organs (additional lack of maternal product causes severe segmentation defects)	16

[a] Genes have been grouped according to classification of phenotypes. For complete references, see Akam (1987) and Nüsslein-Volhard *et al.* (1987).

[b] (*1*) Frohnhöfer and Nüsslein-Volhard (1986), (*2*) Schüpbach and Wieschaus (1986), (*3*) Frohnhöfer and Nüsslein-Volhard (1987), (*4*) Nüsslein-Volhard *et al.* (1987), (*5*) Lehmann and Nüsslein-Volhard (1986), (*6*) Frohnhöfer (1987), (*7*) Lehmann and Nüsslein-Volhard (1987a), (*8*) Wieschaus *et al.* (1984a), (*9*) Jürgens *et al.* (1984), (*10*) Strecker *et al.* (1986), (*11*) Petschek *et al.* (1987), (*12*) Nüsslein-Volhard *et al.* (1984), (*13*) Wakimoto *et al.* (1984), (*14*) Lewis (1981), (*15*) Jürgens and Weigel (1988), and (*16*) Macdonald and Struhl (1986).

uperantia (*exu*) and *swallow* (*swa*) embryos show defects in the posterior abdominal region in addition to the head defects. Embryological and molecular studies revealed that the products of *swa* and *exu* are necessary for the localization of *bcd* activity, which itself acts as a morphogen (Frohnhöfer and Nüsslein-Volhard, 1986, 1987; Driever and Nüsslein-Volhard, 1988a,b).

The group of posterior organizers comprises seven genes [among them *vasa* (*vas*), *staufen* (*stau*), *oskar* (*osk*), and *nanos* (*nos*)]. They have a common mutant phenotype characterized by the deletion of the entire abdomen. Transplantation experiments and genetic studies show that *nos* encodes the actual posterior organizing activity whereas the other genes of the posterior group are involved in transport and localization of *nos* activity (Lehmann and Nüsslein-Volhard, 1987b; Nüsslein-Volhard *et al.*, 1987). Since *osk* is the gene that has been most studied so far, this group has been referred to as the *"osk"* group of genes.

The group of terminal organizers comprises five genes [among them *torso* (*tor*) and *torsolike* (*tsl*)], characterized by a common mutant phenotype: deletion of the two nonsegmented regions of the embryo, the acron (at the anterior pole) and the telson (at the posterior pole). Several lines of genetic evidence indicate *tor* to be the crucial gene for organizing the terminal pattern (Nüsslein-Volhard *et al.*, 1987; Klingler *et al.*, 1988).

If the cuticle defects of *bcd, nos,* and *tor* embryos are projected onto the blastoderm fate map, we can plot the regions where the activities of these genes are required at the blastoderm stage (see Fig. 4): *bcd* would be required in the anterior half of the blastoderm embryo, *nos* in the posterior at 20–50% EL, and *tor* in the two terminal regions, that is, in the regions which give rise to the acron and telson. The spatial distribution of only one maternal product, *bcd,* has been measured to date; these experiments revealed a graded distribution of *bcd* protein along the anteroposterior axis, with the peak at the anterior and an exponential decrease toward the posterior (Driever and Nüsslein-Volhard, 1988a). For *nos* and *tor* such data are not yet available.

Since the regions of requirement are largely nonoverlapping, the gene products of *bcd, nos,* and *tor* can be regarded as organizing their respective realms autonomously. However, one effect that is evidence of gene interaction requires mention: *bcd* modifies the action of the terminal organizer genes in the anterior terminal region. *bcd*-deficient em-

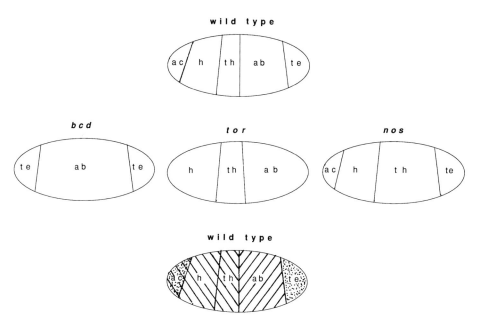

FIG. 4. Maternal organizer genes: schematic representations of the mutant phenotypes and of the regions in which their activities are required for normal development, as projected onto the blastoderm fate map. ac, Acron; h, head; th, thorax; ab, abdomen; te, telson. Regional requirements for maternal organizers are as follows: (⧄⧄) *bcd,* (⊡) *tor,* and (▧) *nos.*

bryos not only lack all anterior structures of the embryonic body pattern, but these structures are replaced by an inverted telson, that is, the posterior terminal structure. This means that, depending on whether *bcd* is present, the terminal organizer genes promote the development of either acron or telson in the anterior. All the maternal organizer genes have a pronounced influence on the expression patterns of both *hb* and *Kr*.

B. MATERNAL CONTROL OF *hunchback* EXPRESSION

Regarding the maternal expression of *hb,* neither the activity of *bcd* nor the activity of the terminal organizers is necessary for the formation of the normal pattern. Only the genes of the posterior organizer group exert a negative influence, restricting the maternal *hb* protein to the anterior of the embryo. In posterior organizer mutant embryos the *hb* protein is uniformly and ubiquitously expressed until the onset of the syncytial blastoderm stage (Tautz, 1988). The importance of this early regulation of the maternal *hb* distribution has been demonstrated by a number of experiments. The paternal rescue of the *hb* phenotype had indicated that the ubiquitous early presence of *hb* RNA is not necessary for normal development (Lehmann and Nüsslein-Volhard, 1987a). We now know that persistent expression of *hb* protein in the posterior region causes defects in the abdominal pattern, indicating that the removal of *hb* activity by the posterior organizers may be an essential prerequisite for normal abdominal development (Hülskamp *et al.,* 1989).

The zygotic expression of *hb* is regulated quite differently from the maternal expression. Both the anterior and the terminal organizers are necessary for the formation of the normal zygotic *hb* pattern, and both exert a positive influence on *hb* expression. The terminal organizer genes act positively on the posterior domain of *hb*; neither cap nor stripe forms in the posterior of terminal organizer mutant embryos. The *bcd* protein acts as a positive regulator of *hb* expression in the anterior. In *bcd*-deficient embryos, the anterior *hb* domain is absent and is replaced by a stripe that can be identified as a duplication of the posterior *hb* stripe (Tautz, 1988). This corresponds with the phenotype observed in *bcd*-deficient embryos, in which the telson is duplicated in the anterior as a result of altered terminal organizer activity. Double-mutant embryos, in which both *bcd* and a terminal organizer gene are defective, show no zygotic expression of *hb* at all, confirming that the stripe found in the anterior of *bcd*-deficient embryos is indeed produced by the action of terminal organizers (U. Gaul, unpublished).

The regulation of *hb* by *bcd* has been studied at the molecular level. From *in situ* hybridization with transcript-specific probes it is known that the (solely zygotic) 2.9-kb transcript of *hb* is the one which forms the "anterior domain" (see Fig. 3). The regulatory sequence regions necessary for the normal expression of this transcript have been determined by using *hb* promoter–*lacZ* fusion gene constructs in combination with germ-line transformation; they could thus be delimited to a region of approximately 300 bp, lying upstream of the site of transcription initiation (Schröder *et al.*, 1988). Subsequently, the homeobox-containing protein of *bcd* has been shown to bind to several sites upstream of the transcription start of this *hb* transcript. Using transient expression assays in embryos with chloramphenicol acetyltransferase (CAT) as a reporter gene, three binding sites in the promoter region (−50 to −300 bp) were demonstrated to be necessary and sufficient for the activation of zygotic *hb* expression. The three regulatory sites are separated by 100 bp from each other and act in a cooperative manner to activate *hb* expression (Driever and Nüsslein-Volhard, 1989).

C. MATERNAL CONTROL OF *Krüppel* EXPRESSION

The regulation of the *Kr* protein domain by maternal genes is very different from that of *hb*. *Kr* receives exclusively negative inputs, and it does so from all three groups of organizers (see Fig. 5). In posterior organizer mutants the *Kr* protein domain extends toward the posterior, but the pole region (0–30% EL) does not show any *Kr* expression (Fig. 5a). In weak *bcd* alleles, the embryos show a *Kr* protein domain that is both enlarged and shifted toward the anterior without reaching the anterior pole (Fig. 5b). Under amorphic conditions for *bcd,* the *Kr* protein domain is similarly shifted but not enlarged (Fig. 5c) (Gaul and Jäckle, 1987b). We think that this discrepancy between hypomorphic and amorphic *bcd* alleles reflects the influence of *bcd* on the terminal organizers (see discussion below).

The repression of *Kr* at the poles in posterior organizer mutants and in *bcd* mutants results from terminal organizer activity. However, the negative effect of the terminal organizers on *Kr* protein expression is not revealed in terminal organizer single-mutant embryos, in which *Kr* expression looks almost normal (Fig. 5d). It is revealed, though, in double-mutant embryos in which both posterior and terminal or both *bcd* and terminal organizer genes are defective. In such embryos, *Kr* expression reaches the posterior or the anterior pole, respectively (Fig. 5f,g). Additional evidence for terminal organizer influence on *Kr* expression is that in the case of gain-of-function alleles of *tor* which cause

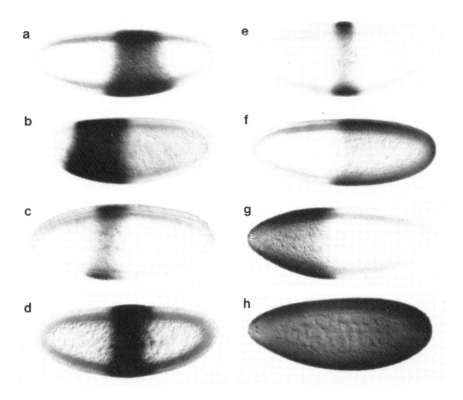

FIG. 5. Effect of maternal mutations on *Kr* protein expression at midblastoderm, as visualized by antibody staining. (a) *osk* embryo, (b) *bcd* (hypomorphic allele) embryo, (c) *bcd* (amorphic allele) embryo, (d) *tor* embryo, (e) *RL3* (gain-of-function allele of *tor*) embryo, (f) *osk tsl* embryo (posterior and terminal organizers defective), (g) *bcd tsl* embryo (*bcd* and terminal organizers defective), (h) *bcd osk tsl* embryo (*bcd*, posterior, and terminal organizers defective). For wild-type expression of *Kr*, see Fig. 2.

aberrant function of *tor* throughout the embryo, *Kr* expression is suppressed (Fig. 5e; Klingler *et al.*, 1988). If the terminal organizers do have a negative effect on *Kr* expression, why does their absence not lead to a marked alteration of the *Kr* expression pattern in terminal organizer single-mutant embryos? The most straightforward explanation is that, independent of the terminal organizers, *bcd* and posterior organizers act to repress *Kr* in the anterior and posterior terminal regions, respectively. Thus, even if the terminal organizer activity is removed, anterior and posterior organizers could function to repress *Kr* at the poles, masking the lack of the terminal organizer.

Going one step further, in a triple-mutant embryo (*bcd osk tsl*), in which all three organizers are defective, *Kr* expression spans the entire length of the embryo (Fig. 5h) (Gaul and Jäckle, 1989).

We noted above that *Kr* expression reaches less anterior positions in amorphic *bcd* alleles than it does in weak *bcd* alleles. This, we believe, is a reflection of the loss of the "modifying" influence of *bcd* on terminal organizer activity. As a result of the complete lack of *bcd* activity, the terminal organizer activities promote telson instead of acron development in the anterior of *bcd*-deficient embryos, and they apparently also span a larger region than in weak *bcd* alleles, resulting in an extension of the region of *Kr* repression.

The data above support the following model for the regulation of *Kr* expression (see Fig. 6). Since all maternal genes organizing the anteroposterior axis have an exclusively negative influence on *Kr* expression, it is most parsimonious to assume that *Kr* expression is constitutive and becomes restricted to the central region of the embryo as a result of the activities of anterior, posterior, and terminal organizer genes. Expression of the *Kr* gene would be part of the "ground state" of any blastoderm cell if it were not for the repression by the maternal organizers acting from the poles. Such a regulatory mechanism requires that the negative influences of the maternal activities cease in the central region. For *bcd,* as we have noted, a graded distribution of the protein showing a maximum at the anterior pole and an exponential decrease in concentration toward the posterior has been demonstrated. For the posterior factors, however, no molecular data are available yet.

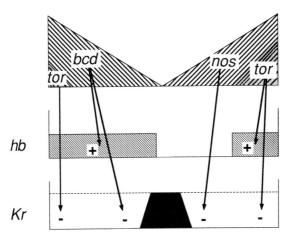

FIG. 6. Schematic representation of the influences of maternal organizer activities on the zygotic expression of *Kr* and *hb*.

One might have expected that a region-specific activator switches on *Kr* transcription in the central domain, as *bcd* does in the case of *hb*. Not only do the data just discussed argue against such a concept, but it can be dismissed on phenotypic grounds alone. The mutant phenotype of a region-specific activator of *Kr* would have to show region-specific deletions in the middle portion of the segment pattern, that is, it would have to resemble the *Kr* phenotype. As yet, no maternal loss-of-function mutant with such a phenotype is known. The only mutant that shows this kind of defect is *RL3,* and the defects in the segment pattern of *RL3* are caused by aberrant *tor* activity (Klingler *et al.,* 1988). Yet another mechanism for activation of *Kr* expression is outlined in a segmentation model proposed by Meinhardt (1986). It assumes that a maternal morphogen gradient spans the entire length of the egg. In different concentration ranges of the morphogen, different gap genes become activated. According to this model, *Kr,* as it is expressed in the central region of the embryo, should be activated by low concentrations of the morphogen, and consequently mutants lacking this morphogen should not express *Kr* at all. Our data show that the regulatory input to *Kr* is negative throughout, therefore ruling out such a mechanism.

The analysis of the regulation of *hb* and *Kr* expression by maternal genes shows that no general principle governs the establishment of the domains of the two gap genes. There is no common scheme, either with respect to the maternal components involved or with respect to the mode of control (activation/repression). The anterior and terminal organizers provide positive input for the establishment of zygotic *hb* expression, whereas all three maternal organizers provide negative input for the establishment of the *Kr* domain (see Fig. 6).

D. GAP GENE INFLUENCES ON THE EXPRESSION OF *hunchback* AND *Krüppel*

The gap genes receive input not only from maternal gene activities but also from other gap genes. This complicates matters and, in particular, raises the question to what extent the maternal gene effects are mediated by gap genes and are thus indirect.

Both the anterior and the posterior domains of zygotic *hb* expression are influenced by other gap genes (see Fig. 7). The anterior *hb* domain extends posteriorly in *Kr*-deficient embryos (Fig. 7f and g). However, differences between wild-type and *Kr* mutant embryos first become visible in late blastoderm stages. We do not yet know whether it is the promoter for the "anterior domain" (2.9-kb transcript) or the promoter for the posteriormost "anterior stripe" (3.2-kb transcript) that is subject to influence from *Kr*. In any case, the effect occurs too late to be relevant

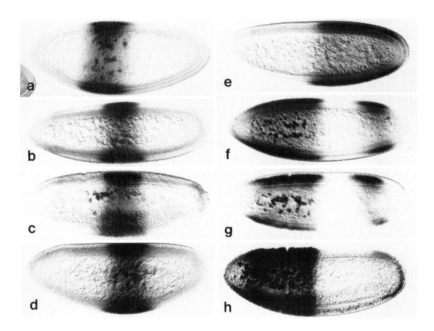

FIG. 7. Effect of gap gene mutations on *Kr* and *hb* protein expression, as visualized by antibody staining (stages are midblastoderm, unless stated otherwise). (a–e) *Kr* protein expression: (a) in *hb* mutant embryos, in *kni* mutant embryos (b) at late blastoderm and (c) at gastrulation, (d) in *tll* embryos, (e) in *osk tll* embryos. (f–h) *hb* protein expression: in *Kr* mutant embryos at (f) late blastoderm and (g) at the onset of gastrulation, (h) in *tll* embryos at the onset of gastrulation. For the wild-type pattern of *Kr* and *hb* expression, see Fig. 2.

for the establishment of the anterior domain of *hb,* and we therefore believe that *bcd* is the primary factor responsible for the establishment of this *hb* domain. If *bcd* alone is responsible, how does the *hb* domain become (relatively) sharply defined at its posterior boundary, given that *bcd* shows a graded distribution? As we have noted, molecular analysis of the activation of *hb* by *bcd* has shown that *bcd* binds in a cooperative manner to the upstream transcriptional control region of *hb*. This cooperativity in activation would allow for a strong alteration in *hb* expression within a small range of *bcd* concentration; it would suffice to create relatively sharp borders of *hb* expression (Driever and Nüsslein-Volhard, 1989).

The posterior domain of *hb* receives a positive input from *tll* activity. In *tll* mutant embryos, a "cap" of *hb* expression is found in the posterior

(Fig. 7h); the "posterior stripe," however, never develops. Hence, formation of the stripe requires *tll* activity, whereas formation of the cap could be caused either by the maternal terminal organizer activity (*tor*) itself or by some yet unidentified zygotic factor that depends on it. Thus, unlike the case for the anterior domain, the very establishment of the posterior domain of *hb* requires zygotic input.

The most profound zygotic contribution to the expression pattern of *Kr* comes from *tll*, which acts as a repressor of *Kr* in almost the entire posterior terminal region of the embryo. The effect can be visualized in *osk tll* double-mutant embryos. In *osk tll* embryos, the *Kr* domain extends almost to the posterior pole (well beyond the border of *Kr* expression in *osk* single-mutant embryos); a small cap at the pole, however, still lacks *Kr* expression (Fig. 7e). The effect is similar to the effect that the terminal organizers have on *Kr* expression. This, along with the similarity of the mutant phenotypes, suggests that *tll* is activated by the maternal terminal organizers and mediates their influence on *Kr*. This view is corroborated by an independent set of experiments. The detrimental effect of *tor* gain-of-function mutations on segmentation can largely be neutralized by removing *tll* activity, suggesting that, to a large extent, *tor* acts through *tll* (Klingler *et al.,* 1988). The lack of *Kr* expression at the extreme pole in *osk tll* mutant embryos indicates, however, that mediation by *tll* is not instrumental in the entire region of maternal terminal organizer activity. We do not yet know whether the maternal organizers themselves or some zygotic factor that depends on them are responsible for the residual repression. There is evidence, however, that at least one zygotic factor other than *tll* is required for organizing the terminal region (i.e., *huckebein*; D. Weigel, unpublished), which could well be the "missing link" not only with respect to *Kr* but also with respect to *hb* expression in the posterior polar region (see above).

In the posterior region of the embryo, *Kr* expression receives a negative influence from *kni* (Jäckle *et al.,* 1986; Gaul and Jäckle, 1987b; Harding and Levine, 1988). At midblastoderm, the *Kr* domain in *kni* mutant embryos extends slightly toward the posterior (Fig. 7b). With the onset of gastrulation the difference between wild-type and *kni* mutant embryos is more striking, for by then the *Kr* domain is about twice the size as in the wild type, owing to the fact that it fails to narrow as it does in the wild type (Fig. 7c, compare with Fig. 3h). Thus, the effect of *kni* on the pattern of *Kr* protein expression is similar to that of the maternal posterior organizers, but it is also weaker. This finding could again be explained by assuming that the negative influence the posterior organizers exert on *Kr* expression is partially mediated by *kni*; either the maternal organizer itself or another zygotic factor (which is activated by the maternal organizer at a more posterior position) would repress *Kr* if *kni* is absent.

Finally, in the anterior region, *hb* has a negative effect on *Kr* expression (Jäckle *et al.*, 1986; Gaul and Jäckle, 1987b; Harding and Levine, 1988). In *hb* mutant embryos, the *Kr* domain extends anteriorly and is also somewhat shifted toward the anterior, that is, the posterior border of the *Kr* domain lies at a more anterior position (Fig. 7a). The effect of *hb* on *Kr* protein expression is thus similar to the effect of its maternal activator *bcd,* but, again, the influence of the maternal activity is stronger [compare hypomorphic *bcd* alleles with amorphic *hb* alleles (Figs. 5b and 7a)].

We want to examine the regulation of *Kr* expression in the anterior more closely. The findings reported raise two questions: How are the regulatory effects of *hb* and *bcd* on *Kr* expression related? If *hb* exerts a negative influence on *Kr* expression, how is it that the domains of *hb* and *Kr* expression overlap in wild-type embryos? Regarding these questions, the inspection of *hb* and *Kr* expression in two maternal mutant combinations, *stau exu* and *vas exu,* has been instructive. We would like to discuss these mutants in more detail at this point, since they also yield information relevant to other sections of this article.

E. PATTERNS OF *hunchback* AND *Krüppel* IN *stau exu* AND *vas exu* EMBRYOS

The two combinations of maternal organizer mutations have the following features in common. The anterior pattern organizer *bcd* is affected by a reduction in the amount of its product and a change in its spatial distribution (*exu*); in addition, posterior organizer genes are defective (*stau, vas*). The combination of these mutations leads to a uniform distribution of low amounts of *bcd* activity in the entire egg (Berleth *et al.*, 1988; Gaul and Jäckle, 1989). The terminal organizers, however, are intact in both combinations. The phenotypes of these maternal mutant combinations show body patterns that lack most of the wild-type pattern elements, and the residual elements are arranged in mirror image symmetry. The phenotype of *stau exu* embryos is characterized by a lack of head, gnathal, and abdominal structures. At the poles of the embryo two telsons of opposite orientation develop, while segments of thoracic identity form in the middle (Fig. 8g) (Schüpbach and Wieschaus, 1986). In contrast, the only cuticular structures formed in *vas exu* embryos are of gnathal origin (maxillary mouth hooks and cirri); in addition, posterior midgut invaginations form at both ends of *vas exu* embryos during gastrulation, indicating that posterior terminal structures do develop (Fig. 8n) (Schüpbach and Wieschaus, 1986).

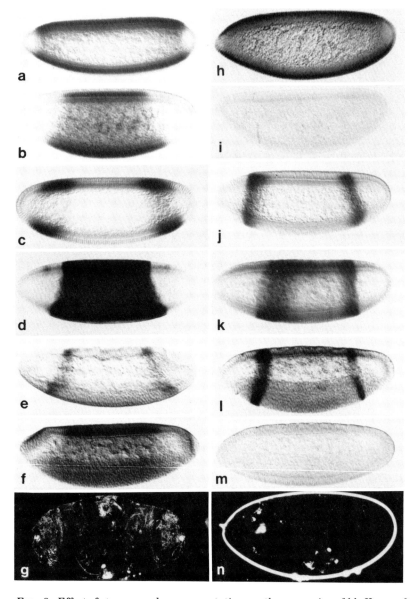

FIG. 8. Effect of *stau exu* and *vas exu* mutations on the expression of *hb, Kr, eve, ftz, en,* and *Antp*, as visualized by antibody staining, and on the cuticle phenotype. (a–g) *stau exu* embryos: (a) *hb*, (b) *Kr*, (c) *eve*, and (d) *ftz* expression in late blastoderm; (e) *en* and (f) *Antp* expression in gastrulating embryos; (g) cuticle phenotype. (h–n) *vas exu* embryos: (h) *hb*, (i) *Kr*, (j) *eve*, and (k) *ftz* expression in late blastoderm; (l) *en* and (m) *Antp* expression in gastrulating embryos; (n) cuticle phenotype.

The expression patterns of *hb* and *Kr* are quite different in these two mutant combinations. In *stau exu* embryos, *hb* is expressed along the entire length, at levels indistinguishable from the wild type, shortly before blastoderm and thereafter (Fig. 8a). This omnipresence of *hb* is due to the ubiquitous activity of *bcd*. *Kr* is expressed in a very broad domain covering most of the egg, also at levels indistinguishable from the wild type (Fig. 8b). *Kr* is not expressed, however, at the pole regions, owing to the presence of terminal organizers and their zygotic "mediators." So we find that, in the *stau exu* mutant combination, *hb* and *Kr* are codistributed within a very large region of the embryo. In *vas exu* embryos, *hb* is again expressed at wild-type levels in the entire embryo (Fig. 8h). In contrast, no expression of *Kr* is detected during blastoderm (Fig. 8i) (Gaul and Jäckle, 1989).

The broad codistribution of *hb* and *Kr* observed in *stau exu* embryos has two important implications. First, it must be due to genuine coexpression, therefore suggesting that the relatively narrow overlap of *hb* and *Kr* protein observed in wild-type embryos similarly results from an overlap of expression territories rather than from diffusion. Second, and more specifically, it indicates that *hb* by itself is not sufficient to repress *Kr*. That *hb* cannot play a crucial role in *Kr* expression is also suggested by the comparison of *stau exu* and *vas exu* embryos: while the *hb* distribution is the same, *Kr* is expressed in the former and repressed in the latter.

What is the crucial difference between the two mutant combinations that accounts for the difference regarding *Kr* expression? The pivotal fact is apparently that the spatially uniform levels of *bcd* are higher in *vas exu* embryos than they are in *stau exu* embryos. *bcd* has been shown to act as a morphogen that determines "positions" in the anterior half of the embryo. Increases or decreases in *bcd* protein levels in a given region of the embryo cause a corresponding shift of the anterior anlagen to the anterior or to the posterior, respectively (Driever and Nüsslein-Volhard, 1988b). The cuticular structures formed in *vas exu* embryos derive from more anterior positions in the fate map than those formed in *stau exu* embryos: the middle region produces gnathal elements instead of thorax. Since higher levels of *bcd* determine structures lying more anteriorly in the fate map, we can infer that higher levels of *bcd* are present in *vas exu* mutants.

The analysis of the patterns of *Kr* and *hb* expression in *stau exu* and *vas exu* mutant embryos thus suggests that *bcd* (or a *bcd*-dependent factor other than *hb*) acts as a repressor of *Kr* in a concentration-dependent manner. Low levels of *bcd* would allow *Kr* to be expressed (*stau exu*); higher levels would cause its repression (*vas exu*). Since *hb* is activated indiscriminately by *bcd* in these two mutants, we infer that the threshold level of *bcd* for activating hb is lower than that for repressing *Kr*.

This difference in threshold levels provides an explanation for the overlap between *hb* and *Kr* in the wild type. In wild-type embryos, *bcd* shows a graded distribution with its peak at the anterior pole and an exponential decrease in concentration posteriorly (Driever and Nüsslein-Volhard, 1988a). Under these conditions, the difference in the threshold levels for *hb* activation, on the one hand, and *Kr* repression, on the other, causes the simultaneous expression of *Kr* and *hb* in a narrow region where the level of *bcd* is too low to repress *Kr* but still high enough to activate *hb*. Only where *bcd* levels are sufficiently high will *Kr* be repressed.

So how do we envisage the control of *Kr* expression in the anterior of the embryo? In particular, what is the role of *hb*? It has a definite role in controlling *Kr* expression, since, as we have seen, in *hb* mutant embryos the *Kr* domain is expanded toward the anterior. We see two mechanisms by which *bcd* and *hb* could both be necessary for *Kr* repression. A first possibility is that *bcd* acts as a repressor of *Kr* but needs help from *hb* as a cofactor at lower levels of concentration to achieve *Kr* repression. It is only at higher levels that *bcd* can act by itself. Such a mechanism implies that *bcd* alone is not a good repressor for *Kr*, that *hb* acts as an "enhancer," and that it can do so only if *bcd* is already present at some threshold level.

A second possibility is that *hb* is only indirectly involved in *Kr* control. One idea employs the fact that the factor to which *hb* shows homology, TF IIIA, binds not only to DNA but also to RNA; thus, *hb* may be involved in positive posttranscriptional regulation of *bcd*. Under such conditions, *bcd* would be the only repressor of *Kr*, and an effective one. The lack of *hb* would cause less *bcd* protein to be produced, and the concentrations necessary for *Kr* repression would be reached only at more anterior positions.

Taken together, these studies demonstrate that gap genes exert a considerable influence on the expression of *hb* and *Kr*. We emphasize, in particular, the strong influence of *tll* in the posterior terminal region, where the control of both *hb* and *Kr* expression appears to be entirely zygotic. In the anterior, the maternal factor, that is, *bcd*, appears to be the primary determinant of *hb* and *Kr* expression, whereas the effects of *hb* and *Kr* on each other are less pronounced and subordinate. In the posterior, *kni* has a negative effect on *Kr* expression, but it is also weaker than that of the corresponding maternal organizer.

IV. Influence of Gap Genes on Pair-Rule Genes

Gap gene expression in the blastoderm embryo is necessary for instructing the patterning of the pair-rule genes. As noted earlier, embryos mutant at any pair-rule locus show defects in every other segment, indicating that these genes are key factors in the establishment of the periodicity of the embryo body pattern (Nüsslein-Volhard and Wieschaus, 1980). Of the eight members of the pair-rule gene class, six have been cloned (for review, see Akam, 1987; Ingham, 1988). Each gene shows a unique, highly dynamic pattern of expression, but all patterns conform in gradually resolving into a series of seven (or eight) discrete stripes during cellularization in late blastoderm. Interactions between the pair-rule genes have been shown to be involved in the formation of these patterns.

In gap gene mutant embryos, the patterns of pair-rule genes monitored so far [*fushi tarazu (ftz), even skipped (eve), hairy (h)*, and *runt (run)*] show severe perturbations, indicating that the gap genes organize pair-rule gene activity (Carroll and Scott, 1986; Ingham *et al.*, 1986; Frasch and Levine, 1987; Carroll *et al.*, 1988; Ingham and Gergen, 1988). Note that this is true not only for the "classic" gap genes *hb, Kr,* and *kni,* but also for *tll* and *gt.* (The term "gap gene" has come to be somewhat ambiguous. Originally, when the genes exhibiting the prototypical "gap" mutant phenotype were the only ones known to regulate pair rule gene expression, phenotypic and functional aspects of the definition were in agreement. But meanwhile, genes have been identified that, while not showing the "gap" mutant phenotype, do qualify as gap genes if the emphasis is on function, since they have an effect on pair rule gene expression.)

While it is evident that the gap genes control pair-rule gene expression, the way they control it is poorly understood. The defects exhibited by the pair-rule gene patterns in gap gene mutant embryos are not suggestive of any straightforward interpretation, and it is still one of the most puzzling problems regarding the process of segmentation to understand how the few gap genes, with their comparatively broad, aperiodic domains of expression, provide sufficient information to establish the intricate periodic pair-rule gene patterns.

Some further information can be gained by examining the correlations between the expression patterns of gap genes and pair-rule genes, both in wild-type and in mutant embryos; so far, only the probes *hb* and *Kr* have been available for this purpose.

As we have seen, *hb* and *Kr* show relatively simple, essentially

nonperiodic patterns of expression in blastoderm embryos. Their domains are largely separate: *hb* expression covers the anterior half of the
embryo and forms a stripe in the posterior, and *Kr* is expressed in one
single domain in the middle region. As described above, however, there
is a band of overlap of *hb* and *Kr* expression. Moreover, the concentration profiles of the gene products do not exhibit a step function; rather,
both *Kr* and *hb* staining fade gradually toward the margins of their
domains. Which of these features encode information relevant for the
establishment of the pair-rule gene pattern? In particular, does the
overlap between the *hb* and *Kr* domains have a specific function, or is
it meaningless? In other words, does the set of pair-rule genes that a
blastoderm nucleus expresses vary depending on whether it experiences the expression of only *Kr* (or only *hb*) or of both *Kr* and *hb* at the
same time?

To address this question, we first look at the correlations between gap
gene and pair-rule gene patterns in the wild type (see Fig. 9). To
monitor pair-rule gene expression we used *ftz* antibodies. In a late
blastoderm wild-type embryo, the anterior *hb* domain encompasses the
second *ftz* stripe and fades into the second interstripe at its posterior
border. The *Kr* domain starts with the second *ftz* stripe at its anterior
and ends in the forth interstripe. Thus, the second *ftz* stripe is formed in
the region of overlap between the *Kr* and *hb* domains. These data are
derived from immuno-double-label experiments.**

Subsequently, we studied embryos in which pattern formation is
perturbed because of maternal-effect mutations. We chose mutant combinations in which most of the wild-type pattern elements are lacking
and the residual elements are arranged in mirror image symmetry.
Under such mutant conditions, the patterns of segmentation genes
become much simpler and the reduction in the number of parameters
considerably facilitates the interpretation of the correlations between
the relevant patterns. To monitor pair-rule gene expression we used *ftz*
and *eve* antibodies. In the wild type, *ftz* and *eve* are initially expressed in
the entire region that is to be segmented; later, their patterns resolve
into series of stripes that are roughly complementary.

** The more sensitive staining technique which shows that the overlap of the domains
of *hb* and *Kr* comprises about 8–12 cells (cf. footnote § on p. 246) also reveals that it
encompasses not only the second *ftz* stripe but also at least part of the two adjacent *eve*
stripes. The (second) *ftz* stripe, however, lies in the central portion of the region of overlap,
which is characterized by higher levels of both *hb* and *Kr* products. The broad coexpression of *hb* and *Kr* in the middle region of *stau exu* embryos, which is correlated with
expression of *ftz*, should thus be regarded as corresponding to this central portion of the
overlap (see below).

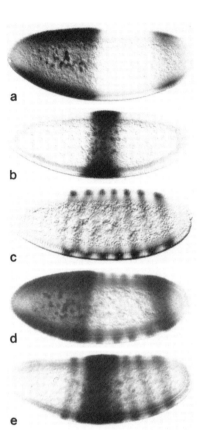

FIG. 9. Correlations between the patterns of *hb, Kr,* and *ftz* protein expression in the wild-type embryos at late blastoderm, as visualized by antibody staining. (a) *hb,* (b) *Kr,* (c) *ftz,* (d) *hb/ftz* double label, (e) *Kr/ftz* double label.

Without a gap gene pattern, that is, if there is no change in the state of expression of gap genes along the anteroposterior axis, no pair rule gene pattern forms. In *bcd osk tsl* embryos, in which all three maternal organizers are defective, *Kr* is the only active gap gene and is expressed throughout the embryo (Fig. 10a). In such embryos, no patterns of pair-rule genes form: *ftz* and *eve* are uniformly expressed throughout the embryo (Fig. 10b and c) (Gaul and Jäckle, 1989). This finding verifies that zygotic gene patterns cannot form *de novo* in the *Drosophila* blastoderm, but rather depend on a prior pattern. In the absence of patterned instruction from maternal organizers, the gap

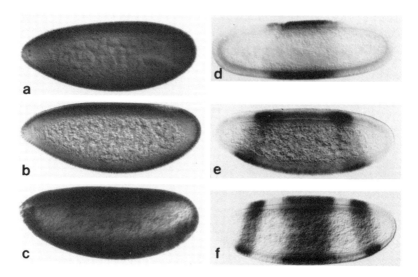

FIG. 10. Effect of *bcd osk tsl* and *bcd osk* mutations on the expression of *Kr, eve,* and *ftz* at the end of blastoderm, as visualized by antibody staining. (a–c) *bcd osk tsl* embryos: (a) *Kr*, (b) *eve*, (c) *ftz* pattern. (d–f) *bcd osk* embryos: (d) *Kr*, (e) *eve*, (f) *ftz* pattern.

genes do not pattern, and this causes a subsequent failure of pair-rule genes to resolve spatial patterns.

In contrast, in *bcd osk* embryos, in which anterior and posterior organizers are defective while terminal organizers are intact, patterns do form. *Kr* is expressed in the middle region of *bcd osk* embryos only (Fig. 10d); the repression of *Kr* expression at the poles is due to the presence there of terminal organizers and their zygotic "targets." (As indicated above, apart from *tll* at least one more zygotic factor is involved in the organization of the terminal region. This factor apparently acts in the terminalmost portion of the embryo to suppress segmentation gene expression.) This constellation of gap gene expression is correlated with the following pair-rule gene pattern. *eve* "tries" to produce two peripheral stripes, but the expression of *eve* persists in the middle and the stripes do not resolve (Fig. 10e). Meanwhile, *ftz* forms a broad band in the central region and one peripheral stripe on either side (Fig. 10f). Hence *ftz* and *eve*, unlike in the wild type, are not mutually exclusive in the middle region of *bcd osk* embryos.

Thus, in *bcd osk* embryos, a pattern of pair-rule gene expression forms in regions where the input from the gap genes changes, namely, where *Kr* goes off and *tll* comes on. In the central region, the uniform expression of *Kr* is accompanied by uniform pair-rule gene expression: *ftz* is on and *eve* is on at a low level (Gaul and Jäckle, 1989). (We cannot specifically explain the pair-rule gene pattern at the very termini at present, since we know too little about the distribution of the zygotic terminal genes.)

In another double-mutant combination, *stau exu,* we find the following patterns. As described above, *hb* and *Kr* are coexpressed in *stau exu* embryos in a broad middle region encompassing about 60% of the embryo (Fig. 8a and b). At the poles, *hb* and the zygotic terminal genes are expressed; the latter is implied by the repression of *Kr* at the termini. In these embryos, *eve* forms two stripes in the terminal region (Fig. 8c). The *ftz* protein exhibits a roughly complementary distribution: a very broad domain is established in the central region of the embryo as well as two small stripes near the poles (Fig. 8d). Thus, in the large region where both *Kr* and *hb* are expressed no further minima or maxima of *ftz* or *eve* expression are found. It is only in the vicinity of the *Kr* borders that the expression patterns resolve. The central *ftz* domain closely resembles the *Kr* pattern, whereas the patterns of *eve* and *Kr* are approximately complementary (Gaul and Jäckle, 1989).

Thus, we find that the combination of *hb* and *Kr* expression is correlated with the presence of *ftz* and the absence of *eve*. This is the same situation as in the wild type: in the region where the domains of *hb* and *Kr* expression overlap, the second *ftz* stripe is formed. By contrast, in the middle region of *bcd osk* embryos, where only *Kr* provides input, a different constellation is generated at the pair-rule gene level, with *ftz* on and *eve* on at a low level. Hence, the difference in the constellations of gap gene activities is reflected in a different expression of pair-rule genes. This result suggests that it does make a difference for pair-rule gene expression what kind of gap gene expression a blastoderm cell experiences. In particular, it indicates that it makes a difference whether a blastoderm cell experiences the expression of *Kr* alone or of both *hb* and *Kr* at the same time. This suggests that the overlap between *Kr* and *hb* expression observed in the wild type indeed provides essential input for the formation of the pair-rule gene pattern.

In all three mutants we find that within an area of uniform gap gene activity no alteration occurs in the constellation of pair-rule gene expression. In the middle regions of *bcd osk* and *stau exu* embryos, *ftz* and *eve* are present or absent in a uniform manner. The pair-rule gene expression alters in a region where the gap gene input changes, namely, around the borders of the *Kr* domains. Hence, unless some differential information is provided by the gap genes, the uniformity of pair-rule gene expression (or nonexpression) is not broken down.

If we take these conclusions seriously they imply that the expression pattern of the pair-rule genes has to be preformed in the pattern of gap genes. The question remains which features of the gap gene expression encode information. So far, we have seen that an overlap of gap gene expression can be functionally relevant. By generalizing, a combinatorial model could be conceived, according to which qualitatively different states of gap gene expression along the anteroposterior axis of the (wild-type) embryo (. . . hb, $hb + Kr, Kr$, . . .) provide different input for the pair-rule genes, and eventually "translate" into different states of pair-rule gene expression. In such a model, however, the informational content provided by the gap genes would not suffice to produce the pattern of pair-rule genes: within the range of Kr expression, for example, at least three stripes and two interstripes of ftz expression have to form, while at most three "states" of gap gene expression ($hb + Kr, Kr, Kr + kni$) would be there to provide the input.

However, more genes might be involved in the generation of the pair-rule gene patterns. In particular, the function of gt remains to be elucidated. In addition, the function of the interactions among the pair-rule genes in the formation of their patterns has to be further clarified.

Apart from searching for further components, however, it may be reasonable to aim at a more quantitative approach to the problem. As a matter of fact, the distribution of hb and Kr is graded toward the margins of their domains. And there is important evidence against simple binary input–output modes in the interaction of gap genes with pair-rule genes. The mutant alleles of gap genes differ in the strength of their effect on segmentation, and a graded phenotypic series from wild-type to loss-of-function phenotypes can be established, reflecting a graded requirement for gap gene activity. Even under merely heterozygous conditions (at least for hb, Kr, and kni) the segment patterns of hatched larvae show abnormalities (Wieschaus et al., 1984a; Lehmann, 1985). The effects of heterozygous gap gene mutants are already visible in the pair-rule gene pattern at blastoderm (Frasch and Levine, 1987). These findings show that gap gene activities are required in a delicate concentration-dependent manner and thus point to the need of a thorough quantitative analysis of the interactions between gap and pair-rule genes (see, e.g., Edgar et al., 1989).

V. Gap Genes Act on Homeotic Selector Genes

Apart from instructing the pair-rule genes, the gap genes, along with other segmentation genes, are thought to be involved in establishing the spatial expression of homeotic selector genes. Thus, besides participating in the establishment of the periodic partitioning of the embryo, the gap genes also have a function in determining segment identity.

The homeotic genes of the Antennapedia (ANT-C) and bithorax (BX-C) complexes are first transcribed just before cellularization of the blastoderm, at about the same time as the pair rule genes. Their spatial patterns change profoundly during development. Initially, they are expressed in broad overlapping domains at rather uniform levels. Then, at the end of blastoderm, they develop a parasegment-specific pattern of expression: for example, high levels of expression of *Sex combs reduced* (*Scr*), *Antennapedia* (*Antp*), and *Ultrabithorax* (*Ubx*) are observed in parasegments 2, 4, and 6, respectively (for review, see Akam, 1987; Ingham, 1988).

The expression patterns of homeotic genes in segmentation gene mutants have been studied by a number of investigators. However, only limited data characterizing the interactions between gap genes and homeotic genes are available so far. *hb* activity appears to repress *Ubx* in both the head and the thorax and to define the posterior border of *Ubx* expression as well (White and Lehmann, 1986). Moreover, *hb* seems to be involved in the activation of *Antp* (P2 promoter) (Harding and Levine, 1988). *Kr* is essential for the activation of one of the *Antp* promoters (P1) (Harding and Levine, 1988), and the posterior determinants dependent on *osk* are thought to be necessary for the initial activation of *Ubx* in the abdomen (Akam, 1987). The lack of these controlling factors causes alterations in the distribution of homeotic transcripts as early as during the blastoderm stage. The pair rule genes also influence homeotic selector gene expression. *ftz* appears to play a key role in coupling the spatial expression of homeotic genes to the evolving segment pattern; *ftz* mutant embryos do not exhibit the early peaks of *Scr, Antp,* and *Ubx* expression at the blastoderm stage (Ingham and Martinez-Arias, 1986).

What is the relative importance of the gap genes in instructing the homeotic selector gene pattern, as compared with the pair rule genes? It is now *opinio communis* that the gap genes are necessary for the early patterning, that is, for the broad, early domains of the homeotic genes, and that the pair-rule genes then set the homeotic gene activity into a parasegmental frame.

The analysis of *stau exu* and *vas exu* embryos supports this view. It illustrates how the same pair-rule gene pattern can be interpreted in different ways, namely, according to the preceding constellation of gap gene activities. As described earlier, the phenotypes of *stau exu* and *vas exu* differ significantly: *stau exu* embryos develop thoracic structures in the middle portion of the embryo, whereas *vas exu* embryos develop gnathal structures. The only major difference in the expression of segmentation genes in these embryos is at the level of the gap genes. While *hb* is expressed in both mutants, *Kr* protein is present in *stau exu* embryos and absent in *vas exu* embryos (Fig. 8a and b, h and i). In contrast, the patterns of the pair-rule genes *eve* and *ftz*, and also that of the segment polarity gene *engrailed* (*en*), are essentially the same in both mutants (see Fig. 8c–e, j–l). In particular, a broad domain of *ftz* expression is established in the middle region of the embryo.

However, the difference in the expression of gap genes is reflected in the expression of homeotic genes. In *stau exu* embryos, *Antp* is expressed in a broad domain that is almost coextensive with the *Kr* domain (Fig. 8f). This expression pattern correlates well with the mutant cuticle phenotype, since, in wild-type embryos, *Antp* is concentrated in the primordia of the thorax (Wirz *et al.*, 1986; Carroll *et al.*, 1986). In contrast, *Antp* expression could not be detected in *vas exu* embryos (Fig. 8m), which likewise corresponds with the phenotype observed.

These data show that *hb* is not sufficient to promote *Antp* expression (see above), whereas *hb* and *Kr* together are. More importantly, they demonstrate the priority of gap gene influence over that of pair-rule genes on homeotic selector gene expression.

In summary, these studies suggest strongly that it is the aperiodic region-specific accumulation of gap gene activities which is crucial for the establishment of the initial region-specific expression domains of the homeotic selector genes. Moreover, as the early domains of homeotic gene expression are instructed by the gap genes and the further refinement of pattern is supervised by the pair rule genes, it is clear that segmentation and control of segment identity are intimately "integrated" processes: the expression of homeotic genes is kept abreast of the evolving state of the segmentation process.

VI. Late Expression Patterns of *hunchback* and *Krüppel* and Their Regulation

Both *hb* and *Kr* are not merely expressed prior to and during the blastoderm stage, that is, at the critical time for the generation of the metameric body pattern. They are also expressed at later stages of embryonic development, showing a complex, dynamic pattern (Gaul *et al.*, 1987). After blastoderm, *hb* is expressed in the neurogenic region of the head and of the germ band. In addition, nuclei in the peripheral nervous system and the nuclei of the yolk sac show *hb* expression.

As for *Kr*, with the onset of gastrulation a new domain of expression is established at the posterior pole, and *Kr* expression in the central domain begins to spread both toward the cephalic furrow and toward the posterior (Fig. 11a,b). After the fast phase of germ band elongation, *Kr* is expressed along the entire length of the germ band, showing a patchy, segmentally repeated pattern that spans the ectodermal and mesodermal layers. In the proctodeal invagination, cells which form the anlage of the Malpighian tubules show *Kr* staining (Fig. 11c). Another group of *Kr*-positive cells, located anterior to the midgut invagination, becomes invaginated as the stomodeum forms.

During the extended germ band stage, *Kr* is expressed in the amnioserosa and the anlage of the Malpighian tubules (Fig. 11d–f). In addition, the neurogenic regions of the head and the germ band show *Kr* expression, the most prominent staining being observed in the group of cells in the stomodeum; these cells may be part of the developing stomatogastric nervous system. During germ band retraction, when the embryonic organs are translocated to the characteristic larval position, the pattern of *Kr* expression changes. The cells of the amnioserosa, which covers the dorsal surface of the embryo, remain *Kr* positive,

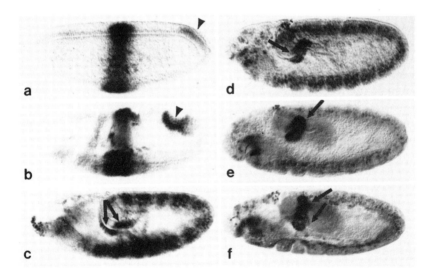

FIG. 11. Expression patterns of *Kr* protein after the blastoderm stage. (a) Onset of gastrulation, (b) end of gastrulation, (c) after the fast phase of germ band extension, (d–f) during the extended germ band stage. *Kr* expression at the posterior pole (arrowhead) becomes restricted to the cells which form the anlagen of the Malpighian tubules (arrows).

whereas the expression of *Kr* in the developing Malpighian tubules ceases. In the central nervous system, a subpopulation of neurons shows *Kr* staining. This staining persists, but the number of *Kr*-positive cells decreases with time. In very old embryos *Kr* becomes expressed in the larval photoreceptor organ (the "Bolwig organ").

We do not know what function *hb* and *Kr* may serve in all the diverse locations and developing tissues, and it is largely unknown which genes control their expression and which they act on in turn. However, a small section of the *Kr* pattern, namely, the expression of *Kr* in the anlage of the Malpighian tubules, has been examined to some extent, with interesting results.

EXPRESSION OF *Krüppel* IN THE ANLAGE OF THE MALPIGHIAN TUBULES

Gloor (1950) first observed that *Kr*-deficient embryos lack Malpighian tubules. In the wild type, the Malpighian tubules develop as outpouchings of the proctodeum at the border between the hind- and midgut. They consist of two pairs of cul-de-sac tubuli floating freely in the body fluid and serve the larva as excretory organs. The lack of Malpighian tubules in *Kr* mutant embryos results from the fact that the cells which are to form the Malpighian tubules in the wild type become transformed and develop as part of the hindgut instead (Harbecke and Janning, 1989). This "transformation" is manifest as early as the extended germ band stage, at which time, in the wild type, the cells of the anlage of the Malpighian tubules begin to form four small pouches at the border between the hindgut and posterior midgut. In *Kr* mutant embryos, these cells do not form the pouches but remain at their original position instead, as demonstrated in embryos mutant for a particular *Kr* allele (Kr^9) which produces a nonfunctional but immunodetectable *Kr* protein (Redemann *et al.*, 1988). Hence, *Kr* seems to be the key gene in promoting Malpighian tubules over hindgut development.

Which genes are responsible for the control of *Kr* expression in the anlage of the Malpighian tubules? As indicated earlier, the *tor* group of maternal genes is necessary for organizing the whole posterior terminal region. Embryos mutant for these genes lack all structures behind the seventh abdominal segment. Apart from these maternal genes, three zygotic genes have been demonstrated to be active in the terminal region: *tll, forkhead* (*fkh*), and *caudal* (*cad*). In *tll* mutant embryos, the telson, the "ectodermal" part of the terminal region, is completely missing. Such embryos, however, will still form a proctodeal invagination, albeit reduced in size (Strecker *et al.*, 1986). In the larvae, no Malpighian tubules can be detected. *fkh* mutant embryos, in addition to showing cuticular defects in the telson, fail to develop the hindgut and Malpighian tubules (Jürgens and Weigel, 1988). Similarly, in *cad* mutant embryos, defects both in the telsonal cuticle and in the Malpighian tubules are observed (Macdonald and Struhl, 1986).

The analysis of the pattern of *Kr* expression in the mutants shows that the activities of the group of terminal organizer genes, of *tll,* and of *fkh* are necessary for the establishment of normal *Kr* expression in the anlage of the Malpighian tubules. In *cad* mutant embryos, in contrast, *Kr* is expressed in the anlage of the Malpighian tubules, and the *Kr* staining shows that the anlage does form pouches (Gaul and Weigel, 1990). The lack of expression of *Kr* in the anlage of the Malpighian tubules in the maternal gene mutants is expected, since the *ftz* patterns in such mutants show that the terminal section of the blastoderm anlage is recruited for segmentation. It is likely, however, that the maternal organizers do not exert any direct influence. The zygotic genes *tll* and *fkh* are better candidates for being involved in direct control of *Kr* expression. Interestingly, in *tll* mutant embryos, the expression of *Kr* at the posterior pole that is observed in wild-type embryos during early gastrulation is still established, and it is only later, at the extended germ band stage, that the *Kr* signal is no longer maintained. In contrast, in *fkh* mutant embryos, the early *Kr* expression at the posterior pole is already missing. These findings suggest that *fkh* may indeed be a positive regulator of *Kr* expression and that the presence of *tll* activity is necessary for maintaining *Kr* expression.

The *fkh* gene has recently been cloned in our laboratory (Weigel *et al.,* 1989), and antibodies have been produced against its protein. Immunochemical labeling shows that the *fkh* protein is localized in the nucleus; hence, it is not unlikely that *fkh* functions as a transcriptional regulator. At blastoderm *fkh* is expressed in the terminal regions of the embryo, covering the first 13% at the posterior pole. At the extended germ band stage, the whole proctodeal invagination, that is, both hindgut and posterior midgut, are *fkh* positive. Therefore, in order to confine *Kr* to the anlage of the Malpighian tubules, factors other than *fkh* must be active.

Taken together, these results demonstrate that the upstream control of the expression of *Kr* in the anlage of the Malpighian tubules differs from the regulation of the central blastoderm domain. While *tll* activity acts to repress *Kr* in the terminal portion of the embryo during early blastoderm, it is needed to maintain the expression of *Kr* in the anlage of the Malpighian tubules. Moreover, *fkh,* which does not play a part in the regulation of the central *Kr* domain in blastoderm, acts as a positive regulator of *Kr* expression in the anlage of the Malpighian tubules (Gaul and Weigel, 1990).

On which genes does *Kr* act to promote the development of the Malpighian tubules? As noted above, *cad* is a good candidate for a "downstream" gene; it becomes expressed in the anlage of the Malpighian tubules after the cells of the anlage have started to grow out and form the actual tubules. Other possible targets of *Kr* activity have not yet been identified. We know, however, that none of the other segmentation genes monitored so far are switched on in the anlage of the Malpighian tubules. Evidently, an entirely different set of genes is subject to *Kr* control in the development of the Malpighian tubules as compared with the segmentation process.

Thus, not only the way *Kr* is regulated but also the genes it acts on are highly context dependent. *Kr* cannot be viewed as a factor that by itself promotes thorax formation or Malpighian tubule development. Rather, other factors have to come in and specify the action of *Kr*. It is evident, however, that the "promiscuity" of *Kr* expression in itself precludes such specificity of function.

Acknowledgments

We are much indebted to Uli Schwarz for encouragement and steady support, and we thank Garrett Odell for many valuable suggestions on the manuscript.

References

Akam, M. (1987). The molecular basis for metameric pattern in the *Drosophila* embryo. *Development* **101,** 1–22.

Berleth, T., Burri, M., Thoma, G., Bopp, D., Richstein, S., Frigerio, G., Noll, M., and Nüsslein-Volhard, C. (1988). The role of localization of *bicoid* RNA in organizing the anterior pattern of the *Drosophila* embryo. *EMBO J.* **7,** 1749–1756.

Carroll, S. B., and Scott, M. P. (1986). Zygotically-active genes that affect the spatial expression of the *fushi tarazu* segmentation gene during early *Drosophila* embryogenesis. *Cell* **45,** 113–126.

Carroll, S. B., Laymon, R. A., McCutcheon, M. A., Riley, P. D., and Scott, M. P. (1986). The localization and regulation of *Antennapedia* protein expression in *Drosophila* embryos. *Cell* **47,** 113–122.

Carroll, S. B., Laughon, S., and Thalley, B. S. (1988). Expression, function and regulation of the *hairy* segmentation protein in the *Drosophila* embryo. *Genes Dev.* **2,** 883–890.

Chan, L. N., and Gehring, W. (1971). Determination of blastoderm cells in *Drosophila melanogaster. Proc. Natl. Acad. Sci. U.S.A.* **68,** 2217–2222.

Driever, W., and Nüsslein-Volhard, C. (1988a). A gradient of *bicoid* protein in *Drosophila* embryos. *Cell* **54,** 83–93.

Driever, W., and Nüsslein-Volhard, C. (1988b). The *bicoid* protein determines position in the *Drosophila* embryo in a concentration-dependent manner. *Cell* **54,** 95–105.

Driever, W., and Nüsslein-Volhard, C. (1989). The *bicoid* protein is a positive regulator of hunchback transcription in the early *Drosophila* embryo. *Nature (London)* **337,** 138–143.

Edgar, B. A., Odell, G. M., and Schubiger, G. (1989). A genetic switch, based on negative regulation, sharpens stripes in *Drosophila* embryos. *Dev. Genet.* **10**, 124–142.

Evans, R. M., and Hollenberg, S. M. (1988). Zinc fingers: Gilt by association. *Cell* **52**, 1–3.

Foe, V. E., and Alberts, B. M. (1983). Studies of nuclear and cytoplasmic behavior during the five mitotic cycles that precede gastrulation in *Drosophila* embryogenesis. *J. Cell Sci.* **61**, 31–70.

Frasch, M., and Levine, M. (1987). Complementary patterns of *even-skipped* and *fushi tarazu* expression involve their differential regulation by a common set of segmentation genes in *Drosophila*. *Genes Dev.* **1**, 981–995.

Frasch, M., Hoey, T., Rushlow, C., Doyle, H., and Levine, M. (1987). Characterization and localization of the *even-skipped* protein of *Drosophila*. *EMBO J.* **6**, 749–759.

Frohnhöfer, H. G. (1987). Maternale Gene und die Anlage des anteroposterioren Musters in *Drosophila* Embryonen. Ph.D. Thesis, Eberhard-Karls-Universität, Tübingen, Federal Republic of Germany.

Frohnhöfer, H. G., and Nüsslein-Volhard, C. (1986). Organization of anterior pattern in the *Drosophila* embryo by the maternal gene *bicoid*. *Nature (London)* **324**, 120–125.

Frohnhöfer, H. G., and Nüsslein-Volhard, C. (1987). Maternal genes required for the anterior localization of *bicoid* activity in the embryo of *Drosophila*. *Genes Dev.* **1**, 880–890.

Gaul, U., and Jäckle, H. (1987a). How to fill a gap in the *Drosophila* embryo. *Trends Genet.* **3**, 127–131.

Gaul, U., and Jäckle, H. (1987b). Pole region-dependent repression of the *Drosophila* gap gene *Krüppel* by maternal gene products. *Cell* **51**, 549–555.

Gaul, U., and Jäckle, H. (1989). Analysis of maternal-effect mutant combinations elucidates regulation and function of the overlap of *hunchback* and *Krüppel* gene expression in the *Drosophila* blastoderm embryo. *Development* **107**, 651–662.

Gaul, U., and Weigel, D. (1990). Regulation of *Krüppel* expression in the anlage of the Malpighian tubules in the *Drosophila* embryo. In preparation.

Gaul, U., Seifert, E., Schuh, R., and Jäckle, H. (1987). Analysis of *Krüppel* protein distribution during early *Drosophila* development reveals posttranscriptional regulation. *Cell* **50**, 639–647.

Gloor, H. (1950). Phänotypen der Heterozygoten bei der unvollständig dominanten, homozygot lethalen Mutante *Kr* (=*Krüppel*) von *Drosophila melanogaster*. *Arch. Julius Klaus-Stift.* **29**, 277–287.

Harbecke, R., and Janning, W. (1989). The segmentation gene *Krüppel* of *Drosophila melanogaster* has homeotic properties. *Genes Dev.* **3**, 114–122.

Harding, K., and Levine, M. (1988). Gap genes define the limits of Antennapedia and Bithorax gene expression during early development in *Drosophila*. *EMBO J.* **7**, 205–214.

Hülskamp, M., Schröder, C., Pfeifle, C., Jäckle, H., and Tautz, D. (1989). Posterior segmentation of the *Drosophila* embryo in the absence of a maternal posterior organizer gene. *Nature (London)* **338**, 629–632.

Ingham, P. (1988). The molecular genetics of embryonic pattern formation in *Drosophila*. *Nature (London)* **335**, 25–34.

Ingham, P., and Gergen, P. (1988). Interactions between the pair rule genes *runt, hairy, even skipped* and *fushi tarazu* and the establishment of periodic pattern in the *Drosophila* embryo. *Development* **104** (Suppl.), 51–60.

Ingham, P. W., and Martinez-Arias, A. (1986). The correct activation of Antennapedia and bithorax complex genes requires the *fushi tarazu* gene. *Nature (London)* **324**, 592–597.

Ingham, P. W., Ish-Horowicz, D., and Howard, K. R. (1986). Correlative changes in homeotic and segmentation gene expression in *Krüppel* mutant embryos of *Drosophila. EMBO J.* **5,** 1659–1665.

Jäckle, H., Tautz, D., Schuh, R., Seifert, E., and Lehmann, R. (1986). Cross-regulatory interactions among the gap genes of *Drosophila. Nature (London)* **324,** 668–670.

Jürgens, G., and Weigel, D. (1988). Terminal versus segmental development in the *Drosophila* embryo: The role of the homeotic gene *forkhead. Wilhelm Roux's Arch. Dev. Biol.* **197,** 354–364.

Jürgens, G., Wieschaus, E., Nüsslein-Volhard, C., and Kluding, H. (1984). Mutations affecting the pattern of the larval cuticle in *Drosophila melanogaster*. II. Zygotic loci on the third chromosome. *Wilhelm Roux's Arch. Dev. Biol.* **193,** 359–377.

Klingler, M., Erdélyi, M., Szabad, J., and Nüsslein-Volhard, C. (1988). Function of *torso* in determining the terminal anlagen of the *Drosophila* embryo. *Nature (London)* **335,** 275–277.

Klug, A., and Rhodes, D. (1987). "Zinc fingers": A novel protein motif for nucleic acid recognition. *Trends Biochem. Sci.* **12,** 464–469.

Knipple, D. C., Seifert, E., Rosenberg, U. B., Preiss, A., and Jäckle, H. (1985). Spatial and temporal patterns of *Krüppel* gene expression in early *Drosophila* embryos. *Nature (London)* **317,** 40–44.

Lehmann, R. (1985). Regionsspezifische Segmentierungsmutanten bei *Drosophila melanogaster* Meigen. Thesis, Eberhard-Karls-Universität, Tübingen, Federal Republic of Germany.

Lehmann, R., and Nüsslein-Volhard, C. (1986). Abdominal segmentation, pole cell formation, and embryonic polarity require the localized activity of *oskar,* a maternal gene in *Drosophila. Cell* **47,** 141–152.

Lehmann, R., and Nüsslein-Volhard, C. (1987a). *hunchback,* a gene required for segmentation of an anterior and posterior region of the *Drosophila* embryo. *Dev. Biol.* **119,** 402–417.

Lehmann, R., and Nüsslein-Volhard, C. (1987b). Involvement of the *pumilio* gene in the transport of an abdominal signal in the *Drosophila* embryo. *Nature (London)* **329,** 167–170.

Lewis, E. B. (1981). Developmental genetics of the bithorax complex in *Drosophila. In* "Developmental Biology Using Purified Genes. ICN–UCLA Symposia" (D. D. Brown and C. F. Fox, eds.), Vol. 23, pp. 189–208. Academic Press, New York.

Macdonald, M., and Struhl, G. (1986). A molecular gradient in early *Drosophila* embryos and its role in specifying the body pattern. *Nature (London)* **324,** 537–545.

Meinhardt, H. (1977). A model for pattern formation in insect embryogenesis. *J. Cell Sci.* **23,** 117–139.

Meinhardt, H. (1986). Hierarchical inductions of cell states: A model for segmentation in *Drosophila. J. Cell Sci.* **4** (Suppl.), 357–381.

Nauber, U., Pankratz, M. J., Kienlin, A., Seifert, E., Klemm, U., and Jäckle, H. (1988). Abdominal segmentation of the *Drosophila* embryo requires a hormone receptor-like protein encoded by the gap gene *knirps. Nature (London)* **336,** 489–492.

Nüsslein-Volhard, C., and Wieschaus, E. (1980). Mutations affecting segment number and polarity in *Drosophila. Nature (London)* **287,** 795–801.

Nüsslein-Volhard, C., Wieschaus, E., and Kluding, H. (1984). Mutations affecting the pattern of the larval cuticle in *Drosophila melanogaster*. I. Zygotic loci on the second chromosome. *Wilhelm Roux's Arch. Dev. Biol.* **193,** 267–282.

Nüsslein-Volhard, C., Frohnhöfer, H. G., and Lehmann, R. (1987). Determination of anteroposterior polarity in *Drosophila. Science* **238,** 1675–1681.

Petschek, J., Perrimon, N., and Mahowald, A. P. (1987). Region specific effects in *l(1)giant* embryos of *Drosophila. Dev. Biol.* **119,** 177–189.

Preiss, A., Rosenberg, U. B., Kienlin, A., Seifert, E., and Jäckle, H. (1985). Molecular genetics of Krüppel, a gene required for segmentation of the Drosophila embryo. Nature (London) 313, 27–33.

Redemann, N., Gaul, U., and Jäckle, H. (1988). Disruption of a putative Cys–zinc interaction eliminates the biological activity of the Krüppel finger protein. Nature (London) 332, 90–92.

Rosenberg, U. B., Schröder, C., Priess, A., Kienlin, A., Côté, S., Riede, I., and Jäckle, H. (1986). Structural homology of the product of the Drosophila Krüppel gene with Xenopus transcription factor IIIA. Nature (London) 319, 336–339.

Schröder, C., Tautz, D., Seifert, E., and Jäckle, H. (1988). Differential regulation of the two transcripts from the Drosophila gap segmentation gene hunchback. EMBO J. 7, 2881–2887.

Schüpbach, T., and Wieschaus, E. (1986). Maternal-effect mutations altering the anterior–posterior pattern of the Drosophila embryo. Wilhelm Roux's Arch. Dev. Biol. 195, 302–317.

Scott, M. P., and O'Farrell, P. (1986). Spatial programming of gene expression in early Drosophila embryogenesis. Annu. Rev. Cell Biol. 2, 49–80.

Simcox, A. A., and Sang, J. H. (1983). When does determination occur in Drosophila embryos? Dev. Biol. 97, 212–221.

Strecker, T. R., Kongsuwan, K., Lengyel, J. A., and Merriam, J. R. (1986). The zygotic mutant tailless affects the anterior and posterior ectodermal regions of the Drosophila embryo. Dev. Biol. 113, 64–74.

Tautz, D. (1988). Regulation of the Drosophila segmentation gene hunchback by two maternal morphogenetic centres. Nature (London) 332, 281–284.

Tautz, D., and Jäckle, H. (1987). Molecular analysis of regulatory genes in Drosophila segmentation. In "Hormones and Cell Regulation European Symposium" (J. E. Dumont and J. Nunez, eds.), 11th Ed., Vol. 153, pp. 125–136. Colloque INSERM/ John Libbey Eurotext Ltd.

Tautz, D., Lehmann, R., Schnürch, H., Schuh, R., Seifert, E., Kienlin, A., Jones, K., and Jaeckle, H. (1987). Finger protein of novel structure encoded by hunchback, a second member of the gap class of Drosophila segmentation genes. Nature (London) 327, 383–389.

Wakimoto, B. T., Turner, F. R., and Kaufman, T. C. (1984). Defects in embryogenesis in Drosophila melanogaster in mutants associated with the Antennapedia gene complex. Dev. Biol. 102, 147–172.

Weigel, D., Jürgens, G., Küttner, F., Seifert, E., and Jäckle, H. (1989). The homeotic gene forkhead encodes a nuclear protein and is expressed in the terminal region of the Drosophila embryo. Cell 57, 645–658.

White, R. A. H., and Lehmann, R. (1986). A gap gene, hunchback, regulates the spatial expression of Ultrabithorax. Cell 47, 311–321.

Wieschaus, E., Nüsslein-Volhard, C., and Kluding, H. (1984a). Krüppel, a gene whose activity is required early in the zygotic genome for normal embryonic segmentation. Dev. Biol. 104, 172–186.

Wieschaus, E., Nüsslein-Volhard, C., and Jürgens, G. (1984b). Mutations affecting the pattern of the larval cuticle in Drosophila melanogaster III. Zygotic loci on X-chromosome and fourth chromosome. Wilhelm Roux's Arch. Dev. Biol. 193, 296–307.

Wirz, J., Fessler, L. I., and Gehring, W. J. (1986). Localization of the Antennapedia protein in Drosophila embryos and imaginal discs. EMBO J. 5, 3327–3334.

ROLE OF THE *zerknüllt* GENE IN DORSAL–VENTRAL PATTERN FORMATION IN *Drosophila*

Christine Rushlow and Michael Levine

Department of Biological Sciences, Columbia University,
New York, New York 10027

I. Introduction

Early in the development of the *Drosophila* embryo, totipotent cleavage stage nuclei migrate to the periphery of the egg and become committed to specific pathways of development during cellularization. Embryonic cells express distinct sets of genes and follow different pathways of development as a result of their locations in the early embryo. Many of the genes that specify this positional information have been identified in past genetic screens based on mutations which disrupt cuticular structures secreted by the epidermis of advanced-stage embryos. The first genetic screens of this sort were restricted to the identification of regulatory genes active in the zygote (Nüsslein-Volhard and Wieschaus, 1980; Jürgens *et al.*, 1984; Nüsslein-Volhard *et al.*, 1984; Wieschaus *et al.*, 1984). These screens were quite extensive and probably revealed nearly every essential gene that causes a specific

ADVANCES IN GENETICS, Vol. 27

disruption of the epidermis when mutated. More recent efforts have led the identification of many of the maternally active genes that control the embryonic pattern (Anderson and Nüsslein-Volhard, 1984a; Boswell and Mahowald, 1985; Perrimon et al., 1986; Schüpbach and Wieschaus, 1989). These latter screens are considerably more difficult than those designed for the identification of zygotically active genes, and it is not clear whether all of the relevant maternal genes have been found.

Over 70 regulatory genes are now known in Drosophila (reviewed in Akam, 1987; Anderson, 1987; Ingham, 1988). About 30 of the genes are maternally active and encode products which are deposited in the unfertilized egg during oogenesis. The other 40 genes are zygotically active and are expressed after fertilization. Both the maternal and zygotic regulatory genes fall into two broad groups: (1) those that control cell identify along the anterior–posterior (AP) embryonic body axis and (2) those that organize the dorsal–ventral (DV) pattern. In both systems, interactions among maternally encoded products lead to the graded expression of one or more regulatory products in the early embryo. These broadly expressed regulatory factors initiate localized expression of early acting zygotic genes, which in turn specify more refined positional identities.

Many of the zygotic regulatory genes control the AP pattern, and mutations in these genes often disrupt the segmentation process. Localized expression of the segmentation genes involves a well-defined hierarchy of interactions, which occurs during a brief period of early development (about 2 to 4 hours after fertilization) and culminates with the precise expression of about 10 different segment polarity genes in specific subsets of cells in each segment primordium. One of the crucial first steps in this process is the localization of maternally encoded bicoid (bcd) transcripts to the anterior pole of the developing oocyte. bcd proteins come to be distributed in a steep gradient along the AP axis, and different thresholds of bcd protein help define the limits of gap gene expression (Driever and Nüsslein-Volhard, 1988a,b, 1989). It has been proposed that the localization of bcd transcripts involves an inherent AP polarity within the egg chamber (Frohnhöfer and Nüsslein-Volhard, 1987). As bcd RNAs are transported from their site of synthesis in nurse cells to the oocyte, they are trapped at their point of entry at the anterior pole of the developing oocyte. A 600-bp portion of the bcd transcript, contained within the untranslated trailer sequence, has been shown to include a signal for localization (Macdonald and Struhl, 1989). Perhaps the bcd RNA interacts with a ubiquitous cytoskeletal element in the oocyte but is restricted to the anterior pole owing to polarized transport from the nurse cells.

The specification of a maternal gradient along the dorsal–ventral (DV) axis appears to involve a fundamentally different mechanism. The initiation of DV polarity depends on a ventral-to-dorsal gradient of a protein encoded by a maternal gene called *dorsal* (Steward, 1987). The graded distribution of this protein appears to depend on a posttranscriptional (and/or posttranslational) process: *dorsal* mRNAs are found at equal levels in both dorsal and ventral regions.

In this article, we consider aspects of the regulatory hierarchy responsible for the asymmetric distribution and function of zygotic regulatory genes along the DV axis of early embryos. Table 1 lists 26 maternal and zygotic genes that are responsible for the differentiation of the dorsal–ventral pattern. This figure could be an underestimate, however, because of the complex phenotypes sometimes observed in advanced-stage DV mutants. Such mutants frequently disrupt early

TABLE 1
Genes Involved in Dorsal–Ventral Pattern Formation

| | Maternal | | |
Gene type	Egg shape and embryonic polarity	Embryonic polarity	Zygotic
Dorsalizing	fs (1) K10 *spire (spir)* *cappucino (capu)*	*dorsal (dl)* *windbeutel (wind)* *gastrulation defective (gd)* *nudel (ndl)* *pipe (pip)* *tube (tub)* *snake (snk)* *easter (ea)* *Toll (Tl)* *spatzle (spz)* *pelle (pll)*	*twist (twi)* *snail (sna)*
Ventralizing	*gurken (grk)* *torpedo (top)*	*cactus (cac)*	*decapentaplegic (dpp)* *shrew (srw)* *screw (scw)* *tolloid (tld)* *zerknüllt (zen)* *twisted gastrulation (tsg)* *short gastrulation (sog)*

embryonic processes, such as gastrulation, germ band elongation, or head involution, which in turn, causes complex secondary alterations in the body pattern of older embryos. These "terminal phenotypes" are not always easily interpreted as alterations in the DV pattern. Thus, while it is likely that all zygotic lethal mutations which alter the embryonic pattern have been identified, mutants showing ambiguous cuticular phenotypes might not be correctly assigned to the DV category. As an example, the *zerknüllt* (*zen*) gene, which is the primary focus of this article, was originally described as a gene required for normal head development (Wakimoto *et al.*, 1984).

We do not intend to catalog regulatory interactions among all of the known DV genes. Rather, we primarily focus on the *zen* gene as a means to raise several current issues regarding DV pattern formation. *zen* is one of four zygotic DV genes that have been cloned and characterized (Doyle *et al.*, 1986; Rushlow *et al.*, 1987a). *zen* and another of the cloned genes, called *snail* (Boulay *et al.*, 1987), encode proteins with known DNA-binding motifs. A third gene, *twist*, shares homology with a newly discovered DNA-binding motif, which is probably associated with an enhancer-binding class of transcription factors that regulate immunoglobulin promoters in mammals (Thisse *et al.*, 1988; Murre *et al.*, 1989). It is likely that these genes control cell fate by modulating gene expression at the level of transcription. Direct evidence for this function has been obtained for *zen*, which encodes a homeobox protein that controls the differentiation of the amnioserosa (Han *et al.*, 1989).

In this article, we consider how *zen* comes to be expressed in a restricted portion of the DV axis, and how localized *zen* products might regulate the differentiation of the amnioserosa and other dorsally derived embryonic tissues. *zen* is one of the first zygotically active DV genes to be expressed in the early embryo, and we review the evidence that the initiation of *zen* expression depends on maternally encoded factors. The study of *zen* expression provides an excellent model system for investigating a central problem in the control of positional information in *Drosophila*. How do discrete patterns of zygotic gene expression emerge in response to crudely localized maternal cues? The maintenance and refinement of the *zen* expression pattern during later periods of development involves regulatory interactions with other zygotically active DV genes, which appear to specify the components of a cell–cell communication pathway. Finally, we present evidence that the *zen* protein is a sequence-specific transcriptional activator and discuss this observation in the context of the regulation of "target" genes responsible for the differentiation of dorsal tissues.

II. Genetic Characterization of the *zerknüllt* Gene

A. IDENTIFICATION OF THE *zerknüllt* GENE

The *zen* gene was identified by Wakimoto *et al.* (1984) during their efforts to saturate the Antennapedia complex (ANT-C) for embryonic lethal complementation groups. Advanced-stage *zen* mutants show a complex phenotype, including a twisting along the germ band and a failure of head involution. Analysis of mutant embryos during earlier periods of development indicated a failure in germ band elongation and the absence of the optic lobe and amnioserosa, structures that derive from the dorsalmost regions of the fate map. However, these developmental defects were not interpreted as resulting from a breakdown in a dorsal–ventral patterning process. Instead, the loss of the optic lobe and the failure of head involution were thought to reflect a segmentation defect. Further support for this view was based on additional observations, including the demonstration of a "synergistic" head defect phenotype in *zen⁻, fushi tarazu (ftz⁻)* double mutants. Moreover, *zen* was shown to map between the homeotic genes *Deformed* and *proboscipedia,* which act on the maxillary and labial segments of the head. As for the bithorax complex, a colinear correspondence had been shown for the ordering of the homeotic genes within the ANT-C and their domains of function along the anterior–posterior axis of embryos and adults. Thus, given the location of *zen* in a proximal region of the ANT-C, the "colinearity rule" suggested a role in head segmentation.

As we discuss below, recent molecular and developmental studies of *zen* activity suggest that it is not a bona fide segmentation or homeotic gene, but instead functions primarily in subdividing the DV axis into domains of diverse developmental potential. However, it is conceivable that *zen* also participates in the specification of anterior portions of the head, which derive solely from dorsal regions of the embryonic fate map.

B. THE MUTANT PHENOTYPE

Previous genetic studies indicated that *zen* function is required for only a brief period during early development (Wakimoto *et al.,* 1984). The phenocritical period of a temperature-sensitive mutation is transient, extending from about 2 to 4 hours after fertilization. Thus, the complex *zen* terminal phenotype can be explained on the basis of a primary lesion that occurs during early development. Specifically, it appears that there is a transformation in cell fate of the dorsalmost ectoderm toward a more ventral pathway of development in gastrulating *zen⁻* embryos (C. Rushlow, unpublished).

Figure 1 compares the gastrulation phenotype of wild-type (Fig. 1a–e) and *zen* mutant embryos (Fig. 1f–j). The position of the cephalic

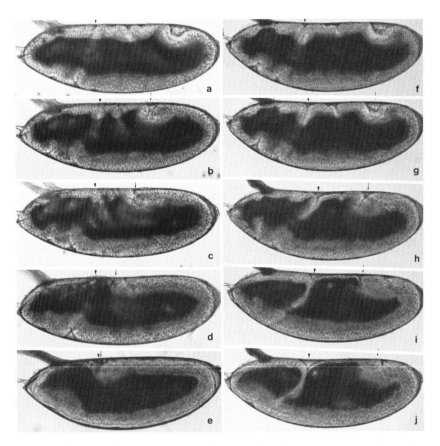

FIG. 1. Gastrulation and germ band elongation in wild-type and *zen⁻* embryos. The embryos are oriented so that anterior is to the left and dorsal is up. (a–e) Time-lapse photomicrographs of a wild-type embryo at successively later stages during germ band elongation. (f–j) Comparable stages for a *zen*^{w36} embryo. The embryos in a and f are at 3–3½ hours postfertilization, and those in e and j have just completed germ band elongation, 5–5½ hours postfertilization. Arrowheads mark the cephalic furrow, while arrows indicate the posterior midgut (PMG), which is at the posterior limit of the germ band. Note the abnormal appearance of the cephalic furrow in the *zen⁻* embryo, as well as the failure of the PMG to expand toward the anterior pole.

furrow has proved to be a useful landmark for identifying shifts in dorsal–ventral cell fates (Anderson and Nüsslein-Volhard, 1984b). In the wild type, the cephalic furrow is derived from an invagination of the lateral ectoderm at 70% egg length (where 0% is defined as the posterior pole). In *zen⁻* embryos, the cephalic furrow arises from a more dorsal region, which is consistent with the idea that the dorsalmost ectoderm has taken on a more ventral fate. Throughout embryonic development the cephalic furrow persists as a deep dorsal fold in *zen⁻* embryos (Fig. 1j). Furthermore, the anterior and posterior transverse furrows (dorsal folds) that normally form on the dorsal surface are virtually absent in *zen⁻* (compare Fig. 1b and g).

During germ band elongation additional abnormalities are observed in *zen⁻* embryos, including a failure of the posterior midgut (PMG) invagination to extend anteriorly along the dorsal surface (compare Fig. 1c–e with h–j). The absence of the amnioserosa might be responsible for these abnormalities, which in turn could cause the severe "twisting" phenotype seen in advanced-stage mutants. Normally, the amnioserosa plays a key role in the elongation process and becomes profusely folded where the posterior tip of the germ band contacts the procephalon and is driven into the proctodeal cavity by the expanding germ band (Campos-Ortega and Hartenstein, 1985). The absence of the amnioserosa might cause the germ band to be twisted and thrown into folds during an abortive elongation process in *zen* mutants.

Support for the ventralization basis of the *zen* phenotype is based on the observation that a similar gastrulation phenotype is observed for embryos which derive from females bearing a dominant allele of the maternal gene *Toll* or females homozygous for a mutant *cactus* allele (Anderson and Nüsslein-Volhard, 1984b; Schüpbach and Wieschaus, 1989). Furthermore, the cuticle phenotype of *zen⁻* embryos is similar to that of weak *cactus* mutants (Fig. 2). Strong *cactus* mutations, like *Toll*^{D} mutations, cause a clear ventralization of the embryo: ventral denticle belts encompass both ventral and dorsal regions of the cuticle in such mutants. An allelic series of progressively weaker mutations ultimately results in a cuticular phenotype quite similar to *zen* (compare Fig. 2b and c).

One problem with this interpretation of the *zen* phenotype, however, is that it is not clear what happens to the transformed cells. The amnioserosa and optic lobe are internal tissues that do not contribute to the external cuticular phenotype. The dorsal lateral regions of the fate map that reside on either side of the *zen* domain give rise to dorsal epidermis and associated cuticular "dorsal hairs" (reviewed in Campos-Oretega and Hartenstein, 1985). If the dorsalmost regions of the fate

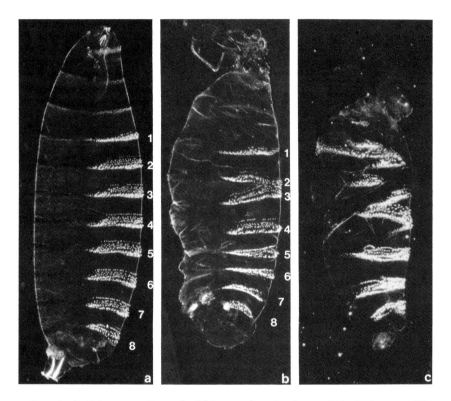

FIG. 2. Cuticle preparations of wild-type and *zen⁻* embryos. Anterior is up, and the ventral lateral regions are displayed. Numbers refer to the eight abdominal segments. (a) Wild-type embryo. (b) Embryo homozygous for the *zen^{w36}* null allele. A correct number of segments is observed, despite occasional disruptions and fusions of some of the denticle belts. Note the failure of head involution and the twisted appearance of the *zen* mutant embryo. (c) Embryo derived from a *cac^{PD74}* female. PD74 is one of the weaker alleles that shows slight ventralization, similar to the phenotype caused by *zen* null mutations.

map differentiate into cell types characteristic of more lateral regions, then advanced-stage *zen⁻* embryos might be expected to possess extra dorsal hairs. It is not clear that this is the case; additional studies will be required to trace the final fate of the transformed cells in *zen* mutants.

The available evidence suggests that *zen* is unique among the homeobox genes in *Drosophila* in that it is not primarily engaged in the process of segmentation or other aspects of AP pattern formation. Mutations in other homeobox genes generally cause disruptions in the segmentation pattern (reviewed in Levine and Harding, 1989). Despite the gross abnormalities seen in advanced-stage *zen⁻* embryos, there is a correct number of body segments, and the cuticular pattern elements associated with each segment appear to be normal.

C. Molecular Cloning and Characterization

zen maps within a proximal region of the ANT-C, which actually contains two closely linked transcription units that contain homeoboxes. Although these units were originally called z1 and z2, it has become clear that z1 corresponds to *zen,* while z2 is a related gene that specifies either a redundant activity or no function at all (Rushlow *et al.,* 1987a; Pultz *et al.,* 1988). *In situ* hybridization studies have shown that *zen* and z2 display virtually identical patterns of expression, which closely match previous genetic studies on the timing and sites of *zen* function.

zen transcripts are among the first zygotically active gene products to appear during embryogenesis, being detected during cleavage cycle 10–11 (Doyle *et al.,* 1986). At this time *zen* shows a broad dorsal on–ventral off pattern of expression, and *zen* products are distributed around 40% of the embryonic circumference in dorsal and dorsal lateral regions. By the onset of gastrulation *zen* products are restricted to the dorsalmost cells, including only 10% of the embryonic circumference. After this refinement in expression, the distribution of *zen* products coincides quite closely with the regions of the embryonic fate map that give rise to the optic lobe and amnioserosa, tissues which are absent in *zen* mutants. The early expression of *zen* transcripts is consistent with the possibility that it is directly regulated by maternal factors deposited in the unfertilized egg.

zen encodes a single, 1.3-kb mRNA which specifies a 39-kDa protein composed of 353 amino acid residues. A full-length *zen* protein was produced in *Escherichia coli* and used to prepare polyclonal antibodies that recognize the native protein in developing embryos (Rushlow *et al.,* 1987b). The distribution of the *zen* protein closely parallels the mRNA pattern, although there is a delay of about 30–45 minutes between the appearance of the mRNA (cleavage stage 10–11) and the first detection of the *zen* protein (cleavage stage 13–14). The protein is restricted to the nuclear regions of cells that express the gene (Fig. 3a), which is consistent with its role in regulating transcription.

The *zen* protein persists during gastrulation and germ band elongation, at which time it is expressed in the differentiating amnioserosa and optic lobe. An unexpected finding is that the *zen* protein also

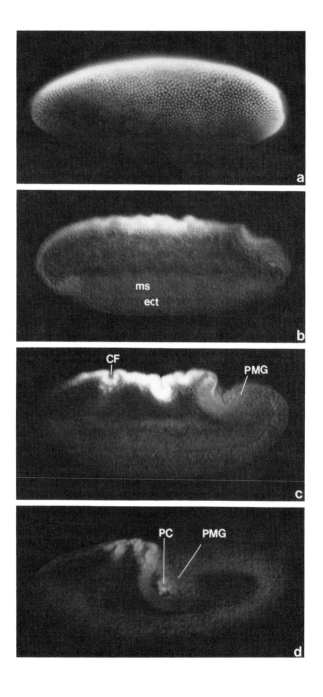

appears in a subset of the pole cells during this time (Fig. 3d). Pole cell transplantation experiments will be necessary to determine whether *zen* functions in these cells. By the completion of germ band elongation the protein disappears and is never again detected during the course of the life cycle. This is unlike the situation observed for a number of early acting segmentation genes, which display novel patterns of expression in advanced-stage embryos (Carroll and Scott, 1985; Frasch *et al.*, 1987; Gaul *et al.*, 1987).

III. Regulation of *zerknüllt* Expression by Maternal Factors

A. ROLE OF THE MATERNAL DORSAL–VENTRAL GENE *Toll*

One of the most interesting aspects of the control of cell fate in the early embryo concerns the problem of how discrete, on–off patterns of zygotic gene expression are established by crudely localized maternal factors, particularly gradients of maternal morphogens. As we discuss below, the regulation of *zen* expression should provide an excellent model system for addressing this issue.

Past genetic studies have identified most or all of the maternally active genes needed for the initiation of the DV pattern, and these genes fall into two classes: (1) early-acting maternal genes, which are responsible for determining DV polarity in the egg chamber during oogenesis (Schüpbach, 1987; Manseau and Schüpbach, 1989); and (2) late-acting genes, which control DV polarity in the early embryo following fertilization (Anderson *et al.*, 1985a,b; Schüpbach and Wieschaus, 1989). The

---

FIG. 3. Wild-type expression of the *zen* protein. Embryos were collected from the *zen*^{w36}/DF (3R)LIN; P-*zen*$^+$ line (Rushlow *et al.*, 1987a), which contains the 4.4-kb *zen* genomic DNA fragment located on the second chromosome that rescues the null phenotype. Whole mount preparations were stained with anti-*zen* antibodies, and the micrographs are oriented so that anterior is to the left and dorsal is up. (a) Precellular cleavage stage 14 embryo. The *zen* protein is broadly distributed along the dorsal half of the embryo and around the anterior and posterior poles. (b) Gastrulating embryo. The *zen* protein is confined to the dorsal surface and is not detected at the poles. No staining is detectable in the mesoderm (ms) or in the ventral ectoderm (ect). (c) Embryo undergoing germ band elongation. Staining is restricted to the differentiating amnioserosa, as well as to other tissues located at the dorsal surface. CF, Cephalic fold; PMG, posterior midgut. (d) Embryo that has completed germ band elongation. The *zen* protein is detected in a subset of the pole cells (PC) within the posterior midgut (PMG) invagination. Staining also persists at low levels in the invaginated amnioserosa.

latter class comprises a total of 12 genes, including 11 "dorsal-group" genes which give similar mutant phenotypes when disrupted. Null mutations in any one of these genes cause the same "dorsalizing" phenotype, whereby all embryonic cells, in both dorsal and ventral regions, follow a dorsal pathway of development. Conversely, as mentioned above, null mutations in the gene *cactus* lead to the formation of ventralized embryos (Schüpbach and Wieschaus, 1989).

It has been proposed that a cascade of interactions among the products of the dorsal-group genes is responsible for the specification of a morphogen gradient that ascribes unique positional identities to the different cells along the DV axis (Nüsslein-Volhard, 1979; Anderson *et al.*, 1985a; DeLotto and Spierer, 1986; Anderson, 1987). According to this model, the ventralmost regions of the early embryo contain the highest concentration of morphogen, with more dorsal regions containing progressively lower concentrations. In the absence of morphogen, which results when any of the 11 dorsal-group genes is eliminated, all cells follow a dorsal pathway, the developmental ground state, by default.

In cross sections, approximately 72 cells make up the circumference of cellular blastoderm stage embryos. An implication of previous proposals is that the ventral-to-dorsal morphogen gradient specified by the dorsal-group genes determines a unique cell fate for each of the 72 cells, depending on the dose of maternal morphogen they come to contain. A more conservative view is that the maternal morphogen establishes localized expression of several zygotic DV genes, which in turn specifies more discrete positional identities along the DV axis. The initial subdivision of the DV axis might involve only four or five distinct cell types, including the following: (1) ventral mesoderm (Simpson, 1983); (2) ventral ectoderm (Mayer and Nüsslein-Volhard, 1988) and the neurogenic domain in ventral lateral regions (Artavanis-Tsakonas, 1988; Campos-Ortega, 1988); (3) dorsal ectoderm regions giving rise to dorsal epidermal structures; and (4) the presumptive amnioserosa, which derives from the dorsalmost regions of the fate map (summarized in Fig. 4).

Anderson has proposed that the dorsal-group gene *Toll* provides the signal for the establishment of the maternal DV morphogen gradient (discussed in Anderson, 1987). There were several reasons for invoking *Toll* as the key gene in the maternal cascade. First, cytoplasmic transplantation experiments indicated that embryos obtained from *Toll⁻* mothers lack inherent DV polarity (Anderson *et al.*, 1985b). When *Toll⁻* embryos are injected with cytoplasm obtained from any portion of wild-type donor embryos, the site of injection defines the location of the

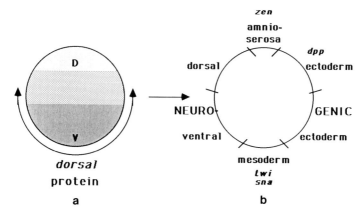

FIG. 4. Zygotic gene response to the maternal dorsal–ventral morphogen *dorsal*. In this model, the *dorsal* protein gradient is depicted by shaded areas with the highest concentration in ventral regions (a). This gradient is interpreted by zygotic genes (b), such as *zen, dpp,* and *twi*, which initially subdivides the DV axis into four or five regions.

ventralmost point of the pattern. Moreover, the entire DV pattern is organized relative to this point. In contrast, the other dorsalizing mutants, including *dorsal,* posses inherent DV polarity. Most of these other mutants can be at least partially "rescued" when injected with wild-type cytoplasm. No matter where the wild-type cytoplasm is injected, ventral regions differentiate into ventral structures and dorsal regions into dorsal structures.

Further support for the importance of *Toll* in establishing a morphogen gradient is the identification of dominant *Toll* (*Toll*^{D}) alleles (Anderson *et al.,* 1985a). While *Toll*$^-$ embryos show the same dorsalizing phenotype as the other dorsal group mutants, *Toll*^{D} mutants display the opposite phenotype. Embryos obtained from mothers that are heterozygous for any one of several *Toll*^{D} alleles are ventralized, such that dorsal regions of the fate map follow a more ventral pathway of development. It has been proposed that *Toll*^{D} alleles specify a constitutively active *Toll*$^+$ product, which promotes the differentiation of ventral structures in both dorsal and ventral regions of mutant embryos.

The nature of the product encoded by *Toll* indicates that it is not the final step in a maternal cascade leading to an active morphogen. *Toll* specifies an integral membrane protein with a large extracytoplasmic domain containing multiple copies of a leucine-rich repeat of 22–26 amino acids (Hashimoto *et al.,* 1988). A protein of similar structure, chaoptin, plays an important role in cell adhesion during the morpho-genesis of *Drosophila* photoreceptor cells (Reinke *et al.*, 1988). Several human proteins, including the α chain of platelet glycoprotein 1b, share structural similarities with the putative *Toll* protein. The extra-cytoplasmic domain of GP1b$_\alpha$ functions as a receptor for thrombin and von Willebrand factor, and the intracellular domain is believed to be associated with actin filaments (Lopez *et al.*, 1987). By analogy to GP1b$_\alpha$, the *Toll* protein may function as a receptor for a diffusible ligand; the binding of this ligand could help define positional identities through cell–cell contact and/or by localizing intracellular maternal factor(s) to appropriate regions along the DV axis of the early embryo. As we discuss below, it is likely that "active" *Toll* products in some way allow *dorsal* proteins to accumulate in ventral regions of early embryos.

B. Is *dorsal* THE MATERNAL MORPHOGEN?

Evidence that *dorsal* encodes the morphogen was first suggested by genetic epistasis studies (Anderson *et al.*, 1985a), which demonstrate that it is one of the last steps in the cascade of interactions among the products specified by the maternal dorsal-group genes (DeLotto and Spierer, 1986; Chasan and Anderson, 1989). Studies on double-mutant combinations suggest that *Toll*^{D} products can promote, to some extent, the formation of ventral structures in the absence of other dorsalizing gene products. Thus, most of the dorsal-group genes appear to influence the DV pattern indirectly, through modulation of *Toll*$^+$ activity. The *dorsal* gene is an important exception to this rule. *Toll*^{D}, *dorsal*$^-$ double mutants display a fully dorsalized phenotype that is indistinguishable from that of *dorsal* mutants. Moreover, there is evidence that *dorsal*$^+$ gene activity becomes progressively restricted to ventral regions of developing embryos. Cytoplasm obtained from any region of a wild-type embryo just after fertilization can rescue *dorsal*$^-$ embryos, but at later stages rescuing activity becomes enriched in ventral regions (Santama-ria and Nusslein-Volhard, 1983). These observations suggest that one of the last, if not the last, steps in the maternal gene cascade is the asymmetric distribution of *dorsal*$^+$ products, which in turn control DV cell identities. This progressive localization does not occur at the level of transcription since *dorsal* RNAs are uniformly distributed throughout the unfertilized egg and embryo, even after *dorsal*$^+$ rescu-ing activity becomes enriched in ventral regions (Steward *et al.*, 1988).

Cloned *dorsal* sequences have been used to prepare polyclonal anti-bodies that selectively recognize the native protein in whole mount preparations of early embryos (Steward *et al.*, 1988). *dorsal* products are first detected during cleavage cycle 11–12. The protein is detected in ventral, not dorsal, regions of early embryos, and it is restricted to the nuclei of expressing cells. A graded distribution of the protein is observed, with peak levels detected on the ventral side and progressively lower levels in lateral and dorsal lateral regions. *dorsal* protein was not detected in the dorsalmost regions of wild-type embryos, which correspond to the initial domains of *zen* expression. The protein was found to persist through cellularization and to disappear rapidly during gastrulation.

The establishment of the ventral-to-dorsal gradient of *dorsal* protein appears to depend on a translational and/or posttranslational process. *dorsal* RNAs are synthesized in the egg chamber and are first detected during previtellogenic stages of oogenesis. As for most of the maternal RNAs stored in the unfertilized egg, *dorsal* RNAs are selectively transported from their site of synthesis in nurse cells to the growing oocyte. These maternal *dorsal* RNAs appear to be translationally arrested since the *dorsal* protein is not detected until well after fertilization.

C. A Regulatory Cascade of Interactions Leads to a Morphogen Gradient

What is the link between *Toll* and other members of the dorsal group genes and the selective appearance of *dorsal* proteins in ventral regions of early embryos? Steward *et al.* (1988) propose that the ventral-to-dorsal gradient of *dorsal* protein involves a selective translation and/or degradation mechanism. Perhaps *dorsal* mRNAs are more efficiently translated in ventral versus dorsal regions of developing embryos. A nonexclusive alternative is that the *dorsal* mRNA is translated throughout the DV axis, but the protein fails to accumulate in dorsal regions because of selective degradation. Null mutations in dorsal group genes might disrupt a selective translation process, thereby leading to an absence of *dorsal* protein. It is also conceivable that these mutations result in the deregulation of a selective degradation process that is normally restricted to dorsal regions. *dorsal* protein might be degraded in both dorsal and ventral regions of such mutants. Previous genetic epistasis studies do not rule out the possibility of parallel processes, whereby both selective translation and protein degradation are important for the distribution of *dorsal* proteins. Perhaps some of the dorsal-group genes participate in selective translation while others are involved in region-specific protein degradation.

In addition to *Toll* and *dorsal*, two other dorsal-group genes have been cloned and characterized, namely, *snake* and *easter* (DeLotto and

Spierer, 1986; Chasan and Anderson, 1989). Interestingly, both of the latter genes appear to encode serine proteases that are structurally similar to the proteases involved in blood clotting (Furie *et al.*, 1982). As for null mutations in each of the other dorsal-group genes, the loss of either of these serine proteases causes an inactivation of *dorsal* function, presumably owing to the lack of *dorsal* protein in both ventral and dorsal regions of mutant embryos.

Until the remaining eight dorsal-group genes are cloned and characterized, it is difficult to envision how a cascade of disparate proteins, such as putative integral membrane protein and serine proteases, lead to the formation of a ventral-to-dorsal gradient of *dorsal* protein. A current guess is that the membrane-bound *Toll* protein is somehow "activated," directly or indirectly, by the serine protease encoded by *easter*. This activation could involve cleavage of one or more of the extracellular leucine repeats. In some way, this activation of *Toll* leads to the selective translation of *dorsal* mRNAs in ventral regions.

It is possible that the ventral activation of *Toll* is brought about by a prelocalization of *easter* activity. Perhaps the *easter* protease is normally active only in ventral regions. This ventral restriction might involve a proteolytic cascade analogous to that in the process of blood clotting, resulting in the selective activation of an inactive precursor form of the *easter* protease that is ubiquitously distributed along the DV axis. Dominant mutations have been identified for *easter* which cause a ventralizing phenotype similar to that observed for certain $Toll^D$ mutants (Chasan and Anderson, 1989). The dominant *easter* mutations might stem from a constitutively active serine protease in both dorsal and ventral regions, leading to the activation of *Toll* in both dorsal and ventral regions. Uniform activation of *Toll* would lead to the translation of *dorsal* mRNAs in both dorsal and ventral regions and, consequently, a ventralizing phenotype.

D. MATERNAL AND ZYGOTIC GENE INTERACTIONS

Whatever the mechanism, once the *dorsal* protein is restricted to the nuclei of ventral cells, it participates in the differential activation and repression of zygotic regulatory genes along the DV axis (see Fig. 4). As discussed above, *zen* is one of at least eight zygotically active DV genes, all of which constitute putative targets for *dorsal* function. Four of these genes have been cloned and characterized, including *zen*, *twist*, *dpp*, and *snail* (Boulay *et al.*, 1987; Padgett *et al.*, 1987; Rushlow *et al.*, 1987a; Thisse *et al.*, 1987a).

A particularly close relationship has been established between two of the genes, *twist* and *snail,* and the maternal gene *dorsal* (Simpson, 1983). Both *twist* and *snail* are required for the differentiation of the ventralmost tissues, including the ventral mesoderm. *twist* encodes a protein that shares homology with the *myc* superfamily of transcription factors, containing the so-called amphipathic helix–loop–amphipathic helix motif which mediates DNA binding and protein dimerization (Thisse *et al.,* 1987b; Murre *et al.,* 1989). Double heterozygotes for *dorsal* and *twist* ($dl^-/+$; $twi^-/+$) show the same lack of ventral structures as observed in *twist* mutants, which suggests that a lower dose of $dorsal^+$ products causes a reduction in the level of $twist^+$ activity.

The genetic interaction between *dorsal* and *twist* (and *snail*) is quite specific, and it is not observed for any of the other dorsalizing maternal mutants. Thus, it appears that one of the ways in which *dorsal* controls DV polarity is by initiating *twist* expression on the ventral side of early embryos. The products of $twist^+$ are normally restricted to the ventralmost regions of gastrulating embryos, and expression is abolished in mutant embryos derived from $dorsal^-$ mothers (Thisse *et al.,* 1987b). *dorsal* might also initiate the expression of *snail,* which encodes a protein that is likely to function as a transcription factor since it contains five copies of the zinc finger DNA-binding motif (Boulay *et al.,* 1987). As discussed below, it is conceivable that *dorsal* not only participates in the activation of *twist* and/or *snail* expression, but functions more intimately in concert with these zygotic products to regulate gene expression via protein–protein interactions.

E. Initiation of *zerknüllt* Expression

The ventral-to-dorsal gradient of *dorsal* protein might contribute to the DV pattern by repressing gene expression, in addition to its role in activating ventrally restricted regulatory genes, such as *twist* and *snail.* As discussed earlier, the initial domain of *zen* expression corresponds to dorsal and dorsal lateral regions where the *dorsal* protein is either absent or present at only low levels. The early, broad dorsal on–ventral off pattern of *zen* expression could involve the direct repression of *zen* by the *dorsal* protein. Null mutations in each of the dorsal-group genes cause the ectopic expression of *zen* in ventral regions (Rushlow *et al.,* 1987b).

Figure 5 compares the transcription pattern of *zen* in embryos derived from wild-type (Fig. 5a) and mutant females (Fig. 5b). Initiation of the pattern is clearly altered. As we discuss below, none of the zygotic DV mutations that have been examined disrupt the normal initiation of the *zen* pattern. Moreover, the early onset of the normal *zen* pattern is

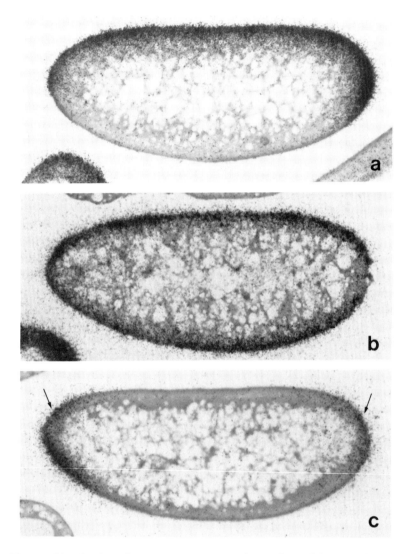

FIG. 5. Distribution of *zen* transcripts in embryos derived from wild-type and mutant females. Embryos of cleavage cycle 11–12 were collected, and sections were hybridized with ^{35}S-labeled *zen* single-stranded RNA probes. (a) Wild-type embryo. (b) *spätzle*[1973]-derived embryo. Equally intense labeling is detected throughout the embryo, in dorsal as well as ventral regions. (c) *cactus*[PD74]-derived embryo. *zen* transcripts are not detected in either dorsal or ventral middle body regions but are expressed at normal levels at the poles (arrows).

consistent with the notion that the gene is directly regulated by maternal factors. *zen* transcripts are first detected during cleavage cycle 10–11, and even at this very early period they are restricted to dorsal and dorsal lateral regions and are absent from ventral regions (see Fig. 5a). This early time of appearance of *zen* products corresponds quite closely to the time when the *dorsal* protein is first detected with antibody probes (Steward *et al.*, 1988). Additional evidence for a relatively direct interaction between *dorsal* and *zen* is the finding that ectopic expression of the *dorsal* protein in dorsal regions can cause a complete repression of *zen* in both dorsal and ventral regions. For example, the ventralizing maternal mutant *cactus* causes an overexpression of the *dorsal* protein so that it is present at high levels in both ventral and dorsal regions (R. Steward, personal communication). This leads to a failure to initiate *zen* expression in either dorsal or ventral tissues (Fig. 5c).

An interesting implication of the *dorsal*-mediated repression of *zen* is that it might involve a threshold response of the *zen* promoter. The *dorsal* protein shows a graded distribution in the dorsal lateral regions where the domain of *zen* expression shows a sharp boundary. It would appear that the *zen* promoter is activated only in cells containing a concentration of *dorsal* protein which is below a certain threshold level. Similar considerations pertain to *decapentaplegic* (*dpp*), another dorsally restricted regulatory gene. Its early domain of expression is quite similar to the initial *zen* pattern (St. Johnston and Gelbart, 1987), which is also likely to involve direct or indirect repression by *dorsal*.

The characterization of the *dorsal* protein suggests that it might regulate the expression of zygotic DV regulatory genes at the level of transcription. *dorsal* exhibits extensive homology with the avian oncogene v-*rel* (Gilmore and Temin, 1986; Steward, 1987). Both *dorsal* and *rel* are located within the nucleus, suggesting a relatively direct modulation of gene expression. The *rel* protein, when introduced into tissue culture cells in transient expression assays, can activate or repress specific mammalian promoters (Gelinas and Temin, 1988). The mechanism of transcriptional regulation is not known, and neither *rel* nor *dorsal* contains a known DNA-binding motif such as a zinc finger or homeobox.

Preliminary studies by K. Han and C. Rushlow (unpublished) suggest that the *dorsal* protein indiscriminately activates the expression of many basal promoters in *Drosophila* tissue culture cells in a non-sequence-specific manner. Moreover, a full-length *dorsal* protein made in bacteria fails to bind *in vitro* specific DNA sequences (C. Rushlow, unpublished), such as the region of the *zen* promoter that is a putative target for *dorsal* function (see below). It is possible that *dorsal* is a promiscuous transcriptional regulatory protein which must interact with sequence-specific DNA-binding proteins in order to modulate the activities of particular promoters. Perhaps the *dorsal* protein interacts with zygotic *twist* and/or *snail* products, which contain known sequence-specific DNA-binding motifs, in order to regulate subordinate genes responsible for the differentiation of the ventral mesoderm. Such protein–protein interactions could explain the tight genetic interaction between *dorsal* and *twist*.

It should be possible to determine how *dorsal* directly or indirectly represses *zen* expression in early embryos. Promoter fusion studies have identified a specific region within the *zen* promoter that is required for transient repression in ventral regions (Doyle *et al.,* 1989). The organization of the *zen* promoter appears to be relatively simple, in that only 1.6 kb of 5′ flanking sequence is required for a fully normal pattern of *zen* expression (Rushlow *et al.,* 1987a; see Fig. 3). This situation differs from the case of the very large promoters of most other homeobox genes; for example, 8 kb of 5′ flanking sequences is not sufficient for normal expression of the pair-rule gene *eve* (Harding *et al.,* 1989). A 200-bp region within the distal half of the *zen* promoter, located between −1.4 and −1.2 kb upstream from the transcription start site, is needed to keep *zen* off in ventral regions during early development. Fusion promoters lacking this sequence are ventrally "derepressed" during early periods of development.

Given the promiscuous regulatory activities of the *dorsal* protein based on transient expression assays, our working model is that early ventral repression of *zen* activity involves the interaction of the *dorsal* protein with a sequence-specific DNA-binding protein. This *dorsal*–factor complex might specifically interact with the distal regulatory element of the *zen* promoter to repress expression. The use of whole nuclear extract binding assays should permit the identification and isolation of the factor(s) that interacts with the distal element. Such biochemical approaches might be needed to supplement previous genetic analyses, in order to identify all the factors responsible for the regulation of a complex developmental promoter such as that of *zen*. If the putative sequence-specific factor is rather ubiquitous and interacts with a variety of promoters, it might be "invisible" to past genetic screens.

In summary, a ventral-to-dorsal gradient of the *dorsal* protein is established in the early embryo by a cascade of interactions among the products encoded by at least 12 maternally active genes. This gradient plays a critical role in the selective expression of different zygotic regulatory genes in discrete regions along the DV axis. Disruption of the *dorsal* gradient alters the expression of the zygotic genes, thereby causing transformations in cell fate. For example, null mutations in most of the maternally active genes result in loss of the *dorsal* protein and ectopic expression of *zen* in ventral regions. *zen* normally participates in the differentiation of dorsal ectodermal tissues, and its misexpression contributes to the "dorsalized" phenotype observed for these mutants. *dorsal⁻*, *zen⁻* double mutants show a less severe phenotype than do *dorsal⁻* embryos (K. Arora, personal communication). Presumably, misexpression of *dorsal* in the maternal ventralizing mutant *cactus* (R. Steward, personal communication) results in the expression of ventral mesoderm genes, such as *twist* and *snail,* over a larger region, thereby transforming dorsal regions of the fate map to more ventral pathways of development.

IV. Regulation of *zerknüllt* Expression by Zygotic Factors

A. Maintenance and Refinement of the *zerknüllt* Pattern

The wild-type pattern of *zen* expression can be divided into two phases: (1) a broad initial pattern, when expression includes 40% of the embryo's circumference in dorsal and dorsal lateral regions, and (2) refinement of the pattern, so that expression becomes restricted to the dorsalmost 10% of the embryo's circumference during gastrulation. This two-tier control of the spatial pattern is similar to the situation observed for a number of segmentation genes, whereby broad, early patterns of expression give rise to sharp stripes during cellularization and gastrulation (Frasch and Levine, 1987). While the early *zen* pattern appears to be subject to maternal control, the refinement of the pattern depends on regulatory interactions with other zygotic DV genes. As discussed below, the second tier refinement process involves proximal regions of the *zen* promoter that are distinct from the distal regulatory element which mediates ventral repression by maternal factors.

We have analyzed the pattern of *zen* expression in many, but not all, zygotic DV regulatory mutants (C. Rushlow, unpublished). At least four genes, *dpp, tolloid (tld), twisted gastrulation (tsg),* and *short gastrulation (sog),* are required for the maintenance and refinement of the *zen* expression pattern during cellularization and gastrulation. The most dramatic disruption of the *zen* pattern is observed in mutants for the *dpp* (Fig. 6) and *tld* genes. Null mutations in *dpp* and *tld* genes cause a

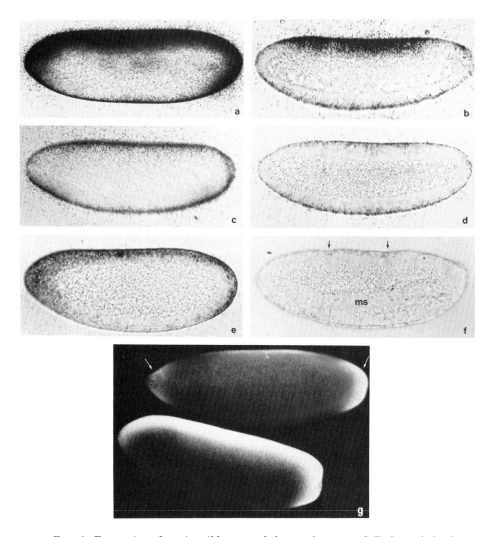

FIG. 6. Expression of *zen* in wild-type and *dpp⁻* embryos. (a–f) Embryos hybridized with *zen* RNA probes, dipped in emulsion, and mounted onto glass slides for
autoradiography. (a) Wild-type precellular, stage 13 embryo. (b) Wild-type cellularized, late stage 14 embryo. (c and e) *dpp⁻* [*dpp*^Hin37/DF (2L)DTD2] precellular, early
stage 14 embryo. An alteration in the *zen* pattern is first observed at this time as a
drastic reduction in the level of *zen* RNAs. All embryos of earlier stages showed the
wild-type pattern. (d) *dpp⁻* cellularized, late stage 14 embryo. (f) *dpp⁻* embryo undergoing gastrulation. By this time, *zen* RNAs are no longer detectable. Note the ventralized phenotype of this embryo: the cephalic fold is shifted dorsally and the dorsal
folds are greatly reduced. ms, Mesoderm. (g) Wild-type and *dpp⁻* (top) embryos
stained with anti-*zen* antibodies. *zen* protein is normally first observed just prior to
cellularization (early stage 14). In *dpp⁻* embryos, *zen* RNA is lost at about this same
time so that *zen* protein is barely detected in the middle body region; however, staining
does appear at the poles.

ventralizing phenotype, similar to that observed for certain maternal mutants, including $Toll^D$ and *cactus* (Jürgens *et al.*, 1984; Nüsslein-Volhard *et al.*, 1984). Ventral denticle belts encircle the circumference of mutant embryos, rather than being restricted to the ventral surface (although strong *tld* mutants show a less severe phenotype than do strong *dpp* mutants). Each of the maternal and zygotic ventralizing mutants shows a nearly complete loss of *zen* expression by the onset of gastrulation. In fact, based on antibody staining, *zen* products are never detected at normal levels in middle body regions of any of these mutants (Fig. 6g). This observation suggested that some of the zygotic DV regulatory genes might be important for the initiation, not just maintenance, of the *zen* pattern. However, *in situ* hybridization experiments indicate that the early, broad *zen* pattern is essentially normal in young (stage 10–13) dpp^- embryos, whereas during cellularization the absence of *dpp* products has a dramatic effect on the maintenance of the *zen* pattern (Fig. 6a–f). As discussed previously, there is a significant lag between the time of appearance of the *zen* mRNA and protein, and, consequently, antibody staining alone is not particularly helpful in addressing the question of initiation.

The regulation of the *zen* pattern by *dpp* and *tld* might involve cell–cell communication mediated by a signal transduction mechanism as *dpp* products have extensive homology with the mammalian transforming growth factor TGF_β (Padgett *et al.*, 1987). Moreover, mosaic analyses have shown that *dpp* gene function is not cell autonomous (Spencer *et al.*, 1982). Together, these results suggest that dpp^+ products are secreted and act on neighboring cells. The initial *dpp* and *zen* expression patterns are quite similar, encompassing about 40% of the embryo's circumference in dorsal and dorsal lateral region (St. Johnston and Gelbart, 1987). It is at this time that *dpp* is required for the maintenance of *zen* expression.

Perhaps the *tld* gene encodes a membrane-bound receptor that interacts with secreted *dpp* products (St. Johnston, 1988). Another candidate for the putative receptor is *screw (scw)*. *scw* mutants exhibit a weak ventralizing phenotype in which the maintenance of the *zen* pattern is also affected (K. Arora, personal communication). The interaction of *dpp* with a receptor present in dorsal regions might lead to enhanced *zen* expression through a signal transduction mechanism. This situation is reminiscent of the interaction between two segment polarity genes, *engrailed (en)* and *wingless (wg)* (DiNardo *et al.*, 1988; Martinez-Arias *et al.*, 1988). *wg* encodes a protein with homology to a secreted mammalian growth factor, called int-1 (Baker, 1987a,b). *wg* and *en* are expressed in neighboring cells, and it has been shown that the loss of *wg* activity causes a premature loss of *en* expression. In the case of *dpp* and *zen,* however, both genes are expressed in a group of cells rather than single cells like *wg* and *en.*

Mutations in the zygotic DV genes *tsg* and *sog* (Wieschaus *et al.,* 1984) result in failure of the refinement process, such that *zen* products do not become restricted to the dorsalmost ectoderm, that is, the presumptive amnioserosa (Fig. 7). *tsg* mutants, like *zen,* show a weak ventralization phenotype; *sog* is a gene which has not yet been classified as a DV gene, but *sog* mutants show defects in gastrulation and germ band elongation (Zusman *et al.,* 1988). Although the effect of the *tsg* and *sog* mutations on the *zen* pattern is similar, it is seen most dramatically in *sog* mutants. At times during germ band elongation when *zen* products are normally restricted to the differentiating aminoserosa (Fig. 7d), *sog* mutants show a greater persistence of the early, broad *zen* pattern (Fig. 7e).

It is possible that the normal loss of *zen* products in dorsal lateral regions (i.e., the presumptive dorsal epidermis) involves cell–cell communication, similar to the *dpp*-mediated maintenance of the *zen* pattern. Mosaic analyses suggest that the *sog* gene is primarily active in ventral, not dorsal tissues (Zusman *et al.,* 1988). Thus, the loss of *sog* function in ventral tissues causes overexpression of *zen* products in dorsal lateral cells.

The cell–cell communication pathway that appears to be responsible for the maintenance and refinement of the *zen* pattern is probably quite distinct from the initiation of *zen* expression during early embryogenesis. In the latter case, a prelocalized transcription factor (*dorsal*) present in the early embryo establishes broad patterns of zygotic gene expression via threshold activation and repression of early target genes. In contrast, the refinement of the *zen* pattern might involve region-specific modulation of general transcription factors that are ubiquitously distributed along the DV axis. Such a mechanism might be more representative of how embryonic cells come to adopt distinct fates during metazoan development. The early *Drosophila* embryo is distinctly unique among most metazoans in containing a number of prelocalized maternal factors such as *bcd* and *dorsal.* In most other systems it appears that the establishment of diverse pathways of morphogenesis involves a gradual series of cell–cell interactions. Thus, the zygotic regulation of *zen* expression would appear to provide an excellent model system for studying a global strategy of embryonic differentiation, namely, the role of cell–cell communication in the specification of cell fate.

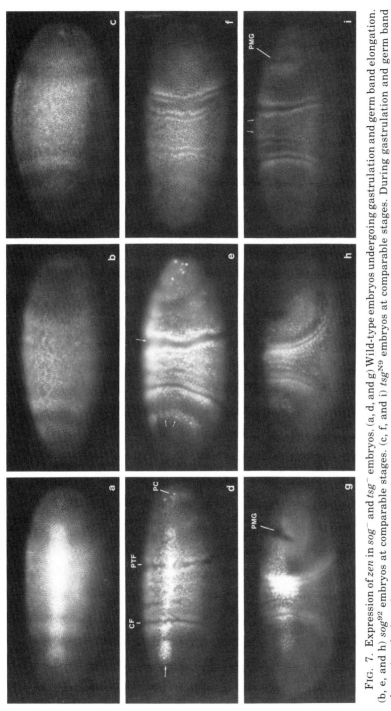

FIG. 7. Expression of *zen* in *sog⁻* and *tsg⁻* embryos. (a, d, and g) Wild-type embryos undergoing gastrulation and germ band elongation. (b, e, and h) *sog⁹²* embryos at comparable stages. (c, f, and i) *tsg^N9* embryos at comparable stages. During gastrulation and germ band elongation the broad *zen* pattern normally becomes localized to a sharp dorsal strip of cells which make up the amnioserosa and optic lobe primordia. This refinement does not occur in *sog* and *tsg* mutant embryos.

B. cis REGULATION

Promoter fusion studies done with *zen* 5′ flanking sequences indicate that the cis-regulatory elements which mediate interactions with the zygotic DV genes are distinct from the distal element responsible for maternal repression (Doyle *et al.,* 1989). The zygotic regulatory sequences that are important for maintaining high levels of *zen* expression during gastrulation and germ band elongation map within the proximal half of the promoter. Promoters lacking these sequences show a normal early *zen* pattern but a premature loss of expression later in development. Two additional regulatory elements have been identified within the proximal promoter. Both are important for repressing expression in dorsal lateral regions and participate in the localization of *zen* products to the dorsalmost ectoderm. Promoters defective in either of these sequences show a persistence of expression in dorsal lateral regions well after the time when *zen* products are normally restricted to the dorsalmost regions.

Thus, there is a close correlation between the cis and trans components of the *zen* expression pattern, based on promoter fusion studies and analysis of *zen* expression in various regulatory DV mutants. The initiation of the *zen* pattern depends on one or more maternal repressors, probably encoded by *dorsal,* that directly or indirectly interact with a distal region of the *zen* promoter. This interaction results in an early; broad dorsal on–ventral off pattern of *zen* expression. *dorsal* may also be responsible for the initiation of other zygotic DV genes, including *twist* and *snail*. These broadly expressed zygotic genes interact with each other to restrict their activities to more discrete spatial boundaries. The zygotic interactions appear to involve primarily cell–cell communication and a signal transduction mechanism. In the case of *zen,* transcription factors present in the early embryo are in some way modulated by these cellular interactions and interact with proximal regions of the *zen* promoter. This regulation of *zen* expression results in the restriction of *zen* products to the dorsalmost ectoderm, which is the primary site of *zen* gene function.

This two-tier mechanism of *zen* expression is similar to the situation observed for the regulation of certain segmentation genes along the AP axis. For example, the segment polarity gene *en* is expressed in a series of 14 stripes; each stripe corresponds to the presumptive posterior compartment and segment border. The initiation of this pattern depends on prelocalized transcription factors, including the homeobox-containing pair-rule genes *eve* (Harding *et al.,* 1986; Macdonald *et al.,* 1986) and *ftz* (Howard and Ingham, 1986). As discussed above, maintenance of the *en* pattern depends on regulatory interactions with other segment polarity genes (i.e., *wg*), which involves a cell–cell signaling process. A possible difference between the regulation of gene expression along the AP versus DV axes might be the time when cell–cell interactions first occur. *dpp⁻* embryos show a premature loss of *zen* expression even before the completion of cellularization, which suggests that interactions occur only about 3 hours after fertilization. In contrast, the *en–wg* interaction is the earliest indication of cell communication along the AP axis, and it is not detected until the completion of germ band elongation (~5 hours after fertilization).

V. *zerknüllt* "Target" Genes

We have primarily restricted this article, to a consideration of how *zen* products come to be restricted to the dorsal ectoderm. Future studies must address the very challenging problem of how localized *zen* products participate in the differentiation of specific cell types such as the amnioserosa. Presumably, the *zen* protein modulates the transcription of "subordinate" genes, which are more directly responsible for cellular phenotype. The number and nature of such subordinate genes are currently unknown. It is conceivable that *zen* directly regulates the many genes responsible for the differentiation of the amnioserosa and optic lobe, structures that are absent in *zen⁻* mutants. Alternatively, *zen* might be the first step in a regulatory cascade involving a series of transcription factors. According to the latter possibility, *zen* coordinates the differentiation of the amnioserosa by regulating the expression of just one or a limited number of transcription factors, which in turn leads to the modulation of the cytodifferentiation genes.

While the identities of the *zen* target genes are currently unknown, considerable information regarding the way in which *zen* controls gene expression has been obtained through the use of a transient expression assay in *Drosophila* tissue culture cells (Han *et al.*, 1989). A full-length *zen* protein made in bacteria binds with high affinity to a 10-nucleotide consensus sequence, TCAATTAAAT. The *zen* protein has been shown to activate strongly the transcription of heterologous promoters containing multiple copies of this sequence. This activation depends on the *zen* homeobox, as well as on a distant region of the protein mapping near the carboxy terminus. The latter activation domain includes a series of serine and threonine residues, which might acquire a net negative charge via phosphorylation. Thus, the *zen* protein appears to be similar to the prototypic yeast transcriptional activators Gal4 and

GCN4 (Hope and Struhl, 1986; Giniger and Ptashne, 1988), which contain a discrete DNA-binding domain and a separable activation domain composed of acidic amino acid residues.

ACKNOWLEDGMENTS

The authors wish to thank Kathryn Anderson, Ruth Steward, and Trudi Schüpbach for providing fly stocks; Eric Wieschaus for the illuminating photomicrographs of gastrulating embryos; Rahul Warrior, Helen Doyle, Rachel Kraut, and Kavita Arora for critical reading of the manuscript and help in gathering reference materials. This work was supported by a grant from the American Cancer Society.

REFERENCES

Akam, M. (1987). The molecular basis for metameric pattern in the *Drosophila* embryo. *Development* **101**, 1–22.

Anderson, K. V. (1987). Dorsal–ventral embryonic pattern genes of *Drosophila*. *Trends Genet.* **3**, 91–97.

Anderson, K. V., and Nüsslein-Volhard, C. (1984a). Information for the dorsal–ventral pattern of the *Drosophila* embryo is stored as maternal mRNA. *Nature (London)* **311**, 223–227.

Anderson, K. V., and Nüsslein-Volhard, C. (1984b). Genetic analysis of dorsal–ventral embryonic pattern in *Drosophila*. *In* "Pattern Formation: A Primer in Developmental Biology" (G. Malacinski and S. V. Bryant, eds.), pp. 269–289. Macmillan, New York.

Anderson, K. V., Jürgens, G., and Nüsslein-Volhard, C. (1985a). Establishment of dorsal–ventral polarity in the *Drosophila* embryo: Genetic studies on the role of the *Toll* gene product. *Cell* **42**, 779–789.

Anderson, K. V., Bokla, L., and Nüsslein-Volhard, C. (1985b). Establishment of dorsal–ventral polarity in the *Drosophila* embryo: The induction of polarity by the *Toll* gene product. *Cell* **42**, 791–798.

Artavanis-Tsakonas, S. (1988). The molecular biology of the *Notch* locus and the fine tuning of differentiation in *Drosophila*. *Trends Genet.* **4**, 95–100.

Baker, N. (1987a). Embryonic and imaginal requirements for *wingless,* a segment polarity gene in *Drosophila*. *Dev. Biol.* **125**, 96–108.

Baker, N. (1987b). Molecular cloning of sequences from *wingless,* a segment polarity gene in *Drosophila*: The spatial distribution of a transcript in embryos. *EMBO J.* **6**, 1765–1774.

Boswell, R. E., and Mahowald, A. P. (1985). *tudor,* a gene required for assembly of the germ plasm in *Drosophila melanogaster*. *Cell* **43**, 97–104.

Boulay, J. L., Dennefeld, C., and Alberga, A. (1987). The *Drosophila* developmental gene *snail* encodes a protein with nucleic acid binding fingers. *Nature (London)* **330**, 395–398.

Campos-Oretega, J. (1988). Cellular interactions during early neurogenesis of *Drosophila melanogaster*. *Trends Neurosci.* **11**, 400–405.

Campos-Ortega, J., and Hartenstein, V. (1985). "The Embryonic Development of *Drosophila melanogaster,*" pp. 31–36. Springer-Verlag, Berlin and New York.

Carroll, S. B., and Scott, M. P. (1985). Localization of the *fushi tarazu* protein during *Drosophila* embryogenesis. *Cell* **43**, 47–57.

Chasan, R., and Anderson, D. V. (1989). The role of *easter,* an apparent serine protease, in organizing the dorsal–ventral pattern of the *Drosophila* embryo. *Cell* **56,** 391–400.

DeLott, R., and Spierer, P. (1986). The gene required for the specification of dorsal–ventral pattern in *Drosophila* appears to encode a serine protease. *Nature (London)* **323,** 688–692.

DiNardo, S., Sher, E., Heemskerk-Jongens, J., Kassis, J. A., and O'Farrell, P. H. (1988). Two-tiered regulation of spatially patterned *engrailed* gene expression during *Drosophila* embryogenesis. *Nature (London)* **332,** 604–609.

Doyle, H. J., Harding, K., Hoey, T., and Levine, M. (1986). Transcripts encoded by a homeobox gene are restricted to dorsal tissues of *Drosophila* embryos. *Nature (London)* **323,** 76–79.

Doyle, H. J., Kraut, R., and Levine, M. (1989). Spatial regulation of *zerknüllt:* A dorsal–ventral gene in *Drosophila. Genes Dev.* **3,** 1518–1533.

Driever, W., and Nüsslein-Volhard, C. (1988a). A gradient of *bicoid* protein in *Drosophila* embryos. *Cell* **54,** 83–93.

Driever, W., and Nüsslein-Volhard, C. (1988b). The *bicoid* protein determines position in the *Drosophila* embryo in a concentration-dependent manner. *Cell* **54,** 95–104.

Driever, W., and Nüsslein-Volhard, C. (1989). The *bicoid* protein is a positive regulator of *hunchback* transcription in the early *Drosophila* embryo. *Nature (London)* **337,** 138–143.

Frasch, M., and Levine, M. (1987). Complementary patterns of *even-skipped* and *fushi tarazu* expression involve their differential regulation by a common set of segmentation genes in *Drosophila. Genes Dev.* **1,** 981–995.

Frasch, M., Hoey, T., Rushlow, C., Doyle, J., and Levine, M. (1987). Characterization and localization of the *even-skipped* protein of *Drosophila. EMBO J.* **6,** 749–759.

Frohnhöffer, H. G., and Nüsslein-Volhard, C. (1987). Maternal genes required for the anterior localization of *bicoid. Nature (London)* **324,** 120–125.

Furie, B., Bing, D. H., Feldmann, R. J., Robison, D. J., Burnier, J. P., and Furie, B. C. (1982). Computer-generated models of blood coagulation factor Xa, factor IXa, and thrombin based upon structural homology with other serine proteases. *J. Biol. Chem.* **257,** 3875–3882.

Gaul, U., Seifert, E., Schuh, R., and Jäckle, H. (1987). Analysis of *Krüppel* protein distribution during early *Drosophila* development reveals posttranscriptional regulation. *Cell* **50,** 639–647.

Gelinas, C., and Temin, H. (1988). The v-*rel* oncogene encodes a cell-specific transcriptional activator of certain promoters. *Oncogene* **3,** 349–356.

Gilmore, M., and Temin, H. (1986). Different localization of the product of the v-*rel* oncogene in chicken fibroblasts and spleen cells correlates with transformation by REV-T. *Cell* **44,** 791–800.

Giniger, E., and Ptashne, M. (1988). Cooperative DNA binding of the yeast transcriptional activator GAL4. *Proc. Natl. Acad. Sci. U.S.A.* **85,** 382–386.

Han, K., Levine, M., and Manley, J. (1989). Synergistic activation and repression of transcription by *Drosophila* homeobox proteins. *Cell* **56,** 573–583.

Harding, K., Rushlow, C., Doyle, J. J., Hoey, T., and Levine, M. (1986). Cross-regulatory interactions among pair-rule genes in *Drosophila. Science* **233,** 953–959.

Harding, K., Hoey, T., Warrior, R., and Levine, M. (1989). Autoregulatory and gap gene response elements of the *even-skipped* promoter of *Drosophila. EMBO J.* **8,** 1205–1211.

Hashimoto, C., Judson, K. L., and Anderson, K. V. (1988). The *Toll* gene of *Drosophila,* required for dorsal–ventral embryonic polarity, appears to encode a transmembrane protein. *Cell* **52,** 269–279.

Hope, I. A., and Struhl, K. (1986). Functional dissection of a eukaryotic transcriptional activator protein, GCN4 of yeast. *Cell* **46**, 885–894.

Howard, K., and Ingham, P. W. (1986). Regulatory interactions between the segmentation genes *fushi tarazu, hairy,* and *engrailed* in the *Drosophila* blastoderm. *Cell* **44**, 949–957.

Ingham, P. W. (1988). The molecular genetics of embryonic pattern formation in *Drosophila. Nature (London)* **335**, 25–34.

Jürgens, G., Wieschaus, E., Nüsslein-Volhard, C., and Kluding, H. (1984). Mutations affecting the pattern of the larval cuticle in *Drosophila melanogaster. Wilhelm Roux's Arch. Dev. Biol.* **193**, 283–295.

Levine, M., and Harding, K. (1989). A regulatory hierarchy governing segmentation in *Drosophila. In* "Frontiers in Molecular Biology, Genes and Embryos" (D. Glover and B. Hames, eds.), pp. 39–94. IRL, London.

Lopez, J. A., Chung, D. W., Fujukawa, K., Hagen, F. S., Papayannopoulou, T., and Roth, G. J. (1987). Cloning of the α chain of human platelet glycoprotein Ib: A transmembrane protein with homology to leucine-rich α₂-glycoprotein. *Proc. Natl. Acad. Sci. U.S.A.* **84**, 5615–5619.

Macdonald, P. M., and Struhl, G. (1989). *Cis*-acting sequences responsible for anterior localization of *bicoid* mRNA in *Drosophila* embryos. *Nature (London)* **336**, 595–598.

Macdonald, P. M., Ingham, P. W., and Struhl, G. (1986). Isolation, structure, and expression of *even-skipped*: A second pair-rule gene of *Drosophila* containing a homeobox. *Cell* **47**, 721–734.

Manseau, L. J., and Schüpbach, T. (1989). *cappucino* and *spire:* Two unique maternal-effect loci required for both the anteriorposterior and dorsoventral patterns of the *Drosophila* embryo. *Genes Dev.* **3**, 1437–1452.

Martinez-Arias, A., Baker, N., and Ingham, P. (1988). Role of segment polarity genes in the definition and maintenance of cell states in the *Drosophila* embryo. *Development* **103**, 157–170.

Mayer, U., and Nüsslein-Volhard, C. (1988). A group of genes required for pattern formation in the ventral ectoderm of the *Drosophila* embryo. *Genes Dev.* **2**, 1496–1511.

Murre, C., McCaw, P. S., and Baltimore, D. (1989). A new DNA binding and dimerization motif in immunoglobulin enhancer binding, *daughterless, MyoD,* and *myc* proteins. *Cell* **56**, 777–783.

Nüsslein-Volhard, C. (1979). Maternal effect mutations that alter the spatial coordinates of the embryo of *Drosophila melanogaster. In* "Determinants of Spatial Organization" (S. Subtelny and I. R. Koenigsberg, eds.), pp. 185–211. Academic Press, New York.

Nüsslein-Volhard, C., and Wieschaus, E. (1980). Mutations affecting segment number and polarity in *Drosophila. Nature (London)* **287**, 795–801.

Nüsslein-Volhard, C., Wieschaus, E., and Kluding, H. (1984). Mutations affecting the pattern of the larval cuticle in *Drosophila melanogaster.* I. Zygotic loci on the second chromosome. *Wilhelm Roux's Arch. Dev. Biol.* **193**, 267–282.

Padgett, R. W., St. Johnston, R. D., and Gelbart, W. M. (1987). A transcript from a *Drosophila* pattern gene predicts a protein homologous to the transforming growth factor-β family. *Nature (London)* **325**, 81–84.

Perrimon, N., Mohler, D., Engstrom, L., and Mahowald, A. P. (1986). X-linked female-sterile loci of *Drosophila melanogaster. Genetics* **113**, 695–712.

Pultz, M. A., Diederich, R. J., Cribbs, D. L., and Kaufman, T. C. (1988). The *proboscipedia* locus of the Antennapedia complex: A molecular and genetic analysis. *Genes Dev.* **2**, 901–920.

Reinke, R., Krantz, D. W., Yen, D., and Zipursky, S. L. (1988). Chaoptin, a cell surface glycoprotein required for *Drosophila* photoreceptor cell morphogenesis, contains a repeat motif found in yeast and human. *Cell* **52**, 291–301.

Rushlow, C., Doyle, H., Hoey, T., and Levine, M. (1987a). Molecular characterization of the *zerknüllt* region of the Antennapedia gene complex in *Drosophila*. *Genes Dev.* **1**, 1268–1279.

Rushlow, C., Frasch, M., Doyle, H., and Levine, M. (1987b). Maternal regulation of *zerknüllt*: A homeobox gene controlling differentiation of dorsal tissues in *Drosophila*. *Nature (London)* **330**, 583–586.

St. Johnston, R. D. (1988). The structure and expression of the *decapentaplegic* gene in *Drosophila melanogaster*. Ph.D. thesis, Harvard University, Cambridge, Massachusetts.

St. Johnston, R. D., and Gelbart, W. M. (1987). *Decapentaplegic* transcripts are localized along the dorsal–ventral axis of the *Drosophila* embryos. *EMBO J.* **6**, 2785–2791.

Santamaria, P., and Nüsslein-Volhard, C. (1983). Partial rescue of *dorsal,* a maternal effect mutation affecting the dorsoventral pattern of the *Drosophila* embryo, by injection of wild-type cytoplasm. *EMBO J.* **2**, 1695–1699.

Schüpbach, T. (1987). Germ line and soma cooperate during oogenesis to establish the dorsoventral pattern of egg shell and embryo in *Drosophila melanogaster. Cell* **49**, 699–707.

Schüpbach, T., and Wieschaus, E. (1989). Female sterile mutations on the second chromosome of *Drosophila melanogaster*. I. Maternal effect mutations. *Genetics* **121**, 101–117.

Simpson, P. (1983). Maternal–zygotic gene interactions during formation of the dorsoventral pattern in *Drosophila* embryos. *Genetics* **105**, 615–632.

Spencer, F. A., Hoffmann, F. M., and Gelbart, W. M. (1982). *Decapentaplegic*: A gene complex affecting morphogenesis in *Drosophila melanogaster. Cell* **28**, 451–461.

Steward, R. (1987). *Dorsal,* an embryonic polarity gene in *Drosophila* is homologous to the vertebrate protooncogene, c-*rel. Science* **238**, 692–694.

Steward, R., Zusman, S. B., Huang, L. H., and Schedl, P. (1988). The *dorsal* protein is distributed in a gradient in early *Drosophila* embryos. *Cell* **55**, 487–495.

Thisse, B., Stoetzel, C., Messal, M. E., and Perrin-Schmitt, F. (1987a). Genes of the *Drosophila* maternal dorsal group control the specific expression of the zygotic gene *twist* in presumptive mesodermal cells. *Genes Dev.* **1**, 709–715.

Thisse, B., Stoetzel, C., Gorostiza-Thisse, C., and Perrin-Schmitt, F. (1987b). Sequence of the *twist* gene and nuclear localization of its protein in endomesodermal cells of early *Drosophila* embryos. *EMBO J.* **7**, 2175–2183.

Wakimoto, B. T., Turner, F. R., and Kaufman, T. C. (1984). Defects in embryogenesis in mutants associated with the Antennapedia gene complex of *Drosophila melanogaster. Dev. Biol.* **102**, 147–172.

Wieschaus, E., Nüsslein-Volhard, C., and Jürgens, G. (1984). Mutations affecting the pattern of the larval cuticle in *Drosophila melanogaster*. III. Zygotic loci on the X chromosome and fourth chromosome. *Wilhelm Roux's Arch. Dev. Biol.* **193**, 296–307.

Zusman, S. B., Sweeton, D., and Wieschaus, E. F. (1988). *short gastrulation,* a mutation causing delays in stage-specific shape changes during gastrulation in *Drosophila melanogaster. Dev. Biol.* **129**, 417–427.

MOLECULAR AND GENETIC ORGANIZATION OF THE ANTENNAPEDIA GENE COMPLEX OF *Drosophila melanogaster*

Thomas C. Kaufman, Mark A. Seeger, and Gary Olsen

Department of Biology, Program in Genetics,
Indiana University, Bloomington, Indiana 47405

The anatomy of the crab is repetitive and rhythmical. It is, like music, repetitive with modulation. Indeed, the direction from head toward tail corresponds to a sequence in time: in embryology, the head is older than the tail. A flow of information is possible, from front to rear. [Gregory Bateson, *Mind and Nature: A Necessary Unity* (1988), p. 10.]

I. Introduction

A current major thrust in the field of developmental genetics is to define the role of genes and their products in the processes of ontogeny. From a slightly different perspective one would like to understand how a three-dimensional organism is encoded in the essentially linear information of the DNA molecule. Investigations of the genetic "regulation" of development actually fall into two overlapping areas of inquiry: those that ask how certain loci are regulated both temporally and

309

spatially and those that focus on the roles of gene products in ontogeny. By studying a system in which both of these elements can be probed, one can hope to gain greater insight into the genetic control of development. Just such a dual system is offered by the analysis of the genes which regulate the process of segmentation in insects. On the one hand, these genes can be seen to be exquisitely regulated both temporally and spatially (Akam, 1987), and, on the other hand, evidence exists that the genes themselves act as regulators and influence the pattern(s) of development subsequent to their initial time of expression (Scott and O'Farrell, 1986).

With this rationale in mind several laboratories have undertaken analyses of the genes that affect the basic pattern of body segmentation in *Drosophila melanogaster*. Initially the genes affecting this process were identified by mutational analyses. The results of these studies have revealed that segmentation genes can be classified into two categories. The first is made up of the loci that affect segmental enumeration. Mutations in these enumeration genes generally cause embryonic lethality, and they are characterized by loss of blocks of contiguous segments (gap genes), missing alternating segments or alternating portions of adjacent segments (pair-rule mutants), or disruptions in the polarity of pattern elements within each segment (Nüsslein-Volhard and Wieschaus, 1980). The second type of alteration is caused by mutations in the homeotic loci. These mutations do not affect segment number but rather transform one segment into the identity of another. On a superficial level it may appear that these two processes, enumeration and identity, are independent. However, evidence is accumulating in favor of the existence of interrelated regulatory networks for these two groups of genes (Struhl and White, 1985; Ingham and Martinez-Arias, 1986; Jackle *et al.*, 1986; White and Lehmann, 1986; Akam, 1987). Therefore, it is not only of interest to understand each on its own terms, but it is of some value to consider the function of each as it may be influenced by the other.

Homeotic mutations map throughout the genome of *Drosophila* (Ouweneel, 1976; Garcia-Bellido, 1977). However, there are two distinct clusters of this class of mutations in the right arm of the third chromosome (Kaufman, 1983; Raff and Kaufman, 1983; Kaufman and Abbott, 1984; Mahaffey and Kaufman, 1987b). The more distal of these is the Bithorax complex (BX-C). Mutations in the BX-C show transformations in the posterior thorax and abdomen of both the embryo and adult (Lewis, 1963, 1978; Duncan and Lewis, 1982; Bender *et al.*, 1983). Within the BX-C there are three subdivisions, *Ultrabithorax (Ubx)*, *abdominal-A (abd-A)*, and *Abdominal-B (Abd-B)*. Lesions in *Ubx* cause transformation in the third thoracic (T3) and first abdominal (A1) segments, whereas *abd-A* and *Abd-B* affect the second through fourth and fifth through eighth abdominal segments, respectively (Sanchez-Herrero *et al.*, 1985). The type of transformations observed are alterations of more posterior (abdominal) segments toward a more anterior (thoracic) identity. This is most dramatically seen in animals deleted for the entire BX-C in which all of the segments from T3 to A8 resemble the second thorax (T2) (Lewis, 1978).

The sum of the developmental and genetic analyses of the BX-C has led to the view that these genes act as ontogenic switches which select between alternate cellular fates and/or developmental programs (Kaufman *et al.*, 1978). Subsequent molecular analyses of the protein products of several of the segmentation genes has supported this view: these molecules are localized to the nuclei in the cells in which they are expressed (White and Wilcox, 1984; Caberra *et al.*, 1985; Carrol *et al.*, 1986a,b; Wirz *et al.*, 1986; Riley *et al.*, 1987). Moreover, these proteins have been shown to bind DNA (Desplan *et al.*, 1985). Thus, the switch nature of these genes is apparently reflected in the production of "regulatory" gene products.

The more proximal of the two homeotic gene complexes is the Antennapedia complex (ANT-C). This group of genes has been the focus of genetic (Denell *et al.*, 1981; Hazelrigg and Kaufman, 1983; Wakimoto *et al.*, 1984; Mahaffey and Kaufman, 1987a,b; Merrill *et al.*, 1987, 1989), developmental (Denell *et al.*, 1981; Wakimoto *et al.*, 1984; Mahaffey and Kaufman, 1987a,b; Merrill *et al.*, 1987, 1989; Mahaffey *et al.*, 1989), and molecular (Scott *et al.*, 1983; Scott and Weiner, 1984; Weiner *et al.*, 1984; Abbott and Kaufman, 1986; Seeger and Kaufman, 1987; Pultz *et al.*, 1988; Diederich *et al.*, 1989) investigations in our laboratory for several years. Our investigations as well as those carried out in other laboratories have shown that the homeotic loci of the ANT-C are involved in the specification of segmental identity in the anterior thorax and posterior head (gnathocephalic) regions of the embryo and adult (Kaufman, 1983; Kaufman and Abbott, 1984; Mahaffey and Kaufman, 1987b). Specifically we have recovered and characterized mutations that affect the development of the intercalary, mandibular, maxillary, labial, and three thoracic segments.

Figure 1 shows the chromosomal location and the genetic map of the ANT-C and the flanking regions. Mutations that inactivate the *Antennapedia* (*Antp*) locus cause transformations of posterior portions of T1 all of T2 and the anterior of T3 to a more anterior identity. This is most strikingly seen in adult flies in which $Antp^-$ cells in the thorax produce head structures (Struhl, 1981; Schneuwley and Gehring, 1985; Abbott

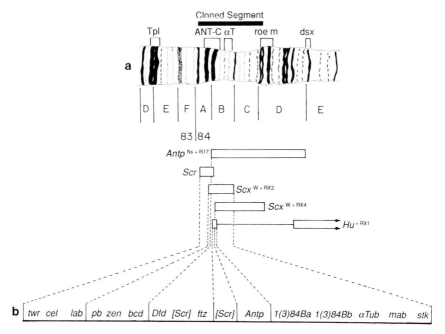

FIG. 1. Genetic map of the Antennapedia complex (ANT-C). (a) Drawing of the portion of the third polytene chromosome on which the ANT-C resides. Open bars below the chromosome show the extent of deletions used to delimit the complex. The genetically defined loci are shown below the deletions: *twisted bristles roughed eye (twr), cell lethal (cel), labial (lab), proboscipedia (pb), zerknüllt (zen), bicoid (bcd), Deformed (Dfd), Sex combs reduced (Scr), fushi tarazu (ftz), Antennapedia (Antp), lethal(3)84Ba* and *lethal(3)84b* [*l(3)84Ba* and *l(3)84Bb*], *α Tubulin (αTub), malformed abdomen (mab),* and *sticking (stk)*. The loci from *lab* to *Antp* inclusive make up the ANT-C. The loci to the left and right are not ostensibly homeotic.

and Kaufman, 1986). Contrasting these loss-of-function mutations at the *Antp* locus are a group of gain-of-function lesions. These mutations produce the opposite effect: thoracic structures (legs) are developed in place of head structures (antennae) (Fig. 2). Thus, the normal function of *Antp* is to repress more anterior head development in the throax, and if this function is inappropriately expressed in the head a thoracic pattern ensues.

Loss-of-function mutations in the *Sex combs reduced (Scr)* locus, on the other hand, cause transformations of T1 toward T2, that is, a posterior directed transformation (Denell *et al.*, 1981; Kaufman and Abbott, 1984; Sato *et al.*, 1985). It is also the case that lesions in the *Deformed (Dfd), proboscipedia (pb),* and *labial (lab)* loci cause adult

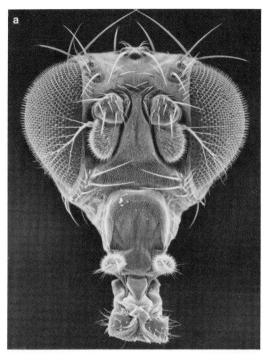

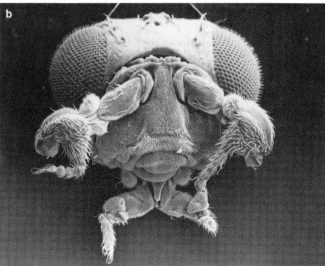

FIG. 2. Scanning electron micrographs of a normal (a) and an $Antp^{73b}$ $pb^4/+$ pb^5 double-mutant fly (b). The $Antp$ mutation transforms the antennae into second thoracic legs, whereas the pb mutation transforms the labial palps into first thoracic legs.

head structures to resemble the thorax (Kaufman, 1978, 1983; Chadwick and McGinnis, 1987; Mahaffey and Kaufman, 1987b; Regulski *et al.*, 1987; Seeger and Kaufman, 1987; Pultz *et al.*, 1988; Merrill *et al.*, 1989). Thus, the ANT-C homeotic loci appear to be playing a dichotomous role: $Antp^+$ is necessary to prevent head development in the thorax, while the other members of the complex are necessary to promote head development and/or repress a thoracic pattern. These genes therefore appear to be exclusionary and antagonistic in their action. What is not clear is the level at which this exclusion takes place and whether other loci are involved in the selection of head versus trunk. What is apparent, however, is that the homeotic loci of the ANT-C are playing both positive and negative roles in the selction process.

Some evidence for the nature of the interaction among components of the ANT-C favoring the possibility that it may be direct comes from the analysis of *Antp* expression in animals carrying lesions in the BX-C. Normally *Antp* does not play a role in the specification of abdominal identity. However, early in embryogenesis there is accumulation of *Antp* RNA and protein in the posterior metameres (Hafen *et al.*, 1984b; Martinez-Arias, 1986). In normal animals this accumulation is transient and by mid-embryogenesis can no longer be detected. In BX-C⁻ animals, however, *Antp* gene products are detected into the later stages of embryogenesis in all of the abdominal segments (Hafen *et al.*, 1984b). Therefore, one of the apparent functions of the BX-C is to "repress" the effect of $Antp^+$ in the posterior end of the animal. The *Antp* locus has a like effect on the expression of *Scr*. The protein product of the *Scr* locus is normally found only in the labial and T1 segments (Mahaffey and Kaufman, 1987a,b). In $Antp^-$ animals, however, *Scr* protein can be found more posteriorly in T2 and T3 (Riley *et al.*, 1987).

Taken together the genetic and molecular results indicate that regulatory hierarchies exist among the different homeotic loci of the ANT-C in the trunk of the animal and that these hierarchies result in the establishment of domains of expression, some of which are clearly exclusive. The presently favored hypothesis is that this exclusivity results from a direct interaction between the product of one locus and the "regulatory" sequences of the other. The above observations are certainly consistent with this interpretation, and direct proof is now being obtained which appears to largely substantiate this view. Therefore, it is quite plausible that this type of cross-regulatory hierarchy or a similar mechanism is responsible for the exclusionary behavior observed in the head versus trunk dichotomy among the ANT-C lesions.

The ANT-C is distinguished from the BX-C not only by its spatial domain of action but by virtue of the fact that it contains loci that are not ostensibly homeotic in nature. Two of these, *fushi tarazu* (*ftz*) and *zerknüllt* (*zen*), have been analyzed extensively and have been shown to affect segment enumeration in the pair-rule category (*ftz*) and to specify certain dorsal structures (*zen*) in the early embryo (Wakimoto *et al.*, 1984). A third nonhomeotic locus is *bicoid* (*bcd*) (Frohnöfer and Nüsslein-Volhard, 1987). Mutations in *bcd* result not directly in zygotic lethality but rather in female sterility. The sterility results from the inability of embryos from *bcd⁻* mothers to produce a normal anterior–posterior polarity. The embryos fail to develop heads, thoraces, and anterior abdominal segments but instead produce structures normally associated with the most posterior (caudal and A8) segments of the embryo in reverse. Interestingly, all three of these genes contain the homeobox (Gehring and Hiromi, 1986) as do the aforementioned homeotic loci. As might be expected, the protein products of the *zen* and *ftz* loci are found in the nuclei of cells expressing these genes (Hiromi *et al.*, 1985; Hiromi and Gehring, 1987; Rushlow *et al.*, 1987a,b).

In addition to the above-mentioned loci, which have been identified by both genetic and molecular means, several other transcription units in the ANT-C have been identified at the molecular level. The first of these is a cluster of what appear to be cuticle-related genes between the *lab* and *pb* loci (see Fig. 12). Analysis of these eight transcription units has revealed sequence similarities to known cuticle genes (Pultz *et al.*, 1988). This same cluster of transcription units has been identified in a screen for ecdysome-regulated loci expressed in imaginal discs (Fechtel *et al.*, 1988). All of the transcripts show sequence similarity (Pultz, 1988) and have apparent homologs elsewhere in the genome (D. Cribbs and M. Pultz, unpublished). Whether these "genes" are homologs of true cuticle genes remains to be shown. What we do know, however, is that all eight of these transcription units can be deleted without affecting the viability of the fly (Pultz *et al.*, 1988). Therefore, the functional significance of this cluster of transcription units remains elusive except to note that a similar cluster of cuticle-like genes is resident in the same position of the ANT-C of *Drosophila pseudoobscura* (P. Randazzo, personal communication). Since *D. pseudoobscura* and *D. melanogaster* diverged approximately 45 million years ago (Beverly and Wilson, 1984) it would appear that some selective force is maintaining this group of transcription units. Whether selection is acting on the cuticle-like genes directly or on the chromosomal neighborhood remains to be demonstrated. The second molecularly identified "gene" is *amalgam* (*ama*) which, like the cuticle cluster units, contains no homeobox (Seeger *et al.*, 1988).

The remainder of this article is concerned with the functional organizatinn of the ANT-C. We first consider the nonhomeotic members of the complex and then the homeotics themselves. It is the former set of loci, *zerknüllt* (*zen*), *bicoid* (*bcd*), *amalgam* (*ama*), and *fushi tarazu* (*ftz*) that most strikingly distinguishes the ANT-C from the BX-C. Analysis of deletions for all or portions of the BX-C have not revealed the existence of similar loci within the complex (Lewis, 1978; Sanchez-Herrero *et al.*, 1985), nor have molecular investigations shown the presence of such independent transcripts (Bender *et al.*, 1983; Karch *et al.*, 1985). Four of the distinguishing transcription units are found in roughly the middle of the ANT-C, while the fifth (*ftz*) is found more distally and embedded in the regulatory elements of the *Scr* locus.

II. *zerknüllt–bicoid–amalgam* Interval

The *zen–bcd–ama* central region of the ANT-C is genetically defined by a set of five deletions which specify its distal and proximal limits in the genome (Fig. 3). Moreover, the complementation behavior of these deletions has shown that two mutationally defined loci, *zen* and *bcd*, are located here. It was therefore somewhat surprising when Northern blots using genomic fragments from this interval revealed the presence of four nonoverlapping small transcription units (Pultz *et al.*, 1988;

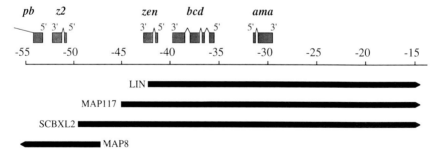

FIG. 3. Molecular map of the *zen–bcd* interval of the Antennapedia complex. The coordinate line is divided into 5-kb intervals and is an extension of the walk initiated by Scott *et al.* (1983) with the *Dfd* locus at a 0 point. Negative values extend proximally along the chromosome. Stippled boxes above the line indicate the relative sizes and positions of the exons of the transcription units; the direction of transcription is also noted for each. Bars below the coordinate line indicate the extent of deletions used to delimit the *zen–bcd* interval. *pb*, *proboscipedia*; *z2, zen2*; *zen, zerknüllt*; *bcd, bicoid*; *ama, amalgam*. This diagram is drawn to scale with others like it in this article (Figs. 9–13).

Seeger, 1989) (Fig. 3). The assignment of two of the transcripts to identified genetic functions has been accomplished by P-element-mediated germ-line transformation. A 4.5-kb *XbaI–SacI* fragment from the *zen* genomic region completely rescues the lethality associated with a *zen* null mutation (Rushlow *et al.*, 1987a,b), and an 8.8-kb *Eco*RI fragment encompassing the *bcd* transcription unit rescues the female sterility associated with *bcd* mutations (Berleth *et al.*, 1988). Neither of these two fragments contains the *z2* or *ama* transcription units (Fig. 3) in their entirety, and therefore neither is necessary for normal *zen* or *bcd* function.

A. *zerknüllt* AND *zen2* LOCI

The *zerknüllt* (*zen*) gene, an embryonic lethal, was identified during saturation mutagensis utilizing two deletion-bearing chromosomes for lethal and visible mutations (Lewis *et al.*, 1980a,b; Wakimoto *et al.*, 1984). Mutations at *zen* result in the loss of several dorsally derived embryonic structures, including the amnioserosa and optic lobe, as well as disruption of germ band extension and the process of head involution (Wakimoto *et al.*, 1984). Examination of an allelic series of *zen* mutations, from weak hypomorphs to amorphic mutations, indicates that the optic lobe and amnioserosa are the structures most sensitive to the loss of *zen*$^+$ activity.

The time of *zen* function, as defined by the analysis of a temperature-sensitive allele, is from 2 to 4 hours of embryogenesis (Wakimoto *et al.*, 1984). Additionally, by using X-ray-induced somatic crossing-over to create *zen*$^-$ clones of cells in adult tissues, it was possible to demonstrate that *zen* activity was unnecessary for postembryonic development (Wakimoto *et al.*, 1984). Observations on the spatiotemporal distribution of *zen* transcript and protein show an excellent correlation with the above developmental and genetic results (Doyle *et al.*, 1986; Rushlow *et al.*, 1987a,b; Seeger, 1989).

The *zen* transcription unit produces a 1300-nucleotide poly(A)$^+$ mRNA that accumulates during early embryogenesis (Rushlow *et al.*, 1987a; Seeger, 1989). The *zen* mRNA is first detected in 0–2 hour embryo collections, peaks at 2–4 hours, and is present at much lower levels in 4–6 hour embryos. *In situ* hybridization has been used to expand these Northern analyses as well as to define the spatial pattern of expression (Fig. 4). The first accumulation of *zen* mRNA is detected during the eleventh to twelfth cell cycle at the syncytial blastoderm stage. By cell cycle 13 substantial *zen* mRNA accumulation is evident (Fig. 4A and B). Prior to and during cell cycle 13, *zen* mRNA is localized

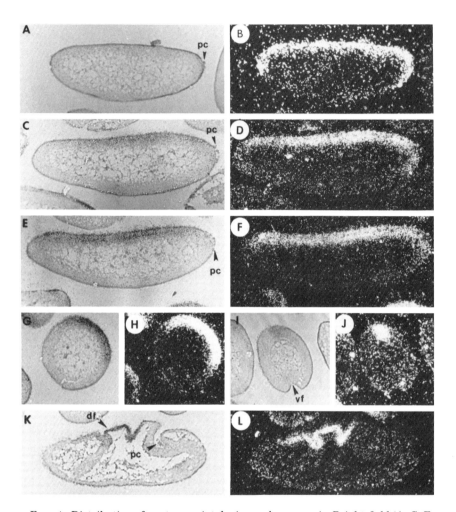

FIG. 4. Distribution of *zen* transcript during embryogenesis. Bright-field (A, C, E, G, I, and K) and corresponding dark-field (B, D, F, H, J, and L) micrographs show sectioned embryos hybridized with a ^{35}S-labeled *zen* mRNA probe. Where appropriate, anterior is left and dorsal is up. (A and B) Sagittal section through a cell cycle 13 embryo. Note the pole cells (pc) at the posterior end. Accumulation of *zen* mRNA is seen over the dorsal surface and around both ends of the embryo. (C and D) Sagittal section through a mid-cycle 14 embryo. Note the relative absence of *zen* mRNA at the anterior end. (E and F) Sagittal section through an older cycle 14 embryo. (G and H) Cross section through an early cycle 14 embryo. (I and J) Cross section through a late cycle 14 embryo where the ventral furrow (vf) has already formed. (K and L) Sagittal section through an early germ band extension stage embryo. Accumulation of *zen* mRNA is seen over the dorsal folds (df).

to the dorsal surface of the embryo and extends around both the anterior and posterior poles as well.

During cell cycle 14 a series of restrictions in the early pattern along the anteroposterior and dorsoventral axes are evident. First, *zen* mRNA no longer accumulates around the anterior pole of the embryo (Fig. 4C and D). Subsequently, *zen* expression around the posterior pole diminishes (Fig. 4E and F). The expression along the dorsoventral axis also becomes more restricted. During early cell cycle 14 *zen* mRNA accumulates over the dorsalmost 30% of the embryonic circumference (Fig. 4G and H), but by the end of cell cycle 14 only the dorsalmost 5% of the cells continue to accumulate *zen* mRNA (Fig. 4I and J). The dorsal cells that continue to accumulate *zen* mRNA are included in the dorsal folds during the process of germ band extension (Fig. 4K and L). These cells will give rise to the amnioserosa as determined by various fate mapping studies (see Campos-Ortega and Hartenstein, 1985). Accumulation has also been observed in a subset of pole cells at the germ band extension stage, but this expression is quite variable among different embryos.

The accumulation of *zen* protein has also been determined using antisera directed against the polypeptide. As mentioned above, examination of embryos stained with the *zen* antisera illustrates the nuclear localization of the *zen* protein product, which is consistent with the presence of a homeobox domain within this protein (Rushlow *et al.*, 1987b; Seeger, 1989).

The first accumulation of *zen* protein is observed over the dorsal surface as well as around the anterior and posterior poles of the embryos during cell cycle 13. The dorsal expression encompasses approximately 30% of the embryonic circumference. This pattern is maintained until early cell cycle 14, after which a series of dynamic changes in the spatial pattern of *zen* protein accumulation are observed that parallel the spatial restriction noted in the *in situ* analysis cited above. These dynamic changes along the dorsoventral and anteroposterior axes culminate in the formation of a middorsal stripe of *zen* accumulation that is roughly 7 cells wide and 70 cells in length at the end of stage 14 just prior to gastrulation.

As gastrulation and germ band extension proceed, the dorsal cells accumulating *zen* protein which lie posterior to the cephalic furrow are dislocated laterally down the sides of the embryo in front of the advancing germ band. These cells will give rise to the amnioserosa, a structure which will continue to express *zen* protein until the end of germ band extension. The cells anterior to the cephalic furrow that stain with *zen* antisera are primordia for the optic lobe and dorsal ridge (see Campos-Ortega and Hartenstein, 1985) and continue to react with *zen* antisera through late germ band extension. Pole cells often begin to accumulate *zen* protein at the start of germ band extension, although this expression is quite variable, with accumulation occurring in 0 to as many as 10 pole cells per embryo.

This elaborate choreography has been shown to be at least partially "regulated" by the set of maternally encoded gene products that specify dorsal–ventral and anterior–posterior polarity in the early embryo (Rushlow *et al.*, 1987b; Seeger, 1989). The nature of the observed interactions and the specifics of *zen* regulation are covered in greater detail in another article in this volume (Rushlow and Levine).

The basic structure of the *zen* locus is shown in Fig. 3. The entire transcription unit is just over 1340 bp in length. There are two exons separated by a 64-bp intron. The open reading frame initiates in the first exon 52 bp downstream of the start of transcription (Rushlow *et al.*, 1987a). In addition to the homeobox domain, the *zen* protein product contains several regions that are enriched for the amino acids serine, threonine, proline, and glutamic acid. These sequences, called PEST regions, have been correlated with proteins that exhibit very short half-lives (Rogers *et al.*, 1986; Rechsteiner, 1987). The presence of such domains within the *zen* protein is entirely consistent with the dynamic changes in the patterns of *zen* protein accumulation seen over very brief periods of time.

The *zen2* (*z2*) locus is located about 10 kb proximally on the chromosome from *zen* (Fig. 3) (Rushlow *et al.*, 1987a; Pultz *et al.*, 1988). The transcription unit is just over 1 kb in length and contains two exons separated by a 67-bp intron. The open reading frame starts in the first exon, and the encoded protein contains a homeobox (Rushlow *et al.*, 1987a; Pultz *et al.*, 1988). The *z2* gene has a pattern of expression nearly identical to that of its neighbor *zen* (Rushlow *et al.*, 1987a,b; Pultz *et al.*, 1988); however, no mutations corresponding to *z2* have been isolated, nor does a deficiency of *z2* exhibit any detectable phenotype (Pultz *et al.*, 1988). Thus, the function of *z2* remains moot. Interestingly, an examination of the homologous interval in *Drosophila pseudoobscura* reveals the presence of a *zen* homolog and the 5′ end of *proboscipedia* but no evidence for a *z2*-like transcription unit (P. Randazzo, personal communication).

B. *bicoid* LOCUS

Mutations in the maternal-effect gene *bicoid* (*bcd*) were isolated in the laboratory of Nüsslein-Volhard during saturation mutagenesis of the *Drosophila* genome for maternal-effect mutations that disrupt the anteroposterior or dorsoventral pattern of the early embryo (Nüsslein-

Volhard *et al.*, 1987). Additionally, two hypomorphic *bcd* alleles were recovered by Lambert (1985) in our laboratory during a screen for female steriles residing within the ANT-C. Subsequently, two strong hypomorphic alleles at the locus were recovered in a screen for additional *bcd* mutations (Seeger, 1989).

As noted earlier, embryos produced from females homozygous for a strong *bcd* mutation exhibit a deletion of all head, thoracic, and anterior abdominal segments with a concomitant duplication of telson structures at the anterior end (Frohnhöfer and Nüsslein-Volhard, 1986; Seeger, 1989). Examination of an allelic series of hypomorphic mutations reveals a graded requirement for *bcd* activity along the anteroposterior axis (Frohnhöfer and Nüsslein-Volhard, 1986). Deletion of anterior head structures is associated with all *bcd* mutations, while further deletions posteriorly are correlated with increasing severity of the mutant allele.

The time of *bcd* function has been determined through the analysis of a temperature-sensitive allele of *bcd*. The experiments indicate that the *bcd* product is required from the time of pole cell formation through cellularization of the blastoderm embryo (Frohnhöfer and Nüsslein-Volhard, 1987).

Finally, cytoplasmic transplantation experiments have led to two interesting observations (Frohnhöfer and Nüsslein-Volhard, 1987). First, the ability of wild-type cytoplasm to rescue phenotypically embryos from *bcd⁻* females is greatest when cytoplasm is removed from the very anterior end of the donor embryo. Cytoplasm taken from 0 to 75% egg length (0% egg length is the posterior pole) does not rescue the *bcd⁻* phenotype, suggesting a highly localized source of functional *bcd* product. Second, donor cytoplasm taken from the anterior end is capable of "inducing" some anterior development regardless of the site of injection in the recipient embryo.

These experiments and others led Frohnhöfer and Nüsslein-Volhard (1986, 1987) to propose that the *bcd* product is localized anteriorly in the form of a protein gradient which then directs the spatially restricted expression of a subset of zygotic segmentation genes. This link between maternal positional information and zygotic gene regulation has recently been demonstrated for *bicoid* and for *hunchback*, a gap class segmentation gene that is required for formation of the labial and thoracic segments (Bender *et al.*, 1987; Lehmann and Nüsslein-Volhard, 1987). Driever and Nüsslein-Volhard (1989) have shown that the *bicoid* gene product, a homeobox-containing protein, binds to *hunchback* regulatory regions and is necessary for the activation of *hunchback* transcription.

The organization of the *bcd* transcription unit (Fig. 3) has been rcvealed by the recovery and sequence analysis of cDNA and genomic clones as well as S1 protection and primer extension studies (Berleth *et al.*, 1988; Seeger, 1989). The sequence data indicate that the *bcd* open reading frame extends from the first through the fourth exon.

Two different *bcd* protein products of 489 and 494 amino acids are produced from the *bcd* transcription unit owing to the use of two different acceptor splice sites at the beginnig of the third exon. Both acceptor splice sites, which are 15 bp apart, are in good agreement with the appropriate consensus sequence. Interestingly, this five-amino acid difference occurs just 5' of the *bcd* homeobox. Similar alternate splicing events just upstream of homeoboxes have also been noted for *Ultrabithorax* (O'Connor *et al.*, 1988), *Antennapedia* (Bermingham and Scott, 1988; Storeher *et al.*, 1988), and *labial* (Mlodzik *et al.*, 1988).

The *bcd* protein product can be organized into at least four discrete domains (Berleth *et al.*, 1988; Seeger, 1989). The first domain is characterized by the repeating combination of amino acid residues His and Pro. Although the function of such a repeat is unclear, it has been found in other *Drosophila* proteins as well (Frigerio *et al.*, 1986). The second domain is the homeobox and represents a DNA-binding domain within the *bcd* protein. As noted previously, Driever and Nüsslein-Volhard (1989) have demonstrated that the *bcd* protein binds to regulatory regions of the *hunchback* gene in a sequence-specific manner. Third, a highly acidic domain is found near the carboxy-terminal end of the *bcd* protein. Driever and Nüsslein-Volhard (1989) have proposed that this acidic domain may provide a transcriptional activator function of the *bcd* protein. The spatial organization of the acidic and homeobox domains is analogous to the separation of DNA-binding and transcription-activating functions that has been found in other eukaryotic transcriptional regulatory proteins (Ma and Ptashne, 1987; Kakidani and Ptashne, 1988). The fourth and last domain of the *bcd* protein is represented by the stretches of poly-Gln or opa (Wharton *et al.*, 1985) that are found in the central regions.

The *bcd* transcription unit produces a 2600-nucleotide poly(A)$^+$ mRNA that accumulates predominantly during early embryogenesis and in adult females. The other minor transcripts, one of approximately 3000 nucleotides which shows the same temporal pattern as the major product and a 1600-nucleotide mRNA which accumulates throughout development (Berleth *et al.*, 1988), are also observed. The possible biological significance of these minor transcripts is not yet clear.

In situ hybridization has revealed that the *bcd* mRNA product is present at the start of embryogenesis and exhibits a striking localization to the anterior pole of the embryo (Fig. 5). In cleavage-stage embryos, *bcd* mRNA accumulation forms a "cap" around the anterior pole (Fig. 5A–D). As nuclei migrate to the periphery during cell cycles 8 through 10 (Foe and Alberts, 1983), the *bcd* mRNA product also becomes concentrated in the cortex (Fig. 5E and F). At the completion of cellularization during cell cycle 14, *bcd* mRNA accumulation has decreased to barely detectable levels (Fig. 5G and H). No accumulation of *bcd* mRNA has been detected at latter stages of embryogenesis.

The asymmetric distribution of *bcd* mRNA product is established during oogenesis when *bcd* mRNA is transcribed in the nurse cells and transported to the oocyte where it becomes "trapped" at the anterior end (Fig. 5I and J). The first accumulation of *bcd* mRNA is detected as early as a stage 8 ovarian chamber (Fig. 5K and L) (King, 1970). In a further analysis, Macdonald and Struhl (1988) identified a cis-acting 625-nucleotide fragment within the 3′ untranslated region of the *bcd* mRNA that is both necessary and sufficient for the observed localization of the *bcd* mRNA product. In addition, the products of two maternal-effect genes, *swallow* and *exuperentia*, were shown to be required *in trans* for proper localization of *bcd* mRNA (Frohnhöfer and Nüsslein-Volhard, 1987). This is demonstrated by the fact that in a *swallow* or *exuperentia* mutant background *bcd* mRNA is more homogeneously distributed throughout the oocyte and early embryo (Frohnhöfer and Nüsslein-Volhard, 1987).

Using antisera to the *bcd* protein, Driever and Nüsslein-Volhard (1988a) found that the *bcd* polypeptide is first observed in the nuclei of cleavage-stage embryos, where it forms a gradient with the highest concentrations located at the anterior end of the embryo. The gradient of protein, however, extends more posteriorly in the animal than does the "trapped" RNA (Driever and Nüsslein-Volhard, 1988a). Alterations of the fate map induced by changes in *bcd* gene dosage or the use of different mutant backgrounds (i.e., *swallow*) can be correlated with corresponding changes in the concentration and shape of the *bcd* protein gradient (Driever and Nüsslein-Volhard, 1988b). These results indicate that anterior positional values are specified by the *bcd* protein product in a concentration-dependent manner. As mentioned previously, a direct role of the *bcd* protein in the specification of anterior identity has been shown by the demonstration that the *bcd* protein acts as a positive regulator of transcription for the zygotic gap gene *hunch-back* (Driever and Nüsslein-Volhard, 1989). These and other experimental results indicate that the *bicoid* protein gradient directly specifies the anterior domain of *hunchback* expression during early embryogenesis, and they have largely substantiated the model of

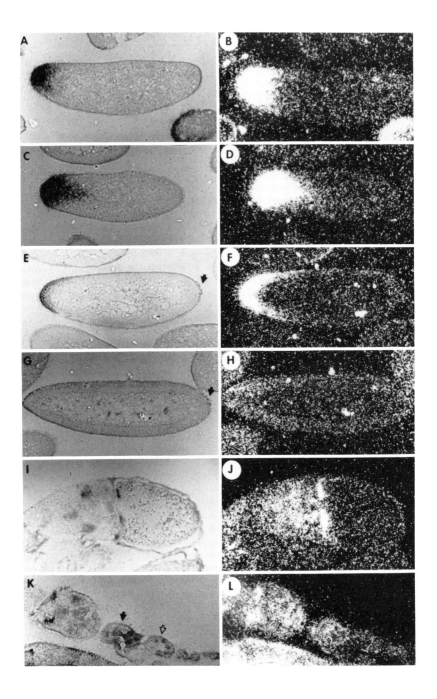

Frohnhöfer and Nüsslein-Volhard (1986, 1987) drawn from genetic and developmental analyses.

C. *amalgam* Locus

The last identified transcription unit in the central region of the ANT-C is *amalgam* (*ama*) (Fig. 3). The *ama* locus is like the cuticle cluster and *z2* loci in that we have not recovered any mutations associated with it, but it is unlike the majority of the genes in the complex (including *z2*) in that the *ama* locus does not contain a homeobox. We have been able to demonstrate, however, that the transcript of the gene accumulates in a highly regulated manner (spatially and temporally) and that an *ama* protein appears to be associated with the membranes of the cells accumulating this product. The latter observation is significant in light of the striking sequence similarity this protein exhibits to members of the immunoglobulin superfamily, a diverse collection of proteins that are involved not only in the immune response but also in various aspects of cell surface recognition (see Williams and Barclay, 1988, for a review).

Using sequence analysis of cDNA and genomic fragments along with S1 protection and primer extension studies we have shown that the *ama* transcription unit is just over 1.8 kb in length and contains two exons (Seeger *et al.*, 1988). The direction of transcription is opposite to that of the other three members of the region (Fig. 3). The sequence data indicate the presence of a long open reading frame which initiates just downstream of the splice site in exon 2 and terminates 281 nucleotides

---

FIG. 5. Spatial distribution of *bcd* transcript. Bright-field (A, C, E, G, I, and K) and corresponding dark-field (B, D, F, H, J, and L) micrographs show paraffin-sectioned embryos and adult female ovaries hybridized with a ^{35}S-labeled *bcd* RNA probe. Embryonic stages are as described by Foe and Alberts (1983), while stages of oogenesis are as described by King (1970). Anterior is left and dorsal is up where appropriate. (A and B) Sagittal section through a cleavage-stage embryo. (C and D) Lateral parasagittal section through a cleavage-stage embryo to illustrate the "caplike" nature of the *bcd* mRNA accumulation pattern. (E and F) Sagittal section through a cell cycle 10 embryo. The arrow points to the pole cells. (G and H) Sagittal section through a cellular blastoderm embryo. The *bcd* mRNA accumulation is barely detectable. The arrow points to the pole cells. (I and J) Section through a stage 10B ovarian chamber. Accmulation of *bcd* mRNA is seen in the nurse cells; in addition, there is a tight band of localization at the anterior end of the oocyte. (K and L) Section through several ovarian chambers at various stages of maturation. The open arrow points to a stage 7 chamber where no *bcd* mRNA accumulation is observed. The solid arrow points to a stage 8/9 chamber where strong accumulation is observed.

upstream of a putative polyadenylation sequence. The identified ORF has a coding potential of 333 amino acids.

The *ama* protein can be divided into five domains based on the analysis of its sequence (Seeger *et al.*, 1988). The first domain includes the amino-terminal 23 amino acids and has the standard characteristics of signal sequences common to secreted and membrane-bound proteins. Consistent with the presence of this signal sequence, the *ama* protein has been initially localized to the extracellular surface by immunostaining of embryos with antisera specific to the *ama* protein (see below).

The fifth domain consists of a short, slightly hydrophobic stretch of amino acids at the carboxy-terminal end of the *ama* protein. Since the *ama* protein contains a signal sequence but no transmembrane domain, it is possible that this fifth domain plays some role in the association of *ama* protein with the extracellular membrane. A growing number of cell surface proteins that lack transmembrane domains have been found to be attached to the membrane via a glycosyl phosphatidylinositol (GPI) anchor covalently linked via ethanolamine to the carboxy terminus of the protein (Ferguson and Williams, 1988). The amino acid composition of the fifth *ama* domain is consistent with the patterns established for GPI attachment; however, experimental testing of this prediction will be necessary.

The sequences that place *ama* in the immunoglobulin superfamily are found in the three internal domains. Each of these domains are approximately 100 amino acids in length and exhibit 22–36% amino acid identity to each other in pairwise comparisons; the greatest identity is centered around the pair of conserved cysteine residues within each domain. The sequence conservation and organization of these domains are typical of most immunoglobulin domains.

Immunoglobulin domains have been divided into three categories: V, C1, and C2, by Williams and colleagues based on structural and sequence criteria (Williams, 1987; Williams and Barclay, 1988). We have analyzed the three *ama* immmunoglobulin domains using these criteria and determined that the first domain is a V-type domain and the second and third domains are representative of the C2 category.

The overall similarity of *ama* to individual members of the immunoglobulin superfamily has been evaluated by comparing the *ama* protein sequence to other members of the superfamily using the FASTP program of Lipman and Pearson (1985). Using this approach, the greatest similarity of *ama* is with vertebrate N-CAM (Barthels *et al.*, 1987), although fasciclin II (Harrelson and Goodman, 1988) and MAG (Lai *et al.*, 1987) also exhibit significant similarity. Interestingly, these three most similar proteins are known or thought to function as cell adhesion molecules (Edelman, 1987; Lai *et al.,* 1987; Salzer *et al.,* 1987; Harrelson and Goodman, 1988).

Among the identified members of the immunoglobulin superfamily, only one other protein, nonspecific cross-reacting antigen (NCA; Neumaier *et al.,* 1988), has the particular combination of domains that are found in the *ama* protein (Killeen *et al.,* 1988; Williams and Barclay, 1988). NCA was first identified from normal lung and spleen tissues as an immunologically highly related form of carcinoembryonic antigen (CEA). DNA sequence analysis of both NCA and CEA strongly supports the hypothesis that CEA evolved from NCA by gene duplication and divergence (see Neumaier *et al.,* 1988). Unfortunately, the function of NCA and CEA remains obscure. Thus, whether *ama* represents a very divergent homolog of an identified vertebrate protein or a new type of protein within the immunoglobulin superfamily is still unclear. Based on sequence similarity and the apparent membrane localization of the protein, however, it is reasonable to propose a role for this molecule in cell adhesion or recognition.

The *ama* transcription unit produces a 1500-nucleotide poly(A)$^+$ mRNA that has two major peaks of accumulation during development, one during mid-embryogenesis and the second during early pupation. The *ama* mRNA is first detectable at 0–2 hours of embryogenesis, peaks at 6–8 hours, and is present at low levels after 12 hours of embryonic development. The second peak of accumulation occurs during early pupal development, with mRNA levels being greatly reduced by late pupation. No *ama* mRNA is detected in adult males or females (Seeger *et al.,* 1988). Using *in situ* hybridization, we have more precisely defined the accumulation pattern of *ama* mRNA in the embryo. The changing patterns of expression of this gene resemble in part the spatial distribution of RNAs derived other members of the ANT-C; this apparent amalgamation of diverse spatial patterns gave the locus its name.

The first accumulation of *ama* mRNA is observed during early cycle 14, before the process of cellularization is complete (Fig. 6A and B). Localization of the mRNA along the dorsoventral as well as the anteroposterior axis is observed. The *ama* mRNA is found over the dorsal- and the ventralmost 30% of the syncytial blastoderm embryo, with little or no accumulation observed along the lateral regions (Fig. 6A and B). Following cellularization, the ventral cells accumulating *ama* mRNA are internalized as a result of ventral furrow formation (Fig. 6A, B, E, and F). These cells represent the mesodermal anlagen as determined by various fate mapping studies (Campos-Ortega and Hartenstein, 1985). The dorsal cells expressing *ama* mRNA are included in the transverse

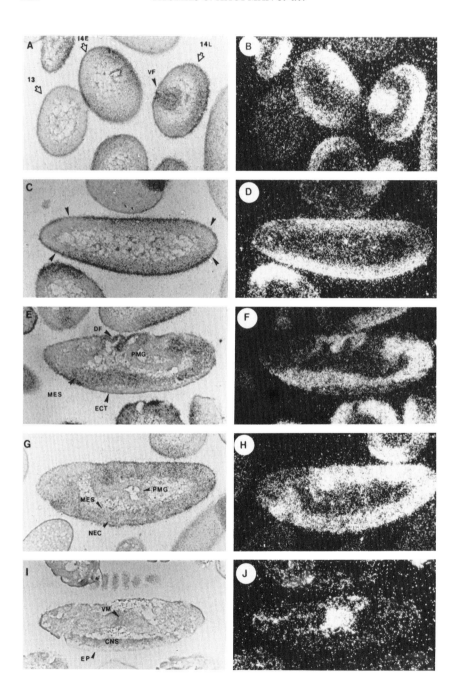

folds that form along the dorsal surface during germ band elongation (Fig. 6E and F). These cells represent the primordia for the amnioserosa and dorsal epidermis (Campos-Ortega and Hartenstein, 1985). Along the anteroposterior axis, *ama* mRNA accumulates from approximately 10 to 90% egg length, with no substantial accumulation observed at the anterior or posterior poles of the embryo (Fig. 6C and D).

The early mesodermal accumulation of *ama* mRNA precedes the first detection of *ama* protein (see below) in these same cells by approximately 30 minutes. In contrast, a major temporal difference is observed for the initial dorsal expression of *ama*. The mRNA accumulation over the dorsal surface precedes the first observed staining of the amnioserosa with *ama* antisera by over 2 hours.

By stage 10 of embryogenesis, *ama* mRNA accumulation is seen throughout the neuroectoderm and mesoderm along the entire extended germ band (Fig. 6G and H). The pattern of expression is quite different by stage 13 of embryogenesis. No accumulation of *ama* mRNA is found within the central nervous system (CNS); however, strong hybridization is observed over the visceral mesoderm (Fig. 6I and J). Weak hybridization over the splanchnopleura is observed occasionally, although the disorganized state of this mesodermal layer combined with the quality of the *in situ* hybridizations make a definitive statement difficult. The pattern during stage 13 contrasts sharply with the observed accumulation of *ama* protein. As shown below, extensive staining with *ama* antisera is seen throughout the CNS (Fig. 7B) as well as the visceral mesoderm and splanchnopleura during stage 13.

FIG. 6. Distribution of *ama* transcript during embryogenesis. Bright-field (A, C, E, G, and I) and corresponding dark-field (B, D, F, H, and J) micrographs show sectioned embryos hybridized with a ³⁵S-labeled *ama* RNA probe. Embryonic stages are based on Campos-Ortega and Hartenstein (1985), and where appropriate anterior is left and dorsal is up. (A and B) Cross sections of three embryos during various stages of cell cycles 13 and 14. 13, Cycle 13; 14E, early cycle 14; 14L, late cycle 14 after cellularization. Accumulation of *ama* transcript is seen over the dorsal and ventral surfaces including the ventral furrow (VF) following cellularization. (C and D) Sagittal section through a cell cycle 14 embryo. Accumulation is seen from approximately 10 to 90% egg length as indicated by the arrowheads. (E and F) Sagittal section through an early stage 8 embryo (early germ band extension). Transcript is restricted to the mesoderm (MES) and dorsal folds (DF). (G and H) Sagittal section through a stage 10 embryo (fully extended germ band). Accumulation is seen throughout the mesoderm (MES) and neuroectoderm (NEC). (I and J) Sagittal section of a stage 13 embryo (shortened germ band). Accumulation is restricted to the visceral mesoderm (VM). Additional abbreviations: CNS, central nervous system; ECT, ectoderm; EP, epidermis; PMG, posterior midgut.

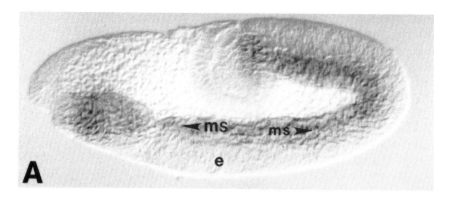

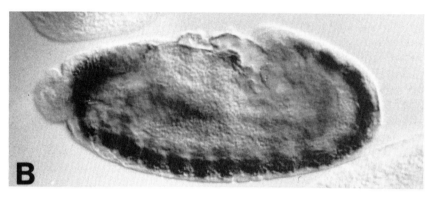

FIG. 7. Distribution of *ama* protein in (A) early and (B) mid-embryogenesis. Anterior is left, dorsal is up. Embryos were stained with an anti-*ama* primary antiserum and a horseradish peroxidase-conjugated secondary antiserum. (A) An early stage 8 embryo (germ band extension) in a midsagittal plane of focus showing that the first detectable accumulation of *ama* protein is in the mesoderm (ms). e, Ectoderm. (B) A stage 12 embryo (germ band shortening), midsagittal plane of focus. The most striking accumulation is seen in the developing nervous system in the ventral nerve cord and the supraesophageal ganglia.

Accumulation of *ama* protein product is first observed during early stage 8 of embryogenesis, shortly after the formation of the three germ layers during the process of gastrulation. This first accumulation is localized exclusively to the mesodermal layer, which is internalized during ventral furrow formation (Fig. 7A). During late stage 9/early stage 10, when neuroblasts first begin to segregate from the ectoderm, *ama* protein begins to accumulate on a row of cells lying at the midline of the extended germ band. This row of small midline cells represents the mesectoderm (Poulson, 1950).

At approximately stage 10 of embryogenesis, weak accumulation of *ama* protein is observed on cells of the amnioserosa. The greatest accumulation is found along the border where amnioserosal cells contact dorsal epidermis and within folded regions of the amnioserosa itself. The weak staining of the amnioserosa persists through germ band shortening the dorsal closure.

Strong accumulation of *ama* protein is also observed on the progeny of neuroblasts beginning at stage 10 and increasing during latter stages (Fig. 7B). The timing of the major neuronal *ama* accumulation coincides temporally with the first appearance of neurons (Hartenstein, 1988). The mesodermal staining observed with *ama* antisera becomes diffuse during stages 11 and 12, corresponding to the reorganization of the mesodermal germ layer into somato- and splanchnopleura (Campos-Ortega and Hartenstein, 1985; Beer *et al.*, 1987). By stage 16 the splanchnopleura has differentiated into somatic musculature, fat body, dorsal vessel, and other mesodermal tissues (Beer *et al.*, 1987); *ama* protein accumulates on a subset of these splanchnopleura derivatives, including the fat body and dorsal vessel, while no accumulation is observed on the somatic musculature.

During stage 13 and extending into later stages, accumulation of *ama* protein is seen throughout the CNS on both neuronal cell bodies and their axons. However, axons leaving the CNS via the segmental and intersegmental nerves do not continue to accumulate *ama* protein on their surfaces. Moreover, axons associated with the peripheral nervous system (PNS) exhibit no accumulation of *ama* protein, while we do observe staining of a subset of cell bodies associated with the PNS. Consistent with the observed sequence similarity between certain vertebrate immunoglobulin family members is a similar expression paradigm. Notably, vertebrate N-CAMs are also expressed predominantly on mesodermal and neural tissues during embryogenesis (Edelman, 1986).

The specific function of *ama* during development remains elusive. It is surprising that no mutations specific to *ama* have been recovered, given its similarity to an important family of cell surface recognition molecules and its broad domain of expression during embryogenesis. The failure to recover mutations occurs in spite of the fact that the ANT-C region has been saturated for ethyl methanesulfonate (EMS)-induced lethal mutations. By comparison, seven EMS-induced lethal alleles have been recovered for *zen,* which produces a protein of similar size to *ama*. The analysis of embryos deficient for *ama* and several adjacent genes does indicate that *ama* is not required for the overall integrity of the embryonic CNS, including the general processes of axon outgrowth and fasciculation (Seeger *et al.*, 1988).

There are several possible explanations for the apparent inability to recover EMS-induced mutations at *ama*. First, the *ama* protein may not be essential for development, and deletion of *ama* may result in either no phenotype or a visible phenotype that has escaped our detection. Second, it is possible that other genes encode functions similar to that of *ama*, and the loss of *ama* protein has no effect because of this functional redundancy.

III. *fushi tarazu* Locus

The *fushi tarazu* (*ftz*) locus is perhaps the most intensively dissected and therefore the mostly widely known member of the ANT-C. We have analyzed the mutant phenotypes of 11 null and hypomorphic as well as 3 temperature-conditional loss-of-function mutations at the *ftz* locus (Wakimoto *et al.*, 1984; T. Kaufman, unpublished). Null alleles act as embryonic lethals and produce a striking mutant phenotype (Fig. 8). Normally the embryonic germ band is divided into 12 trunk segments and 5 head lobes (Fig. 8). In *ftz⁻* animals these numbers are reduced to 6 and 4, respectively, roughly one-half the normal number (Wakimoto *et al.*, 1984) (Fig. 8). Analysis of the partial pattern deletion in the leaky and temperature-sensitive mutations has shown that the mutant phenotype is caused by the deletion of all of the even-numbered parasegments from ps2 to ps1. As noted earlier, *ftz* is not unique but falls into the pair-rule class of segment enumeration loci (Nüsslein-Volhard and Wieschaus, 1980).

Using the temperature-sensitive alleles at the *ftz* locus, we have shown that the temperature-critical period for viability and phenotype falls between 1 and 4 hours of embryogenesis, with a midpoint at 2.5 hours coincident with the blastoderm stage of embryogenesis (Wakimoto *et al.*, 1984). Additionally, using X-ray-induced somatic crossing-over to create *ftz⁻* clones of cells subsequent to the cellular blastoderm stage, it has been possible to demonstrate that *ftz⁺* activity is not necessary for cell viability or pattern formation subsequent to this time (Wakimoto *et al.*, 1984).

In addition to these loss-of-function mutations there are two classes of dominant gain-of-function lesions at the *ftz* locus (Duncan, 1986). Both of these types, *ftz-Regulator of postbithorax-like* and *ftz-Ultraabdominal,* produce mutant phenotypes which mimic lesions in the BX-C. They have been interpreted as showing a regulatory link between the segment enumeration genes and the homeotics (Duncan, 1986).

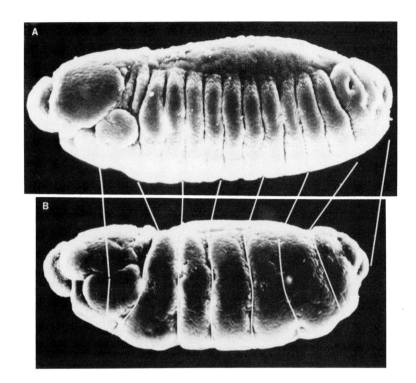

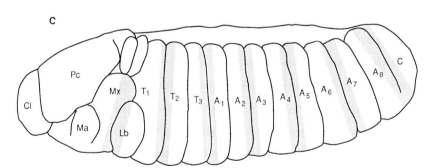

FIG. 8. Summation of defects seen in an early *ftz⁻* embryo. (A) A scanning electron micrograph of a stage 13 wild-type embryo, lateral view. (B) A similarly staged *ftz⁻* animal. White lines between A and B connect the homologous portions of the segmented germ band. (C) The shaded portion shows the parasegments of the germ band that are deleted in the *ftz⁻* animal. Cl, Cypeolabrum; Pc, procephalic; Ma, mandibular; Mx, maxillary; Lb, labial; T_1–T_3, first–third thoracic; A_1–A8, first–eighth abdominal; C, caudal.

The organization of the *ftz* locus (see Fig. 10) has been determined through the analysis of genomic and cDNA clones (Kuriowa *et al.*, 1984; Weiener *et al.*, 1984) and sequencing (Laughon and Scott, 1984). The total transcription unit is small, being just over 2 kb in length, and comprises two exons (800 and 980 bp, respectively) and a single 150-bp intron (Kuriowa *et al.*, 1984; Laughon and Scott, 1984). The open reading frame is 1,239 bp long and initiates in the 5' exon (Laughon and Scott, 1984). The encoded protein (398 amino acid residues) has a predicted molecular weight of 43,000 (Laughon and Scott, 1984). The most prominent motif in the protein is the homeobox (encoded in the second exon); however, like the *zen* protein, the *ftz* protein also contains putative PEST domains (Duncan, 1986). It is likely that these play a role in the resolution of the *ftz* spatial expression pattern in early embryogensis.

Northern analysis shows that the 1.9-kb *ftz* transcript is accumulated early in embryogenesis, starting at about 2 hours, and then declines at about 4 hours (Kuriowa *et al.*, 1984; Weiner *et al.*, 1984). The Northern analysis has been expanded on through *in situ* hybridization of RNA probes and localization of the *ftz* protein product in embryos (Hafen *et al.*, 1984a; Carroll and Scott, 1985). RNA is first observed in a broad band at syncytial blastoderm stage extending from the region of the future cephalic furrow (65% egg length) posteriorly to about 15% egg length. This broad band is resolved into seven transverse stripes which circumscribe the embryo at the cellular blastoderm stage. As gastrulation proceeds, the stripes gradually diminish in the germ band ectoderm and are gone by mid-gastrulation. The earliest detectable protein accumulation is coincident with the blastoderm stage, and, again, seven transverse stripes are observed. Each stripe is initially four cells in width, subsequently narrowing to three cells as germ band elongation proceeds. The position and width of these stripes are coincident with the size and location of the parasegmental anlagen that are missing in *ftz⁻* embryos (Carroll and Scott, 1985).

In addition to this germ band expression of *ftz* in early embryos, there is a subsequent accumulation of *ftz* protein in the CNS of later embryos (Carroll and Scott, 1985). A few cells in each of the segmental ganglia can be seen to express *ftz,* and this expression persists into late embryogenesis.

The identification of this transcription unit as *ftz* has been accomplished through the localization of *ftz*-associated breakpoints in the DNA (Weiner *et al.*, 1984) and P-element-mediated transformation (Hiromi *et al.*, 1985; Hiromi and Gehring, 1987). The latter analyses have included the definition of at least three cis-acting regulatory elements upstream of the *ftz* initiation of transcription. Using *ftz*-derived promoter fragments coupled to a β-galactosidase reporter gene, the above authors showed that an approximately 1-kb fragment just upstream of the CAP site is necessary for the formation of the blastoderm and germ band stripes (the "zebra element"). Just distal to this element another larger fragment is necessary for *ftz* expression in the CNS, and further removed (~6 kb upstream) is an enhancer-like element which is necessary for the maintenance of the striped pattern.

The definition of the cis-acting elements has been augmented by the demonstration that the *ftz* spatial pattern of expression is also affected and likely controlled by other segment enumeration genes (Carroll and Scott, 1986; Carroll *et al.*, 1986b, 1987). The picture emerging is that the segmentation pattern is first broadly defined by maternally expressed genes such as *bcd* which in turn regulates the expression of the gap loci (Driever and Nüsslein-Volhard, 1989). These broad domains are then refined by the action and interaction of the pair-rule and segment polarity genes (Harding *et al.*, 1986; Howard and Ingham, 1986; see Akam, 1987, for a review). In the particular case of *ftz*, the instructive influence of genes upstream of the locus in the segmentation pathway is thought to act both positively and negatively through the proximate "zebra" element to activate or repress *ftz* gene transcription (Hiromi and Gehring, 1987). Once active the *ftz* protein product can autogenously maintain its own expression by interaction with the distal enhancer element (Hiromi and Gehring, 1987). The resolution of the striped pattern results, therefore, through the interaction of *ftz* and the other enumeration genes and its ability to maintain its own expression. Once the striped pattern is established the expression of the *ftz* gene in its proper spatial array is necessary for the proper spatial expression of the homeotic loci (Duncan, 1986), the next gene set in the ontogenic cascade.

IV. Homeotic Loci of the Antennapedia Complex

A. *Antennapedia* LOCUS

We have recovered or obtained and analyzed 60 mutations at the *Antennapedia* (*Antp*) locus. In addition, we have characterized a set of 11 deficiencies which delete distal and proximal portions of the 100-kb *Antp* locus (Scott *et al.*, 1983) (Fig. 9). The results of our genetic and developmental investigations indicated that the *Antp* locus is capable of initiating expression (transcription) from two points (Abbott and Kaufman, 1986). In other words, the *Antp* locus appeared to possess two promoters which could be separately disrupted by mutations. This

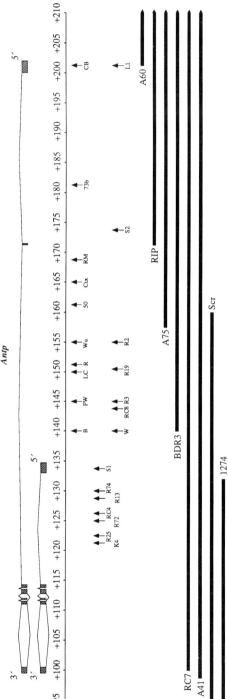

FIG. 9. Diagram of the *Antp* locus of the ANT-C. Coordinates are similar to those of Fig. 3. Bars below the coordinate line indicate the extent of deletions used to characterize the *Antp* locus. The row of arrows closest to the line between coordinates +140 and +205 shows the positions of breakpoints associated with dominant gain-of-function mutations in the locus. The cluster of arrows below and to the left (coordinates +120 to +135) indicates the positions of recessive loss-of-function mutations. The arrows directly below the gain-of-function breaks indicate the position of recessive loss-of-function lesions which do not completely inactivate the gene. The P1-driven transcript initiates at the 5′ exon at coordinate +200, while the P2-driven transcript initiates at coordinate +135.

finding has been subsequently verified at the molecular level by investigators in three other laboratories (Laughon *et al.*, 1986; Stroeher *et al.*, 1986; Jorgensen and Garber, 1987).

Further analyses of the mutations in *Antp* have shown that the distal or P1 promoter can be deleted, preventing the production of transcripts containing the distal two exons of the gene without affecting the function of the proximal P2 promoter. Flies carrying such a lesion do not show any of the *Antp*-associated embryonic or larval phenotypes (Abbott and Kaufman, 1986). These animals survive to the adult stage and exhibit a transformation of dorsal T2 to T1 but no other *Antp*-associated adult phenotypes (e.g., leg → antenna) (Struhl, 1981; Abbott and Kaufman, 1986). We have concluded that the P1-driven transcripts are dispensable for all but a single $Antp^+$ function. It has been shown that the P1- and P2-driven transcripts do have different spatial domains of expression (Frischer *et al.*, 1986). Nevertheless, P1-derived RNAs are found in the embryo despite any obvious phenotypic consequence caused by their absence.

We have also investigated the effects of deletions which proceed through the P1 promoter and include the genomic sequences containing the distal two exons and some of the intron sequences distal to the P2 promoter. Interestingly, these lesions do not behave as the promoter deletions mentioned above; these deletions do cause embryonic transformations. Deletions which remove the distal two exons and portions of the intron show a transformation of posterior T1 to head and anterior T2 to T1. Posterior T2 and anterior T3 are not transformed. Thus, while *Antp* null mutations cause all of these compartments to transform, the partial deletions change only the anterior two. Viewed from another perspective, only the anterior of the two parasegments normally transformed is affected (Martinez-Arias, 1986). Since several chromosome rearrangements in this interval have an identical effect, we concluded that there are regulatory (enhancer-like?) elements proximal to P1 necessary to the full expression of the P2 promoter. What is not clear is whether these elements merely increase the activity of the P2 promoter or are responsible specifically for the expression of P2 in the more posterior parasegmental domain. As would be predicted, deletion of the P2 promoter or the 3' end of the gene abolishes all *Antp* function.

The transcriptional complexity of the *Antp* locus has been further confounded by the discovery of alternate splicing patterns among the exons containing the ORF (Schneuwley *et al.*, 1986). Therefore, in addition to the spatially distinct functions of the P1 and P2 promoters, the possibility exists that alternate protein products of the locus may also be involved in the differential functioning of the *Antp* gene.

Antisera raised to the *Antp* proteins have largely substantiated the spatial pattern of accumulation determined by *in situ* hybridization using antisense RNA probes (Hafen *et al.*, 1983, 1984b). The protein, like the RNA, is seen to be most strongly accumulated in the ventral nerve cord and more weakly in the epidermis and mesoderm (Carroll *et al.*, 1986a; Wirz *et al.*, 1986). When the germ band is extended (stage 10–11), the protein is found starting in the posterior of the first thoracic neuromere and proceeding posteriorly to a region corresponding to the seventh abdominal segment (Carroll *et al.*, 1986a; Wirz *et al.*, 1986). The integumentary staining at this time appears to be more restricted and corresponds to a register approximating parasegments 4 and 5.

As germ band shortening ensues and the ventral nerve cord condenses in late embryogenesis, there is a progressive restriction of *Antp* expression. The thoracic neuromeres corresponding to parasegments 4 and 5 show the highest level of accumulation while the more posterior abdominal regions show a much lower but still detectable amount of protein accumulation (Carroll *et al.*, 1986a; Wirz *et al.*, 1986). As mentioned earlier, this posterior progressive diminishing of *Antp* expression has been shown to be caused by the loci of the BX-C (Carroll *et al.*, 1986a; Wirz *et al.*, 1986) and is thought to be the result of direct transcriptional regulation of the *Antp* locus by the *Ubx* protein product (Boulet and Scott, 1988). Unlike the demonstrated cross-regulatory interaction between *Ubx* and *Antp,* there is no demonstrable effect on the anterior border of the *Antp* domain through the deletion of the *Sex combs reduced* locus (Wirz *et al.*, 1986).

The patterns of *Antp* RNA (Jorgensen and Garber, 1987) and protein accumulation (Wirz *et al.*, 1986) have also been determined in imaginal discs, where they are found in all three leg anlagen, the wing, and the dorsal prothoracic discs. The pattern of accumulation in these regions is entirely consistent with the suite of defects associated with *Antp* lesions in the adult (Struhl, 1981; Abbott and Kaufman, 1986).

The most intriguing lesions at the *Antp* locus are the dominant gain-of-function mutations (Figs. 2 and 9). A rationalization for their effects has been presented earlier (i.e., ectopic expression of $Antp^+$ in the adult head). This view has been substantiated at the molecular level, and the unique (among the homeotics) dual-promoter structure of the gene was shown to play a role. Nearly all dominant lesions at *Antp* are associated with chromosomal rearrangements, and these are broken distal to the P2 promoter (Fig. 9). The rearrangements serve to juxtapose new sequences upstream of the P2 transcriptional initiation site, and they have been shown to cause ectopic *Antp* expression through misregulation of the P2-driven transcript or through production of transcripts bearing novel 5' exons and downstream *Antp* sequences (Frischer *et al.*, 1986; Jorgensen and Garber, 1987; Schneuwley *et al.*, 1987). Since all of the *Antp* open reading frame is contained in the four 3'-most exons (Schneuwley *et al.*, 1986), these alternatively spliced products are clearly capable of producing a normal suite of *Antp* proteins and causing the observed defects.

B. *Sex combs reduced* Locus

We have recovered and characterized 27 mutations at the *Sex combs reduced* (*Scr*) locus (Fig. 10). The *Scr* lesions can be classified into three overlapping categories. The first contains lesions which inactivate the locus entirely and result in embryonic lethality. These lethal individuals exhibit transformations of the embryonic labial and first thoracic segments to a maxillary and second thoracic identity respectively (Sato *et al.*, 1985; Otteson *et al.*, 1990). The second class of recessives is a group of semilethal lesions which affect adult transformations of the labial palps to maxillary and the first thorax to second but which exhibit little or no embryonic transformation (Otteson *et al.*, 1990).

Finally, the third category is a group of dominant gain-of-function lesions. When heterozygous, such mutants exhibit an extra sex combs phenotype indicative of a second and third thoracic to first homeotic transformation (T3, T2 → T1). Each of these lesions has associated with it a recessive lethal or semilethal effect that assigns it to one of the first two categories. One possible explanation for the dominant class is that these mutations represent lesions which cause or allow ectopic expression of the normal Scr^+ product and that Scr^+ activity is necessary to elicit a first thoracic or T1 pattern of development. However, it is also necessary to explain how several of these lesions are also associated with a recessive loss of function and act as Scr^- in the homozygous condition.

The molecular organization of the *Scr* locus is shown in Fig. 10. We have mapped *Scr* breakpoints over an interval of approximately 70 kb extending from the 3' end of the *Antp* locus to 30 kb proximal of the *ftz* gene. Breakpoints associated with the semilethal adult-type lesions all map between *ftz* and *Antp,* including two breakpoints in the *ftz* locus itself. The lethal embryonic mutations all map proximal to *ftz*. The gain-of-function lesions map through the interval (Otteson *et al.*, 1990).

Using genomic fragments and Northern analysis we have found three genomic fragments (all proximal to *ftz*) which identify a single 3.7-kb RNA expressed in embryos, larvae, pupae, and adults. We have recovered cDNAs and shown that these three genomic fragments house

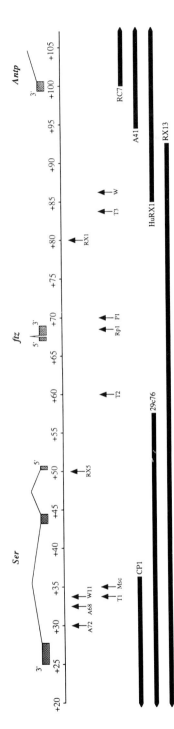

FIG. 10. Diagram of the *Scr* molecular map showing the relative positions of the *Antp* 3'-most exon and the *ftz* transcription unit. Coordinates are as in Fig. 3. Bars indicate the extent of deletions impinging on the *Scr* transcription unit and affecting its function. The top row of arrows indicates the position of breakpoint-associated *Scr* mutations that inactivate the gene (coordinates +30 to +50) or cause partial inactivity (coordinate +80). The bottom row of arrows shows the position of dominant gain-of-function lesions at the locus.

three exons which make up the mature *Scr* transcript (also see LeMotte *et al.*, 1989). Sequencing of the cDNA clones has shown the presence of a homeobox in exon 3 and opa-like repeats in exon 2 (LeMotte *et al.*, 1989; J. Mahaffey and T. Kaufman, unpublished). The open reading frame associated with these two motifs spans exons 2 and 3 but does not extend into exon 1 (LeMotte *et al.*, 1989; J. Mahaffey and T. Kaufman, unpublished). Additionally, a translation stop signal is found about 2 kb upstream of the apparent transcriptional termination site, producing an exceptionally long 3' untranslated tail. The reading frame extends for 1,245 bp and encodes a protein of 45 kDa.

As for the other homeotic loci of the ANT-C we made a chimeric *βgal–Scr* protein (the construct included 198 of the carboxy-terminal amino acid residues of *Scr*). Using the cDNA clones and antisera raised in rabbits to the *βgal–Scr* fusion protein, we determined the spatial and temporal pattern of *Scr* gene product accumulation by *in situ* hybridization and immunostaining (Mahaffey and Kaufman, 1987a). The RNA is first detected in a group of cells just posterior to the cephalic furrow in a pattern reminiscent of *lab* and *Dfd* (also see Martinez-Arias *et al.*, 1987). Similar to *lab* (see below), protein is not detected at this time. Later when the germ band begins to segment, however, protein and RNA can be found in the labial segment (also see Riley *et al.*, 1987; Carroll *et al.*, 1988). Later still RNA and protein also appear in the first thoracic segment. Thus, the spatial domain of *Scr* protein accumulation is consistent with the position of observed defects when this locus is absent (Fig. 14).

During head involution *Scr* protein and RNA accumulate in the labial neuromere of the CNS but do not do so to any appreciable degree in the T1 neuromere. A similar correlation between spatial accumulation of *Scr* protein and defects seen in *Scr*⁻ tissue is seen in the imaginal discs (Mahaffey and Kaufman, 1987a). Protein accumulation is seen strongly in the entire first leg and dorsal prothoracic discs as well as the labial anlagen and a small portion of the eye antennal disc corresponding to the cervical region of the posterior head. We have also shown that the unexpected presence of *Scr* transcripts in adults is accounted for by the expression of this gene in the CNS (labial ganglion) of imagos (Mahaffey and Kaufman, 1987a).

Careful analysis of the limits of expression of the *Scr* locus has revealed that *Scr* protein is accumulated in a segmental pattern in the epidermis and a parasegmental register in the neural portions of the ectoderm (Fig. 14) (Mahaffey *et al.*, 1989). This register in the epidermis is different from that observed for *Antp*, *Ubx*, and the other homeotics expressed in the trunk (see Fig. 14), where a parasegmental register is observed in all ectodermally derived tissues. This shift in *Scr* domain and the failure to observe significant *Scr* protein accumulation in the first thoracic neuromere result in a gap of homeotic protein accumulation between *Scr* and *Antp* in the CNS in a region corresponding to parasegment 3 as well as a small overlap in the *Scr* and *Antp* expression domains in the posterior of the first thoracic epidermis. This small overlap between the two genes is reminiscent of the nested expression patterns observed more posteriorly in the turnk (Fig. 14) but contrasts with the juxtaposition of homeotic gene product accumulation observed more anteriorly.

C. *Deformed* Locus

The molecular organization of the *Deformed* (*Dfd*) locus has been elucidated in the laboratory of Dr. William McGinnis (Regulski *et al.,* 1987). The *Dfd* gene is made up of 5 exons distributed over 11 kb of genomic sequence (Fig. 11). The open reading frame begins in exon 1 and terminates in exon 5 (Regulski *et al.,* 1987). The homeobox is located in exon 4 and the opa repeat in exon 5. The 2.8-kb transcript produced by the locus is accumulated throughout embryogenesis as well as in pupae and adults (Chadwick and McGinnis, 1987).

Our studies on the *Dfd* locus have been almost entirely developmental and genetic. We have recovered and characterized 17 mutations as well as deletions which impinge on the transcription unit identified as *Dfd* (Merrill *et al.,* 1987). Three of the identified alleles are associated

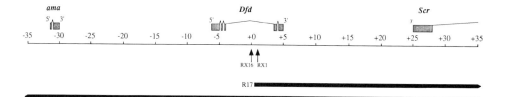

FIG. 11. Molecular map of the *Dfd* region of the ANT-C showing the relative positions of the *Scr* 3′-most exon and the *ama* gene. Coordinates are as in Fig. 3. Breakpoints and deletions are symbolized as in prior figures; all those shown inactivate the *Dfd* locus. Note that the *Dfd* gene is transcribed in a direction opposite to the other homeotics in the ANT-C and that there are no other major transcripts between *ama* and *Scr*.

with chromosomal rearrangements (Fig. 11), and one of them is temperature sensitive. Using both light and electron microscopy, we have shown that Dfd^- animals have defects in the embryonic and larval head and are missing structures derived from the mandibular and maxillary segments. The results of somatic recombination studies have shown that Dfd^+ activity in the ventral aspect of the adult head is required for the development of the vibrissae and the maxillary palps. In the dorsal posterior of the head, cuticular transformations are seen which we have interpreted as head → thorax. Thus, Dfd deficiency results in no obvious homeotic transformation in the embryo or ventral adult head but does cause the development of ectopic structures in the dorsal adult head (Merrill *et al.*, 1987).

Temperature-shift studies on the temperature-sensitive allele have shown that Dfd^+ activity is necessary for viability during two discrete periods of development. The first is in the embryo during segmentation and head involution. The second occurs during the late larval and pupal stages. These two times correlate well with the observed times of Dfd transcript accumulation (Chadwick and McGinnis, 1987; Martinez-Arias *et al.*, 1987). Additionally, the results of the *in situ* hybridization analysis show that Dfd RNA is produced in the embryonic mandibular and maxillary segments and in portions of the eye antennal disc which show defects and transformations in Dfd mutants (Chadwick and Mc-Ginnis, 1987; Martinez-Arias *et al.*, 1987; Merrill *et al.*, 1987).

As is the case for *lab* and *Scr*, accumulation of Dfd transcript is associated with the cephalic furrow of the early gastrula stage embryo, and later accumulation occurs in the subesophageal region of the central nervous system (CNS) (Chadwick and McGinnis, 1987). We have succeeded in producing a *βgal–Dfd* chimeric protein containing a 585-amino acid stretch of the *Drosophila* polypeptide. This protein was used to immunize rabbits, and antisera were obtained. Staining of embryos with the antisera revealed that Dfd protein is accumulated in the same pattern as the RNA, including the cephalic furrow stage (Fig. 14) (Mahaffey *et al.*, 1989; also see Jack *et al.*, 1988). There is a distinct temporal difference in time of initial protein accumulation when one compares *lab* and Dfd (see below). Both genes accumulate RNA at the cephalic furrow (albeit in different populations of cells), but only Dfd protein is present. Later during segmentation of the germ band, both proteins are present in spatially sequential segmental anlagen.

As for *Scr*, we carefully determined the register of Dfd protein accumulation. Using the position of *engrailed* (*en*) expression to mark the segmental/parasegmental boundaries, we found that Dfd behaves in a similar fashion to *Scr*, that is, Dfd protein appears in a segmental register in the epidermis and a parasegmental one in the neural portions of the ectoderm (Mahaffey *et al.,* 1989). This segmental pattern appears to be the first to arise since at the cellular blastoderm stage, when *Dfd* protein is first detectable, all *Dfd*-positive cells which circumscribe the embryo are in this register. A more complete description of *Dfd* expression and regulation will be found in the chapter by McGinnis *et al.* (this volume).

D. *proboscipedia* LOCUS AND CUTICLE CLUSTER

The *proboscipedia* (*pb*) locus maps just distally to the cluster of cuticle-like genes (Fig. 12). A combination of S1 mapping, Northern blotting, and genomic sequencing revealed the presence of a transcribed region extending over 35 kb in this region (Pultz *et al.,* 1988; M. Pultz and D. Cribbs, unpublished). This transcription unit produces a single 4.3-kb RNA which is derived from at least nine exons distributed over the interval (Fig. 12). Exons four and five contain a homeobox which is split by intron four in the same position as the homeobox is split in the *lab* gene (see below) (Mlodzik *et al.,* 1988; Diederich *et al.,* 1989). The *pb* locus differs from *lab,* however, because the opa sequences are downstream (exon 8) of the homeobox rather than 5' to this motif (Fig. 16). We are confident that this transcription unit is *pb* owing to the fact that several *pb*-associated rearrangements and deletions are broken in this chromosomal domain (Fig. 12).

Flies carrying a physical deletion of the *pb* locus survive to adulthood and produce a *pb* null phenotype (labial palps → legs) (Kaufman, 1978; Pultz *et al.,* 1988). Therefore, it was surprising to find, using Northern and *in situ* hybridization, that *pb* RNA is accumulated in embryos in the maxillary and mandibular lobes (Pultz *et al.,* 1988).

This pattern of RNA accumulation has been confirmed and elaborated by the recovery of antisera to a *pb*-derived pepetide (Pultz *et al.,* 1988). A small genomic fragment from exon six (between the homeobox and opa) was cloned into an expression vector and the chimeric βgal–*pb* protein produced. The fragment used had an open reading frame capable of encoding 40 amino acids. Antisera to this peptide used to stain whole mount embryos identifies a nuclear-localized protein which is accumulated in the maxillary and labial segments as well as in cells internal to the mandibular segment in early embryos (Fig. 14). This protein is not present in *pb⁻* embryos. In later stages during head involution the integumentary signal diminishes and a repeating pattern of nuclei expressing the protein in the CNS is seen. The accumulation pattern does not resemble that of the other ANT-C homeotic loci:

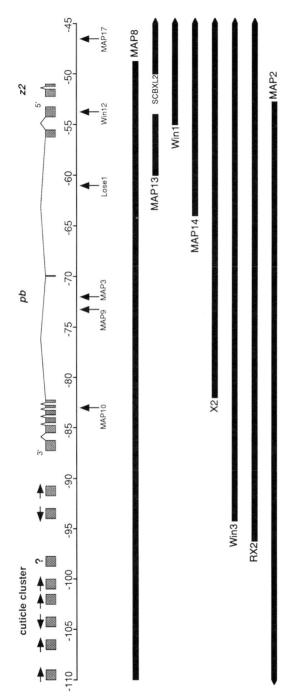

Fig. 12. Molecular organization of the *pb* locus of the ANT-C showing the relative positions of the cuticle cluster and the *z2* transcription unit. Coordinates are as presented in Fig. 3. Bars show the extent of deletions affecting *pb* function. All deletions inactivate the gene, except one (SCBXL2), which shows a partial inactivation of the locus. Arrows show the positions of breakpoints which completely inactivate *pb* (coordinates −85 to −54) or produce a V-type position-effect defect (coordinate −46). Arrows above the cuticlelike genes (▨) indicate the direction of transcription, when known.

a majority of cells in one or two ganglia do not express the protein, rather only a few cells in every ganglia express the product.

The significance of this embryonic expression is not at present clear. As noted above, deletion of the locus produces effects only on adults. It is possible that *pb* deficiency in the embryo causes some barely discernible defect that has escaped notice. However, we do not feel this to be likely; rather, it seems possible that the early *pb* expression represents an evolutionary vestige of the expression pattern of this locus in more primitive insects where head development is not so greatly modified as in the higher diptera. Alternatively, this early accumulation could reflect the mechanism by which *pb* expression in later imaginal development is ensured. The two segments in which *pb* is expressed in the embryo are the logical sites of the progenitor cells of the affected adult structures.

E. *labial* Locus

The *labial* (*lab*) locus is the most proximal member of the ANT-C (Fig. 13). The gene comprises three exons. The proximal two are separated by a 245-bp intron, and they are separated from the 5' exon by a 13.8-kb intron. The homeobox is located in the proximal exons and is split by the small intron (Mlodzik *et al.*, 1988; Diederich *et al.*, 1989). The distal exon contains an opa or M repeat sequence (Wharton *et al.*, 1985). The entire gene (excluding the large intron) has been sequenced; the 5' and

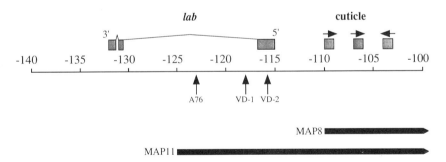

FIG. 13. Molecular organization of the *lab* locus showing the relative position of the cuticle-like genes. Coordinates are as presented in Fig. 3. The MAP11 deletion inactivates *lab*, whereas MAP8 has no effect on *lab* function. Thus, a maximum 5 kb of sequence upstream of the initiation of transcription appears necessary for *lab* expression. The three arrows indicate the positions of breakpoints which inactivate *lab*.

3′ ends and splice junctions have been defined by S1 mapping, primer extension analysis, and recovery of cDNA clones.

A portion of one of the cDNA clones from upstream of the homeobox to the end of translation (270 amino acid residues) was fused into an expression vector and a βgal–lab chimeric protein produced in *Escherichia coli*. This protein was used to raise antisera in rabbits and the antisera used to stain whole embryos. Additionally, we used ^{35}S-labeled antisense RNA probes generated from the cDNA clones for *in situ* hybridization to sectioned animals. The results of these studies have shown that *lab* RNA is accumulated just anterior to the cephalic furrow at the beginning of gastrulation; however, no *lab* protein can be detected at this time. Subsequently during segmentation of the germ band, both *lab* RNA and *lab* protein are detected in the most anterior portions of the gnathocephalic area just anterior and dorsal to the hypopharyngeal lobes. There is also *lab* product detected in a row of cells extending over the top of the gnathocephalic complex in the ventral lateral aspect of the procephalic lobe as well as in the dorsal ridge (Diederich *et al.*, 1989). We have interpreted this accumulation pattern as being indicative of the presence of an intercalary segment and *lab* expression in that metamere.

We analyzed the developmental defects associated with the 15 extant *lab* alleles. Four of these are associated with breakpoints which interrupt the *lab* transcription unit (Fig. 13) and do not produce a *lab* product. Animals deficient for *lab* are nonviable and die at the embryonic first instar boundary. These individuals show defects (missing structures) in derivatives of all of the gnathocephalic (mandibular, maxillary, and labial) segments (Merrill *et al.*, 1989).

Somatic recombination experiments have allowed us to assess the defects associated with *lab* deficiency in the adult. Clones of *lab*⁻ cells fail to develop normally in the ventral (maxillary) portions of the head capsule, and portions of the eye and maxillary palps are absent. Dorsally in the head, portions of the posterior aspect of the head are transformed into an apparent thoracic identity. The regions of the head which are defective overlap substantially with those affected by *Dfd* deficiency (Merrill *et al.*, 1987, 1989). Moreover, fate mapping studies have shown that the mandibular, maxillary, and labial segments are the origin of the structures affected by *lab* deficiency (Jurgens *et al.*, 1986). Thus, we are presented with a seeming paradox. The *lab* gene products are accumulated in a portion of the embryonic head which is not apparently associated with development of the structures affected when the gene is deleted. We have concluded, therefore, that the cuticular defects observed in *lab*⁻ animals at the end of embryogenesis are most readily accounted for as secondary consequences of *lab* deficiency in the intercalary segment. The implication is that this metamere performs some organizational function during head morphogenesis (Diederich *et al.*, 1989; Merrill *et al.*, 1989). Finally, we have observed the accumulation of *lab* RNA and protein in the wall of the midgut. However, animals deleted for *lab* show no obvious defects in gut development.

As for *Scr* and *Dfd*, we have observed the relative positions of *lab* and *en* protein accumulation. The highly modified nature of the *en* pattern in this region of the head has not allowed us to definitively determine whether *lab* is accumulated in a segmental versus parasegmental register. It is apparent, however, that there is no substantial overlap between *lab* and *Dfd* (the next most posteriorly expressed gene). Indeed, in germ band extension (stage 10/11) the cells expressing the two products can be seen to abut. However, later modifications in the *Dfd* pattern of expression in the hypopharyngeal and mandibular lobes and the complex cell movements associated with head involution tend to obscure the initial pattern.

A summary of the relative expression patterns of the anterior (head) and posterior (trunk) homeotics can be seen in Fig. 14. Our analysis indicates two salient differences between the two groups. The trunk loci are clearly expressed in an overlapping set of parasegmentally organized blocks. Anteriorly, however, the register in the epidermis shifts, and the blocks of accumulation become discrete.

Models to account for the role of homeotics in segmental specification have been drawn largely from a paradigm learned in the trunk. Thus, segmental identity is viewed as being instructed in a combinatorial fashion. The identity of any given segment is specified by which subset of several loci are active in that metamere. Our results with the head genes indicate that this paradigm may not be transposable to the anterior of the embryo (Mahaffey *et al.*, 1989). Since only one gene can be seen to be expressed in any given segment, a strict combinatorial code of identity is difficult to envision among the homeotic members of the ANT-C. This is particularly evident considering the *Dfd* locus, which is expressed in two adjacent metameres to the exclusion of the *lab* and *Scr* loci. These two segments, the mandibular and maxillary, develop in unique ways and form distinct structures. What is not clear is how a single regulatory element can distinguish between the two. One possibility lies in the rather different behavior of *Dfd* in the two segments. The *Dfd* protein product is expressed early in the more anterior mandibular segment but does not remain on. Therefore, in the case of *Dfd*, identity specification may have a temporal component as well as a strict spatial demarcation.

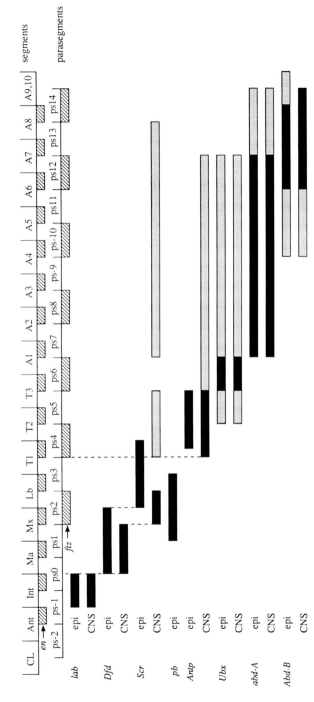

FIG. 14. Summary of the extent of accumulation of protein and/or RNA of the homeotic loci of the ANT-C and BX-C. The coordinate demarcations at top show the segmental (top line) and parasegmental (below) metameric divisions in the embryo. Associated with each line are hatched boxes showing the relative positions of *en* and *ftz* protein accumulation. Solid bars below the register lines show the principal positions of gene product accumulation in the epidermis (epi) and central nervous system (CNS) of the designated homeotic loci. Stippled bars show the extent of minor but detectable accumulation. The patterns of the ANT-C products are summarized from this article, and the *Ubx*, *abd-A*, and *Abd-B* distributions are taken principally from Harding *et al.* (1985), Regulski *et al.* (1985), Akam (1987), and Duncan (1987). Note that the loci expressed in the trunk (T1–A8) show large overlapping patterns of expression, whereas in the head (Int–Lb) a more discrete patterning is observed (dashed lines).

The above discussion assumes that all of the head-specifying homeotics (at least in the gnathal region) have been found. This, of course,
may not be the case, but until others are identified one must consider
the possibility that head development requires an alternate paradigm.

V. Origin and Evolution of the Antennapedia Complex

One of the more popular views of the origin of the homeotic complexes
in *Drosophila* is that they arose as a series of gene duplications and
subsequent divergences of function. This view is at least in part in-
spired by the line of ascent which is thought to have given rise to the
modern higher insect classes such as the diptera. The progenitors of the
insects are likely to have been annelid-like forms with a simple head
followed by identical repeating metameric body segments. A gene or
genes similar to the homeotic selectors can be envisioned as being
involved in establishing the difference between these two types of body
segments. As evolution proceeded through an onycophoran- and
myriapod-like stage to finally give rise to the insects, there was an
increasing elaboration of the originally isomorphous trunk segments
into metamers devoted to specialized mouthparts and bodies designed
to novel types of locomotion and food acquisition. Concomitant with
these changes, the original head/trunk selector would also be elaborated (duplicated) and specialized (diverged) to perform new selector
functions.

The evolutionary time frame of these duplicative and diverging
events was thought to coincide with the above phylogenetic ascent.
However, this classic view has been challenged by some remarkable
discoveries involving homeobox-containing loci in vertebrates (Duboule and Dolle, 1989; Graham *et al.*, 1989). The above authors have
found that the mouse homeobox genes (Hox), like the *Drosophila* homeotics, occur in clusters on the chromosomes. Moreover, like *Drosophila* the order of the genes in the clusters is colinear with the anterior to
posterior order in which the clustered loci are expressed. Additionally,
when one considers the sequence similarity among the homeoboxes
within each respective complex and compares the Hox and *Drosophila*
motifs, it is clear that the order of genes is simlar if not identical in the
mouse and fly. The major difference lies in the fact that the mouse has
four clusters of Hox genes with representatives from both the ANT-C
and BX-C (i.e., the two classes of homeotics are not split up). No evidence is presented for the existence of loci like *ftz, bcd,* and *zen,* but the
major homeotic loci do seem to be represented (Duboule and Dolle,
1989; Graham *et al.*, 1989).

The implications of this conservation over such vast phylogenetic distances are 2-fold. First, there are strong selective and thus likely functional reasons for the grouped organization of these transcription units. Second, the duplication–divergence events outlined above occurred earlier in evolution than envisioned in the classic model, most likely before the protostome–deuterstome divergence in the Precambrian. Therefore, these events were not originally associated with annelid → arthropod specializations; rather, the complex as a unit was adopted and adapted in some fashion within this particular lineage. Any view of the method of change within these loci must therefore consider the nature of the original early diverging events which then will have superimposed specializations unique to the particular phylogenetic lineage in which a given complex is found. As mentioned above, the small nonhomeotic members of the ANT-C may be just such a phylogenetic specialization. Using our one end point *Drosophila,* however, we can begin to gain some insight into the rules governing the way in which change has occurred within the complex. It will be of great interest to find what the structural analysis of the Hox complex genes tells us in the near future.

Figure 15 shows a diagrammatic view of the exon/intron structure of the five homeotic genes of the *ANT-C*. What is immediately apparent from inspection of Fig. 15 is that there is no simple relationship among the five genes in their gross structure which would allow an obvious pattern of structural relatedness. It is clear that the relative position of introns vis-à-vis the ORFs of the five genes is not logically maintained, nor are the number of exons per se.

The difficulty is compounded when one considers the relative positions of the homeobox and opa repeat within each gene (Fig. 16). In the *lab* and *pb* genes an intron splits the homeobox, whereas the other three genes have this motif contained in a single exon. Therefore, one might conclude that the complex contains two subgroups of homeotic genes: *lab*- and *pb*-like on the one hand, *Antp-, Scr-,* and *Dfd*-like on the other. However, in *pb* the opa motif is downstream of the homeobox while in *lab* it is upstream. By this criterion, then, *pb* more closely resembles *Dfd,* and *lab* can be grouped with *Antp* and *Scr.* One possible way to rationalize these observations is to speculate that *pb* arose as a fusion between a *Dfd*-like and *lab*-like ancestor or that *lab* arose by an association of a *pb*-like and an *Scr*-like ancestor. Both scenarios must then also explain the origin of the *Dfd*-like ancestor.

In short, no simple pattern of duplication and divergence will suffice to explain the present structure and organization of the ANT-C. One thing is apparent, however, when one considers the relative juxtapositions of conserved protein domains encoded by these genes. That is, if

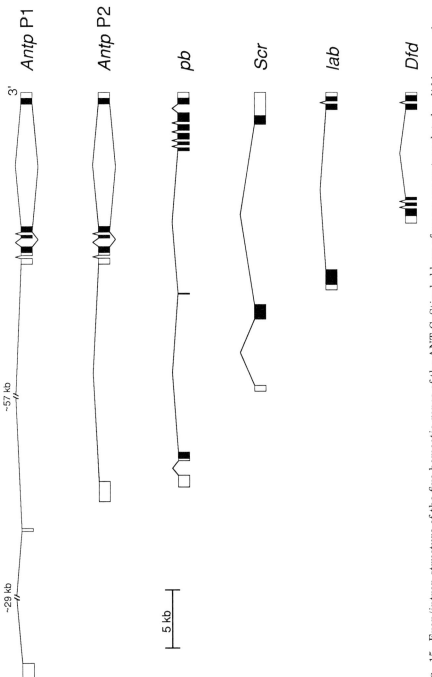

FIG. 15. Exon/intron structure of the five homeotic genes of the ANT-C. Stippled bars of exons are untranslated; solid bars are the open reading frames. The genes are lined up at their 3' ends and are drawn to scale, except for the two 5'-most introns of the *Antp* locus.

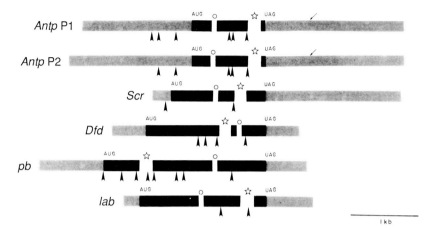

FIG. 16. Processed RNA products of the homeotic genes of the ANT-C. Stippled bars show the 5' and 3' untranslated leader and tail sequences. Solid bars show the ORFs. The relative positions of the homeobox and opa repeats are indicated by stars and circles, respectively. Arrowheads below each RNA show the positions of the splice junctions. The two small arrows above the *Antp* 3' tail show the positions of the internal poly(A) addition sites.

these genes are evolutionary duplicates, one aspect of their divergence has been some form of intron/exon shuffling. What remains to be seen is whether this process is relevant to the functioning of these loci.

VI. Homeobox in the Antennapedia Complex

A further clue to the relatedness of the ANT-C genes can be found through an analysis of the sequence divergence (similarity) in the conserved homeobox. With the exception of *ama* and the cuticle-like loci, all of the genes in the Antennapedia complex possess this element despite the fact that four of them (*ftz, bcd, zen,* and *z2*) are distinct from the homeotics in structure and function. Figure 17 shows a graphic representation of the relatedness of the nine homeoboxes found in the ANT-C. The *bcd* box is only distantly related to those in the rest of the complex. This is perhaps not entirely surprising because of the rather different time and place of expression of this gene (i.e., maternal versus zygotic). The relatedness of the other homeoboxes is, however, somewhat interesting. The *pb* and *lab* genes are more closely related to each other and to *zen* and *z2* than they are to the other homeotics. Moreover,

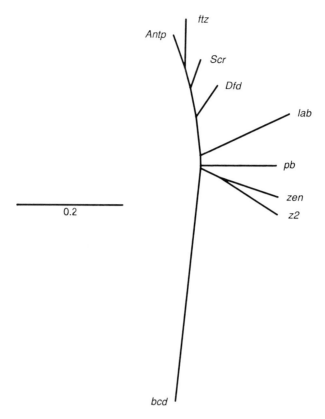

FIG. 17. Family tree of homeoboxes in the ANT-C. The sequence divergence of the homeoboxes listed was calculated with a computer program written by one of the authors (G. Olsen) and the data displayed as a bush. Distances between end points are measured as the total length of the line from tip to tip. The metric is given as 0.2 base changes per position in the total 180 bases of the homeobox.

ftz is more closely related to *Antp* and *Scr* than to *zen*. Overall, the relatedness of the nine homeoboxes more closely predicts the physical proximity on the chromosome of the genes rather than any obvious functional grouping. What is not entirely clear at this point is the significance, if any, of this fact. What is apparent, however, is that sequence similarity among the homeoboxes of the five homeotic loci of the ANT-C should not be used as the sole criterion for predicting evolutionary relatedness among the genes.

VII. Conclusions and Prospects

The molecular and developmental analyses of the Antennapedia complex have shown that it differs from the Bithorax complex in two striking ways. First, interposed among the homeotic loci of the ANT-C is a group of genes that, while important in the ontogenic process, are not homeotic in character. There are no apparent counterparts to these genes in the BX-C. The second and perhaps more interesting difference lies in the spatial patterns of expression of the homeotic genes of the ANT-C that are necessary in the head. The three functional units of the BX-C are expressed in overlapping domains in the trunk of the animal (Akam, 1987). Indeed, a member of the ANT-C, *Antp,* also partially overlaps with the BX-C domains in the anterior of the third thoracic segment. In the head, however, the *lab, Dfd,* and *Scr* genes do not show such an overlap. Rather, they appear to be exclusive in their domains of expression (Mahaffey *et al.,* 1989). The one exception to this rule is *pb,* which does overlap the domains of *Dfd* and *Scr.* This observation is enigmatic, however, because *pb* has no apparent function in the embryo when this concordant expression takes place.

The entirety of the ANT-C has been cloned and antisera raised to the protein products of all of the genes for which mutations exist. With these reagents in hand and with the ease of manipulation of the genome available in *Drosophila,* many questions about the role(s) of these genes in development will be answered. The next years should be very exciting for those interested in the field of developmental genetics.

ACKNOWLEDGMENTS

The authors would like to thank their collaborators in the laboratory Mike Abbott, Scott Chouinard, Bob Diederich, Tulle Hazelrigg, Ricki Lewis, Valerie Merrill, Angela Pattatucci, Mary Anne Pultz, Phil Randazzo, Matt Scott, Barb Wakimoto, and Amy Weiner who have contributed significantly to our understanding of the ANT-C and to the enjoyment of doing the analysis. We thank also Dwayne Johnson and Marie Mazzula for technical assistance and especially thank Drs. David Cribbs and James Mahaffey for technical expertise, insightful comments, and energetic discussions. Finally, we thank Dee Verostko for putting up with all of us and typing the manuscript. The work reported from this laboratory was supported by a National Institutes of Health grant (GM24299) to T. C. Kaufman and an NIH predoctoral fellowship (GM07757) to M. A. Seeger.

REFERENCES

Abbott, M. K., and Kaufman, T. C. (1986). The relationship between the functional complexity and the molecular organization of the *Antennapedia* locus of *Drosophila melanogaster. Genetics* **114,** 919–942.

Akam, M. (1987). The molecular basis for metameric pattern in the *Drosophila* embryo. *Development* **101,** 1–22.

Barthels, D., Santoni, M. J., Wille, W., Ruppert, C., Chaix, J. C., Hirsch, M. R., Fontecilla-Camps, J., and Goridis, C. (1987). Isolation and nucleotide sequence of mouse NCAM cDNA that codes for a M_r 79,000 polypeptide without a membrane-spanning region. *EMBO J.* **6,** 907–914.

Bateson, G. (1988). "Mind and Nature: A Necessary Unity," p. 10. Bantam, New York.

Beer, J., Technau, G. M., and Campos-Ortega, J. A. (1987). Lineage analysis of transplanted individual cells in embryos of *Drosophila melanogaster*. IV. Commitment and proliferative capabilities of mesodermal cells. *Wilhelm Roux's Arch. Dev. Biol.* **196,** 222–230.

Bender, M., Turner, F. R., and Kaufman, T. C. (1987). A developmental genetic analysis of the gene *Regulator of postbithorax* in *Drosophila melanogaster*. *Dev. Biol.* **119,** 418–432.

Bender, W., Akam, M. E., Karch, F., Beachy, P. A., Peifer, M., Spierer, P., Lewis, E. B., and Hogness, D. S. (1983). Molecular genetics of the bithorax complex in *Drosophila melanogaster*. *Science* **221,** 23–29.

Berleth, T., Burri, M., Thoma, G., Bopp, D., Richstein, S., Frigerio, G., Noll, M., and Nüsslein-Volhard, C. (1988). The role of localization of *bicoid* RNA in organizing the anterior pattern of the *Drosophila* embryo. *EMBO J.* **7,** 1749–1756.

Bermingham, J. R., Jr., and Scott, M. P. (1988). Developmentally regulated alternative splicing of transcripts from the *Drosophila* homeotic gene *Antennapedia* can produce four different proteins. *EMBO J.* **7,** 3211–3222.

Beverly, S. M., and Wilson, A. C. (1984). Molecular evolution in *Drosophila* and the higher diptera. II. A time scale for fly evolution. *J. Mol. Evol.* **21,** 1–13.

Boulet, A., and Scott, M. P. (1988). Control elements of the P2 promoter of the Antennapedia gene. *Genes Dev.* **2,** 1600–1614.

Caberra, C., Botas, J., and Garcia-Bellido, A. (1985). Distribution of Ultrabithorax proteins in mutants of *Drosophila* bithorax complex and its trans regulatory genes. *Nature (London)* **318,** 569–571.

Campos-Ortega, J. A., and Hartenstein, V. (1985). "The Embryonic Development of *Drosophila melanogaster*." Springer-Verlag, Berlin and New York.

Carroll, S., and Scott, M. P. (1985). Localization of the *fushi tarazu* protein during *Drosophila* embryogenesis. *Cell* **43,** 47–57.

Carroll, S., and Scott, M. P. (1986). Zygotically active genes that affect the spatial expression of the *fushi tarazu* segmentation gene during early *Drosophila* embryogenesis. *Cell* **45,** 113–126.

Carroll, S. B., Laymon, R., McCuthcheon, M., Riley, P., and Scott, M. P. (1986a). The localization and regulation of Antennapedia protein expression in *Drosophila* embryos. *Cell* **47,** 113–122.

Carroll, S., Winslow, G., Schupbach, T., and Scott, M. (1986b). Maternal control of *Drosophila* segmentation gene expression. *Nature (London)* **323,** 278–280.

Carroll, S., Winslow, G. M., Twombly, V. J., and Scott, M. P. (1987). Genes that control dorsoventral polarity affect gene expression along the anteroposterior axis of the *Drosophila* embryo. *Development* **99,** 327–332.

Carroll, S. B., DiNardo, S., O'Farrell, P. H., White, R. A. H., and Scott, M. P. (1988). Temporal and spatial relationships between segmentation and homeotic gene expression in *Drosophila* embryos: Distributions of the *fushi tarazu, engrailed, Sex combs reduced, Antennapedia* and *Ultrabithorax* proteins. *Genes Dev.* **2,** 350–360.

Chadwick, R., and McGinnis, W. (1987). Temporal and spatial distribution of transcripts from the *Deformed* gene of *Drosophila*. *EMBO J.* **6,** 779–789.

Denell, R. K., Hummels, K., Wakimoto, B., and Kaufman, T. (1981). Developmental studies of lethality associated with the Antennapedia gene complex in *D. melanogaster. Dev. Biol.* **81,** 43–50.

Desplan, C., Thesis, J., and O'Farrell, P. (1985). The *Drosophila* developmental gene engrailed encodes a sequence-specific DNA binding activity. *Nature (London)* **318,** 630–635.

Diederich, R. J., Merrill, V. K. L., Pultz, M. A., and Kaufman, T. C. (1989). Isolation, structure, and expression of *labial,* a homeotic gene of the Antennapedia complex involved in *Drosophila* head development. *Genes Dev.* **3,** 399–414.

Doyle, H. J., Harding, K., Hoey, T., and Levine, M. (1986). Transcripts encoded by a homeobox gene are restricted to dorsal tissues of *Drosophila* embryos. *Nature (London)* **323,** 76–79.

Driever, W., and Nüsslein-Volhard, C. (1988a). A gradient of *bicoid* protein in *Drosophila* embryos. *Cell* **54,** 83–93.

Driever, W., and Nüsslein-Volhard, C. (1988b). The *bicoid* protein determines a position in the *Drosophila* embryo in a concentration-dependent manner. *Cell* **54,** 95–104.

Driever, W., and Nüsslein-Volhard, C. (1989). The *bicoid* protein is a positive regulator of *hunchback* transcription in the early *Drosophila* embryo. *Nature (London)* **337,** 138–143.

Duboule, D., and Dolle, P. (1989). The murine Hox gene network: Its structural and functional organization resembles that of *Drosophila* homeotic genes. *EMBO J.* **8,** 1497–1505.

Duncan, I. (1986). Control of bithorax complex functions by the segmentation gene *fushi tarazu* of *D. melanogaster. Cell* **47,** 297–309.

Duncan, I. (1987). The bithorax complex. *Annu. Rev. Genet.* **21,** 285–320.

Duncan, I., and Lewis, E. (1982). Genetic control of body segment differentiation in *Drosophila. In* "Developmental Order, Its Origin and Regulation: Fortieth Symposium for Development Biology" (S. Subtelney and P. B. Green, eds.), pp. 533–554. Liss, New York.

Edelman, G. M. (1986). Evolution and morphogenesis: The regulator hypothesis. *In* "Genetics, Development, and Evolution" (J. P. Gustafson, G. L. Stebbins, F. J. Ayala, eds.). Plenum, New York.

Edelman, G. M. (1987). CAMs and Igs: Cell adhesion and the evolutionary origins of immunity. *Immunol. Rev.* **100,** 11–45.

Fechtel, K., Natzle, J. E., Brown, E. E., and Fristrom, J. W. (1988). Prepupal differentiation of *Drosophila* imaginal discs: Identification of four genes whose transcripts accumulate in response to a pulse of 20-hydroxyecdysone. *Genes* **120,** 465–474.

Ferguson, M. A. J., and Williams, A. F. (1988). Cell surface anchoring of proteins via glycosyl phosphatidylinositol structures. *Annu. Rev. Biochem.* **57,** 285–320.

Foe, V. E., and Alberts, B. M. (1983). Studies of nuclear and cytoplasmic behaviour during the five mitotic cycles that precede gastrulation in *Drosophila* embryogenesis. *J. Cell Sci.* **61,** 31–70.

Frigerio, G., Burri, M., Bopp, D., Baumgartner, S., and Noll, M. (1986). Structure of the segmentation gene *paired* and the *Drosophila* PRD gene set as part of a gene network. *Cell* **47,** 735–746.

Frischer, L., Hagen, F., and Garber, R. (1986). An inversion that disrupts the *Antennapedia* gene causes abnormal structure and localization of RNAs. *Cell* **47,** 1017–1023.

Frohnhöfer, H. G., and Nüsslein-Volhard, C. (1986). The organization of anterior pattern in *Drosophila* embryos by the maternal gene *bicoid. Nature (London)* **347,** 120–125.

Frohnhöfer, H. G., and Nüsslein-Volhard, C. (1987). Maternal genes required for the anterior localization of *bicoid* activity in the embryo of *Drosophila*. *Genes Dev.* **1**, 880–890.

Garcia-Bellido, A. (1977). Homeotic and atavic mutations in insects. *Am. Zool.* **17**, 613–629.

Gehring, W., and Hiromi, Y. (1986). Homeotic genes and the homeobox. *Annu. Rev. Genet.* **20**, 147–173.

Graham, A., Papalopulu, N., and Krumlauf, R. (1989). The murine and *Drosophila* homeobox gene complexes have common features of organization and expression. *Cell* **57**, 367–378.

Hafen, E., Levine, M., Garber, R. L., and Gehring, W. J. (1983). An improved *in situ* hybridization method for the detection of cellular RNAs in *Drosophila* tissue sections and its application for localizing transcripts of the homeotic *Antennapedia* gene complex. *EMBO J.* **2**, 617–623.

Hafen, E., Kuriowa, A., and Gehring, W. J. (1984a). Spatial distribution of transcripts from the segmentation gene *fushi tarazu* during *Drosophila* embryonic development. *Cell* **47**, 833–841.

Hafen, E., Levine, M., and Gehring, W. (1984b). Regulation of *Antennapedia* transcript distribution by the bithorax complex of *Drosophila*. *Nature (London)* **307**, 287–289.

Harding, K., Wedeen, C., McGinnis, W., and Levine, M. (1985). Spatially regulated expression of homeotic genes in *Drosophila*. *Science* **229**, 1236–1242.

Harding, K., Rushlow, C., Doyle, H., Hoey, T., and Levine, M. (1986). Cross regulatory interactions among pair-rule genes in *Drosophila*. *Science* **233**, 953–959.

Harrelson, A. L., and Goddman, C. S. (1988). Growth cone guidance in insects: Fasciclin II is a member of the immunoglobulin superfamily. *Science* **242**, 700–708.

Hartenstein, V. (1988). Development of *Drosophila* larval sensory organs: Spatiotemporal pattern of sensory neurons, peripheral axonal pathways and sensilla differentiation. *Development* **102**, 869–886.

Hazelrigg, T., and Kaufman, T. (1983). Revertants of dominant mutations associated with the Antennapedia gene complex of *Drosophila melanogaster*: Cytology and genetics. *Genetics* **105**, 581–600.

Hiromi, Y., and Gehring, W. (1987). Regulation and function of the *Drosophila* segmentation gene *fushi tarazu*. *Cell* **50**, 963–974.

Hiromi, Y., Kuroiwa, A., and Gehring, W. J. (1985). Control elements of the *Drosophila* segmentation gene *fushi tarazu*. *Cell* **43**, 603–613.

Howard, K., and Ingham, P. (1986). Regulatory interactions between segmentation genes *fushi tarazu, hairy* and *engrailed* in the *Drosophila* blastoderm. *Cell* **44**, 949–957.

Ingham, P., and Martinez-Arias, A. (1986). The correct activation of Antennapedia and bithorax complex genes requires the *fushi tarazu* gene. *Nature (London)* **324**, 592–597.

Jack, T., Regulski, M., and McGinnis, W. (1988). Pair-rule segmentation genes regulate the expresion of the homeotic selector gene, *Deformed*. *Genes Dev.* **2**, 592–597.

Jackle, H., Tautz, D., Schuh, R., Seifert, E., Lehmann, R. (1986). Cross regulatory interactions among the gap genes of *Drosophila*. *Nature (London)* **324**, 668–670.

Jorgensen, E., and Garber, R. (1987). Function and misfunction of the two promoters of the *Drosophila Antennapedia* gene. *Genes Dev.* **1**, 544–555.

Jurgens, G., Lehmann, R., Schardin, M., and Nüsslein-Volhard, C. (1986). Segmental organization of the head in the embryo of *Drosophila melanogaster*: A blastodem fate map of article structures of the larval head. *Wilhelm Roux's Arch. Dev. Biol.* **195**, 359–377.

Kakidani, H., and Ptashne, M. (1988). GAL4 activates gene expression in mammalian cells. *Cell* **52**, 161–167.

Karch, F., Weiffenbach, B., Peifer, M., Bender, W., Duncan, I., Celniker, S., Crosby, M., and Lewis, E. B. (1985). The abdominal region of the Bithorax complex. *Cell* **43**, 81–96.

Kauffman, S., Shymoko, R., and Trabert, K. (1978). Control of sequential compartment formation in *Drosophila*. *Science* **199**, 259–270.

Kaufman, T. C. (1978). Cytogenetic analysis of chromosome 3 in *Drosophila melanogaster*: Isolation and characterization of four new alleles of the *proboscipedia* (*pb*) locus. *Genetics* **90**, 579–596.

Kaufman, T. C. (1983). The genetic regulation of segmentation in *Drosophila melanogaster*. *In* "Time, Space and Pattern in Embryonic Development" (W. R. Jeffery and R. A. Raff, eds.), pp. 365–383. Liss, New York.

Kaufman, T., and Abbott, M. (1984). Homeotic genes and the specification of segmental identity in the embryo and adult thorax of *D. melanogaster*. *In* "Molecular Aspects of Early Development" (G. M. Malacinski and W. H. Klein, eds.), pp. 189–218. Plenum, New York.

Killeen, N., Moessner, R., Arvieux, J., Willis, A., and Williams, A. F. (1988). The MRC OX-45 nitrogen of rat leukocytes and endothelium is in a subset of the immunoglobulin superfamily with CD2, LFA-3 and carcinoembryonic antigens. *EMBO J.* **7**, 3087–3091.

King, R. C. (1970). "Ovarian Development in *Drosophila melanogaster*." Academic Press, New York.

Kuroiwa, A., Hafen, E., and Gehring, W. J. (1984). Cloning and transcriptional analysis of the segmentation gene *fushi tarazu* of *Drosophila*. *Cell* **47**, 825–831.

Lai, C., Brow, M. A., Nave, K. A., Noronha, A.B., Quarles, R. H., Bloom, F. E., Milner, R. J., and Sutcliffe, J. G. (1987). Two forms of 1B236/mylein-associated glycoprotein, a cell adhesion molecule for postnatal neural development, are produced by alternative splicing. *Proc. Natl. Acad. Sci. U.S.A.* **84**, 4337–4341.

Lambert, L. A. (1985). A genetic and developmental analysis of female sterile loci associated with the Antennapedia complex in *Drosophila melanogaster*. Ph.D. dissertation, Indiana University, Bloomington, Indiana.

Laughon, A., and Scott, M. P. (1984). Sequence of a *Drosophila* segmentation gene: Protein structure homology with DNA-binding proteins. *Nature (London)* **310**, 23–31.

Laughon, A., Boulet, A., Bermingham, J., Laymon, R., and Scott, M. (1986). The structure of transcripts from the homeotic *Antennapedia* gene of *Drosophila*: Two promoters control the major protein coding region. *Mol. Cell. Biol.* **6**, 6476–6481.

Lehmann, R., and Nüsslein-Volhard, C. (1987). *hunchback,* a gene required for segmentation of an anterior and posterior region of the *Drosophila* embryo. *Dev. Biol.* **119**, 402–417.

LeMotte, P., Kuriowa, A., Fessler, L. I., and Gehring, W. J. (1989). The homeotic gene *Sex combs reduced* of *Drosophila*: Gene structure and embryonic expression. *EMBO J.* **8**, 219–227.

Lewis, E. (1963). Genes and develomental pathways. *Am. Zool.* **3**, 33–56.

Lewis, E. B. (1978). A gene complex controlling segmentation in *Drosophila*. *Nature (London)* **276**, 565–570.

Lewis, R. A., Kaufman, T. C., Denell, R. E., and Tallerico, P. (1980a). Genetic analysis of the Antennapedia gene complex (ANT-C) and adjacent chromosomal regions of *Drosophila melanogaster*. I. Polytene chromosome segments 84B–D. *Genetics* **95**, 367–381.

Lewis, R. A., Wakimoto, B. T., Denell, R. E., and Kaufman, T. C. (1980b). Genetic analysis of the Antennapedia gene complex (ANT-C) and adjacent chromosomal regions of *Drosophila melanogaster*. II. Polytene chromosome segments 84A–B. *Genetics* **95**, 383–397.

Lipman, D. J., and Pearson, W. R. (1985). Rapid and sensitive protein similarity searches. *Science* **227**, 1435–1441.

Ma, J., and Ptashne, M. (1987). Deletion analysis of GAL4 defines two transcriptional activating segments. *Cell* **48**, 847–853.

Macdonald, P. M., and Struhl, G. (1988). Cis-acting sequences responsible for anterior localization of *bicoid* mRNA in *Drosophila* embryos. *Nature (London)* **336**, 595–598.

Mahaffey, J. W., and Kaufman, T. C. (1987a). Distribution of the *Sex combs reduced* gene products in *Drosophila melanogaster*. *Genetics* **117**, 51–60.

Mahaffey, J., and Kaufman, T. (1987b). The homeotic genes of the Antennapedia complex and Bithorax complex of *Drosophila*. In "Developmental Genetics of Higher Organisms: A Primer in Developmental Biology" (G. M. Malacinski, ed.). Macmillan, New York.

Mahaffey, J. W., Diederich, R. J., and Kaufman, T. C. (1989). Novel patterns of homeotic protein accumulation in the head of the *Drosophila* embryo. *Development* **105**, 167–174.

Martinez-Arias, A. (1986). The *Antennapedia* gene is required and expressed in parasegments 4 and 5 of the *Drosophila* embryo. *EMBO J.* **5**, 135–141.

Martinez-Arias, A., Ingham, P., Scott, M., and Akam, M. (1987). The spatial and temporal deployment of *Dfd* and *Scr* transcripts throughout development of *Drosophila*. *Development* **100**, 673–683.

Merrill, V. K. L., Turner, F. R., and Kaufman, T. C. (1987). A genetic and developmental analysis of mutations in the *Deformed* locus in *Drosophila melanogaster*. *Dev. Biol.* **122**, 379–395.

Merrill, V. K. L., Diederich, R. J., Turner, F. R., and Kaufman, T. C. (1989). A genetic and developmental analysis of mutations in *labial,* a gene necessary for proper head formation in *Drosophila melanogaster*. *Dev. Biol* **135**, 376–391.

Mlodzik, M., Fjose, A., and Gehring, W. J. (1988). Molecular structure and spatial expression of a homeobox gene from the *labial* region of the Antennapedia complex. *EMBO J.* **7**, 2569–2578.

Neumaier, M., Zimmermann, W., Shively, L., Hinoda, Y., Riggs, A., and Shively, J. E. (1988). Characterization of a cDNA clone for the nonspecific cross-reacting antigen (NCA) and a comparison of NCA and carcinoembryonic antigen. *J. Biol. Chem.* **7**, 3202–3207.

Nüsslein-Volhard, C., and Wieschaus, E. (1980). Mutations affecting segment number and polarity in *Drosophila*. *Nature (London)* **287**, 795–801.

Nüsslein-Volhard, C., Frohnhöfer, H. G., and Lehmann, R. (1987). Determination of anteroposterior polarity in *Drosophila*. *Science* **238**, 1675–1681.

O'Connor, M. B., Binari, R., Perkins, L. A., and Bender, W. (1988). Alternative RNA products from the *Ultrabithorax* domain of the bithorax complex. *EMBO J.* **7**, 435–445.

Otteson, D., Mahaffey, J., and Kaufman, T. (1990). The functional organization of the *Sex combs reduced (Scr)* locus of *Drosophila melanogaster*. In preparation.

Ouweneel, W. J. (1976). Developmental genetics of homeosis. *Adv. Genet.* **18**, 179–248.

Poulson, D. F. (1950). Histogenesis, organogenesis and differentiation in the embryo of

Drosophila melanogaster (Meigen). *In* "Biology of *Drosophila*" (M. Demerec, ed.), pp. 168–274. Wiley, New York.

Pultz, M. A. (1988). A molecular and genetic analysis of *proboscipedia* and flanking genes in the Antennapedia complex of *Drosophila melanogaster*. Ph.D. dissertation, Indiana University, Bloomington, Indiana.

Pultz, M. A., Diederich, R. J., Cribbs, D. L., and Kaufman, T. C. (1988). The *proboscipedia* locus of the Antennapedia complex: A molecular and genetic analysis. *Genes Dev.* **2**, 901–920.

Raff, R. A., and Kaufman, T. C. (1983). "Embryos, Genes, and Evolution." Macmillan, New York.

Rechsteiner, M. (1987). Do *myc, fos* and E1A function as protein phosphatase inhibitors? *Biochem. Biophys. Res. Commun.* **143**, 194–198.

Regulski, M., Harding, K., Kostriken, R., Karch, F., Levine, M., and McGinnis, W. (1985). Homeobox genes of the Antennapedia and Bithorax complexes of *Drosophila. Cell* **43**, 71–80.

Regulski, M., McGinnis, N., Chadwick, R., and McGinnis, W. (1987). Developmental and molecular analysis of *Deformed*; a homeotic gene controlling *Drosophila* head development. *EMBO J.* **6**, 767–777.

Riley, P., Carroll, S., and Scott, M. (1987). The expression and regulation of *Sex combs reduced* protein in *Drosophila* embryos. *Genes Dev.* **1**, 716–730.

Rogers, S., Wells, R., and Rechsteiner, M. (1986). Amino acid sequences common to rapidly degraded proteins: The PEST hypothesis. *Science* **234**, 364–368.

Rushlow, C., Doyle, H., Hoey, T., and Levine, M. (1987a). Molecular characterization of the *zerknüllt* region of the Antennapedia gene complex in *Drosophila. Genes Dev.* **1**, 1268–1279.

Rushlow, C., Frasch, M., Doyle, H., and Levine, M. (1987b). Maternal regulation of *zerknüllt*: A homeobox gene controlling differentiation of dorsal tissues in *Drosophila. Nature (London)* **330**, 583–586.

Salzer, J. L., Holme, W. P., and Colman, D. R. (1987). The amino acid sequences of the myelin-associated glycoproteins: Homology to the immunoglobulin gene superfamily. *J. Cell. Biol.* **104**, 957–965.

Sanchez-Herrero, E., Vernos, I., Marco, R., and Morata, G. (1985). Genetic organization of *Drosophila* bithorax complex. *Nature (London)* **313**, 108–113.

Sato, T., Hayes, P. H., and Dennell, R. E. (1985). Homeosis in *Drosophila*: Roles and spatial patterns of expression of the *Antennapedia* and *Sex combs reduced* loci in embryogenesis. *Dev. Biol.* **111**, 171–192.

Schneuwley, S., and Gehring, W. (1985). Homeotic transformation of thorax to head: Developmental analysis of a new *Antennapedia* allele in *D. melanogaster. Dev. Biol.* **108**, 377–386.

Schneuwley, S., Kuroiwa, A., Baumgartner, P., and Gehring, W. J. (1986). Structural organization and sequence of the homeotic gene *Antennapedia* of *Drosophila melanogaster. EMBO J.* **5**, 733–739.

Schneuwley, S., Kuroiwa, A., and Ghering, W. J. (1987). Molecular analysis of the dominant homeotic Antennapedia phenotype. *EMBO J.* **6**, 201–206.

Scott, M., and O'Farrell, P. (1986). Spatial programming of gene expression in early *Drosophila* embryogenesis. *Annu. Rev. Cell. Biol.* **2**, 49–80.

Scott, M. P., and Weiner, A. J. (1984). Structural relationships among genes that control development: Sequence homology between the *Antennapedia, Ultrabithorax*, and *fushi tarzu* loci of *Drosophila. Proc. Natl. Acad. Sci. U.S.A.* **81**, 4115–4119.

Scott, M. P., Weiner, A. J., Hazelrigg, T. I., Polisky, B. A., Pirrotta, V., Scalenghe, F., and Kaufman, T. C. (1983). The molecular organization of the *Antennapedia* locus of Drosophila. *Cell* **35**, 763–776.

Seeger, M. A. (1989). A molecular and genetic analysis of the *bicoid–zerknüllt* interval of the Antennapedia gene complex in *Drosophila melanogaster.* Ph.D. dissertation, Indiana University, Bloomington, Indiana.

Seeger, M. A., and Kaufman, T. C. (1987). Homeotic genes of the Antennapedia complex (ANT-C) and their molecular variability in the phylogeny of the Drosophilidae. *In* "Development as an Evolutionary Process" (R. Raff and E. Raff, eds.), pp. 179–202. Liss, New York.

Seeger, M. A., Haffley, L., and Kaufman, T. C. (1988). Characterization of *amalgam*: A member of the immunoglobulin superfamily from *Drosophila. Cell* **55**, 589–600.

Stroeher, V., Jorgensen, E., and Garber, R. (1986). Multiple transcripts from the *Antennapedia* gene of *Drosophila. Mol. Cell. Biol.* **6**, 4667–4675.

Stroeher, V. L., Gaiser, C., and Garber, R. L. (1988). Alternative RNA splicing that is spatially regulated: Generation of transcripts from the *Antennapedia* gene of *Drosophila melanogaster* with different protein-coding regions. *Mol. Cell. Biol.* **8**, 4143–4154.

Struhl, G. (1981). A homeotic mutation transforming leg to antenna in *Drosophila. Nature (London)* **292**, 635–638.

Shruhl, G., and White, R. (1985). Regulation of the Ultrabithorax gene of *Drosophila* by other bithorax complex genes. *Cell* **43**, 507–519.

Wakimoto, B. T., Turner, F. R., and Kaufman, T. C. (1984). Defects in embryogenesis in mutants associated with the Antennapedia gene complex of *Drosophila melanogaster. Dev. Biol.* **102**, 147–172.

Weiner, A., Scott, M., and Kaufman, T. (1984). A molecular analysis of *fushi tatazu,* a gene in *D. melanogaster* that encodes a product affecting segment number and cell fate. *Cell* **37**, 843–851.

Wharton, K. A., Yedvobnick, B., Finnerty, V., and Artavanis-Tsakonas, S. (1985). *opa*: A novel family of transcribed repeats shared by the *Notch* locus and other developmentally regulated loci in *D. melanogaster. Cell* **40**, 55–62.

White, R., and Lehmann, R. (1986). A gap gene, *hunchback,* regulates the spatial expression of *Ultrabithorax. Cell* **47**, 311–321.

White, R. A. H., and Wilcox, M. (1984). Protein products of the bithorax complex in *Drosophila. Cell* **39**, 163–171.

Williams, A. F. (1987). A year in the life of the immunoglobulin superfamily. *Immunol. Today* **8**, 298–303.

Williams, A. F., and Barclay, A. N. (1988). The immunoglobulin superfamily—Domains for cell surface recognition. *Annu. Rev. Immunol.* **6**, 381–405.

Wirz, J., Fessler, L., and Gehring, W. (1986). Localization of Antennapedia protein in *Drosophila* embryos and imaginal discs. *EMBO J.* **5**, 3327–3334.

ESTABLISHMENT AND MAINTENANCE OF POSITION-SPECIFIC EXPRESSION OF THE *Drosophila* HOMEOTIC SELECTOR GENE *Deformed*

William McGinnis, Thomas Jack, Robin Chadwick, Michael Regulski, Clare Bergson, Nadine McGinnis, and Michael A. Kuziora

Departments of Molecular Biophysics and Biochemistry and of Biology, Yale University, New Haven, Connecticut 06511

I. Introduction

A. DEVELOPMENT OF POSTERIOR HEAD SEGMENTS DURING EMBRYOGENESIS

The body pattern of *Drosophila* varies a basic segmental morphology at different positions on the anterior–posterior axis to achieve an astonishing variety of related but very different structures. Nowhere are the differences in this basic plan more extreme than in the head region. At early stages of development, the morphological organization of the

ADVANCES IN GENETICS, Vol. 27

posterior head is similar to that of the rest of the body, but thereafter it rapidly diverges. As shown in Fig. 1, during mid-stages of embryogenesis, the posterior head is organized in a segmental manner, superficially similar to that of the thoracic and abdominal regions. The three segmental lobes that comprise most of the posterior head are the mandibular, maxillary, and labial, and they are collectively known as the gnathal segments (Fig. 1).

The morphological divisions that mark the segmental boundaries of the gnathal segments are the first to appear in the visible segmentation of the *Drosophila* body plan (Turner and Mahowald, 1979), arising at about 6.5 hours of development (stage 11 of Campos-Ortega and Hartenstein, 1985). During subsequent developmental stages, rearrangements and fusions of the head segments obscure their segmental origins, and these changes are further complicated by the eventual

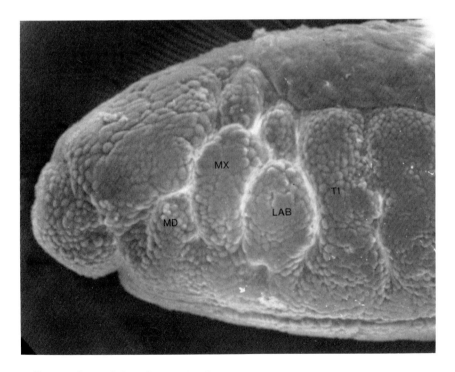

FIG. 1. *Drosophila* embryonic head segments as seen in a scanning electron micrograph of the anterior end of a *Drosophila* embryo at the stage of germ band extension. Dorsal is up. The gnathal segments, mandibular (MD), maxillary (MX), and labial (LAB), are indicated, as is the first thoracic segment (T1).

involution of much of the gnathal region into the stomodeum. Thus, in the first instar larva which emerges at the end of embryonic develop-ment, the cells that were originally located in the gnathal lobes give rise to the lateral walls of the pharynx, the floor of the anterior phar-ynx, and the border of the stomodeum, including a sensory organ-rich region known as the pseudocephalon (Jurgens *et al.*, 1986).

A variety of specific sensory and skeletal features of the larva develop from the gnathal segment cells (Jurgens *et al.*, 1986). Among the most prominent, and most important for the purposes of this article, are the mouth hooks and cirri. The mouth hooks are serrated chitinous struc-tures that function as the jawbones and teeth of the larva. The mouth hooks develop from cells of the maxillary segment, as do the cirri, rows of triangular papillae that flank the opening of the larval mouth (Figs. 2 and 3).

B. Genetic Functions of the *Deformed* Locus in Head Development

One of the original dominant mutant alleles isolated in the labora-tory of Thomas Hunt Morgan was *Deformed* [*Dfd* of Bridges and Mor-gan (1923) here designated *Dfd*^{D}]. Adult flies that carry the *Dfd*^{D} allele

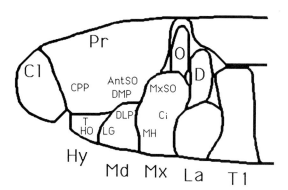

FIG. 2. Fate map of the embryonic head. Anterior is to the left and dorsal up. The subdivisions of the head are as follows: clypeolabrum (Cl), procephalic lobe (Pr), hypopharyngeal lobe (Hy), mandibular (Md), maxillary (Mx), labial (La), first tho-racic segment (T1), optic lobe (O), and dorsal ridge (D). Superimposed on the drawing are the regions in which ablation experiments locate the cellular primordia for the maxillary sense organ (MxSO), cirri (Ci), mouth hooks (MH), dorsolateral papilla (DLP), lateralgraten (LG), T-bars (T), hypopharyngeal organ (HO), cephalo-pharyngeal plates (CPP), antennal sense organ (AntSO), and dorsomedial papilla (DMP).

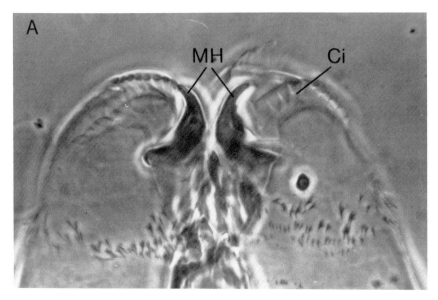

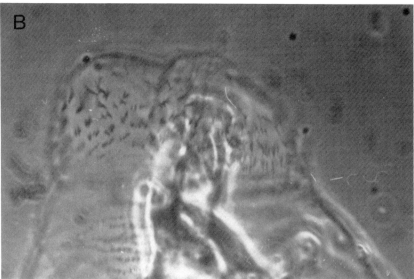

FIG. 3. Morphological phenotype of *Deformed* mutants. (A) Phase-contrast micrograph of the anterior end of a wild-type first instar larva just after emerging from the eggshell. Anterior is up. The mouth hooks (MH) and cirri (Ci) are labeled. (B) Anterior end of a *Dfd* mutant larva, which, among other abnormalities, is missing both mouth hooks and cirri.

are missing ventral eye and orbital tissue. In a genetic analysis of the Antennapedia complex (ANT-C; see article by Kaufman *et al.* in this volume), Hazelrigg and Kaufman (1983) recovered several phenotypic revertants of *Dfd*^D that had lost the dominant adult phenotype but gained a recessive lethal phenotype when combined with alleles of the R3 complementation group of the ANT-C (Lewis *et al.*, 1980). Their genetic results allowed them to conclude that the R3 alleles were lethal, loss-of-function mutations in the *Dfd* locus. Reversion-induced chromosomal breakpoints placed the *Dfd* locus in the middle of the ANT-C and provided both a preliminary definition of sequences required for *Dfd* function, as well as one of the entry points for the isolation of cloned segments of ANT-C DNA (Scott *et al.*, 1983).

The wild-type function of *Dfd* is to assign "head" identity to cells that eventually develop as part of the maxillary and mandibular segments. As described later in this article, *Dfd* can accomplish this head programming function even when expressed far outside its normal positional domain. Embryos homozygous for *Dfd* null mutations develop severe pattern defects in the head region and die without emerging from the eggshell (Wakimoto and Kaufman, 1981; Merrill *et al.*, 1987; Regulski *et al.*, 1987). Thoracic and abdominal segments develop normally. The terminal phenotype of the epidermally derived structures can be observed in cleared cuticle preparations of *Dfd* mutants in which all soft tissues have been dissolved (Van der Meer, 1977). Figure 3 shows the head skeleton of a *Dfd*⁻ larva compared to a wild-type first instar larva. The most conspicuous structures missing in *Dfd*⁻ mutants are the mouth hooks and cirri, both of which are derived from the ventral and ventral posterior regions of the maxillary segment (Fig. 2). Structures derived in large part from other regions of the maxillary segment develop fairly normally, an example being the maxillary sense organ, which is still recognizable although the papillar arrangement is disrupted. At least one structure that derives from the mandibular segment is also missing in *Dfd*⁻ mutants, namely, the dorsal lateral papilla of the maxillary sense organ.

In embryos that are trans-heterozygous for the *Dfd*^RX1 mutation (a chromosomal breakpoint which interrupts the *Dfd* transcription unit, described later) and a deleton for *Dfd* and nearby genes, DF (3R)*Scr,* a homeotic transformation occurs in the larval head (Regulski *et al.,* 1987). In these mutant animals, in addition to the loss of mouth hooks and cirri, the cephalopharyngeal plates (or ventral plates) are partially duplicated in the anterior atrium, replacing the H-piece and lateralgraten. This transformation occurs at a fairly low pentrance

(30–50%) and is not seen in mutants that are point mutants and protein nulls. Why the *Dfd* breakpoint would have an apparently stronger phenotype than a protein null is unknown. A similar phenomenon has been observed at the *Ultrabithorax* (*Ubx*) locus, a homeotic gene whose primary function is to specify identity to the third thoracic and first abdominal segments. When mutants are placed in a allelic series at *Ubx* and then compared with the molecular lesions underlying the mutation, those mutants in the strongest class are deletions or breakpoints in the locus, whereas those in the moderate class are point mutations that result in no *Ubx* protein expression (Weinzierl *et al.,* 1987).

C. MOLECULAR STRUCTURE OF THE *Deformed* LOCUS

DNA containing the *Dfd* locus was isolated both by chromosomal walking in the ANT-C (Scott *et al.,* 1983) as well as by the use of homeobox homology to clone genes that contained sequences closely related to the *Antennapedia* (*Antp*) homeobox (McGinnis *et al.,* 1984a). Two independently isolated chromosomal breakpoints, both lethal loss-of-function *Dfd* mutant alleles, interrupt the homeobox-containing transcription unit and identify it as the *Dfd* transcription unit (Regulski *et al.,* 1987). The *Dfd* transcription unit encodes a single mature mRNA of 2.8 kb, whose five coding exons are spread over 11 kb of genomic DNA (Fig. 4). The 5' end of the *Dfd* transcription unit is approximately 20 kb distal (relative to the third chromosome centromere, left in Fig. 4) to the 3' end of the nearest known adjacent transcription unit of the ANT-C, that of *amalgam* (Seeger *et al.,* 1988). The poly(A) signal of *Dfd* is approximately 6 kb proximal to the Z transcription unit of Kuroiwa *et al.* (1985). The Z transcript has no known genetic function; it has an expression pattern unrelated to that of *Dfd* (M. Regulski, unpublished).

A full-length cDNA for the *Dfd* transcript has been isolated and sequenced (Regulski et al., 1987). It contains a single long open reading frame of 1758 nucleotides. The *Dfd* protein encoded by this open reading frame comprises 586 amino acids, with a calculated molecular weight of 63,500. Untranslated leader and trailer sequences, each approximately 500 nucleotides in length, flank the *Dfd* protein-coding region in the *Dfd* transcript. The protein structure can be divided roughly into three regions. The best understood region in functional terms, the middle third, contains an acidic portion as well as the *Dfd* homeodomain of 66 amino acids (Figs. 5 and 6). Since homeodomains, or domains with even slight similarity to them, act as sequence-specific

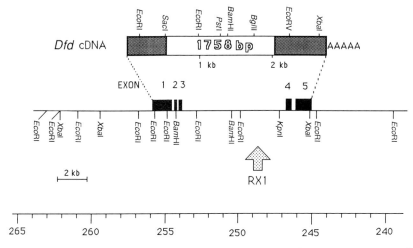

FIG. 4. *Dfd* cDNA and genomic structure. The elongated box at top represents the *Dfd* cDNA (~2.75 kb), which has an open reading frame encoding the *Dfd* protein of 1758 nucleotides. *Dfd* cDNA sequences are encoded in five exons (solid boxes) whose locations are indicated on the line representing the genomic DNA of the *Dfd* locus. The position of the *Dfd*[RX1] breakpoint, a null allele of *Dfd*, is marked by the stippled arrow. The distance in kilobases from the *Humeral* chromosomal breakpoint (Garber *et al.*, 1983) is indicated at the bottom.

DNA-binding domains in transcriptional regulatory proteins (Desplan *et al.*, 1985; Hall and Johnson, 1987; Hoey and Levine, 1988), it is likely that the developmental regulatory functions of *Dfd* are mediated by a DNA-binding, gene regulatory role. As discussed later, one of the genes directly regulated by the *Dfd* protein appears to be the *Dfd* gene itself. The acidic region upstream of the homeodomain in the *Dfd* protein is similar in charge to protein domains that are necessary for gene activation in yeast DNA-binding proteins. The amino-terminal region of the *Dfd* protein contains two glycine-rich regions, and the carboxy-terminal region includes two regions of monotonic amino acid sequence, polyglutamine and polyasparagine. The potential biochemical function of these distinctive protein regions is unknown.

The amino acid sequence of the *Dfd* homeodomain places it in the *Antp* class of *Drosophila* homeobox genes, a group of genes that conserve very similar homeodomain sequences in cross-comparison (70–90% amino acid similarity; McGinnis *et al.*, 1984b; Scott and Weiner, 1984; Regulski *et al.*, 1985). However, the closest structural relatives of the *Dfd* gene are not found in other *Drosophila* homeobox genes, but in homeobox genes of other species whose encoded proteins share

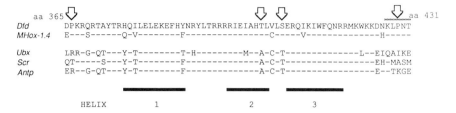

```
     aa 365 ⇩                                    ⇩   ⇩                              ⇩ aa 431
Dfd       DPKRQRTAYTRHQILELEKEFHYNRYLTRRRRIEIAHTLVLSERQIKIWFQNRRMKWKKDNKLPNT
MHox-1.4  E---S------Q-V--------F---------------C-----V-------------H-----

Ubx       LRR-G-QT---Y-T--------T-H---------M--A-C-T-------------L--EIQAIKE
Scr       QT-----S---Y-T--------F-------------A-C-T----------------EH-MASM
Antp      ER--G-QT---Y-T--------F-------------A-C-T----------------E--TKGE


        HELIX           1                 2              3
```

FIG. 5. *Dfd* homeodomain sequence compared to other homeodomains. Using the single-letter code for amino acids, the *Dfd* homeodomain (66 amino acids, residues 365–431 of the *Dfd* protein sequence) is aligned with a mouse *Dfd*-like homeodomain from the *Hox-1.4* (*MHox-1.4*) gene (Rubin *et al.*, 1986) as well as with the homeodomain sequences most similar to *Dfd* in the *Drosophila* genome, namely, those of *Ultrabithorax* (*Ubx*), *Sex combs reduced* (*Scr*), and *Antennapedia* (*Antp*). The positions of the three α-helical regions in the homeodomain are indicated (Otting *et al.*, 1988). Open arrows above the *Dfd* sequence indicate the 8 amino acid residues that are characteristic of *Dfd*-like homeodomains.

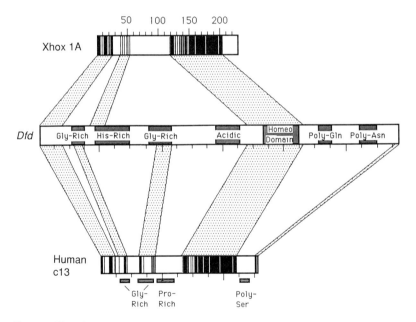

FIG. 6. Vertebrate proteins that are structural homologs of the *Dfd* protein. The Xhox 1A protein of *Xenopus* (Harvey *et al.*, 1986), HHox c13 protein of humans (Mavilio *et al.*, 1986), and *Dfd* protein (Regulski *et al.*, 1987) are indicated by elongated boxes. Solid lines in the schematic Xhox and c13 sequences represent identities with the indicated regions of the *Dfd* protein. Numbers above the *Xenopus* sequence indicate amino acid residues from the amino terminus.

extensive similarity in amino acid sequence both inside and outside of the homeodomain (Mavilio *et al.*, 1986; Rubin *et al.*, 1986; Wolgemuth *et al.*, 1986; Regulski *et al.*, 1987; Graham *et al.*, 1988). For example, the amino acid sequence of the *Dfd* homeodomain has only 8 differences in a 66-amino acid region when the *Dfd* and mouse Hox-1.4 sequences are aligned. In contrast, *Dfd* has 15 differences in the same interval when aligned with the *Sex combs reduced* (*Scr*) homeodomain, its closest *Drosophila* relative, and even more differences when aligned with *Antp* and *Ubx* (Fig. 5).

In addition, when all extant homeodomain sequences are compared, a *Dfd*-like subclass can be identified through the conservation of three distinctive amino acids within the homeodomain and a short peptide sequence just downstream of the conventionally accepted carboxy terminus of the homeodomain (diagrammed in Fig. 5). Using this criterion to define a *Dfd*-like subclass, the *Drosophila* genome has only one member, but the mouse genome has at least four known members, one in each of the Hox (homeobox) gene complexes that contain *Antp* class homeobox genes (Hox-1, Hox-2, Hox-3, and Hox-5; reviewed in Fienberg *et al.*, 1987). The genes of the *Dfd*-like subclass also have considerable amino acid similarity outside of the homeodomain. This is diagrammatically shown in Fig. 6, in which *Xenopus* and human *Dfd*-like proteins are aligned with *Dfd*. The vertebrate homeobox proteins of the *Dfd*-like subclass differ greatly in overall size from *Drosophila Dfd,* an example being *Xenopus* Xhox 1A at 230 amino acids versus *Dfd* at 586 (Harvey *et al.*, 1986). However, much of the *Dfd* protein sequence consists of monotonous amino acid repeats which are missing from homologous positions of the *Xenopus* protein.

The extensive similarity between *Drosophila Dfd* and *Dfd*-like vertebrate homeodomain proteins indicates that a *Dfd*-like gene or genes had evolved and diverged away from other homeobox genes of the *Antp* class prior to the evolutionary divergence between the arthropod and vertebrate lineages (Fig. 7). The common ancestor to present-day *Drosophila* and vertebrates is believed to have existed approximately 600 million years ago (Hadzi, 1963). Though the morphology of the common ancestor is unknown and will probably remain unknowable, it is tempting to speculate that even in this ancient organism, a *Dfd*-like gene functioned to assign head or anterior identity. Although gene function in the common ancestor is not easily tested, the expression pattern and function of *Dfd*-like genes in present-day animals on both sides of the arthropod–vertebrate split will undoubtedly be elucidated. Recent results indicate that some of the *Dfd*-like genes of the mouse are expressed most abundantly in anterior regions (hindbrain and anterior

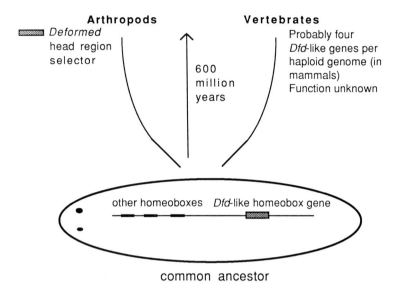

FIG. 7. Evolutionary history of *Dfd*-like genes. As diagrammatically indicated at the bottom, a common ancestor to both present-day arthropods and vertebrates existed about 600 million years ago. Comparison of homeobox gene structure in *Drosophila* and vertebrates strongly suggests that a *Dfd*-like gene or genes had diverged away from other *Antp* class homeobox genes prior to the evolutionary split which eventually gave rise to arthropod and vertebrate lineages.

spinal cord) of the embryonic central nervous system (CNS), as well as in embryonic cervical somites. The genetic function of vertebrate *Dfd*-like genes will provide fascinating insight into whether the structural conservation of the *Dfd*-like structure is paralleled by any conservation of developmental function.

D. *Deformed* POSITIONAL FUNCTION COMPARED TO OTHER HOMEOTIC SELECTOR GENES

As summarized in this article, *Dfd* is one of the *Drosophila* genes which determines cells to assume a position-specific fate on the anterior–posterior axis of the embryo. This function places *Dfd* in the class of homeotic selector genes, whose stable, localized expression is necessary to assign individual identities to head, thoracic, and abdominal segments of the developing embryo. The most obvious realization of this embryonic function is the development of specific cuticular features of the larval exoskeleton. Figure 8 shows that by this criterion, the *Dfd*

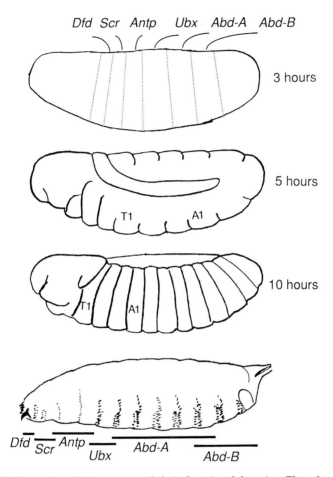

FIG. 8. Homeotic selector genes and their functional domains. The schematic embryo at the top represents the cellular blastoderm stage (~2.75 hours). The order of the expression domains for six homeotic selector genes of the Antennapedia and Bithorax complexes is indicated. Anterior is to the left and dorsal is up for all embryos. At 5 hours of development (extended germ band stage), the segmental primordia are extended around the posterior pole to occupy both ventral and dorsal aspects of the embryo. At 10 hours of development (retracted germ band stage), the segmental primordia have retracted to the ventral aspect of the embryo, and the head segments are beginning to fuse and involute. A simplified view of a first instar larva is shown at the bottom (~22 hours), with the domains of epidermal function of the homeotic selectors indicated on the body plan.

function is most important in the maxillary segment of the head, the *Scr* function most important in the labial segment and first thoracic segments, *Antp* in the thoracic segments, *Ubx* in the third thoracic segment and first abdominal segment, *abdominal-A* (*abd-A*) in the first through seventh abdominal segments, and *Abdominal-B* (*Abd-B*) in the fifth through ninth abdominal segments (Lewis, 1978; Kaufman, 1983; Sanchez-Herrero *et al.*, 1985). All of these homeotic selectors are located in the Antennapedia and Bithorax complexes (the overall structure of the ANT-C is summarized in the article by Kaufman *et al.* in this volume).

The zygotic function of the homeotic selector genes is first required at the cellular blastoderm stage (2.5 hours), at which point the embryo is an oblate spheroid of about 5,000 undifferentiated cells. The initial expression limits of the selectors are largely nonoverlapping, and they comprise circumferential bands of cells at progressively more posterior positions from *Dfd* through *Abd-B* (reviewed in Akam, 1987). The primordia for all three germ layers are included in the bands of cells that exhibit selector expression at this stage. The homeotic selector expression patterns quickly complicate during gastrulation and segmentation stages and continue to change throughout the remainder of embryonic development, as described for *Dfd* in this article. However, for all the indicated homeotic genes, at least some of the initially expressing cells continuously express selector transcripts and proteins throughout most of embryogenesis. The arrangement of the head, thoracic, and abdominal regions at four distinctive embryonic stages is diagrammed in Fig. 8.

II. Establishment of Position-Specific Expression of *Deformed*

A. HIERARCHY OF *Drosophila* DEVELOPMENTAL PATTERNING GENES FOR THE ANTERIOR–POSTERIOR AXIS

The powerful genetic methods available in *Drosophila* have allowed screening protocols with the capability of identifying most or all maternal and zygotic mutants specifically involved in developmental patterning of the anterior–posterior (AP) axis of the embryo (Rice and Garen, 1975; Nüsslein-Volhard and Wieschaus, 1980; Jurgens *et al.*, 1984; Nüsslein-Volhard *et al.*, 1984; Wieschaus *et al.*, 1984; Schupbach and Wieschaus, 1986; and many others). Presumptive patterning genes are detected in these mutant screens by position-specific defects in cuticular structures on the AP axis of the first instar larva. Although most details remain obscure, many of the genes identified in these screens seem to act successively and hierarchically in defining positional information in the developing oocyte and embryo until stable determinative circuits can be imposed at the cellular level.

The positional information genes for the AP axis have been classified into reasonably discrete groups based on their time of action and their mutant phenotypes. The AP coordinate gene class, most acting maternally, appear to determine the overall polarity of the embryo (Nüsslein-Volhard, 1979; Frohnhöfer and Nüsslein-Volhard, 1986; Mohler and Wieschaus, 1986; Schupbach and Weischaus, 1986). The gap genes, which act maternally and zygotically, appear to determine the positional identity of one or two large regions along the AP axis (Nüsslein-Volhard and Wieschaus, 1980). The zygotically acting pair-rule segmentation genes contribute to the determination of segmental number and segmental boundaries by assigning identities to seven or more reiterated blocks of cells at two-segment intervals. The segment polarity genes, zygotically acting, assign the correct AP polarity within each segmental repeat (Nüsslein-Volhard and Wieschaus, 1980). Finally, the homeotic selector genes, all zygotic, act to assign different identities to individual segments or small groups of segments (Lewis, 1978; Kaufman, 1983).

The patterning roles of the various classes of positional information genes are indicated schematically in Fig. 9, which also lists the approximate number of genes presently identified in each class as well as the approximate period of function of each patterning gene class. During all stages prior to the position-specific activation of the homeotic selectors, the embryo is a single, multinucleate cell. One of the underlying assumptions of much of the current experimentation on hierarchies is that a sufficient number of the patterning genes has been identified to allow any direct regulatory relationships between the levels to be identified and characterized. In at least one case, that of *bicoid*-dependent activation of zygotic expression of the gap gene *hunchback,* this assumption has been justified (Driever and Nüsslein-Volhard, 1989).

Dfd, as a homeotic selector gene, lies at the end of the putative regulatory hierarchy. Its position-specific activation (or repression) could conceptually involve the summation of direct inputs from regulatory proteins at all levels, which would be an extreme version of a combinatorial model of position-specific activation. Alternatively, in an extreme hierarchical model, only direct regulatory input from the pair-rule level would be necessary for *Dfd* position-specific activation. The effects of the patterning genes at other levels on *Dfd* expression would all be indirect, by regulating the boundaries of pair-rule expression.

Patterning Genes

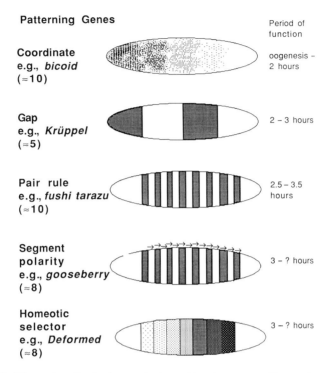

Period of function

Coordinate
e.g., *bicoid*
($\approx$10)

oogenesis – 2 hours

Gap
e.g., *Krüppel*
($\approx$5)

2 – 3 hours

Pair rule
e.g., *fushi tarazu*
($\approx$10)

2.5 – 3.5 hours

Segment polarity
e.g., *gooseberry*
($\approx$8)

3 – ? hours

Homeotic selector
e.g., *Deformed*
($\approx$8)

3 – ? hours

FIG. 9. Hierarchy of anterior–posterior patterning genes. The various classes of anterior–posterior patterning genes are indicated along with their period of function and the approximate number of known members, which is given in parentheses below a representative gene of each class.

Obviously, combinatorial and hierarchical regulatory interactions are not mutually exclusive, and the actual regulatory network will almost certainly involve both types of interactions to activate homeotic selectors in their precise spatial expression domains.

B. INITIATION AND EVOLUTION OF THE *Deformed* EXPRESSION PATTERN

The early and highly restricted nature of its expression pattern make *Dfd* an advantageous gene for studying the regulatory effects of coordinate, gap, and segmentation genes on selector gene expression. After fertilization of the egg, seven or eight synchronous nuclear divisions occur before the nuclei migrate to the egg periphery. Five more nuclear divisions occur during this syncytial blastoderm stage before cell membranes simultaneously enclose the peripheral nuclei. Soon after the thirteenth nuclear division, and prior to cellularization, localized *Dfd* transcripts are detectable in a region at 65–70% embryo length as measured from the posterior pole (Chadwick and McGinnis, 1987; Martinez-Arias *et al.*, 1987). The abundance of *Dfd* transcripts increases rapidly during cellularization, at which stage they are limited to a transverse circumferential band of about six cells (Fig. 10). Fate mapping experiments indicate that these cells comprise the primordia for the ectodermal, mesodermal, and neural tissues of the posterior head segments. The posterior boundary of expressing cells appears to be in flux even at this early stage, as double hybridization of blastoderm tissue sections indicates that the transcripts of the pair-rule gene *fushi*

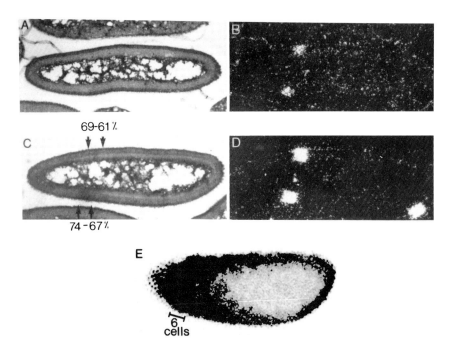

FIG. 10. *Dfd* transcript accumulation in blastoderm stage embryos. (A and B) Bright- and dark-field micrographs of an embryonic tissue section just after the thirteenth nuclear division, hybridized with a *Dfd* cDNA probe. (C and D) Bright- and dark-field views of a cellular blastoderm embryonic section, hybridized with the same probe. Numbers indicate the position of *Dfd* transcript accumulation in percentage egg length from the posterior pole. (E) Bright-field view of a parasagittal section through a cellular blastoderm embryo. A stripe of six cells expresses *Dfd* transcripts. [Reproduced with permission from Chadwick and McGinnis (1987).]

tarazu (*ftz*) and *Dfd* are coexpressed at least transiently in the most posterior two rows of the *Dfd* "stripe" of cells (K. Harding and M. Levine, personal communication). Soon thereafter, at late cellular blastoderm stages, the *ftz* and *Dfd* expression boundaries, as measured with antisera for the respective proteins, appear to be exclusive and abut each other, with the posterior boundary of *Dfd* protein accumulation coincident with the anterior boundary of *ftz* protein accumulation (Jack *et al.*, 1988).

The *Dfd* protein is first detectable at the cellular blastoderm stage, at approximately 2.5 hours (Fig. 11a,b). At this stage *Dfd* protein is expressed at low levels in a stripe of six cells. The protein in this initial stripe of cells is not solely localized in the nucleus, as it is at later stages. Early in gastrulation, *Dfd*-expressing cells are included within the invaginating cells of the cephalic furrow, a site of embryonic cell proliferation, which increases the total number of *Dfd*-expressing cells (Campos-Ortega and Hartenstein, 1985; Jack *et al.*, 1988). At this and

---

FIG. 11. Localization of *Dfd* protein in wild-type embryos. Whole mount embryos were stained with anti-Dfd antiserum in an immunoperoxidase approach. In all embryos anterior is to the left. (a) Wild-type embryo at late cleavage stage 14. *Dfd* expression is confined to a single circumferential band of cells located between 67 and 77% egg length. (b) Close-up of an embryo at the cellular blastoderm stage. *Dfd* is expressed in a stripe of approximately six cells. (c) Lateral view of an embryo in the early stages of germ band extension, approximately 3.5 hours after fertilization. *Dfd* expression at this stage begins to spread in ventral aspects of the embryo, but it remains in a narrow band dorsally, with some *Dfd*-expressing cells inside the disappearing cephalic furrow (arrowhead). (d) Ventrolateral view of a germ band extended embryo. At this stage, the posterior boundary of *Dfd* respects segmental boundaries in lateral cells and parasegmental boundaries along the ventral midline. The arrowhead marks the parasegment 1/2 boundary, the arrow marks the maxillary segment posterior boundary. (e) Close-up of the posterior boundary of *Dfd* expression in a germ band extended embryo. Here the distinction between the parasegmental and segmental frames is evident as the boundary of *Dfd* expression is extended by two rows of cells in lateral positions. The arrowhead marks the parasegmental frame, the arrow marks the segmental frame. (f) Ventral view of an embryo with the germ band fully extended, about 6 hours after fertilization. *Dfd* is expressed in the maxillary (Mx) and mandibular (Mn) segments, which are visible laterally, and in parasegments 0 and 1 ventrally. Also at this stage there is expression of *Dfd* in cells anterior to the mandibular segment in the hypopharyngeal lobe (Hy). (g) Lateral view of an embryo undergoing germ band retraction (10 hours). At this stage, *Dfd* expression disappears in the anterior lateral cells of the maxillary segment. (h) Close-up of the embryo in g. The strong expression of *Dfd* in the cells of the maxillary (Mx) segment that border the labial lobe (La) is clearly visible when compared with the anterior cells of the maxillary segment which no longer express *Dfd*. The expression of *Dfd* in the dorsal ridge (DR) is confined to one or two rows of cells in the anterior portion bordering the optic lobe. T1, First thoracic segment; T2, second thoracic segment. [Reproduced with permission from Jack *et al.* (1988).]

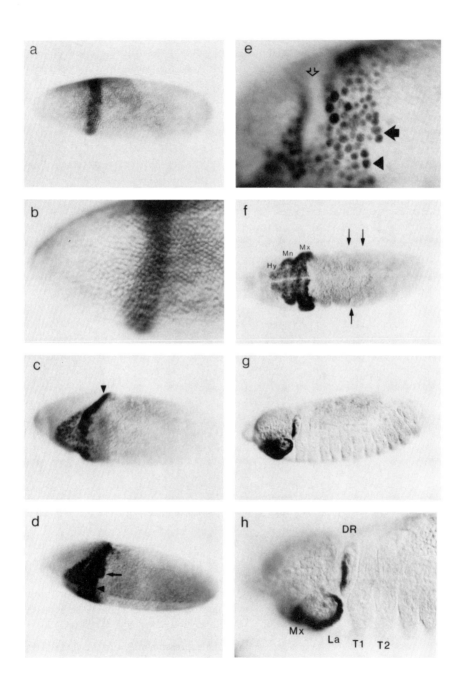

subsequent stages the protein is localized in the nucleus. Although the initial posterior boundary of *Dfd* protein expression, as defined by its relationship to the anterior stripe of *ftz* expression, appears to correspond to the parasegment 1/2 boundary (Martinez-Arias and Lawrence, 1985), this quickly changes as gastrulation proceeds. Before the germ band is fully extended along the dorsal surface, prior to the first morphological signs of segmentation, a more posterior boundary of *Dfd*-expressing cells develops on the lateral aspect of the embryo (Fig. 11d,e). The segmental divisions that arise soon thereafter in the head, as well as double staining for *Dfd* and pair-rule gene expression (Jack *et al.*, 1988), indicate that the lateral expression outlines the posterior boundary of the future maxillary segment, while the ventral posterior boundary remains at the parasegment 1/2 boundary.

During the late germ band extended stage, *Dfd* protein expression further evolves to include cells of the hypopharyngeal region (Fig. 11f). This stage represents the maximal extent of *Dfd* protein expression, with expression occurring in most or all epidermal cells of the maxillary lobe, mandibular lobe, and ventral hypopharyngeal region. During mid to late stages of embryogenesis, *Dfd* protein expression is gradually lost from lateral anterior regions of both the maxillary and mandibular lobes. In addition, a novel site of expression arises in the cells of the dorsal ridge, which are destined to be included within the frontal sac and may represent the primordia of the eye–antennal disc, an imaginal site of *Dfd* expression and function (Chadwick and McGinnis, 1987; Martinez-Arias *et al.*, 1987; Jack *et al.*, 1988). The cells in the ventral and posterior regions of the maxillary segment continue to express *Dfd* protein well after late stages of head involution, during which segmental boundaries are obscured by fusions and rearrangements (Fig. 11g,h) (T. Jack, unpublished).

C. *Deformed* EXPRESSION PATTERNS IN COORDINATE AND GAP MUTANTS

The most striking effect on *Dfd* expression of any single gene mutation in the developmental hierarchy is seen in *bicoid*⁻ embryos. The *bicoid* (*bcd*) gene is a strict maternal whose transcripts accumulate in a highly restricted anterior region of the developing oocyte (Frohnhöfer and Nüsslein-Volhard, 1986; Berleth *et al.*, 1988). The *bcd* protein accumulates in a concentration gradient, with highest amounts near the anterior pole of the syncytial blastoderm embryo (Driever and Nüsslein-Volhard, 1988). Embryos in which the mother has failed to deposit functional *bcd* mRNA do not develop any head or anterior thoracic structures and have extreme posterior structures duplicated (in reverse polarity) in place of the missing head segments. No *Dfd* expression is detected at any stage in *bcd* mutant embryos (T. Jack, unpublished). This result is consistent with a direct effect of *bcd* protein (a homeodomain-containing protein) on *Dfd* expression or an indirect effect mediated by an intervening gene or genes regulated by *bcd*. One candidate for an intermediary is *hunchback,* a gap gene whose zygotic expression is activated by *bcd* (Driever and Nüsslein-Volhard, 1989). *hunchback* (*hb*) is known to be necessary for the normal development of the posterior head; zygotic hb^- mutants are missing the labial and thoracic segments and have defects in abdominal segments 7 and 8 (Lehmann and Nüsslein-Volhard, 1987). Mutant embryos that lack both maternal and zygotic *hb* function have much stronger phenotypes in the head, as no structures from any of the gnathal segments develop.

Different lines of evidence indicate that the regulatory effect of *bcd* on *Dfd* might be direct. First, the removal of zygotic *hb* expression does not affect the *Dfd* expression pattern, indicating that zygotic *hb* is not required for the initiation of position-specific expression of *Dfd* (Jack *et al.,* 1988). Second, *hb* alone cannot account for the activation of *Dfd,* since *bcd* mutant embryos have a normal distribution of maternally deposited *hb* but no *Dfd* expression (Tautz, 1988). Third, some mutant combinations that result in ectopic expression of *bcd* also result in ectopic expression of *Dfd*. For example, in *vas exu* double-mutant embryos, in which *bcd* transcripts are distributed at moderate levels throughout the entire length of the embryos (Berleth *et al.,* 1988),*Dfd* is expressed in both anterior and posterior regions (T. Jack, unpublished).

Although the gap gene *hb* is not sufficient for *Dfd* activation in the absence of *bcd,* it is likely that either its maternal or zygotic expression is necessary. In the absence of both maternal and zygotic *hunchback* expression, posterior head structures (including mouth hooks and cirri) are missing from the embryonic body plan (Lehmann and Nüsslein-Volhard, 1987), indicating that *Dfd* is unlikely to be expressed. We are currently directly testing for *Dfd* expression in *hb* null mutants, which involves the creation of hb^- germ cells in otherwise hb^+ mothers by the use of pole cell transplantation. Normal *Dfd* expression is detected in embryos mutant for the other zygotic gap genes *giant, Krüppel,* and *knirps* (Jack *et al.,* 1988).

D. *Deformed* EXPRESSION PATTERNS IN PAIR-RULE MUTANTS

In contrast to the lack of effect of the gap genes, mutations in eight of the nine pair-rule genes yield unique and obvious changes in the wild-type *Dfd* expression pattern (Jack *et al.,* 1988). In mutants for three of the pair-rule loci, *ftz, odd-skipped* (*odd*), and *hairy* (*h*), there are more

Dfd-expressing cells in mutants compared to the wild type. Thus, these three genes formally appear to act, either directly or indirectly, as negative regulators. In mutants for the five pair-rule loci *engrailed* (*en*), *odd-paired* (*opa*), *runt* (*run*), *paired* (*prd*), and *even-skipped* (*eve*), there are fewer *Dfd*-expressing cells, which implicates them, in a formal sense at least, as positive regulators.

Of the negative regulators, *ftz* appears to be the most important. The *Dfd* misexpression in *h* mutants occurs long after initiation of the *Dfd* stripe and involves only a few cells separated from the normal maxillary–mandibular region. *Dfd* misexpression in *odd* mutants also occurs well after the initial *Dfd* stripe forms and involves an additional two or three rows of cells at the lateral posterior border of the maxillary segment. The *odd* mutant effect on the *Dfd* pattern is likely due to the expansion of expression of *en* (DiNardo and O'Farrell, 1987). As discussed later, expression of *Dfd* at the posterior border of the maxillary segment is dependent on *en* function and may be limited by the *en* expression boundary.

The importance of *ftz* as a negative regulator at the posterior boundary of the initial *Dfd* stripe is supported by three observations. First, the initial expression boundaries of the *Dfd* protein stripe and the *ftz* protein stripe are mutually exclusive, with *ftz* expression defining the anterior boundary of parasegment 2 and *Dfd* the posterior boundary of parasegment 1. Second, even at the earliest stages of *Dfd* protein accumulation in *ftz* mutant embryos, the *Dfd* expression pattern is expanded posteriorly to include one or two additional rows of cells (Fig. 12c,f). Third, when *ftz* protein is misexpressed under the control of a heat-shock promoter, *Dfd* expression is sharply reduced or eliminated throughout the entire *Dfd* stripe (T. Jack, unpublished). This negative regulatory effect is transient and somewhat variable, which is likely due to the instability of *ftz* transcripts and protein and to the lack of high levels of expression from the heat-shock promoter–*ftz* fusion gene (Struhl, 1985; T. Jack, unpublished). The negative regulatory effect of *ftz* on *Dfd* is opposite to its apparent positive regulatory effect on many of the other homeotic selector genes (Ingham and Martinez-Arias, 1986).

The two pair-rule genes that appear to be most important in a positive regulatory sense for the establishment of the *Dfd* stripe of expression are *eve* and *prd*. Mutation of *eve* results in a narrower initial stripe of expression (four rows of cells instead of the normal six; Fig. 12b,e), and embryos mutant for both *eve* and *prd* have no or only one row of the initial *Dfd* stripe of expression (T. Jack, unpublished). The other pair-

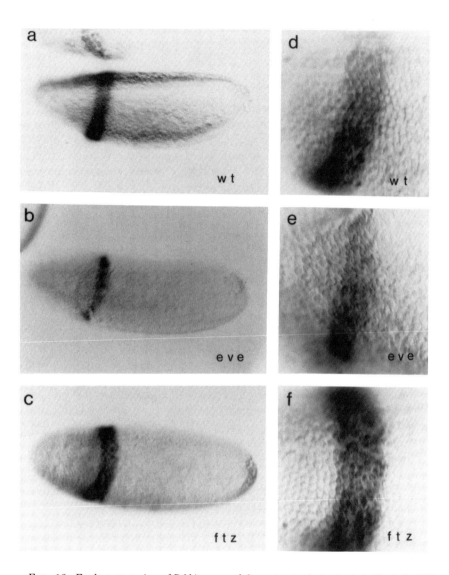

FIG. 12. Early expression of *Dfd* in *eve* and *ftz* mutants. Anterior is to the left. *Dfd* staining of wild-type (wt) (a and d), *eve*[r13] (b and e), and *ftz*[w20] (c and f) cellular blastoderm stage embryos showing the whole embryo (a–c) and close-ups (d–f). In wild-type embryos, *Dfd* protein initially appears as a stripe of five to six cells. In *eve* mutants, the initial stripe of expression is narrower, containing about four cells, whereas in *ftz* mutants, the initial stripe is broader, about seven cells wide. [Reproduced with permission from Jack *et al.* (1988).]

rule gene that may be involved in the establishment of the initial *Dfd* stripe is *opa;* however, *opa* is likely to be required only in the most posterior one or two rows of the initial *Dfd* stripe (Jack *et al.*, 1988), and its precise expression boundaries are unknown, making its role difficult to study.

The overlap in expression domains between the putative pair-rule activators and *Dfd* is consistent with a direct, cell-autonomous regulatory role. The initial anterior expression stripe of *prd* has an anterior boundary that precisely coincides with the anterior *Dfd* expression boundary, consistent with a direct regulatory role (T. Jack, unpublished). The *prd*/*Dfd* expression overlap also coincides with the domain of cells in *prd* mutants which lack *Dfd* expression at later stages (Jack *et al.*, 1988).

The initial anterior stripe of *eve* expression is entirely contained within the initial *Dfd* expression stripe (T. Jack, unpublished), also consistent with a direct role; however, *eve* embryos develop so few morphological markers that the later *Dfd* expression boundaries cannot be accurately assigned. The observation that the *Dfd* expression stripe does not decrease by more than two rows in *eve* or *prd* mutants suggests that the requirement for pair-rule activators is not combinatorial. If, for example, *eve* plus *prd* were required in anterior rows of the stripe, and *eve* plus *opa* in posterior rows of the stripe, then the elimination of *eve* would eliminate nearly all of the initial *Dfd* expression stripe. This is not observed. All the *Dfd* expression pattern changes in pair-rule mutants are more consistent with a redundant role of pair-rule activators. For example, if *prd* is eliminated *eve* can still apparently function to activate *Dfd* in most of the rows of cells in which *prd* is missing.

Other evidence supports the idea that *eve* might have a direct regulatory effect. When *eve* protein is ectopically expressed throughout the embryo by induction of a heat-shock promoter–*eve* cDNA fusion gene, the initial boundaries of the *Dfd* stripe expand dramatically, encompassing as many as 10–12 rows of cells (T. Jack, unpublished). This result is also consistent with the possibility that the anterior boundary of the initial *Dfd* stripe is set not by the expression of an unknown negative regulator but by the absence of a correct combination of positive regulators in this region. The ectopic activation of *Dfd* by *eve* constitutes additional evidence that combinations of pair-rule genes are not required for activation. It seems much more likely that a combinatorial code for *Dfd* activation involves gene products from different levels of the developmental hierarchy.

E. Summary of Regulatory Genes Required for the Initial *Deformed* Stripe

Although the pair-rule genes as a class are important for the initial *Dfd* expression pattern, pair-rule genes alone are obviously not sufficient to explain the initial activation of *Dfd* in a stripe of six cells. The pair-rule genes with positive regulatory effects, for example, *eve*, the best candidate for a direct activator, are expressed in other positions throughout the thorax and abdomen, where *Dfd* expression is never activated (Frasch *et al.*, 1987). So at least one additional factor is necessary to localize *Dfd* expression to the head primordia.

Our current model for the position-specific establishment of *Dfd* expression is outlined in Fig. 13. In this simple model there is a unique

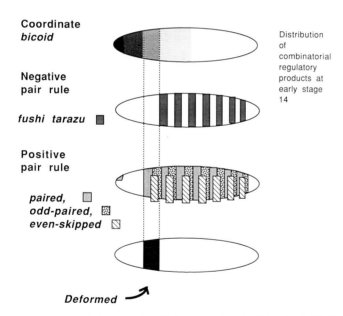

Fig. 13. Summary of the postulated elements involved in the initiation of *Dfd* expression. Postulated positive and negative factors involved in the establishment of *Dfd* expression are indicated in schematic embryos at the syncytial blastoderm stage. We postulate that the pair-rule gene *ftz* negatively regulates *Dfd* and defines the posterior boundary of *Dfd* expression. The *bcd* protein, in combination with either the *prd*, or *opa*, or *eve* proteins, is necessary to activate the initial stripe of *Dfd* expression. [Reproduced with permission from Jack *et al.* (1988).]

group of nuclei on the anterior–posterior axis of the syncytial blastoderm embryo, distinguished by the following criteria: (1) they contain
moderate levels of *bcd* protein, (2) they contain no *ftz* protein, and
(3) they express either *prd* or *eve* or *opa*. Since the expression patterns of
all these genes are constantly in flux, all of the above conditions are
simultaneously present for only a brief period during the syncytial
blastoderm stage, which may be of great importance in assigning the
exact boundaries of the initial stripe. The above model has some easily
tested predictions; for example, expression near the posterior pole
should be inducible if *bcd* expression can be experimentally induced
there at early stages.

Obviously, this model is preliminary, and it us unlikely to describe
every regulatory combination that restricts *Dfd* expression to a single
anterior stripe. The exclusion of *Dfd* expression from the extreme poles
of the embryo in *vas exu* double mutants, even when *bcd* is ectopically
expressed at moderate levels throughout the blastoderm, suggests
that a polar repressor might function to exclude *Dfd* expression
from the anterior and posterior tips. This hypothetical repressor
would also help explain why *prd* expression in the extreme anterior
dorsal patch does not activate *Dfd*, since the boundary conditions
in the above model are satisfied, with *bcd* protein present and *ftz*
absent. In addition, as described previously, either the maternal
or zygotic *hb* protein is very likely to be necessary for *Dfd* expression.

The above model, though crude, proposes that interlevel combinatorial interactions are required to activate position-specific expression of
Dfd. Although the regulatory interactions could be indirect, many of
the gene products that we have defined as initial activators are known
or suspected DNA-binding/gene regulatory proteins. For example, *bcd*
is directly involved in the position-specific activation of *hb*, and *eve* has
been proposed as a direct regulator of *en* activation (Harding *et al.*,
1986; Hoey and Levine, 1988). Both *bcd* and *eve* encode homeodomain-
containing proteins; in a speculative extension of the above model, *bcd*
and *eve* proteins might be unable to bind independently to *Dfd* regulatory sequences, but when both accumulate to suitable levels in the same
nucleus they might cooperatively bind to and activate *Dfd* expression.
The *ftz* protein, another homeodomain-containing protein, which appears able to negate dominantly any positive combination, at least
transiently, might then act either by interfering with the cooperative
interaction between *bcd* and *eve* or by competing for one or both of their
binding sites.

III. Maintenance of Position-Specific Expression of *Deformed*

A. GENETIC REQUIREMENTS FOR MAINTENANCE OF
Deformed EXPRESSION BOUNDARIES

As the *Dfd* pattern of expression changes at nearly every stage of embryogenesis, with cells gaining and losing *Dfd* expression, no complete description of the other factors involved in its persistent expression can be attempted at present. In this section, some of the known genes required to limit *Dfd* expression to maxillary–mandibular segments are described; these act, in a formal sense, as negative regulators of *Dfd* expression in other segments. In addition, at least two genes are known to act as positive regulators of persistent expression within the maxillary segment. One of these positive regulators is the *Dfd* gene itself.

1. Cross-Regulation Due to Other Homeotic Selector Genes

Mutations in other homeotic selector genes of the Antennapedia and Bithorax complexes, which are expressed in approximately the same time period as *Dfd,* have no effect on the maintenance of the normal *Dfd* expression pattern. This stands in sharp contrast to homeotic selectors expressed in the thorax and abdomen, where cross-regulatory interactions have been shown to be required for the correct maintenance of normal expression boundaries. For example, the normal posterior boundary of abundant *Antp* expression is in the third thoracic segment, but in embryos lacking the genes of the Bithorax complex (BX-C), *Antp* is persistently expressed throughout the abdominal segments, suggesting that the BX-C genes repress *Antp* expression in wild-type embryos (Struhl, 1982; Hafen *et al.,* 1984). If *Dfd* were to fit this general rule, then one might predict that when *Scr* and other posteriorly expressed selector genes are missing, *Dfd* expression would spread into posterior segments. However, in *Scr⁻* mutants (as well as *Scr⁻ Antp⁻* double mutants and *Scr⁻ Antp⁻* BX-C⁻ triple mutants) the *Dfd* pattern is normal (Jack *et al.,* 1988).

2. Homeotic Regulatory Genes

Another group of genes that influence the maintenance of homeotic selector expression patterns are the homeotic regulatory genes [e.g., *Polycomb* and *extra sex combs* (Duncan and Lewis, 1982; Struhl, 1983)], which seem to be required in all cells to assure the stability of boundaries of homeotic selector expression. For example, embryos mutant for

Polycomb (*Pc*), exhibit ectopic expression of BX-C genes like *Ubx* and *Abd-B* in anterior segments at late stages of development, where they are normally never expressed (Wedeen *et al.*, 1986). This results in homeotic transformations of the posterior head, thorax, and abdomen to a phenotype similar to that normally developed by the eighth abdominal segment. *Dfd* expression boundaries appear to be less sensitive, at least in epidermal cells, to the effects of homeotic regulatory genes. In *Pc* mutant embryos, *Dfd* is ectopically expressed, but only in reiterated groups of cells in the CNS of the embryo (Wedeen *et al.*, 1986). In *extra sex combs* mutant embryos, *Dfd* is also misexpressed throughout the CNS, as well as in a small group of cells in the posterior labial epidermis (M. Regulski and T. Jack, unpublished). *Dfd* appears to be refractory to the negative regulatory effects of the *Abd-B* gene products in *Pc* mutants, which eventually repress more anteriorly expressed homeotic selectors such as *Antp* and *Ubx*. In *Pc* mutants, expression of *Dfd* in the maxillary segment appears normal even at late stages of embryogenesis (Wedeen *et al.*, 1986).

3. *The Engrailed Gene*

One of the persistent domains of *Dfd* expression in the embryo is in the posterior epidermal cells of the maxillary segment. The initial and persistent expression of *Dfd* in this region requires the action of the *engrailed* (*en*) gene (Jack *et al.*, 1988). *engrailed* is usually classed with the pair-rule genes since its principal mutant phenotype (though variable) involves the loss of alternate segment boundaries (Kornberg, 1981). However, the expression pattern of *en* in the posterior region of each segment is more consistent with a role similar to genes of the segment polarity class. The earliest and most abundant site of *en* protein expression in the embryo is in the posterior compartment of the maxillary segment (*en* stripe 2; DiNardo *et al.*, 1985; Weir and Kornberg, 1985).

Mutants in *en* lack *Dfd* expression in the two lateral posterior rows of cells of the maxillary segment, whereas *Dfd* expression in the mandibular segment (where *en* is also expressed in the posterior compartment) is normal (Jack *et al.*, 1988). The rows of cells that do not express in *en*⁻ mutants are those which make up the lateral offset stripes of cells shown in Fig. 11e. These give rise to the posterior border of the maxillary segment, where wild-type *Dfd* expression is abundant and persists until very late stages of embryogenesis, well after the incorporation of the maxillary segment into the involuted head. In later stage *en*⁻ embryos, the maxillary–labial segment boundary does not form, and cells in the posterior region of the maxillary segment never exhibit *Dfd* expression.

B. AUTOREGULATION OF THE *Deformed* TRANSCRIPTION UNIT

One of the simplest possible mechanisms for the maintenance of position-specific expression of a homeotic selector gene would be for it to regulate its own expression after being activated in a discrete region early in embryogenesis. Considerable evidence now exists that *Dfd* normally uses an autoregulatory circuit to maintain its own expression in epidermal cells in the posterior region of the maxillary segment. The initial indications that *Dfd* is autoregulated arose during a study of the effects of ectopically expressed *Dfd* protein using a heat-shock-inducible promoter attached to a *Dfd* cDNA.

1. *Inducible Ectopic Expression of Deformed Protein*

Kuziora and McGinnis (1988) constructed fly strains carrying a fusion gene consisting of the heat-shock 70 promoter and leader sequences upstream of a near full-length *Dfd* cDNA (Fig. 14). The T58 *Drosophila* strain carries a single homozygous insert of this fusion gene, designated *heat-shock Deformed* (*hsDfd*), on the second chromosome. A mild heat shock at early embryonic stages induces the expression of the *hsDfd* gene and causes the transient accumulation of *Dfd* protein in all or nearly all embryonic nuclei (Fig. 15D). Embryos not heat shocked have no detectable ectopic *Dfd* expression. After 1 hour of recovery from heat shock, virtually all nuclei contain *Dfd* protein, but after 3 hours of recovery little protein remains in lateral cells of the embryo (Fig. 15E). Normal expression is observed in the maxillary and mandibular segments, where *Dfd* expression is directed by the normal chromosomal locus. After 5 hours of recovery, *Dfd* protein is still detected in the ventral or ventral posterior region of each body segment but is undetectable in dorsal and dorsolateral regions. The persistent expression pattern is not dependent on the insertion site of the *hsDfd* fusion gene.

The persistent expression of the *Dfd* protein in the ventral posterior cells of each body segment suggested either that the protein was differentially stable in a position-specific fashion in the embryonic epidermal cells or that a persistent expression circuit was being activated to allow continuous *Dfd* protein production in ventral posterior cells. If the latter explanation for persistent expression were correct, it seems probable that the source of the persistent expression is the normal chromosomal locus of *Dfd,* since the *hsDfd* gene had so little regulatory sequence. Thus, a plausible and testable explanation for the persistent expression of *Dfd* is that the *Dfd* protein produced from the *hsDfd* gene is capable of inducing the expression of the normal chromosomal copy of

Deformed locus

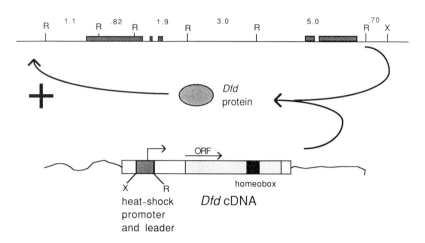

FIG. 14. Autoregulatory circuit in *hsDfd* embryos. The top line shows a simplified restriction map of the genomic DNA of the *Dfd* locus [a more detailed map is found in Regulski *et al.* (1987)]. Stippled bars denote the *Dfd* exons. After an early induction by heat shock, the *hsDfd* fusion gene can activate expression from the endogenous *Dfd* locus either indirectly or (as shown) by direct interaction of the *Dfd* protein with normal *Dfd* regulatory sequences. Thereafter, the endogenous *Dfd* locus can apparently regulate itself either indirectly or (as shown) by direct interaction of the *Dfd* protein produced from the endogenous locus with normal *Dfd* regulatory sequences. Since persistent autoregulation takes place only within posterior ventral ectodermal cells of each segment, there is also a requirement for positive factors which assist *Dfd* in these cells and/or negative factors which prevent autoregulation in other cells.

Dfd, which thereafter could maintain its own expression, at least in ventral posterior cells of each segment (Fig. 14).

As one test of this hypothetical regulatory effect, embryos that carried the *hsDfd* gene but also carried protein null mutations for the normal *Dfd* gene were stained for persistent expression of the *Dfd* protein. As is seen in Fig. 15, although the *Dfd⁻* embryos accumulate a transient ectopic expression of *Dfd* protein, this pattern of expression does not persist. After 3 and 5 hours of recovery from heat shock, essentially no *Dfd* protein is detectable in any segments, including the normal site of expression in the maxillary and mandibular segments.

Direct evidence for ectopic activation of the normal chromosomal (endogenous) *Dfd* transcription unit was obtained by the use of an *in situ* hybridization probe that was specific for transcripts from the normal chromosomal transcription unit. Tissue sections were prepared from embryos that carried both the *hsDfd* gene and the wild-type chromosomal copy (Kuziora and McGinnis, 1988). The endogenous *Dfd*-specific probe hybridizes transcripts that are persistently expressed in the ventral posterior epidermal cells of each body segment, matching the sites of persistent *Dfd* protein accumulation in the *hsDfd* embryos shown in Fig. 15. Thus, the normal chromosomal transcription unit of *Dfd* is being activated and persistently expressed at ectopic sites dependent on the induction of the *hsDfd* gene expression.

The persistent pattern of *Dfd* expression in metamerically reiterated groups of cells suggested that they might be the only cells that support autregulated *Dfd* expression. The alternative hypothesis would be that *hsDfd* expression can transiently activate the normal chromosomal *Dfd* locus in all cells, but that its expression is progressively restricted to the ventral posterior epidermis as development proceeds. The use of the endogenous *Dfd* gene-specific probe on embryos soon after the *hsDfd* expression has been induced indicates that the normal chromosomal *Dfd* locus is never globally activated (M. Kuziora, unpublished). From very early stages, ectopic activation of *Dfd* is limited to metamerically reiterated blocks of cells. Apparently only a subset of cells in each segment, located in the ventral posterior region, is capable of supporting *Dfd* autoregulation. This is presumably due to a segmentally reiterated positive regulatory factor(s) that cooperates with *Dfd* to activate its own transcription, or to the exclusion of negative regulatory factors from the ventral posterior epidermis of each thoracic and abdominal segment.

2. Homeotic Transformations in heat-shock Deformed Embryos

Induction of ectopic expression of *Dfd* in the *hsDfd* embryos is lethal to embryos when administered at early embryonic stages. Most *hsDfd* embryos do continue to develop to the end of embryogenesis but suffer severe morphological transformations at near 100% penetrance if the heat-shock treatment is given precisely at cellular blastoderm (Kuziora and McGinnis, 1988). A similar heat-shock treatment given to embryos not carrying the *hsDfd* fusion gene has no morphological effect.

Cleared cuticle preparations show that the induced *hsDfd* embryos fail to involute head segments and have a partial transformation of other head and thoracic segments toward a maxillary segment identity. As described previously, the principal structures missing in *Dfd* null mutant larvae are the cirri and mouth hooks, both maxillary structures. As shown in Fig. 16, the labial segment of *hsDfd* larvae is strongly transformed toward maxillary as evidenced by the substitution of mouth hooks and cirri for the normal labial structures. The

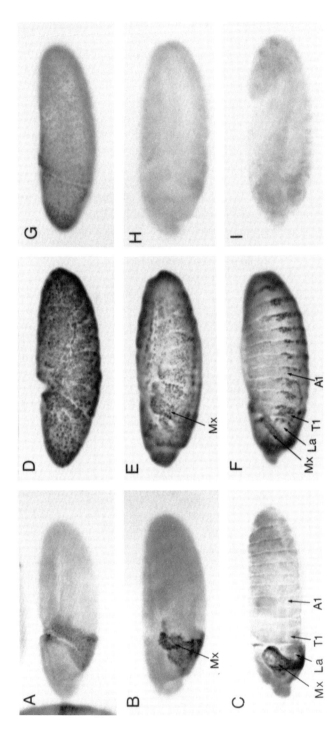

strongest transformations in the thoracic segments are always seen in the first thoracic segment, which often has well-developed cirri and sclerotized cuticle that presumably represents incompletely developed mouth hooks. Cirri are also seen with high penetrance in ventral regions of second and third thoracic segments. These *hsDfd*-induced homeotic transformations are always limited to the head and thorax; transformations of abdominal segments have never been observed. The lack of *hsDfd* homeotic transformations in abdominal segments, even though early and persistent expression is observed in this segments, indicates that persistent *Dfd* expression is not sufficient for head transformation of *Drosophila* embryonic cells. A plausible hypothesis attributes this to phenotypic dominance of the homeotic genes of the Bitho-

FIG. 15. *Dfd* protein expression patterns in *hsDfd* embryos. Embryos were stained with anti-*Dfd* antiserum; all are arranged with anterior to the left and dorsal up. (A) Non-heat-shocked wild-type embryo at approximately 4 hours of development. Stained nuclei are limited to parasegments 0 and 1, the normal *Dfd* expression pattern at this stage (Jack *et al.*, 1988). (B) Non-heat-shocked wild-type embryo at approximately 6 hours of development. The normal *Dfd* expression pattern at this stage is observed, which includes the hypopharyngeal region and the mandibular and maxillary (Mx) segment primordia (Turner and Mahowald, 1979; Chadwick and McGinnis, 1987; Martinez-Arias *et al.*, 1987; Jack *et al.*, 1988). (C) Non-heat-shocked wild-type embryo at approximately 9–10 hours of development. The normal *Dfd* expression pattern in maxillary (Mx) and mandibular segments is present. Note that expression in the maxillary segment at this stage is limited to the ventral and posterior regions. La, Labial lobe; T1, first thoracic segment; A1, first abdominal segment. (D) *hsDfd* embryo at about 4 hours of development, after 1 hour of recovery from a 1-hour heat shock at 2.5–3 hours of development. (E) *hsDfd* embryo at about 6 hours of development, after a 3-hour recovery from a 1-hour heat shock at 2.5–3 hours of development. (F) *hsDfd* embryo at about 9–10 hours of development, after 5 hours of recovery from a 1-hour heat shock at 2.5–3 hours of development. Note the persistent *Dfd* protein staining in the ventral posterior regions of the body segments. Anti-*Dfd* antibody staining of *hsDfd* embryos after 1 hour (G), 3 hours (H), and 5 hours (I) of recovery from a 1-hour heat shock at 2.5–3 hours of development. One-quarter of the embryos in the population stained were expected to be *Dfd* null mutants (Dfd^{RX1}/Dfd^{RX1}). (G) Representative embryo at 1 hour post-heat shock. We cannot identify *Dfd* mutants by any staining or morphological criteria at this stage, but essentially all embryos from the population at this stage stained for *Dfd* protein in nearly all nuclei. (H) Embryo that we classify as a *Dfd* mutant based on the lack of high-level expression in the maxillary and mandibular segments (compare with E). One-quarter of the embryos (10/41) at this stage exhibited such a staining phenotype. (I) Embryo that we classify as mutant for *Dfd* based on the lack of high-level expression in the maxillary and mandibular segments as well as on the different maxillary–mandibular–procephalic morphology characteristic of *Dfd* mutants (Regulski *et al.*, 1987). One-quarter (27/100) of the embryos at this stage exhibited such a staining phenotype. [Reproduced with permission from Kuziora and McGinnis (1988). Copyright held by Cell Press.]

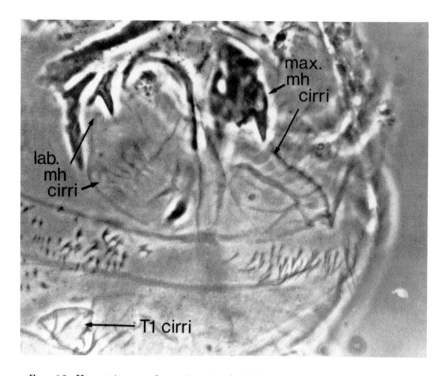

FIG. 16. Homeotic transformations in *hsDfd* larvae. A phase-contrast close-up shows the anterior lateral region of a *hsDfd* first instar larva that was subjected to a 1-hour 37°C heat shock at the cellular blastoderm stage. The head segments have not involuted normally and thus reside on the exterior of the larva. The maxillary (max.) and labial (lab.) segments are labeled, as are the mouth hooks (mh) and cirri, which occur in their normal positions in the maxillary segment and ectopically in the labial segment and first thoracic segment (T1 cirri). [Reproduced with permission from Kuziora and McGinnis (1988). Copyright held by Cell Press.]

rax complex, which are expressed in abdominal epidermal cells and define their identity. This hypothesis is currently being tested by placing the *hsDfd* fusion gene in a BX-C mutant background.

The penetrance of these homeotic transformations is critically dependent on the stage at which *hsDfd* ectopic expression is induced by heat shock. Maximal penetrance is seen at the cellular blastoderm stage, with progressively less penetrance at gastrulation (ventral furrow formation) and early germ band extension stages. By late germ band extension (~5 hours), no transformations can be induced by heat-shock treatment.

The *hsDfd* homeotic transformations are also dependent on the persistent expression of the normal chromosomal *Dfd* gene, which is activated by the autoregulatory effect of *hsDfd*. Strains that carry the *hsDfd* fusion gene and that are mutant for the normal chromosomal *Dfd* gene develop no homeotic transformations of other head and thoracic segments (Kuziora and McGinnis, 1988). Nor does the transient exposure to *Dfd* protein in the normal maxillary and mandibular cells rescue any of the usual patterning defects (missing cirri and mouth hooks) resulting from loss of chromosomal *Dfd* function.

3. *Other Genetic Factors Required for Autoregulation*

As mentioned before, the persistent, autoregulated expression of *Dfd* in the ventral posterior region of each segment suggests that a factor(s) that is persistently expressed in a subsegmental repeat might combine with the *Dfd* protein to achieve autoregulation. Such expression patterns are observed in some genes of the segment polarity class; for example, *en* is expressed in the posterior compartment of each segment. Some of the segment polarity genes have been tested for involvement in the *Dfd* autoregulatory circuit by examining the ability of segment polarity mutant embryos to support persistent expression of *Dfd* in its ventral posterior repeat (N. McGinnis, unpublished). Preliminary results indicate that some of the segment polarity genes are in fact required, though how direct their effect may be is unknown. One of the clearest mutant effects is seen in *hsDfd en⁻* embryos, which persistently express *Dfd* in many fewer cells per repeat.

C. MOLECULAR INTERACTIONS IN THE *Deformed* AUTOREGULATORY CIRCUIT

A simple scheme for autoregulation would involve the *Dfd* protein directly binding to positive regulatory sites at or near the *Dfd* transcription start site. The binding, or the formation of an active complex, presumably would also require the combinatorial input of the segmentally distributed protein(s) that restricts the autoregulatory circuit to the ventral posterior epidermal cells. Recent results support the hypothesis that the autoregulation is in part directly mediated by *Dfd* protein binding to *Dfd* regulatory sequences. An 870-bp *Hind*III fragment that maps approximately 5 kb upstream of the *Dfd* transcription start contains the highest affinity *Dfd* binding sites in a 40-kb DNA region which includes the *Dfd* transcription unit and upstream and downstream flanking sequences (M. Regulski, unpublished). The same *Hind*III fragment is capable, in stably transformed embryos, of conferring maxillary segment-specific expression of β-galactosidase when attached to *lacZ* coding sequences (C. Bergson, unpublished). Experiments are underway to test whether this maxillary expression is dependent on expression of functional *Dfd* protein, whether the *hsDfd* protein can direct ectopic expression of the putative autoregulatory region–marker gene constructs, and whether changes in the *Dfd*-binding sequences will abolish autoregulation in embryonic cells.

D. SUMMARY OF THE MAINTENANCE PHASE OF *Deformed* EMBRYONIC EXPRESSION

During the maintenance phase of *Dfd* expression, from 4 hours (late gastrulation–early germ band extension) to late stages of embryogenesis, its pattern is constantly changing. Thus, it is likely that *Dfd* regulatory sequences respond to a great number of different inputs after the original activation of *Dfd* in a six-cell stripe. Despite this complexity, it seems clear that two genes, *en* and *Dfd* itself, play very important roles in the maintenance of a part of the embryonic *Dfd* expression pattern.

A functional *en* gene is necessary to activate *Dfd* expression in the posterior region of the maxillary segment, a position of *Dfd* expression that persists after most other cells no longer express. *Dfd* epidermal expression in late stage *en* mutant embryos is found in only a few ventral cells, and *en* mutants have defects in maxillary structures (cirri and mouth hooks) normally specified by the morphogenetic function of *Dfd*. *en* is not sufficient for autoregulated *Dfd* expression, as it is expressed in lateral posterior cells of each segment which do not exhibit persistent *Dfd* expression after *hsDfd* induction. Nor is *en* the only positive cofactor necessary to achieve the persistent ectopic expression pattern shown in Fig. 15, since a few cells per segment persistently express *Dfd* even in *en* mutant embryos.

The results summarized previously show that *Dfd* protein can activate expression from its own transcription unit, which in some cells will assure its persistent, autoregulated expression throughout much of the rest of embryogenesis. Although at the moment this autoregulatory circuit is actually best understood in ectopic positions where it is initiated in other segments under the control of *hsDfd*, it presumably is normally used in the posterior region of the maxillary segment. At present, the evidence that *Dfd* autoregulation is a normal mechanism in the maxillary segment is based on the experiments described previously in which *Dfd* upstream regulatory sequences can refer persistent, maxillary segment-specific expression.

The autoregulatory circuit also requires *en* protein, either as a direct or indirect participant in *Dfd* transcription activation (Fig. 17). Both *Dfd* and *en* proteins contain homeodomains and presumably function as

DNA-binding/gene regulators. If *en* acts directly, its binding to *Dfd* regulatory sequences could be either at sites independent from those bound by *Dfd* or concerted (perhaps through the formation of hetero-dimers with *Dfd*) at the same site. In either case, the binding of either *Dfd* or *en* alone would not result in an active complex, but the binding of both would yield an activated transcription unit.

Homeotic selector genes like *Dfd* are distinct from other region-specific patterning genes in that, even though they are expressed from early stages of embryogenesis, their function is apparently required throughout most of development to maintain the identity of cells. Genetic evidence for this requirement has been obtained principally by the generation of somatic clones of imaginal disc cells mutant for a homeotic selector gene in a wild-type background (Lewis, 1965; Struhl, 1982). Even when mutant clones are generated relatively late in development, a homeotic transformation of the mutant cells is observed. The results described in this article on the autoregulated expression of *Dfd* indicate that a similar requirement for persistent expression exists in the embryonic cells whose position-specific fate is determined by homeotic selector expression. A transient exposure to *Dfd* protein at early stages of embryogenesis is not enough to assign head identity to *Drosophila* cells, even in the normal maxillary–mandibular region. Ho-

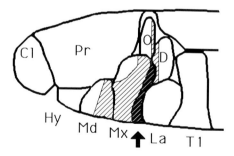

FIG. 17. Genetic functions required for maintenance of *Dfd* expression. The embryonic head is shown schematically with anterior to the left and dorsal up. The subdivisions of the head are indicated as follows: clypeolabrum (Cl), procephalic lobe (Pr), hypopharyngeal lobe (Hy), mandibular (Md), maxillary (Mx), labial (La), first thoracic segment (T1), optic lobe (O), and dorsal ridge (D). Superimposed are the regions of *Dfd* protein expression (hatching) and the region in the posterior maxillary segment where both *en* and *Dfd* are required for persistent expression (shaded region marked by the arrow).

meotic transformations are seen only in embryos in which the autoregulatory circuit can persistently supply *Dfd* protein to ectopic positions. Considering this mechanism of head fate assignment in the conceptual framework of classic developmental biology, the autoregulatory circuit is one way of supplying a long-term cellular memory of the original positional value and, thus, a stable, "determined" identity (Spemann, 1938).

It is possible that other homeotic selector genes may use autoregulation as a mechanism to confer stable, lineage-specific expression. For the homeotic gene *Ubx,* Bienz and Tremml (1988) have shown that *Ubx–lacZ* fusion genes can be persistently expressed in the normal *Ubx* domain in the visceral mesoderm, and that the persistent expression is dependent on functional *Ubx* protein. There is as yet no evidence that other *Drosophila* homeotic selectors are autoregulated in epidermal cells. This may be because some of their expression boundaries are adequately maintained by cross-regulatory relationships. In a more general sense, it is likely that many homeodomain proteins will be part of autoregulatory circuits, since they already carry much of the biochemical machinery for gene regulation, and a simple mechanism for assuring lineage-specific expression is certain to be useful for some of the developmental functions specified by homeobox genes.

ACKNOWLEDGMENTS

Much of this research was funded by the generous support of the National Science Foundation and the National Institutes of Health, both through research grants and a Presidential Young Investigator award to W. McGinnis as well as through training grants and fellowship support. Many people were very generous with antisera and arduously constructed fly stocks; thanks are due particularly to Markus Noll for *prd* antiserum, Manfred Frasch for *eve* antiserum, Gary Struhl for the hs-*eve* and hs-*ftz* stocks, and the laboratories of Nüsslein-Volhard, Wieschaus, Shupbach, Levine, and Kaufman for mutant strains.

REFERENCES

Akam, M. E. (1987). The molecular basis for metameric pattern in the *Drosophila* embryo. *Development* **101**, 1–22.

Berleth, T., Burri, M., Thoma, G., Bopp, D., Richstein, S., Frigero, G., Noll, M., and Nüsslein-Volhard, C. (1988). The role of localization of *bicoid* RNA in organizing the anterior pattern of the *Drosophila* embryo. *EMBO J.* **7**, 1749–1756.

Bienz, M., and Tremml, G. (1988). Domain of *Ultrabithorax* expression in *Drosophila* visceral mesoderm from autoregulation and exclusion. *Nature (London)* **333**, 576–578.

Bridges, C. B., and Morgan, T. H. (1923). The third-chromosome group of mutant characters of *Drosophila melanogaster. Canegie Inst. Washington Publ.* No. 327, p. 93.

Campos-Ortega, J. A., and Hartenstein, V. (1985). "The Embryonic Development of *Drosophila melanogaster.*" Springer-Verlag, Berlin and New York.

Chadwick, R., and McGinnis, W. (1987). Temporal and spatial distribution of transcripts from the *Deformed* gene of *Drosophila. EMBO J.* **6,** 779–789.

Desplan, C., Theis, J., and O'Farrell, P. H. (1985). The *Drosophila* developmental gene, *engrailed,* encodes a sequence-specific DNA binding activity. *Nature (London)* **318,** 630–635.

DiNardo, S., and O'Farrell, P. (1987). Establishment and refinement of segmental pattern in the *Drosophila* embryo: Spatial control of *engrailed* expression by pair-rule genes. *Genes Dev.* **1,** 1212–1225.

DiNardo, S., Kuner, J. M., Theis, J., and O'Farrell, P. H. (1985). Development of embryonic pattern in *D. melanogaster* as revealed by accumulation of the nuclear *engrailed* protein. *Cell* **43,** 59–69.

Driever, W., and Nüsslein-Volhard, C. (1988). A gradient of *bicoid* protein in *Drosophila* embryos. *Cell* **54,** 83–93.

Driever, W., and Nüsslein-Volhard, C. (1989). The *bicoid* protein is a positive regulator of *hunchback* transcription in the early *Drosophila* embryo. *Nature (London)* **337,** 138–143.

Duncan, I., and Lewis, E. B. (1982). Genetic control of body segment differentiation in *Drosophila. In* "Developmental Order: Its Origin and Regulation" (S. Subtelny, ed.), pp. 533–554. Liss, New York.

Fienberg, A. A., Utset, M. F., Bogarad, L. D., Hart, C. P., Awgulewitsch, A., Ferguson-Smith, A., Fainsod, A., Rabin, M., and Ruddle, F. H. (1987). Homeo box genes in murine development. *Curr. Top. Dev. Biol.* **23,** 233–256.

Frasch, M., Hoey, T., Rushlow, C., Doyle, H., and Levine, M. (1987). Characterization and localization of the *even-skipped* protein of *Drosophila. EMBO J.* **6,** 749–759.

Frohnhöfer, H. G., and Nüsslein-Volhard, C. (1986). Organization of anterior pattern in the *Drosophila* embryo of the maternal gene *bicoid. Nature (London)* **324,** 120–125.

Garber, R. L., Kuriowa, A., and Gehring, W. J. (1983). Genomic and cDNA clones of the homeotic locus *Antennapedia* in *Drosophila. EMBO J.* **2,** 2027–2036.

Graham, A., Papalopulu, N., Lorimer, J., McVey, J. H., Tuddenham, E. G. D., and Krumlauf, R. (1988). Characterization of a murine homeo box gene, *Hox-2.6,* related to the *Drosophila Deformed* gene. *Genes Dev.* **2,** 1424–1438.

Hadzi, J. (1963). "The Evolution of the Metazoa." Macmillan, New York.

Hafen, E., Levine, M., and Gehring, W. J. (1984). Regulation of *Antennapedia* transcript distribution by the bithorax complex in *Drosophila. Nature (London)* **307,** 287–289.

Hall, M. N., and Johnson, A. D. (1987). Homeo domain of the yeast repressor alpha 2 is a sequence specific DNA-binding domain but is not sufficient for repression. *Science* **237,** 1007–1012.

Harding, K., Rushlow, C., Doyle, H. J., Hoey, T., and Levine, M. (1986). Cross-regulatory interactions among pair-rule genes in *Drosophila. Science* **233,** 953–959.

Harvey, R. P., Tabin C. J., and Melton, D. A. (1986). Embryonic expression and nuclear localization of *Xenopus* (Xhox) gene products. *EMBO J.* **5,** 1237–1244.

Hazelrigg, T., and Kaufman, T. C. (1983). Revertants of dominant mutations associated with the Antennapedia gene complex of *Drosophila melanogaster:* Cytology and genetics. *Genetics* **105,** 581–600.

Hoey, T., and Levine, M. (1988). Divergent homeo box proteins recognize similar DNA sequences in *Drosophila. Nature (London)* **332,** 858–861.

Ingham, P. W., and Martinez-Arias, A. (1986). The correct activation of Antennapedia and bithorax complex genes requires the *fushi tarazu* gene. *Nature (London)* **324,** 592–597.

Jack, T., Regulski, M., and McGinnis, W. (1988). Pair-rule segmentation genes regulate the expression of the homeotic selector gene, *Deformed. Genes Dev.* **2**, 635–651.

Jurgens, G., Wieschaus, E., Nüsslein-Volhard, C., and Kluding, H. (1984). Mutations affecting the pattern of the larval cuticle in *Drosophila melanogaster.* II. Zygotic loci on the third chromosome. *Wilhelm Roux's Arch. Dev. Biol.* **193**, 283–295.

Jurgens, G., Lehman, R., Schardin, M., and Nüsslein-Volhard, C. (1986). Segmental organization of the head in the embryo of *Drosophila melanogaster. Wilhelm Roux's Arch. Dev. Biol.* **195**, 359–377.

Kaufman, T. (1983). Genetic regulation of segmentation in *Drosophila melanogaster. In* "Time, Space, and Pattern in Embryonic Development" (W. R. Jeffry and R. A. Raff, eds.), pp. 365–383. Liss, New York.

Kornberg, T. (1981). *engrailed:* A gene controlling compartment and segment formation in *Drosophila. Proc. Natl. Acad. Sci. U.S.A.* **78**, 1095–1099.

Kuroiwa, A., Kloter, U., Baumgartner, P., and Gehring, W. J. (1985). Cloning of the homeotic *Sex combs reduced* in *Drosophila* and *in situ* localization of its transcripts. *EMBO J.* **4**, 3757–3764.

Kuziora, M. A., and McGinnis, W. (1988). Autoregulation of a *Drosophila* homeotic selector gene. *Cell* **55**, 477–485.

Lehmann, R., and Nüsslein-Volhard, C. (1987). *hunchback,* a gene required for segmentation of anterior and posterior regions of the *Drosophila* embryo. *Dev. Biol.* **119**, 402–417.

Lewis, E. B. (1965). Genetic control and regulation of developmental pathways. *In* "The Role of Chromosomes in Development" (M. Locke, ed.), pp. 231–252. Academic Press, New York.

Lewis, E. B. (1978). A gene complex controlling segmentation in *Drosophila. Nature* (*London*) **276**, 565–570.

Lewis, R. A., Wakimoto, B. T., Denell, R. E., and Kaufman, T. C. (1980). Genetic analysis of the Antennapedia gene complex (ANT-C) and adjacent chromosomal regions of *Drosophila melanogaster.* II. Polytene chromosome segments 84A–B. *Genetics* **95**, 383–397.

McGinnis, W., Levine, M., Hafen, E., Kuroiwa, A., and Gehring, W. J. (1984a). A conserved DNA sequence found in homeotic genes of the *Drosophila* Antennapedia and Bithorax complexes. *Nature* (*London*) **308**, 428–433.

McGinnis, W., Garber, R. L., Wirz, J., Juroiwa, A., and Gehring, W. J. (1984b). A homologous protein-coding sequence in *Drosophila* homeotic genes and its conservation in other metazoans. *Cell* **37**, 403–408.

Martinez-Arias, A., and Lawrence, P. A. (1985). Parasegments and compartments in the *Drosophila* embryo. *Nature* (*London*) **313**, 639–642.

Martinez-Arias, A., Ingham, P. W., Scott, M. P., and Akam, M. E. (1987). The spatial and temporal deployment of *Dfd* and *Scr* transcripts during development of *Drosophila. Development* **100**, 673–68.

Mavilio, F., Simeone, A., Giampaolo, A., Faiella, A., Zappavigna, V., Acampora, D., Poiana, G., Russo, G., Peschle, C., and Boncinelli, E. (1986). Differential and stage-related expression in embryonic tissues of a new human homeo box gene. *Nature* (*London*) **324**, 664–667.

Merrill, V. K. L., Turner, F. R., and Kaufman, T. C. (1987). A genetic and developmental analysis of mutations in the *Deformed* locus in *Drosophila melanogaster. Dev. Biol.* **122**, 379–395.

Mohler, J., and Wieschaus, E. (1986). Dominant maternal-effect mutations of *Drosophila melanogaster* causing the production of double abdomen embryos. *Genetics* **112**, 803–822.

Nüsslein-Volhard, C. (1979). Maternal effect mutations that alter the spatial coordinates of the embryos of *Drosophila melanogaster*. In "Determinants of Spatial Organization" (S. Subtelny and I. R. Konigsberg, eds.), pp. 185–211. Academic Press, New York.

Nüsslein-Volhard, C., and Weischaus, E. (1980). Mutations and affecting segment number and polarity in *Drosophila*. *Nature (London)* **287**, 795–801.

Nüsslein-Volhard, C., Wieschaus, E., and Kluding, H. (1984). Mutations affecting the pattern of the larval cuticle in *Drosophila melanogaster*. I. Zygotic loci on the second chromosome. *Wilhelm Roux's Arch. Dev. Biol.* **193**, 267–282.

Otting, G., Qian, Y., Muller, M., Affolter, M., Gehring, W., and Wuthrich, K. (1988). Secondary structure determination for the *Antennapedia* homeodomain by nuclear magnetic resonance and evidence for a helix–turn–helix motif. *EMBO J.* **7**, 4305–4309.

Regulski, M., Harding, K., Kostriken, R., Karch, F., Levine, M., and McGinnis, W. (1985). Homeo box genes of the Antennapedia and Bithorax complexes of *Drosophila*. *Cell* **43**, 71–80.

Regulski, M., McGinnis, N., Chadwick, R., and McGinnis, W. (1987). Developmental and molecular analysis of *Deformed*: A homeotic gene controlling *Drosophila* head development. *EMBO J.* **6**, 767–777.

Rice, T. B., and Garen, A. (1975). Localized defects of blastoderm formation in maternal effect mutants of *Drosophila*. *Dev. Biol.* **43**, 277–286.

Rubin, M. R., Toth, L. E., Patel, M. D., D'eustachio, P., and Nguyen-Huu, M. C. (1986). A mouse homeo box gene is expressed in spermatocytes and embryos. *Science* **233**, 63–667.

Sanchez-Herrero, E., Vernos, I., Marco, R., and Morata, G. (1985). Genetic organization of *Drosophila* bithorax complex. *Nature (London)* **313**, 108–113.

Schupbach, T., and Wieschaus, E. (1986). Maternal effect mutations altering the anterior–posterior pattern of the *Drosophila* embryo. *Wilhelm Roux's Arch. Dev. Biol.* **195**, 302–317.

Scott, M. P., and Weiner, A. (1984). Structural relationships among genes that control development: Sequence homology between the *Antennapedia, Ultrabithorax,* and *fushi tarazu* loci of *Drosophila*. *Proc. Natl. Acad. Sci. U.S.A.* **81**, 4115–4119.

Scott, M. P., Weiner, A. J., Hazelrigg, T. I., Polisky, B. A., Pirotta, V., Scalenghe, F., and Kaufman, T. C. (1983). The molecular organization of the *Antennapedia* locus of *Drosophila*. *Cell* **35**, 763–776.

Seeger, M. A., Haffley, L., and Kaufman, T. C. (1988). Characterization of *amalgam*: A member of the immunoglobulin superfamily from *Drosophila*. *Cell* **55**, 589–600.

Spemann, H. (1938). "Embryonic Development and Induction." Yale Univ. Press, New Haven, Connecticut.

Struhl, G. (1982). Genes controlling segmental specification in the *Drosophila* thorax. *Proc. Natl. Acad. Sci. U.S.A.* **79**, 7380–7384.

Struhl, G. (1983). Role of the esc^+ gene product in ensuring the selective expression of segment-specific homeotic genes in *Drosophila*. *J. Embryol. Exp. Morphol.* **76**, 297–331.

Struhl, G. (1985). Near reciprocal phenotypes caused by inactivation or indiscriminate expression of the *Drosophila* segmentation gene *ftz*. *Nature (London)* **318**, 677–680.

Tautz, D. (1988). Regulation of the *Drosophila* segmentation gene *hunchback* by two maternal morphogenetic centers. *Nature (London)* **332**, 281–284.

Turner, R. F., and Mahowald, A. P. (1979). Scanning electron microscopy of *Drosophila melanogaster* embryogenesis. III. Formation of the head and caudal segments. *Dev. Biol.* **68**, 96–109.

Van der Meer, J. M. (1977). Optical clean and permanent whole mount preparation for phase contrast microscopy of cuticular structures of insect larvae. *Drosophila Inf. Serv.* **52,** 160.

Wakimoto, B. T., and Kaufman, T. C. (1981). Analysis of larval segmentation in lethal genotypes associated with the Antennapedia gene complex in *Drosophila melanogaster*. *Dev. Biol.* **81,** 51–64.

Wedeen, C., Harding, K., and Levine, M. (1986). Spatial regulation of *Antennapedia* and *bithorax* gene expression of the *Polycomb* locus in *Drosophila*. *Cell* **44,** 739–748.

Weinzierl, R., Axton, J. M., Ghysen, A., and Akam, M. (1987). *Ultrabithorax* mutations in constant and variable regions of the protein coding sequence. *Genes Dev.* **1,** 386–397.

Weir, M. P., and Kornberg, T. (1985). Patterns of *engrailed* and *fushi tarazu* transcripts reveal novel intermediate stages in *Drosophila* segmentation. *Nature (London)* **318,** 433–439.

Wieschaus, E., Nüsslein-Volhard, C., and Jurgens, G. (1984). Mutations affecting the pattern of the larval cuticle of *Drosophila melanogaster*. III. Zygotic loci on the X-chromosome and fourth chromosome. *Wilhelm Roux's Arch. Dev. Biol.* **193,** 296–307.

Wolgemuth, D. J., Engelmyer, E., Duggal, R. N., Gizang-Ginsberg, E., Mutter, G. L., Ponzett, O. C., Viviano, C., and Zakeri, Z. F. (1986). Isolation of a mouse cDNA coding for a developmentally regulated, testis specific transcript containing homeo box homology. *EMBO J.* **5,** 1229–1235.

MECHANISMS OF A CELLULAR DECISION DURING EMBRYONIC DEVELOPMENT OF *Drosophila melanogaster:* EPIDERMOGENESIS OR NEUROGENESIS

José A. Campos-Ortega

Institut für Entwicklungsphysiologie, Universität zu Köln,
D-5000 Köln 41, Federal Republic of Germany

I. Introduction

The origin of cell diversity is one of the main questions in developmental biology: what are the mechanisms leading to the large number of different cell types found in multicellular organisms? Early neurogenesis in insects, the processes that lead to the formation of a neural primordium, is an appropriate system in which to study the origin of cell diversity, and, indeed, it has received considerable attention since the mid-1980s. In *Drosophila melanogaster,* neighboring cells in the neurogenic region (NR) of the ectoderm adopt one of two alternative fates. From an undifferentiated state during early stages, the cells of the NR will develop either as neuroblasts (NBs, progenitor cells of the central nervous system) or as epidermoblasts (EBs, progenitor cells of the epidermis). Hence, the cells of the NR become subdivided into two different populations.

The study of early neurogenesis in *Drosophila* comprises at least two major problems: (1) what makes the NR different from the other parts of

403

the ectoderm, so that its cells may adopt the neural fate, and (2) what leads to the separation of the presumptive NBs from the EBs within the NR? These two components of neurogenesis are subject to different regulatory influences and certainly have different mechanistic bases (see Campos-Ortega, 1983). A third level of consideration of early neurogenesis in *Drosophila* follows from the fact that, although EBs and NBs are referred to by generic denominations, they actually constitute heterogeneous cell populations giving rise to very different cell lineages (Fig. 1). For example, some EBs include in their progeny elements that form sensory organs, of which there are many different types; others give rise to nonsensorial epidermis only (see Bate, 1978; Technau and Campos-Ortega, 1986). Although little is yet known about the lineages of individual NBs, there is evidence that particular NBs give rise to

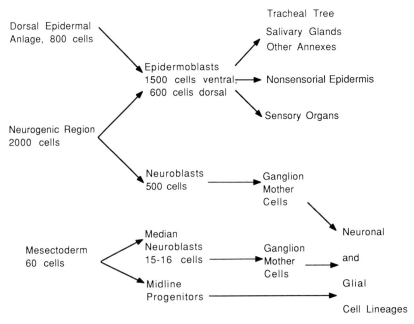

FIG. 1. Major ectodermal cell lineages in the germ band of *Drosophila melanogaster*. The ectodermal layer gives rise to a large variety of cell types. Approximately 600 epidermoblasts develop from the dorsal epidermal anlage, another 1500 from the ventrally located neurogenic region. In addition, the latter region gives rise to approximately 500 neuroblasts. Although here epidermoblasts and neuroblasts are considered generically, both cell types actually constitute heterogeneous cell populations. Only the major cell lineages are indicated.

specific cell lineages and produce different types of neurons, for example, cholinergic or dopaminergic, motorneurons or interneurons (see Thomas *et al.*, 1984; Doe and Goodman, 1985b). Thus, the study of neurogenesis must also seek to explain the origin of the different types of progenitor cells.

Of these three major kinds of problems related to neurogenesis in insects, most of the work done since 1980 has been concerned with the mechanisms of segregation of NBs from the NR of *Drosophila melanogaster* to give rise to the neural and epidermal cell lineages, whereas very little is known about the formation of the NR as such (Campos-Ortega, 1983; Hartenstein and Campos-Ortega, 1984) or about the mechanisms of origin of the different types of NBs and EBs and the lineages to which they give rise (see Doe *et al.*, 1988; Ghysen and Dambly-Chaudiére, 1988; Jiménez, 1988). Therefore, this article deals mainly with the process of separation of the progenitor cells of the epidermal and central neural lineages. Some reference is made to the development of the sensory organs as well, since there is evidence that the same elements and perhaps the same mechanisms are involved in the development of the progenitors of both the central and the peripheral nervous systems. The problems discussed in this article have been approached from the different points of view of embryology, genetics, and molecular biology; therefore, the various aspects of the process are considered in these different contexts.

II. Cellular Aspects of Neurogenesis

The first morphological manifestation of neurogenesis permits the differentiation of the neurogenic region from the other ectodermal regions. In *Drosophila* the NR is subdivided into two parts, one located in the procephalic lobes, giving rise to the supraesophageal ganglia, and the other in the presumptive territory of the trunk, which provides the cells of the subesophageal (gnathal), thoracic, and abdominal neuromeres that fuse to form the ventral cord (see Campos-Ortega and Hartenstein, 1985, for a general introduction to *Drosophila* embryogenesis). Very little is known of the processes that lead to regionalization of the ectoderm, although these processes are likely to depend on the action of genes which control pattern formation along the dorsoventral body axis (see Anderson and Nüsslein-Volhard, 1986; Campos-Ortega, 1983). The following sections discuss morphological aspects of the formation of the NR and of the segregation of the NBs and their further development. The discussion is concerned with the NR of the trunk, neglecting the procephalic portion, our knowledge of which is rather fragmentary.

A. FORMATION OF THE NEUROGENIC ECTODERM

In the wild type, the NR becomes morphologically manifest at stage 8 (embryogenetic stages according to Campos-Ortega and Hartenstein, 1985), prior to the onset of mitotic activity in the ectodermal germ layer. The ectoderm becomes subdivided into a lateral part comprising small cylindrical cells and a medial part made up of large cuboidal cells (Hartenstein and Campos-Ortega, 1984). The lateral region of small cells will give rise to the tracheal placodes and the dorsal epidermis, whereas the medial sector is the NR itself, and gives rise to the ventral cord and the ventral epidermis (Technau and Campos-Ortega, 1985). It is worth emphasizing that in *Drosophila* all cells of the NR, that is, 100 rows on either side of the midline, with approximately 9 cells in each row, enlarge to become conspicuously different from the remaining, nonneurogenic ectodermal cells (Hartenstein and Campos-Ortega, 1984, 1985). During early neurogenesis in grasshoppers, in contrast, only single neuroectodermal cells among groups of several cells enlarge; these cells are the prospective NBs themselves, which will segregate from the remaining ectodermal cells (Bate, 1976, 1982; Doe and Goodman, 1985a).

1. Segregation of Neuroblasts

The segregation of the NBs is a highly dynamic process that is associated with important morphological modifications in the ectodermal germ layer (Poulson, 1950). The process lasts for approximately 3 hours and is discontinuous, proceeding in three discrete pulses (Hartenstein and Campos-Ortega, 1984). In *Drosophila* these three pulses of segregation give rise to three subpopulations of NBs called SI, SII, and SIII NBs. Single cells among the population of large, medial ectodermal cells of the *Drosophila* NR undergo conspicuous shape changes, described in detail by Poulson (1950), and eventually leave the outer layer and migrate internally. The cells remaining in the NR after the segregation of the NBs, namely, the prospective epidermal progenitor cells, also show characteristic changes. Particularly remarkable is the fact that the cells immediately adjacent to each of the SI NBs remain in intimate contact with the NB by means of long basal processes which transiently surround the segregated NB to form a sort of sheath. The ensheathing processes are later retracted, and the prospective epidermal progenitor cells diminish in size to take on an epithelial appearance. Since the NBs segregate in pulses, groups of cells of fairly large size, which give rise to SII and SIII NBs, are present in the NR until late stage 11.

The relationships between the various parts of the NR and mitotic activity in the ectodermal layer are worth mentioning (Hartenstein and Campos-Ortega, 1984, 1985). The first postblastoderm mitosis affects metamerically arranged groups of cells in the middle of the NR. SI NBs segregate from those cells of the NR that flank the mitotic clusters. Therefore, these NBs actually segregate without having divided since the blastoderm stage. Shortly after the segregation of SI NBs, but before the segregation of the SII and SIII NBs, motitic activity spreads from the lateral epidermal anlage into the NR, and some of the cells with neurogenic abilities enter mitosis. This means that some of the SII NBs, as well as most or all SIII NBs, may share common lineages with EBs (see below and Technau and Campos-Ortega, 1986).

It is also remarkable that the proportions of EBs to NBs are fairly reproducible from animal to animal. Roughly 25% (~500) of all cells of the NR adopt the neural fate and develop as NBs, whereas the remaining 75% (~1500 cells) develop as EBs (Hartenstein and Campos-Ortega, 1984). There is no evidence as to the mechanisms that restrict the production of one or the other type of cells. The timing of segregation could be one possible constraint that restricts the final number of each one of the progenitor cell types.

2. Map of Neuroblasts

After segregation, the NBs form a continuous monolayer between ectoderm and mesoderm. In fixed preparations, the three classes of NBs can, in principle, be distinguished from each other at this stage on the basis of their size and location within the array. The distinction between SI and SII NBs is particularly easy, since they are arranged in three longitudinal rows on either side of the ventral midline and form a fairly regular and constant pattern, in which the SII NBs are slightly smaller than the SI NBs (Hartenstein and Campos-Ortega, 1984; Hartenstein et al., 1987). All rows are in register, and in gnathal and thoracic regions each contains four SI and four SII NBs per hemisegment; in abdominal neuromeres, however, each intermediate row exhibits some gaps and contains only two NBs.

Immediately after SI and SII NBs leave the ectodermal layer, they round up and start dividing asymmetrically to produce ganglion mother cells. The increase in cell number and the reduction in the size of the NBs that follows the initiation of mitotic divisions leads to the loss of the regular pattern of SI and SII NBs. Therefore, at the time when SIII

NBs segregate, they encounter a strongly modified NB array, into which they integrate. SI can still be distinguished from SII NBs because of size differences, but this distinction is difficult and therefore not completely reliable. SIII NBs seem to segregate preferentially from medial parts of the NR, and, thus, they become intermingled with the SI NBs in the median row or arranged in a fourth, paramedian row. Hartenstein *et al.* (1987) distinguish a total of eight SIII NBs in each hemineuromere. However, the total complement of NBs in each neuromere is not yet reliably established. Numbers published for the *Drosophila* embryo vary between 22 lateral and 1 median (Hartenstein *et al.*, 1987) and a total of 25 NBs (Doe *et al.*, 1988).

3. Development of the Peripheral Nervous System

The available experimental evidence suggests that, although there are certain formal differences, the origin of the sensory organs, which form one of the main components of the peripheral nervous system (PNS), follows the same general scheme in *Drosophila* as the progenitor cells of the CNS. Therefore, summarized below are just a few aspects of PNS development that are relevant for some of the genetic data discussed in later sections.

In *Drosophila,* the larval sensory organs originate from among the progeny of the EBs (Technau and Campos-Ortega, 1986). In the trunk region, sensory organs are arranged in three groups, dorsal, lateral, and ventral, with a characteristic topological pattern (Campos-Ortega and Hartenstein, 1985; Dambly-Chaudière and Ghysen, 1986; Ghysen *et al.*, 1986; Hartenstein and Campos-Ortega, 1986; Bodmer and Jan, 1987). In insects, most of the epidermal sensory organs are thought to be derivatives of individual progenitor cells that start dividing during late stage 12, when the main bulk of mitotic divisions within the epidermal primordium has finished, to give rise to cell clones; the members of these clones will then group to form each of the corresponding sensilla (reviewed by Bate, 1978).

Two questions arise from this behavior: what determines the appearance of the sensorial progenitor cells, and do sensory organs develop by a clonal mechanism? Recent experimental data have provided some insight into both of these questions. On the one hand, the mitotic divisions that will eventually give rise to the cells of the sensory organs take place during stages 12–13. However, the corresponding progenitor cells are already detectable in stages 10 and 11, that is, several hours before the sensory organs have formed, as they differentiate molecular markers specific for sensory organs in a segmentally specific manner (Ghysen and O'Kane, 1989). These observations indicate that genetic mechanisms determine the appearance of particular cell sets in each segment, which later form a characteristic array of sensory organs. On the other hand, several observations suggest that the cells of a given sensillum do not have to belong to the same cell clone, but may be recruited from neighboring, lineage-independent cells (Technau and Campos-Ortega, 1986; V. Hartenstein, personal communication). Hence, although further work is required, evidence suggests that the cells forming some of the epidermal sensilla may well become sequentially committed to their fates at grouping, that is, they develop in a way similar to the ommatidia of the compound eye (Lawrence and Green, 1979).

B. CELL COMMITMENT IN THE NEUROGENIC ECTODERM

In insects, the decision of the neuroectodermal cells to adopt the epidermal or the neural fate is mediated by cell–cell interactions. Two pieces of experimental evidence support this contention. On the one hand, laser ablation experiments carried out in grasshoppers showed that the cells remaining in the NR after the NBs have segregated are not firmly committed to their fate (Taghert et al., 1984; Doe and Goodman, 1985b). Under normal circumstances, these cells would develop as EBs; however, under the conditions of the experiment they may adopt the neural fate instead. These results led to the proposal of interactions between the prospective NBs and EBs, such that the latter are inhibited by the former from adopting the neural fate (Doe and Goodman, 1985b). On the other hand, results of cell transplantations in *Drosophila* suggest that positive regulatory signals pass between the cells of the neurogenic ectoderm, causing their commitment to a given fate (Technau and Campos-Ortega, 1986; Technau et al., 1988). Below, experiments supporting the existence of two kinds of signals, one with epidermalizing and the other with neuralizing character (Fig. 2), are discussed. Furthermore, data are presented suggesting that cell commitment is not an irreversible event, but that it is experimentally possible to switch between an epidermal and a neural fate.

1. Homotopic and Heterotopic Transplantations of Ectodermal Cells

Upon their homotopic transplantation, ectodermal cells behave in the host in the same way as they would have in the donor (Technau and Campos-Ortega, 1986). Single cells from ventral ectodermal regions of the wild type, that is, from the NR, homotopically transplanted into the NR of a wild-type host embryo adopt either of two different fates: they develop as NBs or as EBs (Fig. 3). Consequently, the transplanted cells

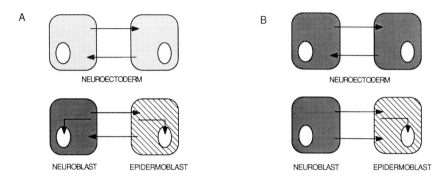

FIG. 2. Cell–cell interactions determine the decision of neuroectodermal cells for the neural or the epidermal fate. Interacting cells of the neuroectoderm are shown schematically; putative regulatory signals are represented by arrows. Two hypotheses are considered. (A) Two regulatory signals occur; one leads to neurogenesis, the other to epidermogenesis. (B) All cells of the neuroectoderm have a primary neurogenic fate; this fate is suppressed in the putative epidermoblasts by the participation of an epidermalizing signal.

give rise to three types of clones: neural cells, epidermal cells, and mixed clones of neural and epidermal cells. Single cells from the dorsal epidermal anlage give rise to only epidermal clones upon homotopic transplantation. However, following heterotopic transplantation, ectodermal cells exhibit a differential behavior. Ventral cells transplanted dorsally develop according to their origin and differentiate either epidermal (more frequently) or neural histotypes (only occasionally). In contrast, dorsal cells transplanted ventrally develop according to their new location, giving rise to either epidermal or neural clones.

This latter observation is actually very striking, for dorsal cells do not develop as NBs, either *in situ* under normal conditions or upon homotopic transplantation. A possible interpretation of this result is that the cells of the dorsal epidermal anlage are normally prevented from developing as NBs by an inhibitory process, and the cells are relieved from this inhibition upon their transplantation into the NR. Another possibility, however, is that the transplanted dorsal cells are actively induced by their neighbors in the NR to adopt a neural fate. Since no intercellular influences that would actively prevent neurogenesis can be experimentally demonstrated in the dorsal region, the results support the existence of a neuralizing signal.

2. Heterochronic Transplantations

Further support for the hypothesis that cellular interactions mediate the segregation of lineages derives from heterochronic transplan-

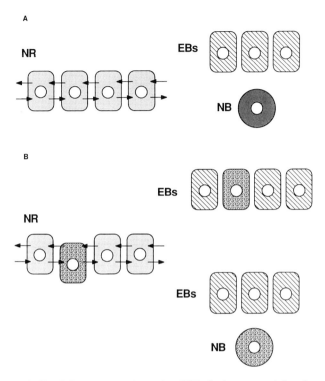

FIG. 3. (A) Cells of the neurogenic region (NR) during normal development. Cell interactions (arrows) lead to the segregation of 75% of the cells as epidermoblasts (EBs) and 25% as neuroblasts (NBs). (B) Situation thought to occur in transplantation experiments. A single HRP-labeled cell is transplanted into the NR; this cell establishes interactions with adjacent cells and, as a result, adopts either an epidermal or a neural fate.

tations of ectodermal cells (Technau *et al.*, 1988). When ectodermal cells of increasingly older donors are transplanted into young hosts, the transplanted cells behave in the same way as younger cells in isochronic transplants. That is to say, the old cells give rise to the same kinds of clones as the cells of the young gastrula stage. The transplantations involved both EBs, namely, cells of the dorsal epidermal anlage, and NBs, namely, cells of the NR of neurogenic mutants (see below), which were aged for up to 170 minutes after gastrulation and had, therefore, divided up to 3 or 4 times, respectively. The frequent switch of fate of the cells upon transplantation observed in these experiments suggests that, under normal circumstances, the segregation of the two types of progenitor cells does not by itself imply their irreversible commitment to the neural or the epidermal fate, for this fate can be changed after experimental manipulation.

III. Genetic and Molecular Aspects of Neurogenesis

The available evidence strongly suggests that the proteins encoded by two groups of genes provide the molecular basis for the regulatory signals which control the process of neurogenesis. The so-called neurogenic (NG) genes, together with a second set of various other genes, including the *achaete–scute* complex (AS-C), *ventral nervous system condensation defective* (*vnd*), *embryonic lethal and visual system defective* (*elav*), and *daughterless* (*da*), are both required for proper segregation of neural and epidermal lineages, whereby the two groups exert apparently opposite effects on the differentiation of epidermal and neural lineages (Campos-Ortega, 1985, 1988).

A. NEUROGENIC GENES

Notch (*N*) generally serves as the prototype NG gene, for it is not only the first NG genes that was discovered (Mohr, 1919; Poulson, 1937) but also one of the most carefully investigated genes in the entire *Drosophila* genome (see Welshons, 1965; Wright, 1970; and Artavanis-Tsakonas, 1988, for reviews). Lehmann *et al.* (1981) studied a collection of embryonic lethals recovered by Nüsslein-Volhard and Wieschaus (1980) and described another four loci, the loss of whose function leads to the same phenotype as N^- mutations. Three of these genes were previously unknown and were called *master mind* (*mam*), *big brain* (*bib*), and *neuralized* (*neu*), whereas the fourth one had already been discovered by Bridges on the basis of wing defects associated with heterozygosity for loss-of-function mutations (see Lindsley and Zimm, 1985) and called *Delta* (*Dl*). The participation of the *Enhancer of split* [*E(spl)*] locus in neurogenesis was discovered by reverting the dominant phenotype of $E(spl)^D$ (Lehmann *et al.*, 1983). E. Wieschaus (unpublished, personal communication) recognized the neurogenic phenotype of embryos derived from females homozygous for the maternal-effect mutation *almondex* (*amx*) (see Shannon, 1972, 1973; Lehmann *et al.*, 1983). Additional NG genes with predominantly maternal effects were described by Perrimon *et al.* (1984) and LaBonne and Mahowald (1985).

Most of the work has been done on the six NG loci with predominantly zygotic expression, namely, *N, bib, mam, neu, Dl,* and *E(spl)*, and this article consequently deals with the zygotic loci. The complete loss of a NG gene function causes the diversion of all the cells of the NR to the neural fate. Hence, approximately 2000 cells initiate neurogenesis in the NG mutants, instead of only 500 in the wild type. This leads to embryonic lethality, associated with massive hyperplasia of the central nervous system (CNS) and an increase in the number of sensory neurons, with concomitant lack of the entire ventrolateral and cephalic epidermis in the mature embryo (different aspects of the complex phenotype of NG mutants are described in Poulson, 1937; Wright, 1970; Lehmann et al., 1981, 1983; Jiménez and Campos-Ortega, 1982; Dietrich and Campos-Ortega, 1984; Hartenstein and Campos-Ortega, 1986).

Since most of the NG genes have both maternal and zygotic expression, the complete loss of gene function is attained only when both components are removed (Jiménez and Campos-Ortega, 1982; Dietrich and Campos-Ortega, 1984). Under these circumstances, all NG genes, with the exception of *bib,* cause an identical mutant embryonic phenotype. *bib* is apparently different from the other NG loci, in that no maternal expression can be detected (Dietrich and Campos-Ortega, 1984). In addition, the phenotypic effects of complete loss of the *bib*$^+$ function are rather weak; the phenotype of *bib*-deficient homozygous embryos attains only an intermediate degree of expressivity (J. Campos-Ortega, unpublished). Besides the neural hyperplasia and the epidermal defects, severe NG mutants are also affected to some extent in all remaining embryonic organs. However, thorough embryological analyses have demonstrated that these are not primary defects, but rather morphogenetic consequences of the neural fate adopted by all the neuroectodermal cells (Wright, 1970; Lehmann et al., 1981, 1983).

A detailed analysis of the epidermal defects associated with homozygosity for any of several NG mutations, using the neural-specific anti-horseradish peroxidase (anti-HRP) antibody (Jan and Jan, 1982), has allowed the recognition of conspicuous hyperplastic defects of the larval sensory organs (Hartenstein and Campos-Ortega, 1986). The severity of the defects varies with the expressivity of the allele. Alleles of weak or intermediate expressivity cause a 2- to 4-fold increase in the number of fully differentiated sensory organs whereby the type of sensillum is preserved, that is, the additional sensilla grow in the immediate neighborhood of normal sensilla and differentiate the same histotypes. In the case of extreme alleles, no organized sensilla are present, but large clusters of anti-HRP-binding neuronal cells with axonal processes are found in dorsal positions instead. These observations indicate that the function of the NG genes is required for the development of the nonsensorial epidermis, as opposed to the fate of the sensory organs, and that this requirement is graded. A functional insufficiency of the NG genes, as in some hypomorphic alleles, leads to the appearance of additional, fully differentiated sensory organs, whereas complete absence of the gene function leads to neural differentiation of all the cells of the epidermal primordium. The observations thus suggest that the amount of NG gene function available may decide the fate of the epidermal cells: a normal function leads to a normal complement of sensilla; a decrease in the function causes the appearance of supernumerary, well-structured sensilla; and the absence of NG gene function causes all cells to develop as neurons.

In summary, the NG genes are required at two different stages during embryonic development, namely, for the segregation of the NBs from the EBs and for the decision of the cells of the epidermal primordium to develop as sensilla as opposed to uninnervated epidermis. At both stages the function of the NG genes appears to be necessary to suppress the neural fate and permit epidermal development (Hartenstein and Campos-Ortega, 1986). We see below that the NG genes seem to perform the same function in the development of the imaginal sensory organs as well.

1. Mosaic Analysis of Neurogenic Mutants

Somatic mosaics have proved extremely useful for studying a large variety of problems in *Drosophila* developmental genetics (see Garcia-Bellido and Ripoll, 1978; and Postlethwait, 1978, for reviews). The following deals with two of these problems: first, the participation of the NG genes in imaginal disc morphogenesis and, second, the cellular autonomy of NG gene expression, both of which can be profitably approached using homozygous mutant cell growing in a background of phenotypically wild-type cells (Stern, 1954, 1968).

Several different techniques have been used to obtain mosaics. Homozygosity for NG mutations has been induced by means of mitotic recombination throughout larval development in cells of various imaginal discs (Dietrich and Campos-Ortega, 1984). The main conclusion was that, in addition to their participation in the process of segregation of the NBs from the EBs during early embryogenesis, the function of the NG genes is required for normal development of the imaginal epidermal cells. Homozygosity for NG mutations, with the exception of *bib*, affects sensory organs, causing morphogenetic defects of variable intensity depending on the allele. In the compound eye, ommatidia with cells homozygous for mutations of weak or intermediate expressivity in the loci *N, mam, neu,* or *Dl* were found to comprise larger numbers of photoreceptor cells and reduced numbers of surrounding pigment cells; within the imaginal epidermis, defects chiefly affected macro- and microchaetae, which either formed conspicuous clumps, when the allele used had weak expressivity (see Shellenbarger and Mohler, 1975, for similar observations on N^{ts1} flies grown at the restrictive temperature), or were absent, in the case of intermediate or severe alleles. Finally, homozygosity for extreme Dl alleles, or for large deletions that comprise the $E(spl)$ locus and several other flanking genes, was found to be cell lethal; lethality of homozygous Dl cells was suppressed by loss-of-function mutations of the *Hairless* (H) locus.

Cell clones homozygous for loss-of-function mutations of the *bib* locus were morphogenetically normal, a result difficult to interpret with the data available. At least two interpretations are possible: first, the *bib* product is not required for imaginal development; or, second, the *bib* product can diffuse freely in such a way that the surrounding heterozygous cells would provide the clone of mutant cells with the gene product, thus allowing the development of a wild-type phenotype.

In summary, most NG mutations affect imaginal sensory organ development. Although several aspects of the observed phenotypes in the imaginal discs could not be satisfactorily resolved, for example, the fate of the sensory cells in clones showing loss of bristles or the cell lethal phenotypes, the current data can be interpreted to mean that functional insufficiency of the NG genes causes the appearance of additional sensilla. Whether the complete lack of NG gene function leads to all imaginal epidermal cells adopting a neural fate still requires experimental demonstration. The graded differences found in the phenotypic expression of the various alleles studied are indeed highly reminiscent of the defects found in the developing larval sensory organs of homozygous mutant embryos discussed above (Hartenstein and Campos-Ortega, 1986).

An additional important conclusion to be drawn from the observations on somatic mosaics is that NG gene products, excepting perhaps that of *bib*, are unable to diffuse, at least over great distances. Furthermore, the results were compatible with cell-autonomous expression of the NG genes (Dietrich and Campos-Ortega, 1984), at least in the imaginal discs. Hoppe and Greenspan (1986) studied gynandromorph embryos formed by N^+ and N^- cells and came to the same conclusion with respect to N in its embryogenetic function.

However, the results of transplanting homozygous mutant ectodermal cells into the NR of wild-type embryos (see Technau and Campos-Ortega, 1987) contradict the conclusion of cell-autonomous expression for all NG genes. We have already seen that individual cells from the

NR of wild-type gastrula stage donors transplanted homotopically and isochronically into wild-type hosts give rise to clonal progenies belonging to three different types: neural, epidermal, and mixed, that is, some cells of the clone differentiate neural and others epidermal histotypes. The results of performing the same experiment, but using homozygous NG mutant embryos as donors, are, with the remarkable exception of $E(spl)^-$ cells, the same as when the donors are wild type, namely, neural, epidermal, and mixed clones in similar proportions to those of the controls. Since the same NG mutant cells would have invariably adopted the neural fate while developing *in situ*, the experimental results indicate that the NG genes under discussion are not cell autonomous in their expression. Granted that single cells lacking any of the NG genes *N, amx, bib, mam, neu,* or *Dl,* develop normally when surrounded by wild-type cells, the mutant cells seem to be capable of receiving and processing the epidermalizing signal with the same apparent efficacy as the wild-type cells and, hence, adopt the epidermal fate in some cases. These results strongly suggest that the corresponding mutants have normal receptor mechanisms but an abnormal signal source.

P. Hoppe and R. Greenspan (personal communication) have reached a similar conclusion with respect to the N locus by using a rather different experiment. They observed that one or two homozygous N^- cells may develop epidermal histotypes within the ventral epidermis of heterozygous animals; in other words, when the mutant cells are completely surrounded by wild-type neighbors they may adopt an epidermal fate.

In striking contrast, cells lacking the $E(spl)$ locus give rise exclusively to neural clones. Therefore, $E(spl)$ is the only NG locus with cell-autonomous expression. Following the same lines of reasoning as before, one could interpret this result to mean that the $E(spl)^-$ cells cannot react to the epidermalizing signal. Thus, the $E(spl)$ locus is a good candidate to encode protein(s) related to the different steps from the receptor to the nucleus, for example, receptor molecules themselves, "second messengers," or transcription factors.

2. Genetics of the Notch Locus

Mutations at the N locus (in band 3C7 of the X chromosome; Slizynska, 1938) lead to a large variety of phenotypic traits (discussed by Wright, 1970). Results of a thorough genetic analysis carried out over several years (Welshons, 1965, 1974a,b; Welshons and Keppy, 1975, 1981), however, indicate that N mutations affect a single gene rather than a gene complex. The phenotype that gave the locus its name, a conspicuous notching of the caudal border of the wings, is associated with haploinsufficient expression and is thus shown by females that are heterozygous for deficiency of the wild-type locus. However, this phenotypic trait exhibits incomplete penetrance and variable expressivity, being largely dependent on modifiers (Lindsley and Zimm, 1985; J. Campos-Ortega, unpublished). In addition to the wing notches, mutants exhibit deltalike thickenings in the veins at the marginal junctions. In particular, vein 3 of heterozygotes with strong N alleles is thicker than in the wild type (Vässin $et\ al.$, 1985). Hemizygous males as well as females homozygous for a deletion of the locus die. Most of the available N^- mutations are indeed noncomplementing, recessive embryonic lethals whereas the visible N^- mutations are very weak hypomorphs. Strong N^- mutations produce the phenotypic NG syndrome (first described by Poulson, 1937) whose main traits were discussed in the previous section, with a variable expressivity depending on the allele. An allelic series can be easily constructed according to the degree of neuralization attained (see Wright, 1970; Lehmann $et\ al.$, 1981, 1983).

N mutations have predominantly zygotic expression. However, the gene is also expressed during oogenesis (Jiménez and Campos-Ortega, 1982). The maternal component of N expression has a clear influence on the severity of the NG phenotype. Embryos homozygous for a complete deletion of the locus that have developed in the absence of maternal expression show a more severe phenotype than those whose mothers carried the wild-type gene. In addition, maternal N^+ expression is necessary for the viability of the embryo, as heterozygous $N/+$ embryos which develop in the absence of the maternal N products are inviable.

The lethality of embryos that lack the zygotic expression of N can be rescued by duplications of the band 3C7. Several N^+ duplications exist which segregate with each of the major chromosomes of $Drosophila$. Three copies of N^+ in females, or two copies in males, cause an irregular thickening of the wing veins known as the Confluens (Co) phenotype (Welshons, 1965). The severity of the Co phenotype increases with the number of N^+ copies. However, the viability of imagos carrying up to five N^+ copies is not noticeably impaired, and no obvious neural defects can be observed in these animals (F. Jiménez and J. Campos-Ortega, unpublished). A third kind of phenotype is that of so-called $Abruptex$ (Ax) mutations, characterized by interruptions in the fourth and/or fifth wing veins. Ax alleles behave as antimorphic or neomorphic mutations (Foster, 1975; Portin, 1975).

Two types of recessive visible N alleles are known (Welshons, 1965) that predominantly affect either the wing or the compound eye. With the exception of fa^{no}, both types are perfectly viable when heterozygous with a deficiency of the N locus; fa^{no} is semilethal with the deficiency. The wings of *notchoid* (*nd*) homozygotes are strongly notched and the veins are thickened; *facet* (*fa*) flies have rough compound eyes (Welshons, 1965). Several *fa* alleles are known (Lindsley and Zimm, 1985) some of which show temperature-sensitive expression (Shellenbarger and Mohler, 1975). The alleles of the group of viable wing mutants do not complement other alleles in the same group; the same applies to the group of compound eye mutants. However, wing mutants complement the phenotype of compound eye mutants and vice versa. *split* (*spl*) is a particularly interesting visible recessive allele. It leads to roughening of the compound eyes, which are smaller than in the wild type, as well as to split or missing bristles. The functional status of *spl* is not clear, for it does not behave as a hypomorph (Welshons, 1965).

Temperature-sensitive recessive lethal N alleles have been recovered (Shellenbarger and Mohler, 1975, 1978), and such mutants are viable at the permissive temperature. The developmental effects of one of these alleles, $l(1)N^{ts1}$, have been studied in some detail. It exhibits three phenocritical periods of lethality, namely, embryonic, larval, and pupal (Shellenbarger and Mohler, 1978); pulses of restrictive temperature during these stages lead either to lethality or to the appearance of a large variety of wing, leg, compound eye, and bristle defects.

Molecular Organization of the Notch Locus.   The N locus comprises approximately 40 kb of genomic DNA to which a large number of N mutations, including both lethal and recessive visible, as well as *Abruptex* alleles, have been mapped (Artavanis-Tsakonas *et al.,* 1983, 1984; Kidd *et al.,* 1983; see Artavanis-Tsakonas, 1988, for review). Southern blot techniques were used to map different N mutants, and their location on the physical map confirms the fine structure genetic map obtained from meiotic recombination studies (Welshons, 1965, 1974a,b). Several of the recessive visible mutations, for example, *facet* mutations, have been found to correspond to insertions of transposable elements, some of them located within a large intron (Artavanis-Tsakonas *et al.,* 1984). There is a striking correlation between the kind of transposon inserted and the mutant phenotype of the corresponding *facet* allele, namely, *facet* versus *facet glossy*-like (Kidd and Young, 1986). Five different *facet* alleles were found to be associated with insertions of the transposon *flea;* all five cause a glossy-like phenotype. Another three alleles were associated with insertions of non-*flea,* probably *copia*-like transposons; they cause a facet rough phenotype. The precise location of the insertion within the locus appears to have little influence.

The genomic DNA to which N mutants have been mapped contains a single, 37-kb large transcription unit (Kidd *et al.*, 1983) which encodes a poly(A)$^+$ RNA of 10.5 kb (Artavanis-Tsakonas *et al.*, 1983). Its expression is temporally regulated, showing a complex pattern (Artavanis-Tsakonas *et al.*, 1983, 1985; Kidd *et al.*, 1983). The RNA is already present in the unfertilized egg, probably corresponding to the maternal expression shown by genetic techniques (Jiménez and Campos-Ortega, 1982), and its concentration increases during the first hours of embryogenesis to reach a maximum at 6–7 hours. The concentration of N RNA decreases steadily during subsequent development and reaches a second peak during the first half of pupation. *In situ* hybridization has shown that, during early embryonic stages, up to stage 11, the N transcript is ubiquitously expressed (Hartley *et al.*, 1987). From stage 11 on, N RNA is still present in most of the embryonic cells, although it is concentrated on the periphery of the CNS in cells which probably correspond to the neuroblasts. These data indicate that the function of the N locus is likely to be required for other processes besides neurogenesis.

The 10.5-kb poly(A)+ RNA is processed from a primary transcript of 37 kb (Kidd *et al.*, 1983) with nine exons (Kidd *et al.*, 1986). The sequence of the transcript has been established both from cDNA clones (Wharton *et al.*, 1985b) and from sequencing part of the genomic DNA (Kidd *et al.*, 1986), and conceptual translation of these sequences reveals the putative N product as a protein of 2703 amino acids. The primary structure of this protein is compatible with a transmembrane location. Its extracellular domain consists mainly of 36 cysteine-rich tandem repeats with homology to various proteins of mammals, among them epidermal growth factor (EGF). There are also three copies of another cysteine-rich repeated motif in the extracellular part, called the Notch repeats (Wharton *et al.*, 1985b). The putative intracellular domain, of approximately 1000 amino acids, exhibits homology to two cell cycle controlling genes in yeast (Breeden and Nasmyth, 1987). In addition, the intracellular domain shows a homopolymeric repeat consisting of glutamine residues, called the opa repeat (Wharton *et al.*, 1985a). Obviously, the structure of the N protein is compatible with its participation in cell communication processes, as suggested by embryological and genetic data. It is remarkable that the protein encoded by the *lin-12* gene of *Caenorhabditis elegans,* which is known to control several developmental cell decisions in the nematode (Greenwald *et al.*,

1983; Sternberg and Horvitz, 1984), exhibits overall similarity to the putative N protein (Greenwald, 1985; Yochem et al., 1988).

3. Genetics of the Delta Locus

Similar to N mutations, lesions in the Dl locus (in the band 92A2 of the third chromosome; Vässin and Campos-Ortega, 1987; see, however, Alton et al., 1988) lead to a large variety of phenotypic traits, pointing to considerable genetic complexity. Dl has haplo-insufficient expression (Lindsley et al., 1972; Vässin et al., 1985); thus, heterozygotes with a deletion of the locus, or any other amorphic Dl mutation, show abnormalities of the wings, compound eyes, and bristles. In the wings of heterozygotes, the veins are irregularly broadened and exhibit deltalike thickenings at the marginal junctions; the compound eyes are rough and smaller than the wild type; and the number of bristles on thoracic segments is increased. Dl deletion homozygotes are embryonic lethal and develop a severe neurogenic phenotype. It is noteworthy that, although Dl is maternally expressed (Vässin et al., 1987), removal of the maternal gene product has no significant effect on the neurogenic phenotype of embryos homozygous for loss-of-function mutations. The phenotype of these Dl mutations is usually comparable to that of N^- embryos that have developed in the absence of maternal and zygotic N expression. Allelic series have been established with respect to the degree of neuralization (Lehmann et al., 1983; Alton et al., 1988).

Three recessive visible Dl alleles (Dl^{via1}, Dl^{via2}, and Dl^{via3}) were recovered by Vässin and Campos-Ortega (1987) which in homozygous flies cause slight deltalike thickenings, rough compound eyes, as well as shortening and fusion of tarsal segments. Approximately 15% of the homozygotes for any of the Dl^{via} alleles die at various stages throughout development. Dead embryos frequently show weak signs of neuralization. Thus, at least to some extent, the NG function of Dl is apparently affected in Dl^{via} alleles as well. However, Dl^{via} alleles are viable when heterozygous with a deficiency of the locus.

Lethal Dl alleles are known to exhibit a complex pattern of heteroallelic complementation (Vässin and Campos-Ortega, 1987; Alton et al., 1988), compatible with the notion that Dl may be a complex locus. Three lethal alleles (Dl^{FE30}, Dl^{FE32}, and Dl^{B107}) have been recovered with clear antimorphic effects, as shown by the fact that animals carrying any of these alleles when heterozygous with a Dl^+ duplication still exhibit wing vein defects.

In contrast to N, Dl function is apparently not sensitive to changes of the dosage. The presence of up to six copies of Dl^+ in the genome does not lead to any apparent neural defect (D. Godt and J. Campos-Ortega, unpublished).

Molecular Organization of the Delta Locus. The *Dl* locus spans a stretch of approximately 25 kb of genomic DNA to which several *Dl* mutations have been mapped by Southern blot analysis (Vässin *et al.*, 1987). The 25-kb genomic DNA encodes one transcription unit that exhibits very complex regulation, reflecting the functional complexity of the *Dl* locus indicated by the available genetic evidence (Vässin and Campos-Ortega, 1987; Alton *et al.*, 1988). It produces three major, largely overlapping poly(A)$^+$ RNAs of 5.4, 4.6, and 3.7 kb (Vässin *et al.*, 1987; M. Haenlin, K. Bremer, B. Kramatschek, and J. Campos-Ortega, unpublished). The 5.4-kb RNA is zygotically expressed, whereas the other two are both maternal and zygotic.

In situ hybridization to embryonic tissue sections shows a distribution of *Dl* transcripts that conforms to expectations for a neurogenic gene (Vässin *et al.*, 1987). I emphasize only two main aspects of this very complex expression pattern: (1) *Dl* is expressed in territories with neurogenic capacities, like the NR or the anlagen of sensory organs; and (2) after an initial phase, during which it is abundantly transcribed in all cells of such territories, the *Dl* RNA becomes restricted to the cells that adopt the neural fate, for example, the NBs or the cells forming sensory organs, and persists in those cells for some time. Among the regions where *Dl* is transcribed, there are only two without known neurogenic abilities. One is the mesodermal layer, where *Dl* is transiently transcribed during stages 9–10 (D. Godt and J. Campos-Ortega, unpublished); the other region is within the anterior half of the hindgut, where a high concentration of *Dl* transcripts is present throughout embryogenesis. It is interesting to note that four of the *E*(*spl*) RNAs, *m4*, *m5*, *m7*, and *m8* (see below and Knust *et al.*, 1987c), are also transcribed in the mesoderm, and their expression starts when the transcription of *Dl* ends. Despite our present inability to give any plausible functional explanation for the presence of *Dl* RNA in mesodermal cells and hindgut, the neurogenic capabilities of the other organs where *Dl* is transcribed is beyond doubt. Preliminary data from *Delta*—β *Gal* fusions (Hiromi *et al.*, 1985) show that the spatial distribution of the protein encoded by *Dl* matches precisely that of the RNA (M. Haenlin, B. Kramatschek, and J. Campos-Ortega, unpublished).

The sequence of the putative protein encoded by the 5.4-kb *Dl* transcript has been deduced from cDNA clones (Vässin *et al.*, 1987) and shows some similarity to the putative *N* protein (Knust *et al.*, 1987a; see

Wharton *et al.*, 1985a,b; Kidd *et al.*, 1986). The sequence of a 4.7-kb cDNA clone, which apparently encompasses all translated sequences of the major 5.4-kb *Dl* transcript, indicates a transmembrane protein with a number of features, among them a putative signal peptide, five potential glycosylation sites, and an extracellular domain comprising nine EGF-like repeats (Knust *et al.*, 1987a). The repeated EGF-like motifs encoded by *N* and *Dl* are very suggestive in light of the assumption that the two genes are involved in cell–cell interactions. The primary structure of the putative *Dl* protein proposed by Vässin *et al.* (1987) on the basis of conceptual translation of the sequence of a cDNA clone (c3.2) has been confirmed by Kopczynski *et al.* (1988). However, the latter authors have sequenced the corresponding genomic DNA and found an open reading frame that encodes only 832 instead of the 880 residues determined by Vässin *et al.* (1987). It remains to be seen whether this discrepancy reflects interstrain differences or rather an artifact of the c3.2 cDNA clone sequenced by Vässin *et al.* (1987).

In view of their homology to EGF, which is synthesized from a larger precursor molecule (Gray *et al.*, 1983), the repeats encoded by *N* and *Dl* might conceivably be cleaved from the cell membrane and diffuse through the intercellular space. Data from genetic mosaics indicate, however, that the products of both *N* and *Dl,* as well as the products of the other NG genes, are unlikely to diffuse over great distances (Dietrich and Campos-Ortega, 1984; Hoppe and Greenspan, 1986). It is more probable that these products mediate protein–protein interactions between neighboring rather than distant cells. The results of transplanting mutant cells (Technau and Campos-Ortega, 1987) further suggest that the proteins encoded by *N* and *Dl* do not act as signal receptors, at least with respect to their role in the neuroepidermal dichotomy of lineages. The data can be interpreted to suggest that both proteins act on the side of the signal source. Since *Dl* shows the appropriate topological specificity in its expression, it seems a better candidate than the *N* protein, which is ubiquitously distributed, to mediate the protein–protein interactions between adjacent cells required for lineage segregation.

Cell ablation experiments in grasshoppers led to the conclusion that NBs inhibit the remaining neuroectodermal cells from producing more NBs (Taghert *et al.*, 1984; Doe and Goodman, 1985b). The presence of the *Dl* protein in NBs suggest that it might be responsible for such an inhibitory action, assuming this inhibition is operative in *Drosophila* as well.

4. Genetics of the Enhancer of split Locus

The $E(spl)$ locus was discovered by means of the mutation $E(spl)^D$ recovered by Green (quoted in Lindsley and Zimm, 1985; see Welshons, 1956). The presence of this mutation in the genome enhances the spl phenotype and renders *spl* expression dominant. Several observations show $E(spl)^D$ to be a rather interesting allele (Knust *et al.*, 1987b). $E(spl)^D$ is homozygous viable and fertile, the homozygotes being virtually wild type in phenotype; without the concomitant presence of *spl* in the genome, $E(spl)^D/E(spl)^D$ flies exhibit merely a slight compound eye roughening. However, 18% of the embryos derived from these homozygotes develop neural hypoplasia defects of variable severity that affect structures of both the CNS and the PNS.

It should be emphasized that females carrying more than two copies of $E(spl)^+$, without the $E(spl)^D$ allele, give rise to a high percentage of embryos with a smaller CNS and PNS. This effect of $E(spl)^+$ is very striking, for an increased dosage of wild-type alleles of any other NG locus, except *neu*$^+$ (see below), does not cause neural hypoplasic defects (F. Jiménez, D. Godt, and J. Campos-Ortega, unpublished). It should be emphasized at this point that this effect of increased dosage of $E(spl)^+$ as well as several of the phenotypic defects associated with $E(spl)^D$ are to a large extent maternal effects (Knust *et al.*, 1987b), and the same applies to the increase of the dosage of $E(spl)^+$. All phenotypic traits of $E(spl)^D$ are enhanced by additional copies of the $E(spl)^+$ allele in the genome. Therefore, $E(spl)^D$ behaves as a gain-of-function mutation in which the gene product is modified rather than absent. However, $E(spl)^D$ causes lethality when heterozygous with a deficiency for the locus (Lehmann *et al.*, 1983). Thus, whereas the modifications of the $E(spl)$ gene product encoded by the $E(spl)^D$ allele are compatible with normal viability of the homozygous $E(spl)^D$ animals, the lethality of hemizygous $E(spl)^D$ mutants suggests that this gene product is also functionally impaired.

There is a single recessive visible $E(spl)$ allele, called *groucho* (*gro*) because the phenotype is associated with clumps of supernumerary bristles at the supraorbital margin that are reminiscent of eyebrows. The bristle defects are rather variable in appearance and expressivity; selected lines can be established in which these traits are expressed in a high proportion of homozygous animals. However, other phenotypic traits, for example, increased pigmentation of the supraorbital border, are expressed with complete penetrance. The severity of the gro phenotype can be considerably increased with heterozygous with a deletion of the locus, suggesting that it is a hypomorph (Knust *et al.*, 1987b). However, the precise functional nature of the *gro* allele is not well established, since under certain conditions it shows dominant traits (H. Schrons, E. Knust, and J. Campos-Ortega, unpublished).

A increasingly large body of experimental evidence (Knust *et al.*, 1987c; Ziemer *et al.*, 1988; Klämbt *et al.*, 1989) indicates that the *E(spl)* locus consists of a cluster of several related genetic functions rather than a single gene. In the following, evidence for complexity of the *E(spl)* locus derived from studies on transmission genetics is considered. An easy way to recover loss-of-function alleles of *E(spl)* is by reverting the dominant effect of *E(spl)*^{D} on *spl* (Lehmann *et al.*, 1983). As a rule, revertants associated with the loss of the *E(spl)* function are embryonic lethals and produce the neurogenic phenotype discussed above to a variable extent, depending on the allele (Knust *et al.*, 1987b; Ziemer *et al.*, 1988). Second-site revertants can also be recovered from the same experiments, and some of them correspond to *Dl*$^-$ or *neu*$^-$ mutations. This fact most probably reflects the functional interrelationships between particular NG genes (see below). The frequency of recovery of loss-of-function *E(spl)* alleles as revertants of *E(spl)*^{D} is relatively low ($1.5–2 \times 10^{-4}$) after irradiation of mature sperm with 5000 R, whereas *E(spl)* mutations can be recovered at a higher frequency ($2–3 \times 10^{-3}$) by other means (F. Jiménez, K. Bremer, A. Ziemer, E. Knust, and J. Campos-Ortega, unpublished). This low frequency, together with the fact that no revertant of *E(spl)*^{D} has been obtained after mutagenesis with ethyl methanesulfonate (EMS) (Ziemer *et al.*, 1988), suggests that the reversion of *E(spl)*^{D} requires a complicated type of molecular lesion (see Section III,A,4,a). Indeed, most of the *E(spl)*^{D} revertants are actually associated with chromosomal aberrations that permit one to locate the neurogenic defects of *E(spl)* loss-of-function mutations to the chromosomal bands 96F8–13.

Animals carrying *E(spl)* mutations associated with inversions or translocations show slight defects in the wing venation pattern; since this phenotypic trait is associated with chromosomal rearrangements but is absent in heterozygtes with deletions of the locus, it is more likely due to positional effects than to haplo-insufficiency. There is an obvious correlation between the severity of the neural phenotype developed by homozygous embryos and the chromosomal aberration they carry: only large deletions lead to severe, fully penetrant neural hyperplasia, whereas homozygosity for inversions or translocations, or any of the other X-ray-induced mutations, produce weak to intermediate phenotypes with incomplete penetrance only (Knust *et al.*, 1987c; Preiss *et al.*,

1988; Ziemer *et al.*, 1988). Expressivity and penetrance of the neurogenic phenotypes of all these mutations can be increased when the corresponding variants are heterozygous with large deletions of the 96F region.

A number of lethal $E(spl)$ alleles were recovered after EMS mutagenesis (Ziemer *et al.*, 1988), and none of them causes fully penetrant neural hyperplasia when homozygous. In fact, the only neural defects associated with EMS-induced $E(spl)$ variants are some slight hypoplastic defects found among homozygotes for the allele $E(spl)^{B7}$, which behaves as a gain-of-function mutation. Thus, with respect to neural hyperplasia, $E(spl)$ alleles behave differently from alleles of the remaining NG loci, for example, N, mam, or Dl, in which several point mutants are known that lead to the amorphic phenotype of the corresponding gene. Consideration of these findings led us to postulate a degree of functional redundancy with respect to the role played by the $E(spl)$ locus in neurogenesis. It appears that several functions have to be eliminated simultaneously in order to abolish the function of the locus completely (Ziemer *et al.*, 1988).

$E(spl)$ mutations exhibit a complex pattern of heteroallelic complementation, with respect to their viability in crosses with lethal alleles and to expression of visible phenotypic traits in crosses with $E(spl)^{D}$ and *gro*. Seven alleles are lethal when heterozygous with any of the remaining lethal alleles; six of these are large deletions, but the polytene chromsomes of the seventh, $E(spl)^{B7}$, do not show any detectable defect. Another 22 alleles are fully viable when heterozygous with at least one other $E(spl)$ mutation. Although a definitive classification of these variants in further complementation groups is premature since the number of alleles of each one of the putative groups is still very low, a preliminary classification can be made. Sixteen of the alleles are lethal when heterozygous with $Df(3)E(spl)^{R-A7.1}$, a 34- to 36-kb deletion characterized at the molecular level (see below, and Knust *et al.*, 1987c), and, therefore, these variants very likely map within the limits of this deletion. The remaining six alleles are fully viable when heterozygous with $Df(3)E(spl)^{R-A7.1}$ and map, therefore, outside the deletion. With respect to their behavior as heterozygotes with the visible alleles *gro* and $E(spl)^{D}$, the variants can be classified into four different complementation groups. All the data are indicative of considerable genetic complexity.

Molecular Organization of the Enhancer of split Locus. Results of the ongoing molecular analysis of the $E(spl)$ locus confirm and extend the conclusion derived from genetic studies (Ziemer *et al.*, 1988), namely, that the $E(spl)$ locus is a gene complex encoding several related functions (Knust *et al.*, 1987c; Klämbt *et al.*, 1989). The extent of the locus is not yet precisely defined. The variant $Df(3)E(spl)^{R-A7.1}$, which lacks 34–36 kb of genomic DNA (see Fig. 4), is defective for some $E(spl)$ functions, as several mutations, including $E(spl)^{D}$, have been mapped to the same stretch of genomic DNA (Knust *et al.*, 1987c; Preiss *et al.*, 1988). It seems probable that homozygous $Df(3)E(spl)^{R-A7.1}$ embryos do not lack the entire $E(spl)$ locus, however, because they do not exhibit the most severe form of the neurogenic phenotype. In fact, embryos homozygous for deletions bigger than $Df(3)E(spl)^{R-A7.1}$ develop a considerably more severe phenotype. Possibly, the $E(spl)$ locus extends still further proximally and/or distally to the region deleted in $Df(3)E(spl)^{R-A7.1}$.

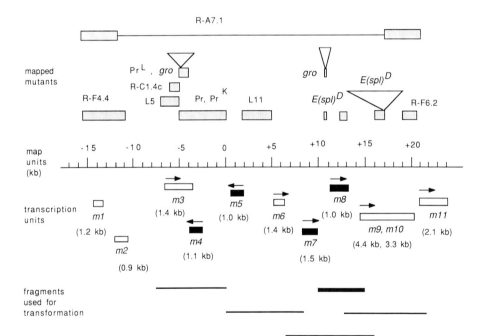

FIG. 4. Molecular map of the $E(spl)$ gene complex. [Modified from Knust *et al.* (1987c).] Several mutants have been mapped to 34–36 kb of genomic DNA; transcription units and directions of transcription are indicated beneath the physical map. Fragments used for transformation experiments (H. Schrons, K. Tietze, E. Knust, and J. Campos-Ortega, unpublished data) are indicated by bars. The thick bar represents the $E(spl)^{D}$ fragment including the *m8* transcription unit of the mutant that causes enhancement of the spl phenotype (Klämbt *et al.*, 1989).

The 34–36 kb of genomic DNA to which $E(spl)$ mutations were mapped contains at least 10 different transcription units and encodes 11 transcripts (one of the units encodes two overlapping RNAs) that have been designated $m1$ to $m11$, in proximodistal direction. All transcripts are temporally regulated and are expressed during embryonic development. There are some suggestive indications that 7 of the 11 RNAs participate in $E(spl)$-related functions. Four of these RNAs are affected in different mutations: $m3$, a 1.4-kb RNA, is missing in $T(3;4)E(spl)^{R-C1.4c}$ embryos and is larger (1.6 kb) in gro embryos; $m9$ and $m10$ are more abundantly expressed in $E(spl)^D$ than in wild-type embryos; and $m8$ is shorter and more abundantly expressed in $E(spl)^D$ than in wild-type embryos (Knust et al., 1987c; Klämbt et al., 1989).

On the other hand, the spatial distributions of $m4$, $m5$, $m7$, and $m8$ RNA during embryogenesis are virtually identical and conform to expectations for the epidermalizing function assumed to be exerted by (genes of) the $E(spl)$ locus. $m4$, $m5$, $m7$, and $m8$ are initially transcribed in cells of the neurogenic primordia, and, after the separation of the neural and epidermal cell lineages, those cells with epidermal fate continue to show transcripts from all four genes for a short time. We cannot exclude the possibility that transcripts $m7$ and $m8$ continue to be present in low concentration in neuroblasts after their separation from the ectoderm, but $m4$ and $m5$ are apparently confined to the epidermoblasts. In addition, all four RNAs are also transcribed in the mesodermal layer in stages 10–11, slightly later than Dl (see above). As a matter of fact, these four RNAs are expressed at early stages in the same cells as Dl. During later stages, however, the behavior of the $E(spl)$ transcripts, at least that of $m4$ and $m5$, is to some extent complementary to that of the Dl RNA. The functional significance of the spatial regulation of Dl and $E(spl)$ expression cannot be explained with the available data. In any case, the similar spatial distribution of $m4$, $m5$, $m7$, and $m8$ suggests that the four RNAs exert similar functions.

Sequence analyses (Klämbt et al., 1988) have uncovered extensive sequence similarity in the putative proteins encoded by the transcription units $m8$, $m7$, and $m5$ and thus substantiate the hypothesis that the various products of the complex perform similar functions. Recall that results of genetic studies had led to the proposal of a degree of redundancy in the various functions of the $E(spl)$ complex (Ziemer et al., 1988). The sequence data support this possibility. In addition, there is a striking similarity between a 57-amino acid domain at the amino-terminal ends of the putative protein products of $m8$, $m7$, and $m5$ and those encoded by four transcripts (T3, T4, T5, and T8) of the achaete–scute complex and by daughterless, all genes required for neural development (see below). This similarity is even more intriguing in light of the fact that the 57-amino acid domain is also present in proteins encoded by members of the *myc* gene family (Battey *et al.*, 1983; Bernard *et al.*, 1983; Watt *et al.*, 1983; DePinho *et al.*, 1987; Stone *et al.*, 1987), in the *MyoD* protein of mice (Davis *et al.*, 1987), and in a protein binding to an enhancer of the immunoglobulin light chain gene (Murre *et al.*, 1989).

Several molecular lesions have been found to be associated with the $E(spl)^D$ allele, including a middle repetitive fragment inserted in the transcription unit *m9–m10* and various deletions and insertions in the coding and in the 5′ regions of the *m8* transcription unit (Knust *et al.*, 1987c; Klämbt *et al.*, 1989). The role played by each of these molecular lesions in the production of the $E(spl)^D$ phenotype is not yet completely understood. However, there is direct evidence for the participation of *m8* in $E(spl)$-related functions. This evidence derives from P-element-mediated transformation experiments in which a mutant *m8* transcription unit, derived from the genome of $E(spl)^D$ animals, was injected into the germ line of wild-type animals. Transgenic flies exhibit one of the properties of the $E(spl)^D$ mutation, namely, the ability to enhance the spl phenotype (Klämbt *et al.*, 1989).

Transformation experiments have been used by Preiss *et al.* (1988) to present evidence that the RNAs *m9–m10* are also part of the $E(spl)$ complex. According to these authors, a fragment encoding both overlapping RNAs is able to rescue (1) the phenotype of *gro,* (2) the lethality of $E(spl)^D$ over $E(spl)^-$ alleles (Lehmann *et al.*, 1983), and (3) part of the neurogenic phenotype caused by a deletion of *m9–m10* and flanking transcription units (Preiss *et al.*, 1988). Hartley *et al.* (1988) have determined the sequence of the protein encoded by the overlapping transcripts *m9–m10* and shown that it is similar to the β subunit of transducin, a G protein of mammals. Along with the sequence similarity of the three small $E(spl)$ proteins *m5, m7,* and *m8* to *myc* proteins, described above, this homology of the putative *m9–m10* protein is a most appealing one, as G proteins are intimately associated with receptors. The results of transplantation experiments using cells homozygous for NG mutations (Technau and Campos-Ortega, 1987), in which only $E(spl)^-$ cells showed an autonomous phenotypic expression, prompted us to propose the participation of $E(spl)$ in functions related to the reception of the epidermalizing signal. These data suggest that the $E(spl)$ locus may encode consecutive steps of this process, from the transduction of the signal at the cell membrane to the regulation of the genetic activity in the receiving cell.

5. *Genetics and Molecular Organization of the master mind Locus*

The *mam* locus (50C23–Dl; Wiegel *et al.*, 1987; Yedvobnick *et al.*, 1988) has been characterized by 41 noncomplementing recessive lethals which were recovered from different mutagenesis programs (Lehmann *et al.*, 1981; Nüsslein-Volhard *et al.*, 1984; H. Schrons, U. Wetter, D. Weigel, U. Dietrich, and J. Campos-Ortega, unpublished). These alleles cause the neurogenic phenotype with variable expressivity, allowing the establishment of an allelic series (Lehmann *et al.*, 1981, 1983; Weigel *et al.*, 1987). The *mam* locus is maternally expressed, and, indeed, the most severe form of the neurogenic phenotype related to *mam* mutants is obtained after elimination of both zygotic and maternal components of *mam* gene expression (Jiménez and Campos-Ortega, 1982).

Flies heterozygous for any of several amorphic *mam* mutations exhibit various defects of the wings, particularly notching at the posterior margin and delta-shaped widenings of the tips of the veins. These phenotypic traits are remarkably similar to those of heterozygotes for *N* or *Dl* amorphic mutations, and, in fact, a second chromosomal mutation known as *Notch-2 of Gallup* (*N-2G;* see Lindsley and Grell, 1968), owing to its resemblance to the *N* phenotype, has turned out to be an allelomorph of *mam* (Lehmann *et al.*, 1983). The expressivity of these phenotypic traits in deficiency heterozygotes is variable (H. Schrons, U. Wetter, D. Weigel, U. Dietrich, and J. Campos-Ortega, unpublished), in a way similar to the case of *N*, being largely dependent on background factors. In fact, there is no correlation between the degree of expressivity of the imaginal defects of heterozygotes and that of embryos homozygous for the various *mam* lethal alleles.

Very little is known about the molecular organization and expression of the *mam* locus. Genomic DNA from the *mam* locus has been cloned and partially characterized by Weigel *et al.* (1987) and Yedvobnick *et al.* (1988). The limits of the locus are not yet well established. Several *mam* mutations have been mapped to a stretch of 45–60 kb of genomic DNA that contains a large number of copies of two different repetitive sequences. Sequence analysis (Yedvobnick *et al.*, 1988) has shown that one of the repeats corresponds to opa (Wharton *et al.*, 1985a) and the other [N (Weigel et al., 1987) or RS repeat (Yedvobnick *et al.*, 1988)] to a CA repeat. The 45-kb stretch encodes two major overlapping RNAs, of approximately 5.0 and 3.9 kb, which show the expected temporal regulation, namely, strong maternal expression and zygotic expression during 3–8 hours of embryonic development.

6. Genetics of the neuralized Locus

The *neu* locus is one of the genetically less well-characterized NG loci. *neu* has been located to 85C4–14 by means of various chromosomal aberrations (A. de la Concha and J. Campos-Ortega, unpublished) and has been further characterized by 15 noncomplementing lethal alleles recovered from EMS and X-ray mutageneses. Ten of the alleles lead to strong neurogenic defects, whereas the other five cause only intermediate or weak phenotypes. The *neu*$^+$ locus is included in *Dp(3;3)Antp*$^{+R8}$, and a few embryos of this strain exhibit neural hypoplastic defects reminiscent of those found among the progeny of females triploid for *E(spl)*$^+$. It is remarkable that increasing the ploidy of *N*$^+$ or *Dl*$^+$ does not cause such embryonic defects, suggesting particular relationships between *neu* and *E(spl)* (see below). *neu* also has a maternal component of expression (Dietrich and Campos-Ortega, 1984).

B. FUNCTIONAL INTERACTIONS BETWEEN NEUROGENIC LOCI

The identical phenotype caused by the loss of any of the NG gene functions is actually one of the most striking features of these loci, for it immediately suggests their products participate in a single functional pathway. The possibility that all of the NG loci contribute to serve the same overall function had already been considered during the first phases of the investigation of these loci, when it was discovered that animals with only one wild-type allele of the *Dl* and *E(spl)* loci are inviable (Lehmann *et al.*, 1983). This effect is associated with maternal expression of the *E(spl)* locus, for it applies only when the *E(spl)*$^-$ mutation is provided by the mother, whereas a few escapers survive from the reciprocal cross (Vässin *et al.*, 1985).

In the meantime, a large number of observations have been made concerning functional interactions between the NG loci (Campos-Ortega *et al.*, 1984; Dietrich and Campos-Ortega, 1984; Vässin *et al.*, 1985; de la Concha *et al.*, 1988). All these observations support the hypothesis that all of the NG loci tested, with the exception of *bib*, are involved in a common function and emphasize the key role played by the *E(spl)* locus. Since mutations in most of the NG loci exhibit dominant traits in their phenotypic expression, for example, wing, bristle, or leg defects, I distinguish below interrelationships that were worked out using heterozygous flies from those using homozygous embryos. As we shall see, different conclusions can be drawn from results of analysis of heterozygotes as compared to that of homozygotes, which may well have important functional significance.

1. Genetic Interactions in Heterozygotes

The observation that animals doubly heterozygous for amorphic *Dl* and *E(spl)* mutations do not survive (Lehmann *et al.*, 1983) was soon extended by the finding that the viability of animals doubly heterozygous for amorphic *N* and *E(spl)* mutations is also highly impaired (Vässin *et al.*, 1985). Some of the doubly heterozygous animals die as embryos; they develop weak neural hyperplasia. Such behavior is particularly striking, as double heterozygotes for *N* and *Dl* mutations with a normal complement of *E(spl)*[+] are fully viable. This observation implies that half-normal levels of expression at either *N* or *Dl* and *E(spl)* simultaneously are not sufficient for normal embryonic development. The phenotypic results of various genetic combinations, including both deficiencies and duplications of *N*+, *Dl*[+], and *E(spl)*[+], led us to postulate a network of reciprocal interactions with opposite character between these three loci (see Fig. 5), the meaning of which remained obscure (Vässin *et al.*, 1985). Functional interactions between *N* and *Dl* are indicated by the fact that two *Dl* alleles have been recently recovered as suppressors of *spl* (M. Brand and J. Campos-Ortega, unpublished). The suppression of the spl phenotype by *Dl* is allele specific

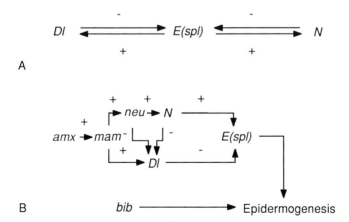

FIG. 5. Network of genetic relationships between the neurogenic genes. (A) Interactions between *E(spl)*, *N*, and *Dl* as worked out from observations on the phenotype of heterozygous animals. [Modified from Vässin *et al.* (1985).] (B) Interactions between NG genes determined from analysis of homozygous mutant embryos carrying extra copies of wild-type alleles of other NG genes. [Modified from de la Concha *et al.* (1988).] The + and − symbols reflect the positive and negative functional influences assumed to be exerted by one gene product on the next. See the text for further details.

because it is not attained by Dl deletions nor by other amorphic alleles, suggesting that the interaction is at the level of the proteins.

Doubly heterozygous animals for amorphic mutations in any of the other NG loci are fully viable. However, a high percentage of lethal embryos were found after combination of N^{55e11}, an amorphic allele, and $mam^{\mu 97}$. It is noteworthy that heterozygosity for $mam^{\mu 97}$ is associated with notches and deltalike wing vein widenings in a high proportion of individuals, and that the wing defects can be completely suppressed by increasing the ploidy of N^+ but very much increased in combinations of $mam^{\mu 97}$ with fa^{nd} (D. Weigel and J. Campos-Ortega, unpublished). These observations suggest the existence of functional interrelationships between N and mam.

Interrelationships between $E(spl)$ and neu are indicated by the fact that neu^- mutations can be recovered as second-site revertants of $E(spl)^D$ (see above). In addition, neu^- mutations considerably reduce the enhancement of the spl phenotype caused by $E(spl)^D$, whereas increasing the ploidy of neu^+ leads to increased enhancement of spl. The described effects of neu on spl care carried out chiefly through $E(spl)$, rather than acting directly on N itself, as neu^- mutations exert very mild effects on spl when the $E(spl)^D$ allele is not present in the genome (A. de la Concha and J. Campos-Ortega, unpublished). However, this finding does not preclude the existence of other direct interactions between neu and N.

Although extensive, the observations on heterozygous animals discussed above are actually incomplete, for not all possible genotypic combinations have been studied yet. The available data, however, suggest that particular NG loci, for example, $E(spl)$ and Dl, $E(spl)$ and N, N and Dl, N and mam, or neu and $E(spl)$, maintain closer functional relationships than others. Some of these interactions are allele specific, suggesting that they occur at the level of the protein products.

2. Genetic Interactions in Homozygotes

The use of duplications of the various NG loci, while focusing on the effects of NG mutations on neurogenesis, has permitted the establishment of epistatic relationships among the loci, with the exception of bib. We observed that the homozygous phenotype of some of the NG mutations can be modified when the genome carries an increased number of copies of the wild-type allele of another NG locus (Campos-Ortega et al., 1984; Vässin et al., 1985; de la Concha et al., 1988). In most cases, the severity of the phenotype of the NG mutant was reduced by the concomitant presence of three copies of the wild-type allele of another NG locus. Although in no case studied was a wild-type phenotype restored, the reduction of phenotypic severity was in all cases significant, for example, from a severe to an intermediate or even in some cases a weak phenotype. In other cases, however, the severity of the NG phenotype was increased by increasing the ploidy for another NG gene. For example, three copies of Dl^+ were found to enhance the phenotype of N^- or of neu^- alleles, whereas three copies of $E(spl)^+$ increased the phenotype of Dl^- alleles. An important aspect of these studies was that the phenotypic modifications became manifest in only some but not all genotypic combinations; in addition, the observed modifications were asymmetrical. For example, whereas three copies of neu^+ do not modify the phenotype of loss of the N function, the reciprocal genotype, that is, three copies of N^+, leads to reduction of the phenotype of loss of the neu function. Such an asymmetry indicates that the state of activity of some NG loci is modified by the activity of other NG loci; in other words, at least with respect to their participation in the segregation of NBs and EBs, as manifested by these genetic crosses, the relationships between NG loci are polarized rather than reciprocal.

The analysis of embryos homozygous either for two different NG mutations or for one NG mutation and carrying a duplication of the wild-type allele of another NG locus has led to the postulate of functional community for six NG loci, whereas the seventh, *bib*, appears to be independent of the others (Fig. 5B). The results are consistent with the six loci being links of a chain or a network of epistatic interactions, the last link of which would be the $E(spl)$ locus.

C. GENES OF SUBDIVISION 1B

The subdivision 1B of the X chromosome contains a number of genes that are required for development of both the CNS and the PNS (see Fig. 6). These genes include the various members of the *achaete–scute* complex (AS-C) and the loci of *ventral nervous system condensation defective* (*vnd*) and *embryonic lethal, abnormal visual system* (*elav*). The participation of the AS-C in neural development was first suggested by Garcia-Bellido and Santamaria (1978) on the basis of the behavior of gynandromorphs for $Df(1)sc^{19}$, which deletes most of the functions of the complex, and other considerations derived from genetic analyses (Garcia-Bellido, 1979). The phenotype of embryos lacking the AS-C was described soon thereafter, and the existence of additional genetic functions mapping adjacent to the AS-C and also being required for neural development was proposed (Jiménez and Campos-Ortega, 1979). White and colleagues identified two of these functions, *vnd* (White, 1980; White et al., 1983) and *elav* (Campos et al., 1985), and

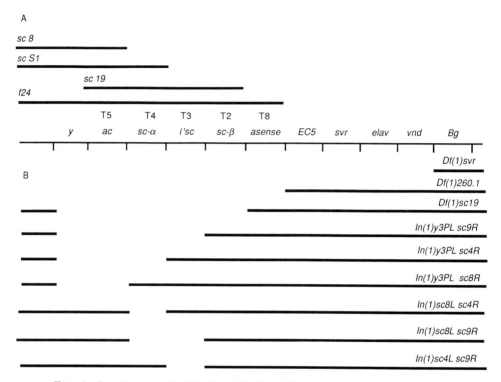

FIG. 6. Genetic map of subdivision 1B of the X chromosome. [Data from Garcia-Bellido (1979), K. White (personal communication), and Ghysen and Dambly-Chaudière (1988).] Bars indicate the extent of chromosomal aberrations used for the genetic dissection of the region. (A) The extent of duplications is represented by the presence of a bar. (B) The extent of deletions is represented by the absence of a bar.

described the main features of the phenotypes associated with their mutations. The participation of genes of the subdivision 1B in the development of epidermal sensilla of the embryo was first noticed by Campos-Ortega and Jiménez (1980) based on the phenotype of *Df(1)svr* hemizygous embryos. A detailed analysis of the PNS defects of AS-C mutants has recently been carried out by Dambly-Chaudière and Ghysen (1987; see below), who also characterized *sc-γ*, an additional gene of the AS-C. This name was later changed by Ghysen and Dambly-Chaudière (1988) to *asense*.

Hemizygous *Df(1)svr* embryos, thus deficient for 1A1 to 1B9–10, exhibit a phenotype opposite to that of NG mutants: they have a highly hypoplastic CNS (Jiménez and Campos-Ortega, 1979, 1987; Campos-Ortega and Jiménez, 1980; White, 1980) and lack all sensory neurons, with the exception of those innervating chordotonal organs and a few multidendritic neurons (Dambly-Chaudière and Ghysen, 1987). The origin of this phenotype has not yet been firmly established. However, two different processes are certainly involved. On the one hand, the complement of NBs is highly defective in these mutants (F. Jiménez and J. Campos-Ortega, unpublished), with neurogenesis being initiated by fewer NBs than in the wild type; on the other hand, there is increased cell death in the primordia of both the CNS and PNS (Jiménez and Campos-Ortega, 1979). Nonneural organs of the mutants are not affected directly.

1. Genetic Organization of Subdivision 1B

As we see in the discussion to follow, the number of genes in the subdivision 1B that are essential for neural development is not yet definitively established. Genetic studies by Muller and Prokofjeva (1935) had recognized the existence of three genes related to bristle development and viability, namely, achaete (ac), scute (sc), and lethal of scute (l'sc). With respect to bristle development, a detailed analysis by Garcia-Bellido and Santamaria (1978) and Garcia-Bellido (1979), using the left–right inversion–recombination test invented by Muller (Muller and Prokofjeva, 1935) led to subdivision of the sc gene into two different functions flanking the l'sc gene, called sc-α and sc-β. All these genes, namely, ac, sc-α and sc-β, and l'sc, are deleted by the $Df(1)sc^{19}$ mutation (Fig. 6).

The current phenotypic analysis of the subdivision 1B mutants (Jiménez and Campos-Ortega, 1987) indicates that CNS defects can be related to the deletion of l'sc, on the one hand, and vnd, on the other hand (see below). l'sc hemizygous embryos have a hypoplasic ventral cord, and very thin connectives and commissures, indicating that they carry fewer axons than wild-type controls. With respect to its neurogenic functions, l'sc seems to interact at least with ac and sc-α. Thus, the deletion of the three genes causes a phenotype of increased severity. There is no indication that sc-β takes part in functions necessary for CNS development; however, I should mention that the genetic identity of sc-β is not understood, as sc-β mutations are actually associated with defective transcription of sc-α (Campuzano et al., 1985; see below). Judging from the severity of the phenotype of embryos lacking genes further proximal to the AS-C, it is probable that, besides vnd and elav, to be considered later, there may be an additional gene located adjacent to sc-β which is necessary for CNS development in a similar way as ac and sc-α, that is to say, by interacting with l'sc (Jiménez and Campos-Ortega, 1987). On topological grounds, this gene is likely to correspond to sc-γ (or asense; Ghysen and Dambly-Chaudière, 1988).

With respect to PNS development, Dambly-Chaudière and Ghysen (1987) studied the effects of eliminating particular genes of the AS-C on the development of the embryonic sensilla. They found that, although there are some overlapping effects, particular genes affect particular subsets of sensory organs. The loss-of-function mutation of the *asense* gene, identified by Dambly-Chaudière and Ghysen (1987), has little effect on adult bristle development (see also Garcia-Bellido, 1979, for observations on phenotypes probably related to *asense*).

On the genetic map of subdivision 1B elaborated by K. White and colleagues (unpublished, personal communication; see Fig. 6), the loci *elav* and *vnd* are still further proximal, separated from the AS-C by three complementation groups that do not seem to function in neural development, *EC4, EC5,* and *svr*. The *elav* locus has been characterized by Campos *et al.* (1985) with noncomplementing embryonic lethals. A gynandromorph analysis permitted mapping of the focus of lethality to the ventral blastoderm region, compatible with a location in the neuro-ectoderm. Homozygotes for the *elav*[ts1] allele raised at the restrictive temperature during postembryonic development exhibit important structural defects of the brain and compound eyes. Similar defects can be observed in clones of cells homozygous for *elav*[1], a stronger allele. Embryos homozygous for the same allele show slight neural hypoplasia and, in particular, lesions of the connectives and commisures (Jiménez and Campos-Ortega, 1987). The phenotypic defects of embryos lacking the *vnd* locus are more severe (White, 1980; White *et al.*, 1983), comparable to those of *l'sc* homozygotes. No other genetic function of the subdivision 1B required for neural development has been identified further proximal to *vnd*.

Hairy wing (*Hw*) mutations are dominant and cause the appearance of supernumerary chaetes on the head, notum, and wing blade (Lindsley and Zimm, 1985). *Hw* mutations behave genetically as gain-of-function mutations of *ac* and *sc* (Garcia-Bellido, 1979, 1981; Garcia Alonso and Garcia-Bellido, 1986). However, the duplication of *Hw*[+] does not cause any clear phenotypic effect apart from the suppression of *h* (Botas *et al.*, 1982).

2. Molecular Organization of Subdivision 1B

a. Molecular Analysis of the achaete–scute Complex. The genomic DNA of the AS-C has been cloned and characterized by Modolell and colleagues (Carramolino *et al.*, 1982; Campuzano *et al.*, 1985, 1986; Ruiz-Gómez and Modolell, 1987; Balcells *et al.*, 1988). Most genes of the AS-C, as defined proximally by the *yellow* gene and distally by the breakpoint of *T(1;2)sc*[19], are contained within approximately 85–90 kb of genomic DNA (Carramolino *et al.*, 1982; Campuzano *et al.*, 1985).

The region encodes a large number of transcripts, three of which, T5, T4, and T3, have been tentatively identified as corresponding to *ac, sc,* and *l'sc* functions (Campuzano *et al.,* 1985; see Fig. 6). Sequence analysis has shown that the putative proteins encoded by the three genes share three conserved domains. One of these domains is acidic; the other two are basic and exhibit some similarity to proteins of the *myc* gene family (Villares and Cabrera, 1987; Alonso and Cabrera, 1988; J. Modolell, personal communication). Various other transcription units are scattered throughout this region, the significance of which is not clear (Campuzano *et al.,* 1985). However, the sequence of one of them, T8, which maps proximal to the breakpoint of $T(1;2)sc^{19}$ and corresponds to the T1a transcript of Balcells *et al.* (1988; J. Modolell, personal communication), has been established recently. The conceptual translation of this DNA sequence has revealed the existence of the two basic domains present in T5, T4, and T3 (Alonso and Cabrera, 1988; J. Modolell, personal communication). Thus, it is likely that T8 (T1a) corresponds to one of the AS-C functions. Although its genetic identity is not yet well established, there are reasons to believe that it corresponds to *asense* (J. Modolell, personal communication).

The spatial distribution of the T3, T4, and T5 transcripts has been studied in sections of staged embryos using *in situ* hybridization (Cabrera *et al.,* 1987; Romani *et al.,* 1987). The distribution is very similar for all three transcripts and shows a high degree of correlation with the processes of NB segregation (see Fig. 7) and development of the sensory organs and stomatogastric nervous system. During early neurogenesis, the three transcripts are expressed in partially overlapping clusters of cells within the neuroectoderm; they seem to appear in three waves that parallel the three pulses of NB segregation (Hartenstein and Campos-Ortega, 1984). During each wave, the RNA of T4 and T5 becomes restricted to the NBs, whereas that of T3 continues to be present in both NBs and EBs until later stages (Cabrera *et al.,* 1987). Since their domains of expression partially overlap, some NBs may contain all three RNAs, and their products, whereas other NBs contain only one or two of them. This pattern of transcription and the correspondence between deletion of some AS-C genes and defects in particular subsets of sensory organs (Dambly-Chaudière and Ghysen, 1987; Ghysen and Dambly-Chaudière, 1988) are suggestive of specific roles for the AS-C genes during neurogenesis. Cabrera *et al.* (1987) proposed that the AS-C genes serve to provide the NBs with specific identities, based on the combination of products expressed in each cell.

Romani *et al.* (1989) analyzed the pattern of transcription of AS-C genes in the imaginal wing disc and found that *ac* (T5) and *sc*-α (T4) are only sparsely expressed in the CNS but quite abundantly in the wing

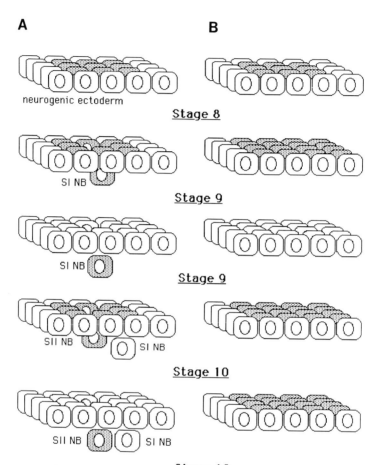

FIG. 7. Spatial distribution of T5 transcripts in (A) wild-type embryos (Romani *et al.*, 1987; Cabrera *et al.*, 1987) and (B) neurogenic mutants (Brand and Campos-Ortega, 1989). An array of 25 neuroectodermal cells is shown at different stages in both cases. In stage 8, nine of these cells transcribe the T5 RNA; in stage 9, one of them segregates as a neuroblast (NB) and continues transcribing T5. At the end of stage 9, transcription of T5 reappears in a cluster of nine cells, of which one will segregate as an SII NB that continues transcribing T5. In the neurogenic mutants, the number of cells transcribing T5 is comparable to the wild type in early stages but increases at the end of stage 9; since all neuroectodermal cells develop as NBs, no segregation of lineages occurs in the neurogenic mutants. Transcription of T5 is interrupted until the end of stage 9, when it reappears in a large cluster of cells. See the text.

disc. Contrarily, *l'sc* (T3) and *ase* (T1a) are almost exclusively expressed within the CNS. Careful planimetric reconstruction of serial sections allowed the authors to correlate precisely the pattern of distribution of RNAs with various sensory organs of the imago.

Hw mutations have been found to correspond to different molecular lesions distributed throughout the transcription units T4 and T5 (*sc-α* and *ac*), which arise from insertions of transposable elements, chromosomal breakages, etc. (Campuzano *et al.*, 1986). Three *Hw* alleles associated with insertions of *gypsy* or *copia* (Campuzano *et al.*, 1986) and another two with chromosomal aberrations (Balcells *et al.*, 1988) have been characterized at the molecular level. They correspond to truncated T4 or T5 transcripts which, in addition, are made in larger amounts than in the wild type. Apparently, the presence of foreign DNA in the neighborhood of the transcription units promotes overexpression of the genes. It is remarkable that in the *Hw* mutants these RNAs are expressed in territories of the wing blade that correspond to the regions where ectopic chaetae grow (Balcells *et al.*, 1988).

b. Molecular Analysis of embryonic lethal and visual system defective Locus. The *elav* locus has been cloned by White and colleagues (Campos *et al.*, 1987; Robinow and White, 1988); a 13.5-kb genomic DNA fragment from the pertinent chromosomal walk was sufficient to rescue phenotypic abnormalities associated with amorphic *elav* mutations (Campos *et al.*, 1987). This fragment encodes a 2.3-kb transcript which is truncated in the amorphic mutation $elav^{G3}$ and expressed exclusively within neural tissue during all developmental stages (Robinow and White, 1988).

The neural specificity of the *elav* gene product indicated by the *in situ* hybridization experiments (Campos *et al.*, 1987; Robinow and White, 1988) is confirmed by the discovery of Bier *et al.* (1988; Y. N. Jan, personal communication) that the antigen recognized by the monoclonal antibody Mab44c11 is present in the nuclei of all nerve cells. Cloning of the DNA encoding the antigen has revealed that it is identical to the *elav* coding sequence. It is noteworthy that the absence of such a protein, located in the nucleus of all *Drosophila* neurons, causes only slight hypoplastic defects in the pattern of commissures and connectives (Jiménez and Campos-Ortega, 1987).

D. *daughterless* Locus

The gene *da* (Bell, 1954) has been known for some time to be required for sex determination and dosage compensation (Cline, 1976, 1980;

Lucchesi and Skripsky, 1981). The requirements for da^+ for normal neural development is a surprising finding (Caudy et al., 1988a). The locus is expressed during both oogenesis and embryonic development. There is evidence that maternal expression is relevant at early blastoderm stages for correct dosage compensation and differentiation in female embryos, whereas zygotic expression is essential for PNS development in both sexes (discussed in Caudy et al., 1988a).

Loss of da^+ function leads to embryonic lethality. The dead embryos show a predominantly neural phenotype: they lack all sensory neurons, and, in addition, the CNS is smaller than normal, the ventral cord being frequently fragmented in several pieces. Nonneural organs are relatively normal, although some minor defects can be seen in the pattern of muscles and in the gut. The phenotype is unique, in that da is the only mutation that affects sensory organs in a global manner (Jan et al., 1987).

The da locus has recently been cloned (Caudy et al., 1988b). Five da mutations have been mapped within an interval of approximately 5 kb of genomic DNA that encodes a single transcription unit with two overlapping RNAs of 3.2 and 3.7 kb. The conceptual translation of the corresponding cDNA sequences uncovers an interesting similarity, namely, the conserved region present in myc, in the proteins encoded by the AS-C transcripts T3, T4, T5, and T8, and in the proteins encoded by the E(spl) transcripts m5, m7, and m8. Whether this finding has functional significance is unclear, although the similarity of the phenotype of mutations in the Drosophila genes makes it very probable.

E. Interactions between Neurogenic Genes and Genes of the achaete–scute Complex and daughterless

Genetic interactions between the AS-C genes and da have been described by Dambly-Chaudière et al. (1988), indicating that these genes are involved in the same function. On the other hand, Brand and Campos-Ortega (1989) have obtained evidence for interactions of the NG genes with the AS-C genes and with da. Observations on double mutants show that the severity of the phenotype of homozygous NG mutants can be considerably reduced if a mutation of the AS-C or of da is present in homo- or hemizygosity in the same genome. This reduction of the phenotypic severity of the double mutant affects both the epidermis, which is larger, and the neural tissue, which is less hyperplastic.

Mapping experiments using the left–right inversion recombinants which delete various amounts of the AS-C (Muller, 1935; Garcia-Bellido, 1979) show that removal of the l'sc function is necessary in order for the modifications of the NG phenotype to be observed to a significant extent. However, removal of the *l'sc* function alone leads to only a moderate attenuation of the phenotype. To obtain the full attenuation, other functions of the AS-C, for instance, the *sc-α* function, have to be removed simultaneously with *l'sc. asense* (Ghysen and Dambly-Chaudière, 1988) does not seem to play any important role. Double-mutant combinations with the *Df(1)svr* demonstrate that the interaction of at least *Dl* occurs specifically with the functions encoded by the AS-C and not with *elav* and *vnd*.

Pattern of Transcription of achaete–scute Complex Genes in Neurogenic Mutants

At least some of the interactions between NG and AS-C genes are likely to involve an influence on the pattern of transcription of these genes (Fig. 7). Changes of the pattern of transcription of the genes T3 and T5 (*l'sc* and *ac*) have been observed in embryos carrying any of several NG mutations (Brand and Campos-Ortega, 1989). In these embryos, T3 and T5 are expressed in more cells than in the wild type. However, the early pattern of expression, up to stage 9, of T3 and T5 in NG mutants is indistinguishable from the wild type. Hence, transcriptional interactions seem to operate, or at least to become evident, at the time when the segregation of lineages is taking place. In the wild type, a restriction of T5 transcription occurs from an initial group of about nine ectodermal cells to a few NBs as they segregate from the EBs (Cabrera *et al.* 1987). In NG mutants, the size of the territories of hybridization is expanded, indicating that this restriction fails to occur. Moreover, the total number of T5-expressing cells per cluster is greater than nine cells. In contrast to this finding, no significant modification of the pattern of transcription of NG genes is observed in *Df(1)sc*[19] embryos, which lack most of the AS-C genes (Brand and Campos-Ortega, 1989). Therefore, the polarity of functional relationships between the two gene groups is likely to be from the NG to the AS-C genes, and not vice versa. These results suggest that cellular interactions mediated by the NG genes are responsible for the refinement of the territories of T3–T5 expression in the wild type and that the NG genes exert this function by suppressing the transcription of T3 and T5 in some of the neuroectodermal cells.

The expansion of the territories of transcription of the AS-C genes found in NG mutants raises the question as to whether this phenomenon is causally related to the misrouting into neurogenesis of all neuroectodermal cells characteristic of NG mutations. If in the wild type a combination of the products of AS-C, and other similar genes, were necessary in a particular neuroectodermal cell for this cell to take on the fate of a NB, one could imagine that the increase in the number of cells transcribing AS-C genes may be responsible for the appearance of additional NBs.

IV. Conclusions and Prospects

In the following, I would like to draw some conclusions and to specu-late on functional implications of the data discussed in this article (see Fig. 8), and, at the same time, I also emphasize several weak points in our current view. The picture of the process of lineage segregation presented below is based on the hypothesis that the proteins encoded by the NG genes directly mediate the postulated cell interactions. However, although there is no reason to doubt that the NG genes are in one way or another involved in this process, the evidence that the products they encode are indeed the basic elements of the postulated cell communication is rather indirect. Yet, this hypothesis best ex-plains all the available data, at least with respect to the decision of cells to follow epidermogenesis.

Figure 8 shows two interacting cells of the NR, after the segregation of lineages has taken place. In Fig. 8 the arrangement of the different genes in both cells is based on data from *in situ* hybridizations and on the assumption that the proteins encoded by these genes show a distri-bution similar to that of the RNA. *N* (Hartley *et al.*, 1987), some of the transcripts of the *E(spl)* complex (*m9–m10, m3;* Knust *et al.*, 1987c; Hartley *et al.*, 1988), and *da* (M. Brand, personal communication) are expressed in all embryonic cells throughout neurogenesis. Prior to lineage segregation, *Dl*, the transcripts of the *E(spl)* gene complex *m4, m5, m7,* and *m8,* and the AS-C transcripts T3, T4, and T5, although restricted almost exclusively to neurogenic territories, are also ex-pressed in both putative EBs and NBs. After lineage segregation, however, these RNAs continue to be transcribed in either the neural or the epidermal cells, in the way indicated in Fig. 8. Thus, for the pur-poses of discussion and with respect to the epidermalizing signal, that is, irrespective of the participation of these genes in other processes, I would like to propose that the various proteins are functionally active in the interacting cells as indicated in Fig. 8.

Sequencing data strongly suggest that the proteins encoded by *N* and *Dl* are located in the membrane of the cells of the NR. It is conceivable that one, or both, of these proteins interact via the EGF-like motifs with protein(s) encoded by *E(spl)*. The results of transplantation experi-

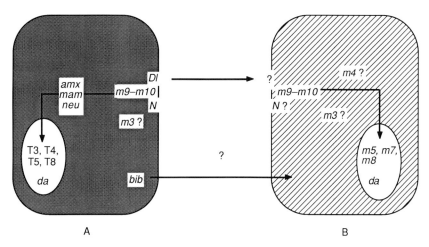

FIG. 8. Schematic of two cells of the neurogenic region [(A) neuroblast and (B) epidermoblast] after the segregation of lineages, showing the distribution of the products of the neurogenic and other genes. Arrows indicate the proposed flow of information. *m3, m4, m5, m7, m8,* and *m9–m10* refer to six transcription units of the *E(spl)* complex. Transcription unit *m9–m10* encodes two different RNAs (Knust *et al.,* 1987c; Hartley *et al.,* 1988). The function of *m3* and *da* is placed on both cells only on the basis of the ubiquitous expression of these genes. The identity of the receptor protein is unknown. *da* is assumed to act chiefly on PNS development. See the text for further discussion.

ments point to a role for both *N* and *Dl* at the source of the epidermalizing signal (Technau and Campos-Ortega, 1987). The allele-specific suppression of the spl phenotype by *Dl* mutations (M. Brand and J. Campos-Ortega, unpublished observations) constitutes weak evidence in favor of direct relationships between *N* and *Dl*. If the proposed association of the proteins of *N* and *Dl* holds true, it is likely that *Dl* provides on the side of the signal source the topological specificity required for normal lineage segregation, as *N* is ubiquitously expressed (Yedvobnick *et al.,* 1985; Hartley *et al.,* 1987) whereas *Dl* expression becomes restricted to the neural cells (Vässin *et al.,* 1987). Therefore, the protein encoded by *Dl* is a good candidate for the ligand to an as yet unknown receptor (see below). This ligand–receptor interaction between adjacent cells would mediate the epidermalizing signal. On the other hand, molecular defects associated with the *spl* mutation (Hartley *et al.,* 1987; Kelley *et al.,* 1987) also suggest such a protein–protein relation between *N* and *E(spl)*.

I further propose that the various proteins encoded by the $E(spl)$ gene complex are responsible for receiving, transducing, and/or relaying the signal further to the genome of the receiving cell. The main support for this claim derives from transplantation experiments with mutant cells (Technau and Campos-Ortega, 1987), indicating a function of the $E(spl)$ locus in signal reception and response. Since $E(spl)$ is a gene complex, encoding a still unknown number of functions, its function is also assumed to be complex. Results of the ongoing molecular analysis lend support to the view that the various proteins of the $E(spl)$ locus serve different steps of the chain which leads the signal from the membrane to the genome of the receiving cell. First, the sequence of a putative protein encoded by transcription unit $m9-m10$ (Knust et al., 1987c) is similar to part of a G protein (Hartley et al., 1988), suggesting that this protein is associated with the membrane. Second, the nearly identical pattern of expression of the four genes $m4$, $m5$, $m7$, and $m8$, complementary to that of Dl, is consistent with that expected for the assumed epidermalizing function of $E(spl)$ (Knust et al., 1987c). Third, the putative sequence of the proteins encoded by $m5$, $m7$, and $m8$, which all share a region similar to one another, is present in proteins of the myc family (Klämbt et al., 1989).

It is tempting to speculate about the possibility that the $m5$, $m7$, and $m8$ proteins act as transcription factors in the nucleus of the receptor cell to regulate the expression of genes of the epidermal pathway. This speculation receives support from the recent finding of other proteins containing a fragment highly similar to the myc box that have been shown to be capable of binding to an enhancer of the immunglobulin light chain gene (Murre et al., 1989). With respect to $m4$, there is no indication yet that would allow ascribing a differential role to this protein. Thus, several data support the notion of a role for the $E(spl)$ locus in processing the regulatory signal.

However, among the various genes of the $E(spl)$ complex studied so far, none is yet known whose sequence suggests that the corresponding protein functions as a membrane receptor. Thus, the identity of the putative receptor protein itself is still obscure. Hartley et al. (1988) suggest that the protein encoded by N interacts directly with the putative G protein encoded by transcription unit $m9-m10$. Thus, the N protein might act as a receptor for the epidermalizing signal. Since both N and the $m9-m10$ transcription unit of the $E(spl)$ complex are expressed ubiquitously, the proposed association between both proteins on either side of the signal is a very likely and appealing possibility (Fig. 8). However, the nonautonomous expression of the N phenotype found in transplantation experiments (Technau and Campos-Ortega,

1987) and in somatic mosaics (P. Hoppe and R. Greenspan, personal communication) argues against the N protein acting exclusively as a receptor. The present data, therefore, do not allow a resolution of the role played by N in the signal-receiving cell.

The data available on the structure of *amx, mam,* and *neu* are insufficient to permit any reasonable prediction about their location in this functional scheme. Transplantation experiments indicate a role for the three genes at the signal source. Genetic analysis shows that *amx, mam,* and *neu* are hypostatic to *Dl, N,* and *E(spl)*, thus suggesting that the former genes may regulate the expression of the latter. Whether their products are also acting on the side of the receiving cell is not yet known. A similar problem concerns the role played by the *bib* locus, which seems to act independently of the other six NG genes. Since *bib* expression is not cell autonomous, its function is assumed to be required on the side of the signal source. The molecular analysis of *amx, mam, neu,* and *bib* will certainly contribute to our understanding of the function of these genes.

One of the least understood aspects of the proposed flow of information is related to the neuralizing signal and to the process of acquisition of the neural fate itself. We do not know whether a neuralizing signal is operative during normal development, for its existence is suggested only by heterotopic transplantations. Several pieces of evidence, derived, for example, from normal embryology (Hartenstein and Campos-Ortega, 1984) or from the phenotype of neurogenic mutants (Lehmann *et al.,* 1983), point to neurogenesis as the primary fate for all the neuroectodermal cells. In principle, such a primary neurogenic fate would make a neuralizing signal superfluous, although it would not preclude the possibility that cell communication processes are necessary to permit NB development. It is conceivable that those cells that finally adopt the neural fate require a neuralizing signal in order to reinforce their primary neurogenic fate.

In any case, it is probable that the genes of the AS-C, and perhaps also some of the neighboring loci in the subdivision 1B (Garcia-Bellido, 1979; Jiménez and Campos-Ortega, 1979; White, 1980; Carramolino *et al.,* 1982; Ghysen and Dambly-Chaudière, 1988), as well as *da* vis-à-vis PNS development (Caudy *et al.,* 1988b), take part in the neuralizing pathway. Indeed, these genes have been termed "proneural" by Romani *et al.* (1989) on the basis of their assumed function in promoting neural development.

I would like to suggest that the proteins encoded by the AS-C genes, and in the case of PNS development by *da,* are part of the proposed neuralizing function, acting as regulators of transcription in the genome of the presumptive NBs and progenitor cells of sensory organs. Four of the AS-C transcripts, T3, T4, T5, and T8 (Villares and Cabrera, 1987; Alonso and Cabrera, 1988), and *da* (Caudy *et al.*, 1988b) encode proteins that carry the *myc* box with DNA-binding capacity (Murre *et al.*, 1989). In addition, the distribution of the transcripts of T3, T4, and T5 is intimately correlated both with NB segregation (Cabrera *et al.*, 1987; Romani *et al.*, 1987) and with development of the imaginal sensory organs (Romani *et al.*, 1989). Regulatory interactions have been demonstrated between the NG genes and the genes under discussion (Brand and Campos-Ortega, 1989), suggesting that the activity of the AS-C genes may be driven by the products of the NG genes. I should mention that, although for the purposes of the present discussion I have dealt with the AS-C genes as a group, clear differences can be observed in the patterns of expression of these genes (Cabrera *et al.*, 1987; Romani *et al.*, 1987, 1989; Alonso and Cabrera, 1988), and it seems probable that functional differences exist between them (Romani *et al.*, 1989). All these questions require further clarification and are presently under investigation.

ACKNOWLEDGMENTS

I would like to thank Volker Hartenstein, Fernando Jiménez, Elisabeth Knust, Harald Vässin, and Gerd Technau, who were involved at different stages of the work reported here; I also thank Elisabeth Knust and Paul Hardy for constructive criticisms on the manuscript. The research reported here was supported by grants of the Deutsche Forschungsgemeinschaft (DFG).

REFERENCES

Alonso, M. C., and Cabrera, C. V. (1988). The *achaete–scute* gene complex of *Drosophila melanogaster* comprises four homologous genes. *EMBO J.* **7**, 2585–2591.
Alton, A. K., Fechtel, K., Terry, A. L., Meikle, S. B., and Muskavitch, M. A. T. (1988). Cytogenetic definition and morphogenetic analysis of *Delta*, a gene affecting neurogenesis in *Drosophila melanogaster*. *Genetics* **118**, 235–245.
Anderson, K. V., and Nüsslein-Volhard, C. (1986). Dorsal-group genes of *Drosophila*. In "Gametogenesis and the Early Embryo." (44th Symp. Soc. Dev. Biol.) (J. G. Gall, ed.), pp. 177–194. Liss, New York.
Artavanis-Tsakonas, S. (1988). The molecular biology of the *Notch* locus and the fine tuning of differentiation in *Drosophila*. *Trends Genet.* **4**, 95–100.
Artavanis-Tsakonas, S., Muskavitch, M. A. T., and Yedvobnick, B. (1983). Molecular cloning of *Notch*, a locus affecting neurogenesis in *Drosophila melanogaster*. *Proc. Natl. Acad. Sci. U.S.A.* **80**, 1977–1981.
Artavanis-Tsakonas, S., Grimwade, B. G., Harrison, R. G, Makopoulou, K., Muskavitch, M. A. T., Schlesinger-Bryant, R., Wharton, K., and Yedvobnick, B. (1984). The *Notch* locus of *Drosophila melanogaster:* A molecular analysis. *Dev. Genet.* **4**, 233–254.

Balcells, L. I., Modolell, J., and Ruiz-Gomez, M. (1988). A unitary basis for different *Hairy-wing* mutations of *Drosophila melanogaster. EMBO J.* **7,** 3899–3906.

Bate, C. M. (1976). Embryogenesis of an insect nervous system. I. A map of the thoracic and abdominal neuroblasts in *Locusta migratoria. J. Embryol. Exp. Morphol.* **35,** 107–123.

Bate, C. M. (1978). Development of sensory systems in arthropods. *In* "Handbook of Sensory Physiology" (M. Jacobson, ed.), Vol. 9, pp. 1–53. Springer-Verlag, Berlin and New York.

Bate, C. M. (1982). Development of the neuroepithelium. *Neurosci. Res. Bull.* **20,** 803–806.

Battey, J., Moulding, C., Taub, R., Murphy, W., Stewart, T., Potter, H., Lenoir, G., and Leder, P. (1983). The human c-*myc* oncogene. Structural consequences of translocation into the IgH locus in Burkitt lymphoma. *Cell* **34,** 779–787.

Bell, A. E. (1954). A gene in *Drosophila melanogaster* that produces all male progeny. *Genetics* **39,** 958–959.

Bernard, R., Dessain, S. K., and Weinberg, R. A. (1983). Sequence of the murine and human cellular *myc* oncogenes and two modes of *myc* transcription resulting from chromosome translocation in B lymphoid tumors. *EMBO J.* **2,** 2375–2383.

Bier, E., Ackerman, L., Barbel, S., Jan, L., and Jan, Y. N. (1988). Identification and characterization of a neuron-specific nuclear antigen in *Drosophila. Science* **240,** 913–916.

Bodmer, R., and Jan, Y. N. (1987). Morphological differentiation of the embryonic peripheral neurons in *Drosophila. Wilhelm Roux's Arch. Dev. Biol.* **196,** 69–77.

Botas, J., Moscoso del Prado, J., and Garcia-Bellido, A. (1982). Gene–dose titration analysis in the search of trans-regulatory genes in *Drosophila. EMBO J.* **1,** 307–310.

Brand, M., and Campos-Ortega, J. A. (1989). Two groups of interrelated genes regulate early neurogenesis in *Drosophila melanogaster. Wilhelm Roux's Arch. Dev. Biol.* (in press).

Breeden, L., and Nasmyth, K. (1987). Similarity between cell-cycle genes of budding yeast and fission yeast and the *Notch* gene of *Drosophila. Nature (London)* **329,** 651–654.

Cabrera, C. V., Martinez-Arias, A., and Bate, M. (1987). The expression of three members of the *achaete–scute* gene complex correlate with neuroblast segregation in *Drosophila. Cell* **50,** 425–433.

Campos, A. R., Grossman, D., and White, K. (1985). Mutant alleles at the locus *elav* in *Drosophila melanogaster* lead to nervous system defects. A developmental-genetic analysis. *J. Neurogenet.* **2,** 197–218.

Campos, A. R., Rosen, D. R., Robinow, S. N., and White, K. (1987). Molecular analysis of the locus *elav* in *Drosophila melanogaster:* A gene whose embryonic expression is neural specific. *EMBO J.* **6,** 425–431.

Campos-Ortega, J. A. (1983). Topological specificity of phenotype expression of neurogenic mutations in *Drosophila. Wilhelm Roux's Arch. Dev. Biol.* **192,** 317–326.

Campos-Ortega, J. A. (1985). Genetics of early neurogenesis in *Drosophila melanogaster. Trends Neurosci.* **8,** 245–250.

Campos-Ortega, J. A. (1988). Cellular interactions during early neurogenesis of *Drosophila melanogaster. Trends Neurosci.* **11,** 400–405.

Campos-Ortega, J. A., and Hartenstein, V. (1985). "The Embryonic Development of *Drosophila melanogaster.*" Springer-Verlag, Berlin and New York.

Campos-Ortega, J. A., and Jiménez, F. (1980). The effects of X-chromosomal deficiencies on neurogenesis in *Drosophila. In* "Development and Neurobiology of *Drosophila*" (O. Siddiqi, P. Babu, L. Hall, and J. H. Hall, eds.), pp. 201–222.

Campos-Ortega, J. A., Lehmann, R., Jimenez, F., and Dietrich, U. (1984). A genetic analysis of early neurogenesis in *Drosophila*. *In* "Organizing Principles of Neural Development" (S. C. Sharma, ed.), pp. 129–144. Plenum, New York.

Campuzano, S., Carramolino, L., Cabrera, C. V., Ruiz-Gomez, M., Villares, R., Boronat, A., and Modolell, J. (1985). Molecular genetics of the *achaete–scute* gene complex of *Drosophila melanogaster*. *Cell* **40,** 327–338.

Campuzano, S., Balcells, L., Villares, R., Carramolino, L., Garcia Alonso, L., and Modolell, J. (1986). Excess function *Hairy-wing* mutations caused by the gypsy and copia insertions within structural genes of the *achaete–scute* locus of *Drosophila melanogaster*. *Cell* **44,** 303–312.

Carramolino, L., Ruiz-Gomez, M., Guerrero, M. C., Campuzano, S., and Modolell, J. (1982). DNA map of mutations at the *scute* locus of *Drosophila melanogaster*. *EMBO J.* **1,** 1185–1191.

Caudy, M., Grell, E. H., Dambly-Chaudière, C., Ghysen, A., Jan, L. Y., and Jan, Y. N. (1988a). The maternal sex determination gene *daughterless* has zygotic activity necessary for the formation of peripheral neurons in *Drosophila*. *Genes Dev.* **2,** 843–852.

Caudy, M., Vässin, H., Brand, M., Tuma, R., Jan, L. Y., and Jan, Y. N. (1988b). *daughterless,* a gene essential for both neurogenesis and sex determination in *Drosophila,* has sequence similarities to *myc* and the *achaete–scute* complex. *Cell* **55,** 1061–1067.

Cline, T. W. (1976). A sex-specific temperature sensitive maternal effect of the *daughterless* mutation of *Drosophila melanogaster*. *Genetics* **84,** 723–742.

Cline, T. W. (1980). Maternal and zygotic sex-specific gene interactions in *Drosophila*. *Genetics* **96,** 903–926.

Dambly-Chaudière, C., and Ghysen, A. (1986). The sense organs in the *Drosophila* larva and their relation to the embryonic pattern of sensory organs. *Wilhelm Roux's Arch. Dev. Biol.* **195,** 222–228.

Dambly-Chaudière, C., and Ghysen, A. (1987). Independent subpatterns of sense organs require independent genes of the *achaete–scute* complex in *Drosophila* larvae. *Genes Dev.* **1,** 297–306.

Dambly-Chaudière, C., Ghysen, A., Jan, L. Y., and Jan, Y. N. (1988). The determination of sense organs in *Drosophila:* Interactions of *scute* with *daughterless*. *Wilhelm Roux's Arch. Dev. Biol.* **197,** 419–423.

Davis, R. L., Weintraub, H., and Lassar, A. B. (1987). Expression of a single transfected cDNA converts fibroblasts to myoblasts. *Cell* **51,** 987–1000.

de la Concha, A., Dietrich, U., Weigel, D., and Campos-Ortega, J. A. (1988). Functional interactions of neurogenic genes of *Drosophila melanogaster*. *Genetics* **118,** 499–508.

DePinho, R. A., Hatton, K. S., Tesfaye, A., Yancopoulos, G. D., and Alt, F. W. (1987). The human *myc* gene family: Structure and activity of *L-myc* and an *L-myc* pseudogene. *Genes Dev.* **1,** 1311–1326.

Dietrich, U., and Campos-Ortega, J. A. (1984). The expression of neurogenic loci in imaginal epidermal cells of *Drosophila melanogaster*. *J. Neurogenet.* **1,** 315–332.

Doe, C. Q., and Goodman, C. S. (1985a). Early events in insect neurogenesis. I. Development and segmental differences in the pattern of neuronal precursor cells. *Dev. Biol.* **111,** 193–205.

Doe, C. Q., and Goodman, C. S. (1985b). Early events in insect neurogenesis. II. The role of cell interactions and cell lineages in the determination of neuronal precursor cells. *Dev. Biol.* **111,** 206–209.

Doe, C. Q., Hiromi, Y., Gehring, W. J., and Goodman, C. S. (1988). Expression and function of the segmentation gene *fushi tarazu* during *Drosophila* neurogenesis. *Science* **239,** 170–175.

Foster, G. G. (1975). Negative complementation at the *Notch* locus of *Drosophila mela-nogaster. Genetics* **81**, 99–120.

Garcia Alonso, L., and Garcia-Bellido, A. (1986). Genetic analysis of *Hairy-wing* muta-tions. *Wilhelm Roux's Arch. Dev. Biol.* **195**, 259–264.

Garcia-Bellido, A. (1979). Genetic analysis of the *achaete–scute* system of *Drosophila melanogaster. Genetics* **91**, 491–520.

Garcia-Bellido, A. (1981). From the gene to the pattern: Achaete differentiation. *In* "Cellular Controls in Differentiation" (C. W. Lloyd and D. A. Rees, eds.), pp. 281–304. Academic Press, New York.

Garcia-Bellido, A., and Ripoll, P. (1978). Cell lineage and differentiation in *Drosophila. In* "Results and Problems in Cell Differentiation" (W. Gehring, ed.), Vol. 9, pp. 119–156. Springer-Verlag, Berlin and New York.

Garcia-Bellido, A., and Santamaria, P. (1978). Developmental analysis of the *achaete–scute* system of *Drosophila melanogaster. Genetics* **88**, 469–486.

Ghysen, A., and Dambly-Chaudière, C. (1988). From DNA to form: The *achaete–scute* complex. *Genes Dev.* **2**, 495–501.

Ghysen, A., and O'Kane, C. (1989). Neural enhancer-like elements as specific cell mark-ers in *Drosophila. Development* **105**, 35–52.

Ghysen, A., Dambly-Chaudière, C., Aceves, E., Jan, L. Y., and Jan, Y. N. (1986). Sensory neurons and peripheral pathways in *Drosophila* embryos. *Wilhelm Roux's Arch. Dev. Biol.* **195**, 281–289.

Gray, A., Dull, T. J., and Ullrich, A. (1983). Nucleotide sequence of epidermal growth factor cDNA predicts a 128,000-molecular weight protein precursor. *Nature (Lon-don)* **303**, 722–725.

Greenwald, I. (1985). *lin-12*, a nematode homeotic gene, is homologous to a set of mamma-lian proteins that includes epidermal growth factor. *Cell* **43**, 583–590.

Greenwald, I. S., Sternberg, P. W., and Horvitz, H. R. (1983). The *lin-12* locus specifies cell fates in *Caenorhabditis elegans. Cell* **34**, 435–444.

Hartenstein, V., and Campos-Ortega, J. A. (1984). Early neurogenesis in wild-type *Dro-sophila melanogaster. Wilhelm Roux's Arch. Dev. Biol.* **193**, 308–325.

Hartenstein, V., and Campos-Ortega, J. A. (1985). Fate mapping in wild-type *Drosophila melanogaster*. I. The pattern of embryonic cell divisions. *Wilhelm Roux's Arch. Dev. Biol.* **194**, 181–195.

Hartenstein, V., and Campos-Ortega, J. A. (1986). The peripheral nervous system of mutants of early neurogenesis in *Drosophila melanogaster. Wilhelm Roux's Arch. Dev. Biol.* **195**, 210–221.

Hartenstein, V., Rudloff, E., and Campos-Ortega, J. A. (1987). The pattern of prolifera-tion of the neuroblasts in the wild-type embryo of *Drosophila melanogaster. Wilhelm Roux's Arch. Dev. Biol.* **196**, 473–485.

Hartley, D. A., Xu, T., and Artavanis-Tsakonas, S. (1987). The embryonic expression of the *Notch* locus of *Drosophila melanogaster* and the implications of point mutations in the extracellular EGF-like domain of the predicted protein. *EMBO J.* **6**, 3407–3417.

Hartley, D. A., Preiss, A., and Artavanis-Tsakonas, S. (1988). A deduced gene product from the *Drosophila* neurogenic locus *Enhancer of split* shows homology to mamma-lian G-protein β subunit. *Cell* **55**, 785–795.

Hiromi, Y., Kuroiwa, A., and Gehring, W. J. (1985). Control elements of the *Drosophila* segmentation gene *fushi tarazu. Cell* **50**, 963–974.

Hoppe, P. E., and Greenspan, R. J. (1986). Local function of the *Notch* gene for embryonic ectodermal choice in *Drosophila. Cell* **46**, 773–783.

Jan, L. Y., and Jan, Y. N. (1982). Antibodies to horseradish peroxidase as specific neuronal markers in *Drosophila* and grasshopper embryos. *Proc. Natl. Acad. Sci. U.S.A.* **79**, 2700–2704.

Jan, Y. N., Bodmer, R., Ghysen, A., Dambly-Chaudiere, C., and Jan, L. Y. (1987). Mutations affecting the peripheral nervous system in *Drosophila. J. Cell. Biochem., Proc. UCLA Symp. Mol. Entomol.*, 45–56.

Jiménez, F. (1988). Genetic control of neuronal determination in insects. *Trends Neurosci.* **11**, 378–380.

Jiménez, F., and Campos-Ortega, J. A. (1979). A region of the *Drosophila* genome necessary for CNS development. *Nature (London)* **282**, 310–312.

Jiménez, F., and Campos-Ortega, J. A. (1982). Maternal effects of zygotic mutants affecting early neurogenesis in *Drosophila. Wilhelm Roux's Arch. Dev. Biol.* **191**, 191–201.

Jiménez, F., and Campos-Ortega, J. A. (1987). Genes in subdivision 1B of the *Drosophila melanogaster* X-chromosome and their influence on neural development. *J. Neurogenet.* **4**, 179–200.

Kelley, M. R., Kidd, S., Deutsch, W. A., and Young, M. W. (1987). Mutations altering the structure of epidermal growth factor-like coding sequences at the *Drosophila Notch* locus. *Cell* **51**, 539–548.

Kidd, S., and Young, M. W. (1986). Transposon-dependent mutant phenotypes at the *Notch* locus of *Drosophila. Nature (London)* **323**, 89–91.

Kidd, S., Lockett, T. J., and Young, M. W. (1983). The *Notch* locus of *Drosophila melanogaster. Cell* **34**, 421–433.

Kidd, S., Kelley, M. R., and Young, M. W. (1986). Sequence of the *Notch* locus of *Drosophila melanogaster:* Relationship of the encoded protein to mammalian clotting and growth factors. *Mol. Cell. Biol.* **6**, 3094–3108.

Klämbt, C., Knust, E., Tietze, K., and Campos-Ortega, J. A. (1989). Closely related transcripts encoded by the neurogenic gene complex *Enhancer of split* of *Drosophila melanogaster. EMBO J.* **8**, 203–210.

Knust, E., Dietrich, U., Tepass, U., Bremer, K. A., Wiegel, D., Vässin, H., and Campos-Ortega, J. A. (1987a). EGF-homologous sequences encoded in the genome of *Drosophila melanogaster* and their relation to neurogenic genes. *EMBO J.* **6**, 761–766.

Knust, E., Bremer, K. A., Vässin, H., Ziemer, A., Tepass, U., and Campos-Ortega, J. A. (1987b). The *Enhancer of split* locus and neurogenesis in *Drosophila melanogaster. Dev. Biol.* **122**, 262–273.

Knust, E., Tietze, K., and Campos-Ortega, J. A. (1987c). Molecular analysis of the neurogenic locus *Enhancer of split* of *Drosophila melanogaster. EMBO J.* **6**, 4113–4123.

Kopczynski, C. C., Alton, A. K., Fechtel, K., Kooh, P. J., and Muskavitch, M. A. T. (1988). *Delta,* a *Drosophila,* neurogenic gene, is transcriptionally complex and encodes a protein related to blood coagulation factors and epidermal growth factor of vertebrates. *Genes Dev.* **2**, 1723–1735.

LaBonne, S. G., and Mahowald, A. P. (1985). Partial rescue of embryos from two maternal-effect neurogenic mutants by transplantation of wild-type ooplasm. *Dev. Biol.* **110**, 264–267.

Lawrence, P. A., and Green, S. M. (1979). Cell lineage in the developing retina of *Drosophila. Dev. Biol.* **71**, 142–152.

Lehmann, R., Dietrich, U., Jimenez, F., and Campos-Ortega, J. A. (1981). Mutations of early neurogenesis in *Drosophila. Wilhelm Roux's Arch. Dev. Biol.* **190**, 226–229.

Lehmann, R., Jimenez, F., Dietrich, U., and Campos-Ortega, J. A. (1983). On the phenotype and development of mutants of early neurogenesis in *Drosophila melanogaster. Wilhelm Roux's Arch. Dev. Biol.* **192**, 62–74.

Lindsley, D. L., and Grell, E. H. (1968). Genetic variations of *Drosophila melanogaster*. *Carnegie Inst. Washington Publ.*

Lindsley, D., and Zimm, G. (1985). The genome of *Drosophila melanogaster*. Part 1: Genes A–K. *Drosophila Inf. Serv.* **62**, 1–227.

Lindsley, D. L., Sandler, L., Baker, B. S., Carpenter, A. T. C., Denell, R. E., Hall, J. C., Jacobs, P. A., Miklos, G. L., Davis, B. K., Gethman, R. C., Hardy, R. W., Hessler, A., Miller, S. M., Nozawa, H., Parry, D. M., and Gould-Somero, M. (1972). Segmental aneuploidy and the genetic structure of the *Drosophila* genome. *Genetics* **71**, 157–184.

Luchesi, J. C., and Skripsy, T. (1981). The link between dosage compensation and sex differentiation in *Drosophila melanogaster*. *Chromosoma* **82**, 217–227.

Mohr, O. (1919). Character changes caused by mutation of an active region of a chromosome in *Drosophila*. *Genetics* **4**, 275–282.

Muller, H. J. (1935). The origination of chromatin deficiencies as minute deletions subject to insertion elsewhere. *Genetica* **17**, 237–252.

Muller, H. J., and Prokofjeva, A. A. (1935). The individual gene in relation to the chromomere and the chromosome. *Proc. Natl. Acad. Sci. U.S.A.* **21**, 16–26.

Murre, C., McCaw, P. S., and Baltimore, D. (1989). A new DNA binding and dimerization motif in immunoglobulin enhancer binding, *daughterless, MyoD,* and *myc* proteins. *Cell* **56**, 777–783.

Nüsslein-Volhard, C., and Wieschaus, E. (1980). Mutations affecting segment number and polarity in *Drosophila*. *Nature (London)* **287**, 795–801.

Nüsslein-Volhard, C., Wieschaus, E., and Kluding, H. (1984). Mutations affecting the pattern of the larval cuticle of *Drosophila melanogaster*. I. Zygotic loci on the second chromosome. *Wilhelm Roux's Arch. Dev. Biol.* **193**, 267–282.

Perrimon, N., Engstrom, L., and Mahowald, A. P. (1984). Analysis of the effects of zygotic lethal mutations on the germ line functions in *Drosophila*. *Dev. Biol.* **105**, 404–414.

Portin, P. (1975). Allelic negative complementation at the *Abruptex* locus of *Drosophila melanogaster*. *Genetics* **81**, 121–123.

Postlethwait, J. H. (1978). Clonal analysis of *Drosophila* cuticular patterns. In "Biology of *Drosophila*" (M. Ashburner and T. R. F. Wright, eds.), Vol. 2c, pp. 359–441. Academic Press, New York.

Poulson, D. F. (1937). Chromosomal deficiencies and embryonic development of *Drosophila melanogaster*. *Proc. Natl. Acad. Sci. U.S.A.* **23**, 133–137.

Poulson, D. F. (1950). Histogenesis, organogenesis and differentiation in the embryo of *Drosophila melanogaster*. In "Biology of *Drosophila*" (M. Ashburner and T. R. F. Wright, eds.), Vol. 2c, pp. 268–274. Academic Press, New York.

Preiss, A., Hartley, D. A., and Artavanis-Tsakonas, S. (1988). The molecular genetics of *Enhancer of split,* a gene required for embryonic neural development in *Drosophila*. *EMBO J.* **7**, 3917–3928.

Robinow, S., and White, K. (1988). The locus *elav* of *Drosophila melanogaster* is expressed in neurons at all developmental stages. *Dev. Biol.* **126**, 294–303.

Romani, S., Campuzano, S., and Modolell, J. (1987). The *achaete–scute* complex is expressed in neurogenic regions of *Drosophila* embryos. *EMBO J.* **6**, 2085–2092.

Romani, S., Campuzano, S., Macagno, E. R., and Modolell, J. (1989). Expression of *achaete* and *scute* genes in *Drosophila* imaginal discs and their function in sensory organ development. *Genes Dev.* **3**, 997–1007.

Ruiz-Gomez, M., and Modolell, J. (1987). Deletion analysis of the *achaete–scute* locus of *Drosophila melanogaster*. *Genes Dev.* **1**, 1238–1246.

Shannon, M. (1972). Characterization of the female sterile mutant *almondex* of *Drosophila melanogaster. Genetica* **43**, 244–256.

Shannon, M. (1973). The development of eggs produced by the female sterile mutant *almondex* of *Drosophila melanogaster. J. Exp. Zool.* **183**, 383–400.

Shellenbarger, D. L., and Mohler, J. D. (1975). Temperature-sensitive mutations of the *Notch* locus in *Drosophila melanogaster. Genetics* **81**, 143–162.

Shellenbarger, D. L., and Mohler, J. D. (1978). Temperature-sensitive periods and autonomy of pleiotropic effects of *l(1)Nts*, a conditional *Notch* lethal in *Drosophila. Dev. Biol.* **62**, 432–446.

Slizynska, H. (1938). Salivary chromosome analysis of the *white-facet* region of *Drosophila melanogaster. Genetics* **23**, 291–299.

Stern, C. (1954). Two or three bristles. *Am. Sci.* **43**, 213–247.

Stern, C. (1968). "Genetic Mosaics and Other Essays." Berkeley Univ. Press, Berkeley, California.

Sternberg, P. W., and Horvitz, H. R. (1984). The genetic control of cell lineage during nematode development. *Annu. Rev. Genet.* **18**, 489–524.

Stone, J., DeLange, T., Ramsey, G., Jakobovits, E., Bishop, J. M., Varmus, H., and Lee, W. (1987). Definition of regions in human c-*myc* that are involved in transformation and nuclear localization. *Mol. Cell. Biol.* **7**, 1697–1709.

Taghert, P. H., Doe, C. Q., and Goodman, C. S. (1984). Cell determination and regulation during development of neuroblasts and neurones in grasshopper embryos. *Nature (London)* **307**, 163–165.

Technau, G. M., and Campos-Ortega, J. A. (1985). Fate mapping in wild-type *Drosophila melanogaster.* II. Injections of horseradish peroxidase in cells of the early gastrula stage. *Wilhelm Roux's Arch. Dev. Biol.* **194**, 196–212.

Technau, G. M., and Campos-Ortega, J. A. (1986). Lineage analysis of transplanted individual cells in embryos of *Drosophila melanogaster.* II. Commitment and proliferative capabilities of neural and epidermal cell progenitors. *Wilhelm Roux's Arch. Dev. Biol.* **195**, 445–454.

Technau, G. M., and Campos-Ortega, J. A. (1987). Cell autonomy of expression of neurogenic genes of *Drosophila melanogaster. Proc. Natl. Acad. Sci. U.S.A.* **84**, 4500–4504.

Technau, G. M., Becker, T., and Campos-Ortega, J. A. (1988). Reversible commitment of neural and epidermal progenitor cells during embryogenesis of *Drosophila melanogaster. Wilhelm Roux's Arch. Dev. Biol.* **197**, 413–418.

Thomas, J. B., Bastiani, M. J., Bate, M., and Goodman, C. (1984). From grasshopper to *Drosophila:* A common plan for neuronal development. *Nature (London)* **310**, 203–207.

Vässin, H., and Campos-Ortega, J. A. (1987). Genetic analysis of *Delta,* a neurogenic gene of *Drosophila melanogaster. Genetics* **116**, 433–445.

Vässin, H., Vielmetter, J., and Campos-Ortega, J. A. (1985). Genetic interactions in early neurogenesis of *Drosophila melanogaster. J. Neurogenet.* **2**, 291–308.

Vässin, H., Bremer, K. A., Knust, E., and Campos-Ortega, J. A. (1987). The neurogenic locus *Delta* of *Drosophila melanogaster* is expressed in neurogenic territories and encodes a putative transmembrane protein with EGF-like repeats. *EMBO J.* **6**, 3431–3440.

Villares, R., and Cabrera, C. V. (1987). The *achaete–scute* gene complex of *Drosophila melanogaster:* Conserved domains in a subset of genes required for neurogenesis and their homology to *myc. Cell* **50**, 415–424.

Watt, R., Stanton, L. W., Marcu, K. B., Gallo, R. C., Croce, C. M., and Rovera, G. (1983). Nucleotide sequence of cloned cDNA of human c-*myc* oncogene. *Nature (London)* **303**, 725–728.

Weigel, D., Knust, E., and Campos-Ortega, J. A. (1987). Molecular organization of *master mind,* a neurogenic gene of *Drosophila melanogaster. Mol. Gen. Genet.* **207,** 374–384.

Welshons, W. J. (1956). Dosage experiments with *split* mutants in the presence of an enhancer of split. *Drosophila Inf. Serv.* **30,** 157–158.

Welshons, W. J. (1965). Analysis of a gene in *Drosophila. Science* **150,** 1122–1129.

Welshons, W. J. (1974a). The cytogenetic analysis of a fractured gene in *Drosophila. Genetics* **76,** 775–794.

Welshons, W. J. (1974b). Genetic basis for two types of recessive lethality at the *Notch* locus of *Drosophila. Genetics* **68,** 259–268.

Welshons, W. J., and Keppy, D. O. (1975). Intragenic deletions and salivary band relationships in *Drosophila. Genetics* **80,** 143–155.

Welshons, W. J., and Keppy, D. O. (1981). The recombinational analysis of aberrations and the position of the *Notch* locus on the polytene chromosome of *Drosophila. Mol. Gen. Genet.* **181,** 319–324.

Wharton, K. A., Kedvobnick, B., Finnerty, V. G., and Artavanis-Tsakonas, S. (1985a). *opa:* a novel family of transcribed repeats shared by the *Notch* locus and other developmentally regulated loci of *Drosophila melanogaster. Cell* **40,** 55–62.

Wharton, K. A., Johansen, K. M., Xu, T., and Artavanis-Tsakonas, S. (1985b). Nucleotide sequence from the neurogenic locus *Notch* implies a gene product that shares homology with proteins containing EGF-like repeats. *Cell* **43,** 567–581.

White, K. (1980). Defective neural development in *Drosophila melanogaster* embryos deficient for the tip of the X-chromosome. *Dev. Biol.* **80,** 322–344.

White, K., DeCelles, N. L., and Enlow, T. C. (1983). Genetic and developmental analysis of the locus *vnd* in *Drosophila melanogaster. Genetics* **104,** 433–488.

Wright, T. R. F. (1970). The genetics of embryogenesis in *Drosophila. Adv. Genet.* **15,** 262–395.

Yedvobnick, B., Muskavitch, M. A. T., Wharton, K. A., Halpern, M. E., Paul, E., Grimwade, B. G., and Artavanis-Tsakonas, S. (1985). Molecular genetics of *Drosophila* neurogenesis. *Cold Spring Harbor Symp. Quant. Biol.* **50,** 841–854.

Yedvobnick, B., Smoller, D., Young, P., and Mills, D. (1988). Molecular analysis of the neurogenic locus *mastermind* of *Drosophila melanogaster. Genetics* **118,** 483–497.

Yochem, J., Weston, K., and Greenwald, I. (1988). The *Caenorhabditis elegans lin-12* gene encodes a transmembrane protein with overall similarity to *Drosophila Notch. Nature (London)* **335,** 547–550.

Ziemer, A., Tietze, K., Knust, E., and Campos-Ortega, J. A. (1988). Genetic analysis of *Enhancer of split,* a locus involved in neurogenesis in *Drosophila melanogaster. Genetics* **119,** 63–74.

INDEX

A

Agglutinins, in yeast mating, 36
amalgam (ama) gene, *Drosophila*
  mRNA, distribution during
    embryogenesis, 327–329
  protein product
    association with cell membrane, 325
    localization during embryogenesis,
      330–332
    organization in five domains, 326
    sequences similar to members of
      immunoglobulin family,
      326–327
  structure, position in ANT-C, 316,
    325–326
Amnioserosa, differentiation, *Drosophila*
  absence in *zen⁻* mutants, 281, 283, 285,
    317
  zen target genes and, 303
Anchor cell (AC), *C. elegans*
  generation, *lin-12* role, 79, 81,
    88–89, 92
  interactions with
    VPCs, 75–78
    VU cells, 66, 81
Aneuploid genotypes, viability,
  Drosophila, 213–215
Antennapedia complex (ANT-C),
  Drosophila
  distinction from BX-C genes, 355
  genetic map, 311–312
  homeotic loci, head-specifying,
    335–350; *see Antennapedia,*

Deformed, labial, proboscipedia,
  Sex combs reduced genes
  exon/intron structures, 351, 352
  expression pattern, comparison with
    trunk-specifying, 348–349
  homeobox and opa repeat positions,
    351, 353
  origin and evolution
    duplication–divergence events
      and, 350–351
    similarity to mouse homeotic
      genes, 350
  protein products, comparison with
    BX-C genes, 349
  nonhomeotic loci, 315-335; *see*
    amalgam, bicoid, fushi tarazu,
    zerknüllt genes
  relatedness of nine homeoboxes,
    353–354
Antennapedia (Antp) gene, homeotic,
  Drosophila
  dominant gain-of-function mutations,
    313, 336, 338–339
  expression during embryogenesis, 338
  promoters P1 and P2, distinct
    functions, 335, 337
    activation by gap genes, 267, 268
  protein product, homeotic sequences,
    370, 371
  structure, 335–337
Anterior–posterior (AP) axis, *Drosophila*
  homeotic gene functional domains,
    372–374
  maternal-effect gene roles, 246–249

lin-12 protein of *C. elegans,* 419–420
pb, during embryogenesis, *Drosophila*
RME1 (meiosis inhibition), yeast
in α and **a** cells, 41
absence in **a**/α cells, 42
Scr, Drosophila embryo
accumulation during development, 341–342, 349
homeodomain sequences, 370, 371
STE group, in yeast mating, 35–36, 41
STE12, pheromone response and, 53–54
Toll, structure and function, *Drosophila,* 289–290
Ubx, homeodomain sequences, *Drosophila,* 370, 371
Xhox 1A, *Xenopus,* structural homology to *Dfd* protein, 370, 371
zen, during embryogenesis, *Drosophila,* 285–287, 319
PRTF, *see* Pheromone and receptor transcription factor

dorsal, selective translation in ventral region of embryo, 291, 292
E(spl), multiple units, spatial distribution, 427
ftz, during embryogenesis, 334
lab, during embryogenesis, 347, 348
mam, two major units, expression, 429
N, structure and processing, 419
pb, during embryogenesis, 344, 349
Scr, during embryogenesis, 341, 349
zen, during embryogenesis, 285
spaciotemporal distribution, 317–319
in wild-type and mutant, 293–295
pre-mRNA, sex-specific splicing
dsx, 200–201
regulation, 202, 204
Sxl, 199–200, 205
tra, 200
regulation, 202, 203
tra-2, 201–202
regulation, 202–203

R

Restriction fragment length polymorphisms (RFLP), mapping, *C. elegans,* 73
Rhizobium, σ^{54} promoter, in nitrogen metabolism regulation, 13, 26–27
RNA, *Drosophila*
mRNA
ama, accumulation during embryogenesis, 327–329
Antp, accumulation during embryogenesis, 338
AS-C multiple transcripts, spatial distribution, 437–439
in neurogenic mutants, 438, 441–442
bcd, localization
in embryo during development, 322–325
in oocyte, 278
Dfd, during embryogenesis, 343, 376–378
Dl, structure and distribution, 421

S

Saccharomyces cerevisiae, see Yeast
Salmonella typhimurium
flagellar genes, regulation, 19, 25
nitrogen metabolism genes, regulation, 26
Segment polarity genes *(gooseberry),* positional function along AP axis, *Drosophila,* 375, 376
Serine proteases, encoded by *snake* and *easter,* in embryo, *Drosophila,* 291–292
Serum response factor (SRF), mammalian, homology to yeast MCM1, 50–51
Sex comb, control by *dsx* and homeotic genes, *Drosophila,* 217–218, 220
Sex combs reduced (Scr) gene, homeotic, *Drosophila*
molecular map, position in ANT-C, 339–441
mRNA accumulation in embryo, 341, 349